tionship

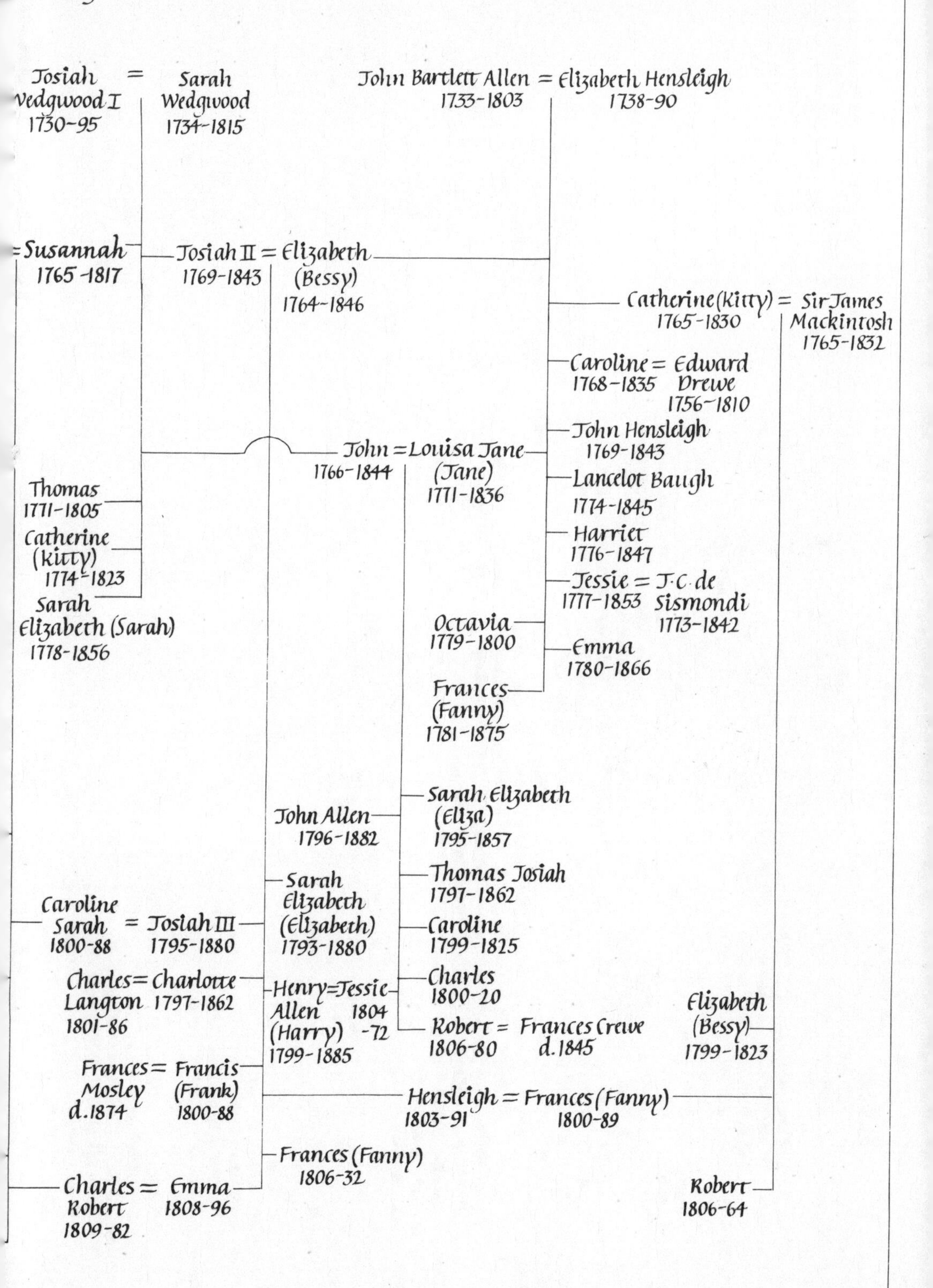

Josiah Wedgwood I 1730-95
=
Sarah Wedgwood 1734-1815
John Bartlett Allen = Elizabeth Hensleigh
1733-1803
1738-90
=Susannah 1765-1817
Josiah II = Elizabeth (Bessy)
1769-1843
1764-1846
Catherine (Kitty) = Sir James Mackintosh
1765-1830
1765-1832
Caroline = Edward Drewe
1768-1835
1756-1810
John Hensleigh
1769-1843
John = Louisa Jane (Jane)
1766-1844
1771-1836
Lancelot Baugh
1774-1845
Thomas 1771-1805
Catherine (Kitty) 1774-1823
Harriet 1776-1847
Jessie = J.C. de Sismondi
1777-1853
1773-1842
Sarah Elizabeth (Sarah) 1778-1856
Octavia 1779-1800
Emma 1780-1866
Frances (Fanny) 1781-1875
Sarah Elizabeth (Eliza) 1795-1857
John Allen 1796-1882
Thomas Josiah 1797-1862
Sarah Elizabeth (Elizabeth) 1793-1880
Caroline Sarah = Josiah III
1800-88
1795-1880
Caroline 1799-1825
Charles 1800-20
Charles Langton = Charlotte
1801-86
1797-1862
Henry Allen (Harry) = Jessie
1799-1885
1804-72
Robert = Frances Crewe
1806-80
d.1845
Elizabeth (Bessy) 1799-1823
Frances Mosley = Francis (Frank)
d.1874
1800-88
Hensleigh = Frances (Fanny)
1803-91
1800-89
Frances (Fanny) 1806-32
Charles Robert = Emma
1809-82
1808-96
Robert 1806-64

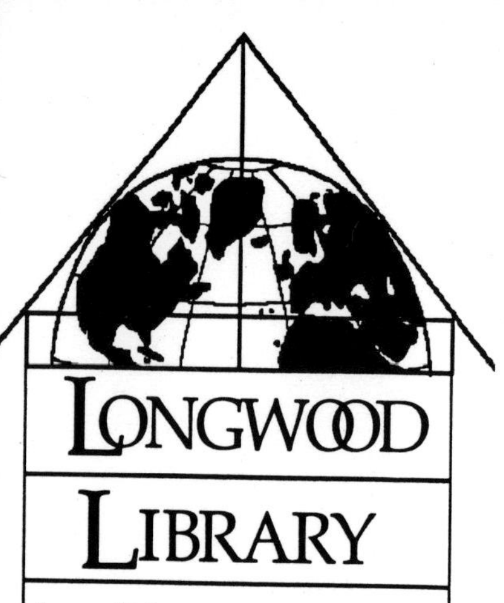

THE CORRESPONDENCE OF
CHARLES DARWIN

Editors

FREDERICK BURKHARDT SYDNEY SMITH

Associate Editors

JANET BROWNE DAVID KOHN

WILLIAM MONTGOMERY

Managing Editor

STEPHEN V. POCOCK

Assistant Editors

CHARLOTTE BOWMAN MARSHA L. RICHMOND

ANNE SECORD

Research Associates

ANNE SCHLABACH BURKHARDT

NORA CARROLL STEVENSON

This edition of the Correspondence of Charles Darwin is sponsored by the American Council of Learned Societies. Its preparation is made possible by the co-operation of the Cambridge University Library and the American Philosophical Society.

The principal sources of funds for editing this volume have been the National Endowment for the Humanities, the National Science Foundation, and the Andrew W. Mellon Foundation. The National Endowment's grants (Nos. RE-23166-75-513, RE-27067-77-1359, RE-00082-80-1628, and RE-20166-82) were from its Program for Editions; the National Science Foundation's support of the work was under grants Nos. SOC-75-15840 and SES-7912492. Any opinions, findings, conclusions or recommendations expressed in this publication are those of the authors and do not necessarily reflect the views of the grantors.

Joseph Dalton Hooker. Lithograph by T. H. Maguire, 1851.
(Courtesy of the Council of the Linnean Society of London.)

THE CORRESPONDENCE OF
CHARLES DARWIN

VOLUME 3 1844–1846

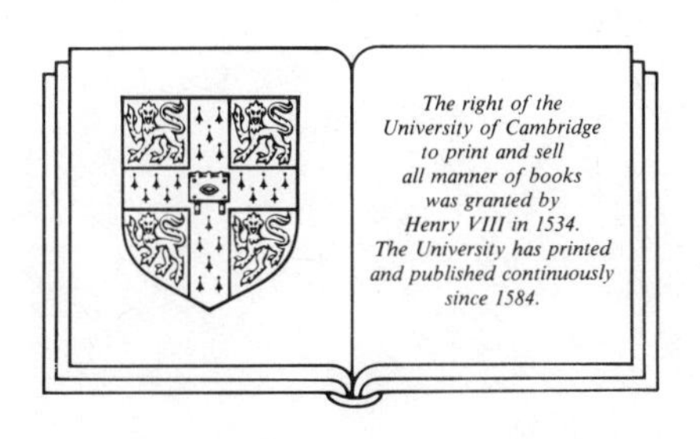

The right of the
University of Cambridge
to print and sell
all manner of books
was granted by
Henry VIII in 1534.
The University has printed
and published continuously
since 1584.

CAMBRIDGE UNIVERSITY PRESS

CAMBRIDGE

NEW YORK NEW ROCHELLE

MELBOURNE SYDNEY

Published by the Press Syndicate of the University of Cambridge
The Pitt Building, Trumpington Street, Cambridge CB2 1RP
32 East 57th Street, New York, NY 10022, USA
10 Stamford Road, Oakleigh, Melbourne 3166, Australia

First published 1987

Printed in Great Britain at the University Press, Cambridge

British Library cataloguing in publication data

Darwin, Charles
The correspondence of Charles Darwin.
Vol. 3: 1844–1846
1. Darwin, Charles 2. Naturalist—
England—Biography
I. Burkhardt, Frederick H.
II. Smith, Sydney, *1911–*
575′.0092′4 QH31.D2

Library of Congress cataloguing in publication data

Darwin, Charles, 1809–1882.
The correspondence of Charles Darwin.
I. Darwin, Charles, 1809–1882. 2. Naturalists—
England—Correspondence. I. Burkhardt, Frederick,
1912– . II. Smith, Sydney, 1911– . III. Title.
QH31, D2A33 1985 575′.0092′4[B] 84-45347

ISBN 0 521 25589 9

CONTENTS

ILLUSTRATIONS

LIST OF LETTERS

The following list is in the order of the entries in the *Calendar of the correspondence of Charles Darwin*. It includes all those letters that are dated in the *Calendar* within the date range covered by this volume of the *Correspondence*. Alongside the *Calendar* numbers are the current dates ascribed to every item. Some letters have been redated since publication of the *Calendar* so this list is necessary to enable users of the *Calendar* to locate such letters in the *Correspondence*. In the list, a date printed in italic type indicates that that item either appears as a summary only in this volume or has been omitted from it entirely.

646. [28 January or 4 February 1846]
678. [before 1 October 1844]
694. 9 [May 1846]
726. [3 June 1844]
727. [May 1844 – 1 October 1846]
728. [3 January 1844]
729. [11 January 1844]
730. 13 January – 12 February 1845
731. 20 January [1844]
732. 25 January 1844
733. [27 January 1844]
734. 29 January 1844
735. [3–17 February 1844]
736. 23 February [1844]
737. [23 February – 6 March 1844]
738. [6 March 1844]
739. 9 March 1844
740. 11 March [1844]
741. 14 March 1844
742. 16 March [1844]
743. [27 March 1844]
744. 31 March [1844]
745. 5 April 1844
746. [17 April 1844]
747. 20 April [1844]
748. 26 April 1844
749. [after 26 April 1844]
750. 5 May [1844]
751. 23 May [1844]
752. 28 May 1844
753. [11 or 18 December 1844]
754. 1 June [1844]
755. 3 June [1844]
756. [before 8 June 1844]
757. 11 June [1844]
758. 15 June 1844
759. 29 [June 1844]
760. 4 July [1844]
761. 5 July 1844
762. 11 July 1844
763. [14 July 1844]
764. 22 July [1844]
765. [25 July 1844]
766. 25 July 1844
767. [30 July 1844 – 1 October 1846]
768. [25 July – 29 August 1844]
769. 9 August [1844]
770. 12 August [1844]
771. 29 August [1844]
772. 29 [August 1844]
773. [1 September 1844]
774. [*c.* 3 September 1844]
775. 5 September [1844]
776. [8 September 1844]
777. [before 14 September 1844]
778. [before 14 September 1844]
779. 1 October 1844
780. 10 October [1844]
781. 11 October [1844]
782. 12 October [1844]

INTRODUCTION

The third volume of Darwin's correspondence covers a period in his career when the scientific results of the *Beagle* voyage still dominated his working life, but during which he broadened his continuing investigations into the nature and origin of species and varieties. In contrast to the received image of Darwin as a recluse in Down, the letters published here show him to be an established and confident naturalist at the heart of British scientific society, travelling often to London and elsewhere to attend meetings and confer with colleagues, and involved in the social and political activities of the community of savants as well as in its philosophical and scientific pursuits. At home, time was filled with copious natural history work, writing, and gathering information from an ever-expanding network of correspondents. Down House was altered and extended to accommodate Darwin's growing family and the many relatives and friends who came to stay; and, with his father's advice, Darwin began a series of judicious financial investments to ensure a comfortable future for all those under his care.

In these years, Darwin published two books on geology, *Volcanic islands* (1844) and *Geological observations on South America* (1846), which completed his trilogy on the geological results of the *Beagle* voyage, and extensively revised his *Journal of researches* for a second edition in 1845, having already provided corrections in 1844 for a German translation of the first edition. He continued as an officer of the Geological Society of London, acting as one of four vice-presidents in 1844 and remaining on the council from 1845 onwards; he was a conscientious member of the Royal Geographical Society and the Royal Society; he regularly attended meetings and refereed papers for all these organisations. Between 1844 and 1846 Darwin himself wrote ten papers, six of which related to the *Beagle* collections. Among these were some studies of invertebrates that at first had been intended for publication in *The zoology of the voyage of H.M.S. Beagle* (1838–43) but were deferred when the Government grant was exhausted (*Correspondence* vol. 2, letter to A. Y. Spearman, 9 October 1843, n. 1). In addition, Darwin threw himself into analysing the results emerging from the examination of *Beagle* plant specimens by the young botanist and traveller, Joseph Dalton Hooker. This volume of Darwin's correspondence contains 107 letters between Darwin and Hooker, fully documenting the beginnings of their lifelong friendship.

Darwin's earlier scientific friendships were not neglected either, as the correspondence with Charles Lyell, George Robert Waterhouse, John Stevens Henslow, Leonard Horner, Leonard Jenyns, Edward Forbes, and Richard Owen shows. These friends, with the addition of Hooker, were important to Darwin for—among other things—they were the first people he turned to when he wished to discuss the problems and various scientific issues that arose out of his work on species. This volume shows that Darwin discussed his ideas on species mutability with Hooker, Horner, Jenyns, Lyell, Owen, and Charles James Fox Bunbury; he may well have broached the subject with others. Only two months after their first exchange, early in 1844, Darwin told Hooker that he was engaged in a 'very presumptuous work' which had led to the conviction that 'species are not (it is like confessing a murder) immutable' (letter to J. D. Hooker, [11 January 1844]). Nine months later, in his letter of 12 October [1844], he explained to Jenyns:

> I have continued steadily reading & collecting facts on variation of domestic animals & plants & on the question of what are species; I have a grand body of facts & I think I can draw some sound conclusions. The general conclusion at which I have slowly been driven from a directly opposite conviction is that species are mutable & that allied species are co-descendants of common stocks. I know how much I open myself, to reproach, for such a conclusion, but I have at least honestly & deliberately come to it.

It is clear from the correspondence that his close friends were not outraged by Darwin's heterodox opinions and later in the year both Jenyns and Hooker were invited to read a manuscript essay on his species theory (DAR 113; *Foundations*, pp. 57–255), an expanded version, completed on 5 July 1844, of a pencil sketch he had drawn up some two years earlier. But although eager for the views of informed colleagues, Darwin was naturally protective of his untried theory and seems to have shied away from the risk of pushing it too early into the open. In the event, it was not until the beginning of 1847 that Hooker was given a fair copy of the essay of 1844 to read (see *Correspondence* vol. 4, letter to J. D. Hooker, 8 [February 1847]). Darwin can be seen as a cautious strategist, sometimes confident, but often uneasy about his work, and always attempting to gauge the kind of response that his theory of transmutation would generate. In particular, he anxiously watched the controversy seething around an evolutionary book, *Vestiges of the natural history of creation*, published anonymously in 1844. His old friend Adam Sedgwick attacked the work vehemently in the *Edinburgh Review* (1845), while other colleagues like Edward Forbes ridiculed the theories employed there, caring only to join in the popular guessing-game about the identity of the author. One candidate, known to be working on species and varieties, was Darwin himself: as he told his cousin William Darwin Fox in a letter of [24 April 1845], he felt he ought to be both 'flattered & unflattered' to hear that other naturalists attributed the book to him. But, as his letters to Hooker show, Darwin carefully considered and then rejected almost all of the contents of *Vestiges*, and he feared that the reaction to his own work would be prejudiced by the arguments aroused by its skilful but scientifically unsound reasoning.

Perhaps the most interesting letter relating to Darwin's species theory, which also bears on his concern for the future, is that addressed to his wife Emma, dated 5 July 1844, just after Darwin had completed the final draft of his essay on the subject. He asked her to ensure that the essay would be published in the event of his death and stipulated a sum of money to be bequeathed, together with his extensive library and portfolios of notes on species, to an editor who would undertake to see the work through the press. Darwin also listed possible editors: at first he proposed any one of Lyell, Henslow, Edward Forbes, William Lonsdale, Hugh Edwin Strickland, or Owen—the last with the caveat that he would probably not wish to take on the work. But the list was subsequently altered after Darwin's second, and possibly third, thoughts on the choice of the right person. The names of Lonsdale, Forbes, and Owen were deleted, Henslow's was queried, and J. D. Hooker's was added. Much later, by the autumn of 1854 when Darwin began sorting out his notes in preparation for writing up his 'big book' on species (*Natural selection*), he had decided that Hooker was by far the best man for the task and added a note on the cover to that effect.

The full consideration that Darwin gave to the future editing and publication of his essay, and the way in which he wrote to colleagues and friends about his work, show clearly his intention to publish his theory. His instructions to Emma may, perhaps, as some scholars have thought, indicate a reluctance to take the responsibility for publishing upon himself, but, more plausibly, they portray a man faced with the task of establishing a theory and its consequences, and fearful lest both the energy and time necessary to achieve this end should be denied him. After prolonged illnesses in 1841 and 1842, years poorly represented in the *Correspondence* because he was for much of the time too ill even to write letters, Darwin felt that his life was only too likely to be cut short. Moreover, even when at his best, Darwin could never work as intensively as he felt he ought to, or needed to, for fear of inducing another breakdown in his health.

Darwin's published work during this period secured his position as one of Britain's foremost naturalists. His study of the volcanic islands visited during the *Beagle* voyage was based on a wide range of rock and mineral specimens, including his own, and considerable research into contemporary theories of volcanic activity, mountain formation, and the elevation of extensive tracts of land relative to the sea. Darwin put forward a new explanation of the origin of so-called 'craters of elevation', which formed the basis of discussions with Charles Lyell and Leonard Horner in letters in this volume. His observations on the lamination of volcanic rocks prompted an exchange with James David Forbes on the analogous structure of glacier-ice. In *South America* he proposed that the tension generated in molten rock before final consolidation, which he believed gave rise to this lamination, could also explain and link the widespread phenomena of cleavage and foliation, observable in some metamorphic rocks. His description and explanation of cleavage and foliation in the clay-slates and schists of South America benefitted from the mathematical expertise of William Hopkins and aroused the interest of Daniel Sharpe, whose subsequent work led to the general acceptance of Darwin's views. *South America* drew together all the geological and palaeontological results of Darwin's travels through that area and, like *Volcanic islands*, demonstrated how the structure of the land could best be explained by elevation. Darwin presented a wholeheartedly Lyellian picture of the geology of this vast area, reflecting the influence of Lyell's *Principles of geology* (1830–3) and a commitment to Lyell's idea of gradual geological change taking place over immensely long periods of time; a commitment that transcended Darwin's purely geological thought and influenced his speculations in all fields of natural history. But despite this clear and acknowledged debt, Darwin's independence of mind was never in doubt and is well evidenced by the skilled and determined defence of his theories he invariably made against rivals of whatever standing. Through the pages of *South America* Darwin pursued an argument against the French palaeontologist Alcide d'Orbigny, insisting that the vast pampas formation could not have been laid down at a single moment through the action of a great *débâcle*, as Orbigny proposed. Darwin not only used his personal notes and records but, by letter, marshalled the resources of experts such as palaeontologists Edward Forbes and George Brettingham Sowerby, and the German naturalist Christian Gottfried Ehrenberg, to support his own opinion that the pampas formations had been deposited successively under mostly brackish or estuarine conditions.

In addition to writing up his geology, Darwin undertook the revision of his *Journal of researches* for a second edition in 1845. At Lyell's recommendation, arrangements were made for the rights of the work to be transferred from Henry Colburn, the original publisher, to John Murray, and throughout 1845 Darwin worked hard to provide manuscript copy to be published in three parts during the year. Though the text was reduced in volume, Darwin went to considerable trouble to add the latest descriptions of the *Beagle* collections, to alter and expand some of his previous suggestions about the causes of extinction, and to supplement the original account of the three Fuegians carried on board the *Beagle* back to Tierra del Fuego. By 1845, Darwin was in full command of a sophisticated theory of species transmutation and there is much interplay between the information supplied in letters to Darwin, the contents of the new edition of the *Journal of researches*, and his species work.

The botany of the *Beagle* voyage was a topic still relatively unexplored by Darwin, even though he had collected plants extensively. Henslow, who had undertaken to describe the collection, was overwhelmed by ever-increasing parish and local concerns in Cambridge and Hitcham and apparently relieved to hand over Darwin's plants to Hooker, who had just returned from accompanying James Clark Ross's Antarctic surveying expedition and who hoped to publish a detailed account of the flora of the Southern Hemisphere. Darwin was quick to spot in Hooker a man he judged could become the 'first authority in Europe on that grand subject, that almost key-stone of the laws of creation, Geographical Distribution' (letter to J. D. Hooker, [10 February 1845]) and quick to make use of the young man's already large fund of botanical knowledge and his extensive connections with other British and European botanists. Darwin's questions challenged Hooker to apply his particular knowledge to more general problems, always relating, directly or indirectly, to the question of the origin and nature of species. There is little in contemporary botany and botanical systematics that is not touched upon in their correspondence. Hooker's observations on classification provided Darwin with a professional judgment on the plant world to place beside that of Waterhouse with respect to the animal kingdom. Hooker was also ready to discuss contemporary ideas on transformism in Britain and France and was a constant source of useful references and books. Some indication of the intellectual value that both men placed on their correspondence is found in the fact that they independently kept practically all the letters received from each other. The letters also document aspects of Hooker's life: his search for a paid position, involving an unsuccessful campaign for the chair of botany at Edinburgh University and a period of half-hearted work with the Geological Survey of Great Britain. Like Darwin, he obtained Government aid to publish the results of his own four-year voyage and struggled to keep up to the time-table. And like Darwin, he was deeply committed to philosophical natural history.

It was also Hooker who helped Darwin in the first stages of his barnacle work, a study commenced towards the end of 1846, at the close of this volume. Hooker, ready with advice on microscopes and microscopic technique, assisted Darwin with drawings of his first dissection. The barnacle—'M[r] Arthrobalanus' in Hooker's and Darwin's letters—was a minute, aberrant species collected by Darwin in the Chonos Archipelago, off southern Chile, which lived inside the shell of the mollusc, *Concholepas*. Unusual sexual dimorphism, with the male virtually a parasite on the female, a complex life-cycle, and difficult taxonomic considerations, combined to intrigue Darwin, and he launched himself into a survey of related species to elucidate some of the problems presented by the animal. The cirripedes were to remain central to Darwin's working life for the next eight years.

ACKNOWLEDGMENTS

The editors are grateful to Mr George Pember Darwin for permission to publish the Darwin letters and manuscripts. They also thank the Syndics of the Cambridge University Library and other owners of the manuscript letters who have generously made them available.

The research and editorial work for this edition as a whole has been supported by grants from the National Endowment for the Humanities and the National Science Foundation. The Sloan Foundation and the Andrew W. Mellon Foundation provided grants to match NEH funding, and the Mellon Foundation in 1981 and 1984 awarded grants to Cambridge University that have made it possible to put the entire Darwin correspondence into machine-readable form. Timely research assistance has been furnished by grants from the Royal Society of London and the British Academy, and the Royal Society also helped to meet the cost of publication. Bern Dibner and the late Mary S. Hopkins provided financial assistance for the purchase of books and other necessary project expenses.

The Cambridge University Library and the American Philosophical Society have generously made working space and many services available to the editorial staff.

Since the project began in 1975 we have been fortunate in benefiting from the interest, experience, and practical help of many people in many places, and the editors hope that they have adequately expressed their thanks to them individually as the work proceeded. There are some, however, who have helped over so long a period that it would be ungracious not to thank them personally in the work which has profited so much from their co-operation.

Over the years, George F. Farr Jr, assistant director of the Division of Research Grants of NEH, and Ronald J. Overmann, program director for History and Philosophy of Science of NSF, have given far more time and attention to the project than their formal duties required. The editors also appreciate the interest and encouragement they received from John E. Sawyer, president of the Andrew W. Mellon Foundation and his colleague, James Morris.

Without the expert help of John L. Dawson of the Literary and Linguistic Computing Centre of Cambridge University the computerisation of the correspondence would not have been possible, for the work on both the *Calendar* and the *Correspondence* required the solution of many novel technical problems.

The computer of the Cambridge University Library, as well as that of the University, was used in the preparation of this edition. W. D. S. Motherwell, head of automation at the Library, arranged for the installation of terminals in the project's office in the Manuscripts Department and solved the operational problems that arose.

This book was typeset at the Oxford University Computing Service on a Monotype Lasercomp. We are grateful to Susan Hockey for enabling the typesetting of the volume to be carried out, and to the other Oxford staff involved.

The late Sir Hedley Atkins and Philip Titheradge of the Darwin Museum at Down House, Downe, Kent welcomed the editors on numerous visits and responded most generously to frequent requests for information.

Professor R. D. Keynes kindly made available Emma Darwin's diaries, thus enabling a fuller chronology of Darwin's life to be drawn up and helping to remove some dating difficulties.

Libraries all over the world have given help that was literally indispensable by making available photocopies of Darwin correspondence and other manuscripts in their collections. The institutions and individuals that furnished copies of letters for this volume are listed on pp. xxii–xxiii. To all of them the editors are extremely grateful. We are also grateful to the many people who have transmitted information regarding the whereabouts of particular letters and, in many cases, have generously provided copies of such letters for the project's use.

Among the librarians whose help has been of especial importance, the foremost is Peter J. Gautrey who has charge of the Darwin Archive at the Cambridge University Library. His great knowledge of Darwiniana has always been readily shared with the editors, as it has with countless other Darwin scholars. We are also most grateful to the University librarian, Frederick W. Ratcliffe, and all those members of the staff of the University Library, Cambridge, who have helped us in our work. In particular the following have frequently responded to the editors' needs: Arthur Owen, Elizabeth Leedham-Green, Margaret Pamplin, Jayne Ringrose, Gerry Bye, Godfrey Waller, Janice Fairholm, Roger Fairclough, Cynthia Webster, Louise Aldridge, Shona Johnson, and Richard Flood.

At the American Philosophical Society Library a splendid collection of Darwiniana and works in the history of science has been continuously available to the editorial staff since the project was started. Whitfield J. Bell Jr, secretary of the Society until 1983, has served on the U.S. Advisory Committee for the project and has done his utmost to further its work. The editors have also benefited on a daily basis from the co-operation of Edward Carter II, Murphy D. Smith, Roy C. Goodman, Stephen Catlett, Willman Spawn, Carl F. Miller, Elizabeth Carroll Horrocks, and Bertram Dodelin, all of the APS Library.

Douglas W. Bryant, formerly of the Harvard University Library, Rodney Dennis of the Houghton Library, Constance Carter of the Science Division of the Library of Congress, James Henderson, Walter Zervas, and Joseph Mask of the New York Public Library have all been exceptionally helpful in providing material from their great collections.

The editors often had recourse to assistance from Maldwyn J. Rowlands, Rex E. R. Banks, Anthony P. Harvey, Mrs Valerie C. Phillips, and Dorothy Norman of the British Museum (Natural History) Library; Gina Douglas, librarian of the Linnean Society of London; Mrs C. Kelly, archivist of the Royal Geographical Society; John Thackray of the Geological Society Archives; Victor T. H. Parry, Sylvia FitzGerald, Irene Smith, and Miss Thompson of the Library of the Royal Botanic Gardens, Kew; Anthony M. Carr, Miss Williams, and Mrs Ion of the Local Studies Library, Shrewsbury; Anthony P. Shearman, City librarian, of Edinburgh; Christine Fyfe, archivist of Keele University.

Among the scholars who have been consulted in editing the letters are Joyce Adler, Paul H. Barrett, P. Thomas Carroll, Ralph Colp Jr, the late Richard B. Freeman, Julius Held, Desmond King-Hele, Jack Morrell, Jane Oppenheimer, Martin Rudwick, Silvan S. Schweber, James A. Secord, David Stanbury, Frank J. Sulloway, Garry J. Tee, and Hugh Torrens. Judith Butcher, Fredson Bowers, and G. Thomas Tanselle were frequently called upon for advice on editorial method. Mario di Gregorio prepared translations of letters from European correspondents and generously made available the results of his labours on Darwin's marginalia.

Since work began on the Darwin correspondence a number of able assistants have helped with research, checking, filing, and other essential tasks. We are grateful to: Doris E. Andrews, Pamela J. Brant, Deborah Fitzgerald, Jane Mork Gibson, Dorothy Hoffman, Thomas Horrocks, Christine M. Joyner, Joan W. Kimball, Barbara A. Kimmelman, Nancy Mautner, John A. Reesman, and Edith Stewart.

Special thanks are due to Hedy Franks and Alison Soanes for their skill and patience in deciphering and keyboarding complex manuscript material, and to Heidi Bradshaw for valuable research assistance and diligent proofreading.

Finally and with much regret we must record the deaths of two members of the U.K. Advisory Committee during the preparation period for this volume. John Gilmour and Richard Freeman will be missed by many. We are grateful for the time and expertise that they, and all members of the Advisory Committees, have donated to the project.

LIST OF PROVENANCES

The following list gives the locations of the original versions of the letters printed in this volume. The editors are grateful to the individuals and institutions listed for allowing access to the letters.

American Philosophical Society, Philadelphia, Pa., USA
Amherst College Archives, Amherst, Mass., USA
Auckland Public Library, Auckland, New Zealand
Bath Reference Library, 18 Queen Square, Bath, England
Beinecke Library, Yale University, New Haven, Conn., USA
British Library, Great Russell Street, London, England
British Museum (Natural History), South Kensington, London, England
Brown University Library, Providence, R.I., USA
Burgerbibliothek Bern, Bern, Switzerland
Cambridge University Library, Cambridge, England
Christ's College Library, Cambridge, England
Cleveland Health Sciences Library, Cleveland, Ohio, USA
George Clive (private collection)
CUL *see* Cambridge University Library
DAR *see* Cambridge University Library
Down House, Downe, Kent, England
L. D. Edmondston (private collection)
Emma Darwin (publication)
Fitzwilliam Museum, Cambridge, England
Gardeners' Chronicle and Agricultural Gazette (publication)
Geological Society Archives, Burlington House, London, England
Houghton Library, Harvard University, Cambridge, Mass., USA
Imperial College Archives, Imperial College of Science and Technology, London, England
Keele University Library, Keele, Staffordshire, England (Wedgwood/Mosley papers courtesy of the Trustees of the Wedgwood Museum, Barlaston, Stoke-on-Trent, England)
Kew *see* Royal Botanic Gardens, Kew
Lincolnshire Record Office, The Castle, Lincoln, England
Missouri Botanical Garden Library, Saint Louis, Mo., USA
Museum für Naturkunde der Humboldt-Universität zu Berlin, German Democratic Republic

Muséum National d'Histoire Naturelle, Paris, France
John Murray Archive, John Murray (Publishers), London, England
New York Botanical Garden Library, Bronx, N.Y., USA
New York Public Library, Astor, Lenox and Tilden Foundations, New York, N.Y., USA
Norwich Castle Museum, Norwich, England
Proceedings of the Royal Society of Edinburgh (publication)
Quarterly Journal of the Geological Society of London (publication)
Sir Tom Ramsay (private collection)
Hans Rhyn (private collection)
Royal Botanic Gardens, Kew, Richmond, Surrey, England
Royal Geographical Society, Kensington Gore, London, England
Royal Society, London, England
Smithsonian Institution, Washington, D.C., USA
Sotheby's, London, England (dealers)
J. A. Stargardt, Marburg, Federal Republic of Germany (dealers)
Ulster Museum, Botanic Gardens, Belfast, Northern Ireland
University of Michigan Library, Ann Arbor, Mich., USA
University of Oklahoma Library, Norman, Okla., USA
University of Rochester Library, Rochester, N.Y., USA
University of Saint Andrews Library, Saint Andrews, Fife, Scotland
University of Texas at Austin, Harry Ransom Humanities Research Center, Austin, Tex., USA

A NOTE ON EDITORIAL POLICY

The first and chief objective of this edition is to provide complete and authoritative texts of Darwin's correspondence. Insofar as it is possible the letters have been dated, arranged in chronological order, and the recipients or senders identified. Darwin seldom wrote the full date on his letters and, unless the addressee was well known to him, usually wrote only 'Dear Sir' or 'Dear Madam'. In the 1840s, after the adoption of adhesive postage stamps, the separate covers that came into use with them were usually not preserved and thus the dates and the names of many recipients of Darwin's letters have had to be derived from other evidence. The notes made by Francis Darwin on letters sent to him for his editions of his father's correspondence have been helpful, as have matching letters in the correspondence, but many dates and recipients have had to be deduced from the subject-matter or references in the letters themselves. These tasks, together with the deciphering of Darwin's handwriting, have been the most troublesome problems.

Whenever possible transcriptions have been made from manuscript. If the manuscript was inaccessible but a photocopy or other facsimile version was available, that version has been used as the source. Other copies and published texts have been transcribed when they have provided the only known version of a letter.

The method of transcription employed in this edition is adapted from that described by Fredson Bowers in 'Transcription of manuscripts: the record of varients', *Studies in Bibliography* 29 (1976): 212–64. This system is based on accepted principles of modern textual editing and has been widely adopted in literary editions.

The case for using the principles and techniques of this form of textual editing for historical and non-literary documents, both in manuscript and print, has been forcefully argued by G. Thomas Tanselle in 'The editing of historical documents' in *Studies in Bibliography* 31 (1978): 1–56. The editors of the *Correspondence* have followed Dr Tanselle in his conclusion that a 'scholarly edition of letters or journals should not contain a text which had editorially been corrected, made consistent, or otherwise smoothed out' (p. 48), but they have not wholly subscribed to the statement made earlier in the article in which he says, 'In the case of notebooks, diaries, letters and the like, whatever state they are in constitutes their finished form, and the question of whether the writer "intended" something else is irrelevant' (p. 47). The editors have preserved the spelling, punctuation, and grammar of the original, but they have found it impossible to set aside entirely the question of authorial intent. One obvious reason is that in reading Darwin's writing, there must necessarily be reliance upon both context and intent. Even when Darwin's general intent is clear, there are cases in which alternative readings are, or may be, possible and therefore the transcription decided upon must to some extent be conjectural. In this work, when the editors were uncertain of their transcription they have enclosed the doubtful text in italic square brackets.

It was also necessary to consider Darwin's intent in coping with the transcription problem posed by his practice in the early letters of sprinkling his sentences with points. These would normally be interpreted as full stops, but their number and position make it clear that they could not possibly have been intended to function as such. At an early stage in their work the editors interpreted these points as 'pen rests' marking places where Darwin paused to take thought. This may, indeed, be the explanation for some of them. Where the points may have been intended to act as commas they have been silently changed; those for which no grammatical function could be conjectured have been omitted, with the omissions recorded in the Manuscript alterations and comments section of the apparatus.

This treatment of a particular and rather exceptional problem is consistent with the major editorial decision to adopt the so-called 'clear-text' method of transcription, which so far as possible keeps the text free of brackets recording deletions, insertions, and other alterations in the places at which they occur. These changes are recorded in the Manuscript alterations together with any comment the editors have to make regarding the original text. The alteration notes are keyed to the printed text by paragraph and line number. All lines above the first paragraph of the letter (i.e., date, address, or salutation) are referred to as paragraph '0'. Separate paragraph numbers are used for subscriptions and postscripts. These notes enable the reader who wishes to do so to reconstruct the manuscript versions of Darwin's letters while furnishing printed versions that are uninterrupted by editorial interpolations. They record all alterations made by Darwin in his letters and any editorial amendments made in transcription. For copies and drafts of Darwin letters included in the correspondence no attempt has been made to record systematically all alterations to the text, but ambiguous passages in copies and major revisions of drafts are noted. The editors believe it would be impracticable to attempt to go further without reliable information about the texts of the original or final versions of the letters involved. The letters to Darwin have been transcribed without recording the writers' alterations unless they reflect significant changes in substance; in such cases footnotes will bring them to the reader's attention.

Misspellings have been preserved, even when it is clear that they were unintentional as, for instance, 'lawer' for 'lawyer'. Such errors often indicate excitement or haste, and over a series of letters may exhibit, in the aggregate, a habit of carelessness in writing to a particular correspondent or about a particular subject.

Capital letters have also been transcribed as they occur except in certain cases, such as 'm' and 'c', which are frequently written somewhat larger than others as initial letters of words. In these cases the normal practice of the writers has been followed. If there is doubt about Darwin's intention, these letters have been transcribed as capitals.

In some instances that are not misspellings in a strict sense, editorial corrections have been made. In his early manuscripts and letters Darwin consistently wrote 'bl' so that it looks like 'lb' as in 'albe' for 'able', 'talbe' for 'table'. Because the form of the letters is so consistent in different words, the editors consider that this is most unlikely to be a misspelling but must be explained simply as a peculiarity of Darwin's handwriting. Consequently, the affected words have been transcribed as normally spelled and no record of any alteration is given in the apparatus. Elsewhere, though, there are misformed letters that the editors have recorded because they do, or could, affect the meaning of the word in which they appear. The main example is the occasional inadvertent crossing of 'l'. When the editors are satisfied that the intended letter was 'l' and not 't', as, for example, in 'stippers' or 'istand', then 'l' has been transcribed, but the actual form of the word in the manuscript has been given in an alteration note.

Editorial interpolations are in square brackets. Italic square brackets enclose conjectured readings and descriptions of illegible passages. To avoid confusion, in the few instances in which Darwin himself used square brackets, they have been altered by the editors to parentheses, with the change recorded.

Material that is irrecoverable because the manuscript has been torn or damaged is indicated by angle brackets; any text supplied within them is obviously the responsibility of the editors.

Words and passages that have been underlined for emphasis are printed in italics, in accordance with conventional practice. When the author of a letter has indicated greater emphasis by underlining a word or passage two or more times then the greater emphasis is indicated by printing the text in bold type.

Paragraphs are often not clearly indicated in the letters. Darwin and others sometimes marked a change of subject by leaving a somewhat larger space than usual between sentences; sometimes Darwin employed a longer dash. In these cases and in very long stretches of text, when the subject is clearly changed, a new paragraph is started by the editors without note. The start of letters, valedictions, and postscripts are also treated as new paragraphs regardless of whether they appear as such in the manuscript. Special manuscript devices delimiting sections or paragraphs, e.g., blank lines and lines drawn across the page, are treated as normal paragraph indicators and are not specially marked or recorded unless their omission leaves the text unclear.

Additions to a letter that run over into the margins or are continued at its head or foot, are transcribed at the point in the text at which the editors believe they were intended to be read. The placing of such an addition is recorded in a footnote if it seems to the editors to have some significance or if the position at which it should be transcribed is unclear.

Occasionally, punctuation marking the end of a clause or sentence is not present in the manuscript, but the author has made his or her intention clear by allowing, for example, extra space or a line break to function as punctuation. In such cases the editors have inserted an extra space following the affected sentence or clause to set it off from the following text.

Some Darwin letters and an occasional letter to Darwin are known only from entries in the catalogues of book and manuscript dealers or mentions in other published sources. Whatever information these sources provide about the content of such letters has been reproduced.

For every Darwin letter, the text that is available to the editors is always given in full. Some other items, however, are not printed in their entirety. Most memoranda and other documents that are not letters but are relevant to the correspondence have been summarised, as have those letters to Darwin that the editors consider can be presented adequately in shortened form.

The format in which the transcriptions are printed in the *Correspondence* is as follows:

1. *Order of letters.* The letters are arranged in chronological sequence. A letter that can be dated only approximately is placed at the earliest date on which the editors believe it could have been written. The basis of a date supplied by the editors is given in a footnote, unless the date is conjectured solely from the relationship of a letter to those that immediately surround it, or unless the date is derived from a postmark, watermark, or endorsement as recorded in the physical description of the letter (see section 4, below). Letters that have the same date are printed in alphabetical order of their senders or recipients unless the contents of the letters dictate a clear alternative order.

2. *Headline.* This gives the name of the sender or recipient of the letter and its date. The date is given in a standard form, but those elements not taken directly from the manuscript are supplied in square brackets.

3. *The letter text.* The transcribed text follows as closely as possible the layout of the source, although no attempt is made to produce a type facsimile of the manuscript; that is, word-spacing and line-division in the running text are not adhered to. Dates and addresses given by authors are transcribed as they appear, except that if both date and address are at the head of the letter they are always printed on separate lines with the address first, regardless of the exact manuscript order. If no address is given anywhere in the letter the editors have supplied one, when able to do so, in square brackets at the head of the letter. Addresses on printed stationery are transcribed in italics, often in a simplified form. Addresses, dates, and valedictions have been run into single lines to save space, but the positions of line-breaks in the original are marked by vertical bars.

4. *Physical description.* All letters are signed autograph letters (ALS) unless otherwise described. A letter that is signed by Darwin but written in the hand of an amanuensis is noted by the abbreviation 'LS'. If possible the writer is identified. If Darwin has made corrections or additions in his own hand, the

abbreviation used is 'LS(A)'. If the source of the text is other than manuscript, details are given. Postmarks, endorsements by recipients and others, and watermarks are recorded only when they are evidence for the date or address of the letter.

5. *Source.* The final item provides the provenance of the text. Some sources are given in abbreviated form (e.g., DAR 140: 18), but all are listed in full in the List of provenances, unless the source is a published work. References to published works are given in author–date form, with full titles and publication details supplied in the bibliography.

6. *Darwin's annotations.* Darwin frequently made notes in the margins of the letters he received, scored significant passages, and crossed through details that were of no further interest to him. These annotations are transcribed or described following the letter text together with details of their location on the manuscript and the writing medium. They are keyed to the letter text by paragraph and line numbers. Most notes are short, but occasionally they run from a paragraph to several pages, often written on separate sheets appended to the letter. Extended notes relating directly to a letter and physically associated with it are transcribed whenever practicable.

Darwin's notes occasionally take the form of a draft reply. In such cases, if the final form exists in a preserved letter, the draft is not reproduced, but its existence and significant differences from the final form are noted. If the final form has not been found, the draft reply is transcribed in lieu of the missing letter.

Quotations from Darwin manuscripts in footnotes and elsewhere, and the text of his annotations and notes on letters, are transcribed in 'descriptive' style. In this method the alterations in the text are recorded in brackets at the places in which they occur. For example:

'See Daubeny ['vol. 1' *del*] for *descriptions of volcanoes in [*interl*] S.A.' *ink*

means that Darwin originally wrote in ink 'See Daubeny vol. 1 for S.A.' and then deleted 'vol. 1' and inserted 'description of volcanoes in' after 'for'. The asterisk before 'descriptions' marks the beginning of the interlined phrase, which ends at the bracket. The asterisk is used when more than one word is interlined. The final text can be read simply by skipping the material in brackets. Descriptive style is also used in the alteration notes in the transcription of deleted passages from Darwin's letters.

Volumes of the *Correspondence* are published in chronological order. Each volume is self-contained, having its own index, bibliography, and biographical notes. A comprehensive index is planned for the final volume. References are supplied for all persons and subjects mentioned, even though some repetition of material in earlier volumes is involved.

If the name of a person mentioned in a letter is incomplete or incorrectly spelled, the full correct form is given in a footnote. Brief biographies of everyone mentioned in a volume and dates of each correspondent's letters to and from Darwin are given in the Biographical register and index to correspondents.

The editors use the abbreviation 'CD' for Charles Darwin throughout the notes and appendixes. A list of all abbreviations used by the editors in this volume is given on p. xxxii. For references to Darwin's books, articles, and collections of his letters, short titles are used (e.g., *Descent*, *Collected papers*, *LL*). Short titles are also used for some standard reference works (e.g., *Alum. Cantab.*, *Wellesley index*, *DNB*). For all other works author–date references are used. The full titles of all the books referred to are given in the bibliography where the short titles and author–date references are listed in alphabetical order.

The Wedgwood and Darwin

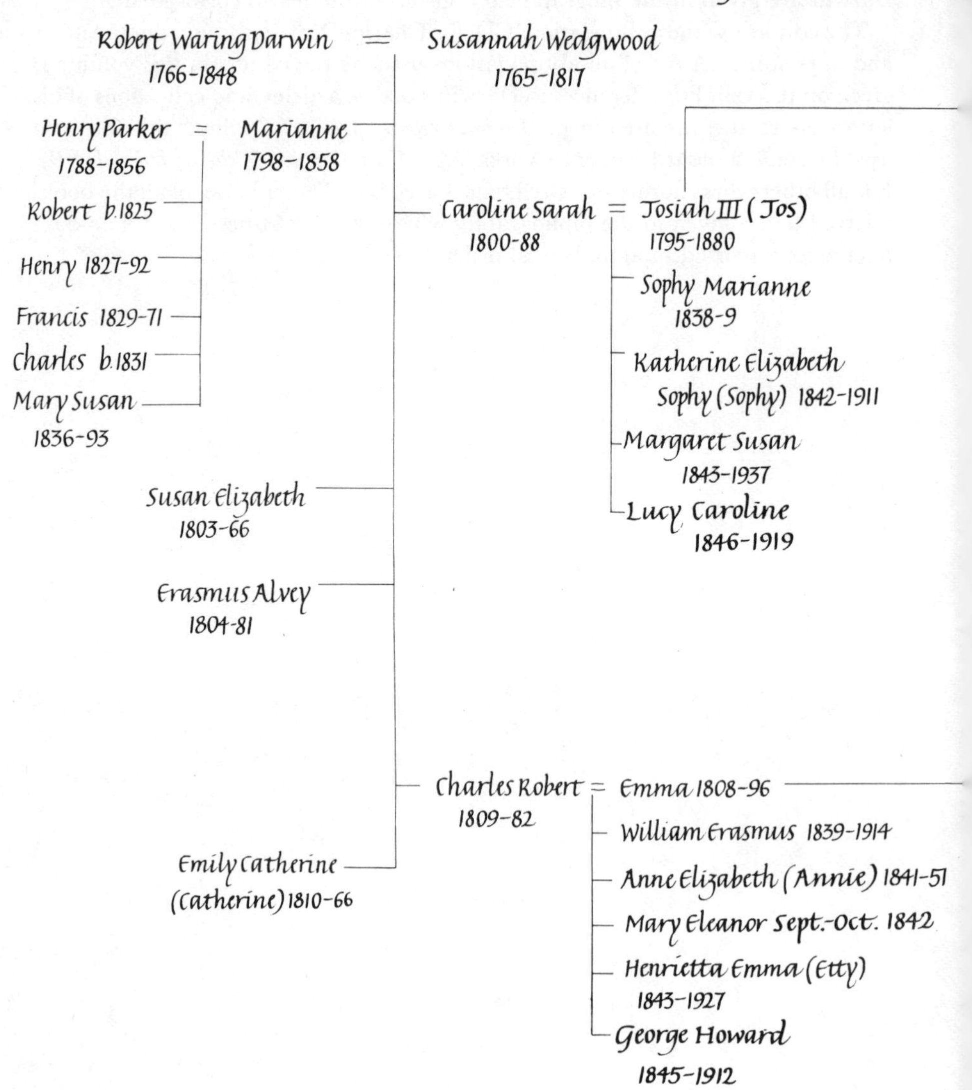

Robert Waring Darwin = Susannah Wedgwood
1766-1848
1765-1817
Henry Parker = Marianne
1788-1856
1798-1858
Robert b.1825
Henry 1827-92
Francis 1829-71
Charles b.1831
Mary Susan
1836-93
Caroline Sarah = Josiah III (Jos)
1800-88
1795-1880
Sophy Marianne
1838-9
Katherine Elizabeth
Sophy (Sophy) 1842-1911
Margaret Susan
1843-1937
Lucy Caroline
1846-1919
Susan Elizabeth
1803-66
Erasmus Alvey
1804-81
Charles Robert = Emma 1808-96
1809-82
William Erasmus 1839-1914
Anne Elizabeth (Annie) 1841-51
Mary Eleanor Sept.-Oct. 1842
Henrietta Emma (Etty)
1843-1927
George Howard
1845-1912
Emily Catherine
(Catherine) 1810-66

Families up to 1846

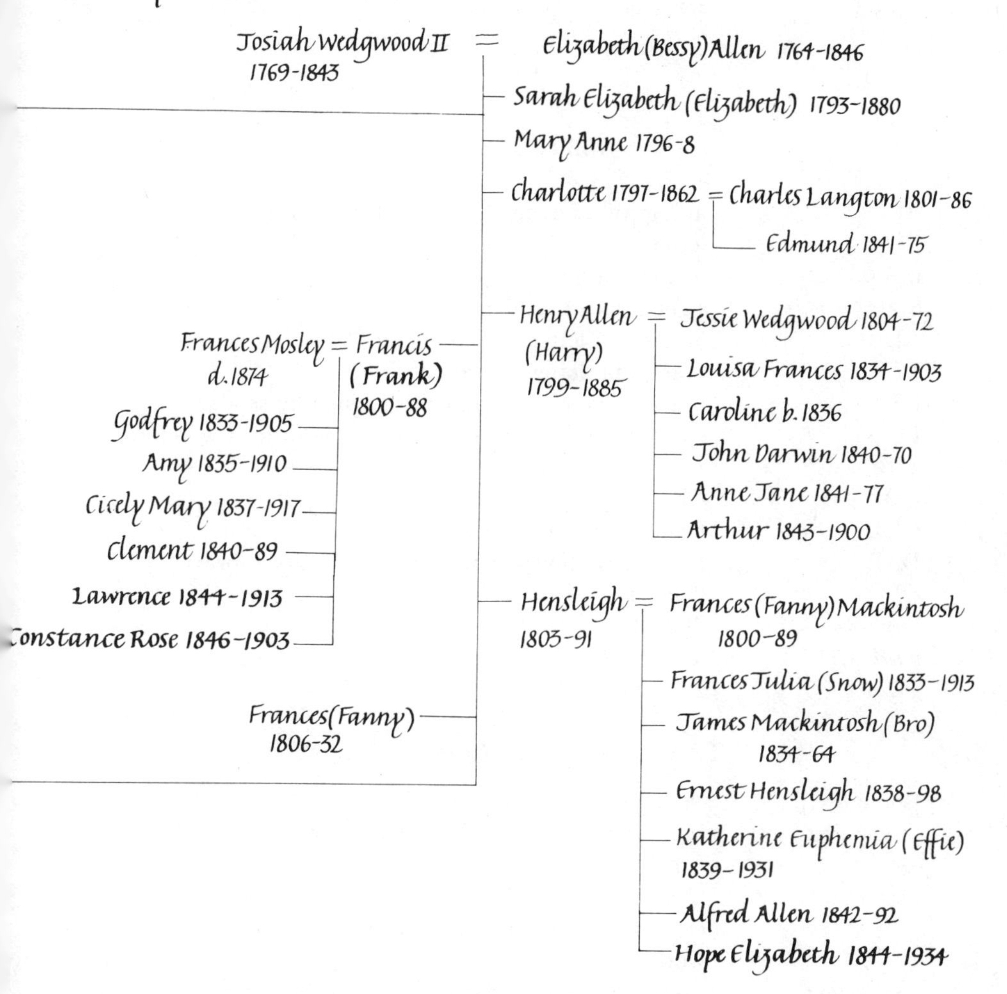
Josiah Wedgwood II
1769-1843
=
Elizabeth (Bessy) Allen 1764-1846
Sarah Elizabeth (Elizabeth) 1793-1880
Mary Anne 1796-8
Charlotte 1797-1862 = Charles Langton 1801-86
Edmund 1841-75
Frances Mosley = Francis
d. 1874
(Frank)
1800-88
Godfrey 1833-1905
Amy 1835-1910
Cicely Mary 1837-1917
Clement 1840-89
Lawrence 1844-1913
Constance Rose 1846-1903
Henry Allen = Jessie Wedgwood 1804-72
(Harry)
1799-1885
Louisa Frances 1834-1903
Caroline b. 1836
John Darwin 1840-70
Anne Jane 1841-77
Arthur 1843-1900
Hensleigh = Frances (Fanny) Mackintosh
1803-91
1800-89
Frances Julia (Snow) 1833-1913
James Mackintosh (Bro)
1834-64
Ernest Hensleigh 1838-98
Katherine Euphemia (Effie)
1839-1931
Alfred Allen 1842-92
Hope Elizabeth 1844-1934
Frances (Fanny)
1806-32

ABBREVIATIONS

A	autograph, i.e., in the hand of the sender
L	letter
mem	memorandum
S	signed by the sender
AL	autograph letter
ALS	autograph letter signed
LS	letter in hand of amanuensis, signed by sender
LS(A)	letter in hand of amanuensis with additions by author
CD	Charles Darwin
CUL	Cambridge University Library
DAR	Darwin Archive, Cambridge University Library
del	deleted
illeg	illegible
interl	interlined
underl	underlined

TRANSCRIPTION CONVENTIONS

[some text]	'some text' is an editorial insertion
*[*some text*]*	'some text' is the conjectured reading of an ambiguous word or passage
[some text]	'some text' is a description of a word or passage that cannot be transcribed, e.g, *3 words illeg*
⟨ ⟩	word(s) destroyed
⟨some text⟩	'some text' is a suggested reading for a destroyed word or passage
⟨*some text*⟩	'some text' is a description of a destroyed word or passage, e.g., *3 lines excised*

THE CORRESPONDENCE OF CHARLES DARWIN
1844–1846

To Geological Society of London [3 January 1844][1]

Down Bromley | Kent
Wednesday

Dear Sir

The Geological map of S. America was not sent me with D'Orbigny's work—[2] would you please send it me *and a card of the Evening meetings*[3]—either by Post, or if *there are any parcels waiting for me*, by "*Down Coach*", from Bolt-in-Tun Fleet St.—[4]

Yours faithfully | C. Darwin

Coach starts daily at 3 olock.

Geological Society Archives (LR8/92)

[1] Dated from surrounding letters in the Geological Society Letter Book, all of which are dated in the first week of January 1844.
[2] Orbigny 1835–47, vol. 3, pt 3: *Géologie*.
[3] CD was vice-president of the Society in 1844.
[4] A tavern from which coaches started for the south of England.

To Joseph Dalton Hooker [11 January 1844]

Down. Bromley Kent
Thursday

My dear Sir

I must write to thank you for your last letter;[1] I to tell you how much all your views & facts interest me.— I must be allowed to put my own interpretation on what you say of "not being a good arranger of extended views"—which is, that you do not indulge in the loose speculations so easily started by every smatterer & wandering collector.— I look at a strong tendency to generalize[2] as an entire evil—

What limit shall you take on the Patagonian side—has d'Orbigny published, I believe he made a large collection at the R. Negro, where Patagonia retains its usual forlorn appearance;[3] at Bahia Blanca & northward the features of Patagonia insensibly blend into the savannahs of La Plata.— The Botany of S. Patagonia (& I collected *every* plant in flower at the season when there) would be worth comparison with the N. Patagonian collection by d'Orbigny.— I do not know anything about King's plants, but his birds were so inaccurately habitated, that I have seen specimen from Brazil, Tierra del & *the Cape de Verde Is*^{d} all said to come from the St. Magellan.—[4] What you say of M[r] Brown is humiliating; I had suspected it, but cd not allow myself to believe in such heresy.— FitzRoy gave him a rap in his Preface, & made me very indignant, but it seems a much harder one w[d] not have been wasted.[5] My crptogamic collection was sent to Berkeley; it was not large; I do not believe he has yet published an account, but he wrote to me some year ago that he had described & mislaid all his descriptions.[6] W[d] it not be well for you to put yourself in communication with him; as otherwise some things will perhaps be twice laboured over.— My best (though poor) collection of the Crptogam. was from the Chonos Islands.—[7]

Would you kindly observe one little fact for me, whether any species of plant, *peculiar* to any isl[d], as Galapagos, St. Helena or New Zealand, where there are no large quadrupeds, have hooked seeds,—such hooks as if observed here would be thought with justness to be adapted to catch into wool of animals.—

Would you further oblige me some time by informing me (though I forget this will certainly appear in your Antarctic Flora) whether in isl[d] like St. Helena, Galapagos, & New Zealand, the number of families & genera are large compared with the number of species, as happens in coral-isl[d], & as I *believe*? in the extreme Arctic land. Certainly this is case with Marine shells in extreme Arctic seas.— Do you suppose the fewness of species in proportion to number of large groups in *Coral-islets*., is owing to the chance of seeds from all orders, getting drifted to such new spots? as I have supposed.—[8]

Did you collect sea-shells in Kerguelen land, I sh[d] like to know their character.?

Your interesting letters tempt me to be very unreasonable in asking you questions; but you must not give yourself any trouble about them, for I know how fully & worthily you are employed.

Besides a general interest about the Southern lands, I have been now ever since my return engaged in a very presumptuous work & which I know no one individual who w[d] not say a very foolish one.— I was so struck with distribution of Galapagos organisms &c &c & with the character of the American fossil mammifers, &c &c that I determined to collect blindly every sort of fact, which c[d] bear any way on what are species.— I have read heaps of agricultural & horticultural books, & have never ceased collecting facts— At last gleams of light have come, & I am almost convinced (quite contrary to opinion I started with) that species are not (it is like confessing a murder) immutable. Heaven forfend me from Lamarck nonsense of a "tendency to progression" "adaptations from the slow willing of animals" &c,—but the conclusions I am led to are not widely different from his—though the means of change are wholly so— I think I have found out (here's presumption!) the simple way by which species become exquisitely adapted to various ends.—[9] You will now groan, & think to yourself 'on what a man have I been wasting my time in writing to.'— I sh[d], five years ago, have thought so.— I fear you will also groan at the length of this letter—excuse me, I did not begin with malice prepense.

Believe me my dear Sir | Very truly your's | C. Darwin

Postmark: JA 11 1844
DAR 114.1: 3

[1] See *Correspondence* vol. 2, letter from J. D. Hooker, [12 December 1843 – 11 January 1844]. Hooker, who was preparing his *Flora Antarctica* (J. D. Hooker 1844–7), had agreed to describe CD's *Beagle* plants.

[2] The original reads 'generatize', which raises the possibility that CD intended to coin a word for the tendency of some taxonomists to proliferate genera. However, CD often crossed 'l's unintentionally. It occurs in words with both 'l' and 't', as well as in words like 'istand' and 'stippers' (see *Correspondence* vol. 1, Manuscript alterations and comments for letters to Susan Darwin, [6 September 1831] and 17 [September 1831]). The context appears to favour a meaning of hasty or over-generalisation, whereas 'generatize' never occurs again.

[3] See *Correspondence* vol. 1, letter to J. S. Henslow, [c. 26 October –] 24 November [1832]. Alcide Charles Victor Dessalines d'Orbigny published descriptions of the Cryptogamia of Patagonia and Bolivia in 1839, and the palms of Paraguay and Bolivia in 1847 (Orbigny 1835–47, vol. 7).

[4] The collections made by Phillip Parker King during the first surveying expeditions of the *Beagle* and *Adventure* to South America, 1826–30.

[5] Robert Brown, keeper of the botanical collections in the British Museum. In *Narrative* 1: x, Robert FitzRoy wrote: 'Captain King took great pains in forming and preserving a botanical collection . . . He placed this collection in the British Museum, and was led to expect that a first-rate botanist would have examined and described it; but he has been disappointed.'

[6] Miles Joseph Berkeley described parts of CD's cryptogamic collection in 1839, 1842, and 1845. No letters from Berkeley during this period have been traced, but see *Correspondence* vol. 2, letters to M. J. Berkeley, [26 November 1840] and [March 1841].

[7] CD recorded: 'in these islands, within the forest, the number of species, and great abundance of mosses, lichens, and small ferns, is quite extraordinary' (*Journal of researches*, p. 349).

[8] *Journal of researches*, p. 541.

[9] CD first formulated his theory of natural selection in autumn 1838 (*Notebook D*: 134e–5e).

To Henry Denny 20 January [1844]

Down near Bromley | Kent
Jan. 20th.—

Dear Sir

It would give me great pleasure to assist you in your valuable scientific labours.—[1]

I cannot remember whether I collected many lice; I think I did some; but part of my collection was lost & from ill-health & my extreme wish to finish one part of my materials (viz Geology) has prevented me for a long time going through my zoological collection.— I intend, however, doing this soon; & you may **rely** on it, that I will put on one side, everything I find, which is in your department, & will then communicate with you.—[2] I fear the result will be small, if any.—

Everything I collected was properly ticketed & if you ask Mr. Waterhouse,[3] I think he can tell you (if the specimens came from me) at least what country, & I shd think the specimens wd have had a separate number, which wd tell the bird, by reference to me.—

With my best wishes for your success.— Believe me | Yours very faithfully | C. Darwin

American Philosophical Society

[1] Denny had been approached by a British Association for the Advancement of Science committee, consisting of William Jardine, William Yarrell, and Edwin Lankester, to help in preparing a report on the exotic species of *Anoplura*. Denny later recorded that he wrote to several naturalists for assistance and specimens. His illustrations were exhibited at the British Association meeting in York in September 1844 (*Report of the 14th meeting of the British Association for the Advancement of Science held at York in 1844*, p. 392).

[2] CD sorted his collections in July 1844 and provided Denny with specimens. See 'Journal' (Appendix II) and letter to Henry Denny, 12 August [1844].

[3] George Robert Waterhouse, who described part of CD's entomological collection from the *Beagle* voyage.

To Ernst Dieffenbach 25 January 1844[1]

[Down]

'. . . I am delighted at the thoughts of being able to contribute any Infusoria to your great countryman Ehrenberg. . .'[2]

J. A. Stargardt, Marburg (catalogue 574, 11–13 November 1965)

[1] Date taken from the Stargardt catalogue.

[2] The Stargardt catalogue description notes that the letter was accompanied by a list of eight substances from CD's collection, described in detail, which might be useful in Christian Gottfried Ehrenberg's research on the Infusoria.

To J. D. Hooker [27 January 1844][1]

Down. Bromley Kent
Saturday night.

My dear Sir

I hope you will excuse me troubling you again, I have heard through a friend[2] from Ehrenberg, that he wants some earth from the Galapagos to hunt for Infusoria,[3] & he suggests that perhaps a little may hang to the roots,—I fear not, but would you kindly look, & if you can shake any off, send it me.— Ehrenberg further is anxious for any earth or more especially peat from T. del. Fuego or the Falkland Islands.— I had specimens of peat, showing the process of its formation, by plants like Astelia &c &c, I do not know whether they were sent to you— I fear they are probably lost,—but I daresay a little peaty earth could be shaken off some of the little peat-loving plants of T. del. Fuego.—

I know you will not grudge some little trouble for so great a naturalist as Ehrenberg.

Pray believe me | Yours very sincerely | C. Darwin

DAR 114.1: 4

[1] Dated from the letter to Ernst Dieffenbach, 25 January 1844, and on the assumption that CD would have written to Hooker immediately.
[2] Ernst Dieffenbach.
[3] Eventually described in Ehrenberg 1853.

From J. D. Hooker 29 January 1844

West Park Kew
January 29th. 1844

My dear Sir

The arrival of your letter this morning put me to shame, for not having answered ere this your former very kind one. You take so much notice of me, that I am almost afraid of saying too much, & of destroying the illusory character you give of my little notes—

Those French works of Voyages are so unsatisfactory, until finished, they come out so irregularly, & are transmitted to our houses so incomplete that as yet I have not been able to make out from D'Orbignys work what particular plants he has described; I find a great deal in scattered parts, but am at a loss as to what countries he intends to fully illustrate.[1] The Botany of Patagonia is entirely cut off as you have remarked from that of Fuegia, & the Chilian coast, in so much so that I thought of considering its Flora seperately, & not connecting it with that of the latter country, further than is avoidable from having a few of its types in the Falkland Islds. No circumstance is so remarkable as the comparative abundance of Leguminosæ along the Patagonian coast, & their almost total absence on the Chilian & Fuegian. This fact your collection illustrates better than any other as your collection of that order from Rio Negro & other Patag. coasts is good. It is however from the absence of most vegetable forms, common to similar Latitudes that the characters of those singular plains should be drawn. Strictly speaking my flora should contain *no* Patagonian plants, as they belong to Northern types, hardly found to the Southward of the straits, but which increase in number of species as the Southern tropic is approached, as far as the Banda Oriental,[2] where they blend the Brazilian; but the maximum of the Fuegian & highly southern types of vegetation will be found in about 45 S., or perhaps 40, they decrease as Botanical features from thence to Valdivia; but not so southward.—

I cannot but consider the Beeches as important a Bot. feature & guide in S. Am, as the Chestnut forests are on Madeira, the belt of pines in North Asia (vid Map to Wrangel's tour)[3] the Birch in N. America, or any other genus or species which defines the geographical limits of a large amount of other species, of plants. Were such a term allowable, as a name to a book, I should prefer calling such a flora, that "of the Beech districts" of S. America.

The Eastern limit to agree with this definition in the Straits of Magal^s should coincide with the change in Geolog. formation, & with the Beeches, cease at Cape Negro; as however the Patagonian flora is the Negative one, & characterized by the absence of *Beech district* forms; some of its peculiar plants are probably also found in the latter & I should include therefore those found on the immediate boundary, as those of Elizabeth Isld & Cape Gregory Bay. The Geological features of K. Charles S. land to the Northd of Admiralty Sound & all to the Westd of that Latitude is I suppose Patagonian in its flora, as well as its Geology. & had I its plants they too should be excluded..— Mr Berkeley was with us shortly before your letter arrived, & I gave him all my Fungi, he has returned me for Mr Henslow[4] your Lichens, which he had; but said nothing about the other orders in his possession, I shall write to him about them.. I found a third species of the curious genus *Cyttaria* on the deciduous Beech in Hermite Isld. a much smaller species with only 4 cells—

I am now examining the Galapago's plants & shall soon duly report to you on the state of the seeds as to arming, &c.[5] I have begun with the Cryptogamia. *Fungi* one species, found all over the world. *Lichens* 2 species both gerontogeous[6] & English; I believe the Tortoise to be the only animal which eats the *Usnea plicata*; the most common Lichen decidedly in the world. There are no Algæ "hiatus valde deflendus" but the Filices with which I am now working, are very nice, most of them are of course common to S. Am & the W. Indies especially, a few more particularly to the W. coast; a few also are quite new; the difficulty of determining such long neglected tribes is however difficult & I do not speak confidently. I hope soon to present all the Cryptog in a written form to the L. Soc. with descript. of n. sp^s.[7] Amongst these the number of genera to sp. is very large, as in all circumscribed portions of land.[8] Were there any tree ferns among them? i.e. with short stems?.—

I have notes on the comp. number of sp to gen in various places but they are at the Admiralty, for though I have been 4 months at home they have not yet returned me the notes drawings &c Botanical & others which I gave up as per order. When I receive them I shall tell you what little I have done; the results I think were curious regarding Arctic forms.— These lists are troublesome to make, as species are so loosely described. For instance of the arbor. comp.[9] of St Helena in one genus 5 species are made out of 3, & in another 4 out of 2, which makes a vast difference in so small a flora— on the other hand, the French in D'Urville's last voyage have made 2 genera out of the 2 Antart. Beeches! & neither of them *Fagus*; true species of which they assuredly both are, as far as the limits of genera can be at all defined.—[10]

You ask me whether I suppose the small proportion of sp to genera in Coral Islets, arises from chance of seeds &c? I cannot answer this, I should say perhaps not:— if genera or small groups are truly natural they are supposed to contain many characters in common, it is but right to assume that the character of *transportable seeds* should hence be common to some groups above others, the inference I need not state. The seeds of Cruciferous plants do not keep well & this I believe to be a character of the group. Yet Cruciferæ are found all over the world. The presence of some most remarkable plants in several remote isolated spots, staggers all my notions of the migration of species— The Kerguelen's Land cabbage is found only in that Island & is the most remarkable plant of its whole Nat Ord in the whole S. Hemisphere there is nothing at all like it any where else; yet almost all the other Kerg. land plants are Fuegian.[11] Vegetation was doubtless once very different on the same spot to what it is now. Nor do I see that we have any chance of solving the question that relates to the existence of certain plants on Islands created, (we suppose,) before the time of man. That there was a beginning to the creation of plants on our globe is very true, we can hardly suppose that we have now only the remains of that original stock or why should not the said cabbage grow on lands we suppose older than Kerg Land, or the Seychelle double cocoa nut on older formations than they are— There may in my opinion have been a series of productions on different spots, & also a gradual change of species. I shall be delighted to hear how you think that this change may have taken place, as no presently conceived opinions satisfy me on the subject.—

The shells of Antarct regions I should have thought were more proportionally abundant in species than genera— In Kerg Land I procured 3 limpets (no Fissurella) 3 Muscles & one land shell, I think only 2 other sea shells were found; but I had them not, I speak from recollection— In all Southern regions it appeared to me that there were more of those genera Mussels & limpets in proportion to others, than of any one genus of Seaweeds to another; & that by very far; the fact always struck me. All the facts of Nat. Hist. that tend to illustrate insular flora's are to me most interesting, & will prove the most valuable of any toward satisfying enquiry about the adaptation of species to various ends— You are I daresay aware of the fact that there is no reason to believe that plants can be artificially acclimated to any extent— Gardeners have hardly made any plant hardy, either by growing it from seeds of an introduced live specimen which did *but just ripen*, or by grafting on allied hardier species.—

Many plants seem made to live every where & others no where but where they *seem* to have been generated.. You may depend upon my best exertions to name the Galapago plants carefully: it is a slow business but I like it much.—

Incomplete
DAR 100: 5–6

CD ANNOTATIONS
1.1 The arrival . . . illustrate. 2.5] *crossed pencil*
3.1 I cannot . . . arming, &c. 5.2] *crossed pencil*
5.4 I believe . . . world 5.5] *scored pencil*
5.6 Filices . . . at all defined.— 6.11] *crossed pencil*
5.12 Were there . . . them?] *scored pencil*
7.4 it is but . . . state. 7.6] *scored pencil*

7.5 the inference I need not state. 7.6] 'Does it not lead to inference that more transported than grow'[12] *added pencil*
7.8 most remarkable . . . Ord in 7.11] *scored pencil*
8.1 The shells . . . recollection— 8.4] *scored pencil*
8.7 All the facts . . . like it much.— 9.3] *crossed pencil*
Top of first page: 'Geology' *pencil, circled pencil*

[1] Orbigny 1835–47. See letter to J. D. Hooker, [11 January 1844], n. 3.
[2] The South American name for the state of Uruguay: literally, the eastern shore of the River Uruguay. CD also used this name, see *Journal of researches*, p. 169.
[3] Ferdinand Petrovich Wrangel. See Wrangel 1840.
[4] John Stevens Henslow, who had originally received CD's *Beagle* plant collections. The plants were subsequently passed to Hooker in the autumn of 1843 (see *Correspondence* vol. 2, letter from J. D. Hooker, 28 November 1843).
[5] See CD's query on hooked seeds in his letter to J. D. Hooker, [11 January 1844].
[6] Belonging to the Old World.
[7] Hooker did not publish descriptions of CD's Galápagos cryptogams separately. Rather they formed part of J. D. Hooker 1845d, published by the Linnean Society.
[8] J. D. Hooker 1844–7, p. 217 n.:

> I may remark, that species in isolated islands are generally well defined; this is in part the natural consequence of another law which I have observed, that genera in islands bear a large proportion to the species, or in other words, that genera are small, seldom containing more than two or three species, and very frequently solitary representatives.

[9] Arborescent Compositae.
[10] The beeches were divided into *Calucechinus* and *Calusparassus* by Jacques Bernard Hombron in the atlas of plates to Dumont d'Urville [1841–54]. Hooker reallocated Hombron's species in J. D. Hooker 1844–7, pp. 345–9. Later volumes in Dumont d'Urville [1841–54] acknowledge Hooker's corrections, see Decaisne 1853, pp. 7–10.
[11] *Pringlea antiscorbutica*. See J. D. Hooker 1844–7, pp. 238–41, for a fuller description of the Kerguelen Land cabbage. See also plate facing p. 288.
[12] A pencil note, written on cream-coloured, unwatermarked paper, 20cm wide, is bound with this letter (DAR 100: 7). The paper seems to have been cut from a sheet of CD's typical 1840s stationery. The note may be a continuation of the annotations to paragraph seven, as it is written in a similar strong hand, pertains to the same passage, and continues to discuss the relation between struggle and diversity:

> Explanation of fewness of species & diversity of genera, I think must be partly accounted for that plants of diverse groups c^d^ subsist in greater numbers, *& interfere less with each other. [*above del illeg*] This must be explanation of Arctic Regions.— How are Alpine Plants— Several genera?

If contemporary, this note is of interest given the closeness of these views to CD's principle of divergence, as explained in a letter to Asa Gray, 5 September [1857], and *Origin*, pp. 111–26.

To J. D. Hooker [3–17 February 1844]

Down. Bromley Kent
Saturday

My dear Sir

I write a line merely to acknowledge & thank you for your long & to me most agreeable letter & to tell you that I am in communication with Ehrenberg to find out more definitely, what objects he wishes for, and I will let you know in time for you to send me any likely objects to contain infusoria. I know thus far that his chief present object is the geographical range of infusoria, so that I cannot doubt, of all things, he would most value specimens from the Antarctic regions.—

Would not floating sea-weed probably still contain some attached to it— I am astonished at your description of the number of Infusoria in the far-antarctic seas.—

Once again I thank you for your letter, & I can hardly tell you, how much all your facts & opinions interest me.—

In Haste | Believe me | Most truly yours | C. Darwin

PS. | Dr. Dieffenbach, the New Zealand traveller,[1] (who has translated my Journal into German)[2] (& I must with *unpardonable vanity* boast to you, that it was at the instigation of Liebig[3] & Humboldt[4]) wrote to me about the Infusoria at the request of Ehrenberg & to him I have written some further questions.—

I cannot doubt, Ehrenberg would value all your notes & drawings whether imperfect or perfect.

DAR 114.1: 5

1 His travels were described in Dieffenbach 1843.
2 Dieffenbach trans. 1844.
3 Justus von Liebig, professor of chemistry in Giessen and patron of Ernst Dieffenbach.
4 Alexander von Humboldt. For his favourable opinion of CD's *Journal of researches* see *Correspondence* vol. 2, letter from Alexander von Humboldt, 18 September 1839.

To Charles Wicksted 13 February [1844?][1]

Feb. 13th.

My dear Sir

I trust to your kindness in allowing me to trouble you with a question on a point, communicated to me by Mr. Tollet at Betley,[2] & which interests me greatly. Mr Tollet told me that the late Mr. Botfield[3] had a Harrier remarkably good for recovering the scent in paths or roads, & that you sent a bitch to this dog, & that he believed that one of the puppies inherited this good quality. If you would be so kind as to take the trouble to give me a few more particulars, I should be greatly obliged: I do not ask out of quite idle curiosity, as I have for several years been collecting facts on the variation of plants & animals, & all cases of inheritable qualities of body & mind come into this subject. Is the faculty of recovering scent in roads to the extent of Mr Botfield's dog very unusual? did it appear as conspicuously in the puppy as in the parent? did only one puppy inherit it? Have you ever bred from this puppy?.— I have seen it stated that young fox-hounds naturally evince different propensities or qualities, so that one is good to find his fox, another to make casts, one is apt to run stragging & another compact &c &c, & that these qualities often reappear in the offspring. As no one, I suppose, has had better opportunities of judging for yourself, if you do not much dislike the trouble of writing, would you be so very kind as to tell me, whether you observed any such & what cases.

I feel that I have much cause to apologise for thus troubling you, I can rely only on your kindness to excuse me, and I beg to remain

Yours very faithfully & obliged | Charles Darwin

Down Bromley Kent

George Clive

[1] The 'Down Bromley Kent' address indicates date ranges of 1843–6 or 1854 or later. George Tollet, referred to in the letter, died in 1855 so the date cannot be later than that, and the unperforated postage stamp on the cover suggests the earlier period. The precise date is conjectured from the details in the letter relating to the habits of foxhounds which closely match details given by CD in his essay of 1844 (*Foundations*, p. 114). CD was working on his species essay in February 1844 (see 'Journal'; Appendix II).

[2] George Tollet of Betley Hall was the father of Charles Wicksted and a friend of the Wedgwood family.

[3] Probably Thomas Botfield, who died on 17 January 1843.

To J. D. Hooker 23 February [1844]

Down Bromley Kent
Feb. 23^d^

Dear Hooker.

I hope you will excuse the freedom of my address, but I feel that as co-circum-wanderers & as fellow labourers (though myself a very weak one) we may throw aside some of the old-world formality.— Absence from home has prevented me sooner answering your note.[1] If you send the earth &c &c by any public conveyance, w^d^ you send it to "43 Grt. Marlborough St",[2] if otherwise either the Geolog. Soc. or Athenæum will do perfectly.— Of course you will put Lat: & Long: to the Antarctic mud or iceberg infusoria.— I have not yet heard from Ehrenberg.— I have just finished a little volume on the volcanic isl^d^. which we visited;[3] I do not know how far you care for dry simple geology, but I hope you will let me send you a copy.— I suppose I can send it from London by common coach conveyance.—

I am quite ashamed of myself that I omitted to thank in words, but not in my mind, Sir. W. Hooker, for his kind invitation to myself to M^rs^. Darwin to visit the gardens at Kew;[4] I fear it is not likely we shall be able, though it w^d^ give us great pleasure.

I am going to ask you some *more* questions, though I daresay, without asking them, I shall see answers in your work, when published, which will be quite time enough for my purposes. First for the Galapagos, you will see in my Journal, that the Birds, though peculiar species, have a most obvious S. American aspect:[5] I have just ascertained the same thing holds good with the sea-shells.—[6] Is it so with those plants, which are peculiar to this archipelago; you state that their numerical proportions are continental (is not this a very curious fact?) but are they related in forms to S. America.— Do you know any other cases of an Archipelago, with the separate islands possessing distinct representative species? I have always intended, (but have not yet done so) to examine Webb & Bert: on the Canary Is[d] for this object.[7] Talking with M[r] Bentham,[8] he told me that the separate isl[ds] of the Sandwich Arch: possessed distinct representative species of the same genera of Labiatæ: would not this be worth your enquiry? How is it with the Azores; to be sure the heavy West: gales w[d] tend to diffuse the same species over that group.—

I hope you will (I daresay my hope is quite superfluous) attend to this general kind of affinity in isolated islands; though I suppose it is more difficult to perceive this sort of relation in plants, than in birds or quadrupeds, the groups of which are, I fancy, rather more confined. Can St. Helena be classed, though remotely, either with Africa or S. America? From some facts, which I have collected, I have been led to conclude, that the Fauna of mountains are *either* remarkably similar (sometimes in the presence of the same species & at other times of same genera) *or* that they are remarkably dissimilar;[9] and it has occurred to me, that possibly part of the peculiarity of the St. Helena & Galapagos Floras may be attributed to a great part of these two Floras, being mountain Floras.— I fear my notes will hardly serve to distinguish much of the habitats of the Galapagos plants, but they may in some cases;[10] most if not all of the green, leafy plants come from the summits of the islands, & the thin, brown leafless plants come from the lower arid parts: would you be so kind as to bear this remark in mind, when examining my collection.

I will trouble you with only one other question. In discussion with M[r] Gould,[11] I found that in most of the genera of birds, which range over the whole or greater part of the world, the individual species have wider ranges: thus the Owl is mundane, & many of the species have very wide ranges.[12] So I believe it is with land & fresh-water shells—& I might adduce other cases. Is it not so with crptogamic plants; have not most of the species wide ranges, in those genera which are mundane— I do not suppose that the converse holds viz—that when a species has a wide range, its genus also ranges wide:— Will you so far oblige me by occasionally thinking over this. It w[d] cost me vast trouble to get a list of mundane phanærogamic genera & then search how far the species of these genera are apt to range wide in their several countries; but you might occasionally in the course of your pursuits, just bear this in mind, though perhaps the point may long since have occurred to you or other Botanists.[13] Geology is bringing to light interesting facts, concerning the ranges of shells; I think it is pretty well established, that according as the geographical range of a species is wide, so is its persistence or duration in time.—[14]

I hope you will try to grudge as little as you can the trouble of my letters, & pray believe me, very truly your's, | C. Darwin

P.S. I should feel extremely obliged for your kind offer of the sketch of Humboldt; I venerate him, & after having had the pleasure of conversing with him in London,[15] I shall still more like to have any portrait of him.—

DAR 114.1: 6

[1] This note has not been found.
[2] The home of Erasmus Alvey Darwin, CD's brother.
[3] *Volcanic islands.*
[4] William Jackson Hooker, Hooker's father, was director of Kew Gardens.
[5] *Journal of researches*, p. 461.
[6] Eventually recorded in *Journal of researches* 2d ed., pp. 390–1.
[7] Webb and Berthelot, 1835–50.
[8] George Bentham.
[9] According to his 'Journal' (Appendix II), CD was expanding the 'pencil sketch' of his species theory written in 1842 (*Foundations*, pp. 1–53). Alpine floras are referred to in the expanded essay of 1844 (see *Foundations*, pp. 163–4).
[10] D. M. Porter 1980b, pp. 87–8, gives an extract from a *Beagle* specimen notebook 'Printed Numbers Nr. 1426—3342' (Down House MS), which lists Galápagos plants and very spare habitat descriptions.
[11] John Gould. He had described CD's ornithological specimens from the *Beagle* voyage in *Birds*.
[12] CD had been interested in such information for some years. See *Notebook B*: 104–5, 'No doubt in birds; mundane genera are birds, (bats, foxes, Mus) that are apt to wander and of easy transportal.— Waders and waterfowl—scrutinize genera and draw up tables.—' See also the essay of 1844 (*Foundations*, pp. 155–6).
[13] On the last page of this letter Hooker wrote out a table that formed the basis of his reply to CD, see next letter.
[14] Described in C. Lyell 1830–3, 3: 48, 55–6. CD's copy of this work is in the Darwin Library–CUL. CD referred to this generalisation in *Notebook B*: 200e. Edward Forbes discussed the same subject in a lecture delivered at the Royal Institution on the day that this letter was written (E. Forbes 1844, pp. 324–5).
[15] CD had met Alexander von Humboldt at Roderick Impey Murchison's house on 29 January 1842 (see *Correspondence* vol. 2, Appendix II). For CD's notes on the meeting with Humboldt see letter to J. D Hooker, [10–11 November 1844], n. 7.

From J. D. Hooker [23 February – 6 March 1844]

Acot. & Moncot are done,[1] as also a few Dicot.— I expect to come to a rule about several groups of Islds. lying to the W. of large continents of Land in the S. Hemisphere; & hope to prove that they contain a vegetation analogous to that of those continents 20 degrees nearer the adjacent Pole. The Flora of the Galapagos is most allied to that of the S. United States & to that of S. Brazil partially,.— That of St. Helena to the Cape—Remotely it is true, but to none other:—of Tristan d'Acunha to what one would suppose the Cape to be if produced to 50 S., where the Antarctic forms would appear— Though the Galapogean Flora is essentially S. American, the proportions of the Nat: Ords to one another is remarkably different, as is the vast quantity of Arborescent Compos. & particularly of *Euphorbiaceæ* (at which I am now working) & which order has no less than 19 representatives, 16 of them *entirely* new, they are however of very common genera, *Euphorbia*, *Acalypha Croton* & *Phyllanthus*. Now it is very remarkable, that in an order so poorly represented as *Graminæa* (& where genera are so mundane), there should be a new genus, & no less than 6 old ones,.—7 genera for 11 species; while in *Euphorbiaceæ* there are only 4 genera for 19 species—I think in the paucity of grasses there will be a strong analogy to other Tropical Isld^{s};—[2] the reverse holds good in the Islands of higher Latitudes. I wish that the Admiralty had returned me my notes that I could draw out some proportions to send you—

With regard to the dissimilarity between the Flora of the several Islds of the group, that is too extraordinary a circumstance for me to offer any remarks upon, until the *florula* is drawn up, the further I proceed the more I wonder. Had the collections been all made by yourself I should have attributed it to accident, but Macræ's collections[3] are large from Albemarle Isld, & the only 3 plants I have hitherto examined from (Malden Isld (which is it?) are different from those of the other Islets— I was not aware of the analogous fact with regard to the Sandwich group; nor have I yet examined the Canary Isld. Campbell Isld, 2 degrees further S. than Auckland, contains several species not found in the former, though the latter is the smallest & furthest South— I should not think however that it would hold with Islds in the more temperate zones generally, as the Azores which have not very many peculiar plants I shall however sift this subject with my friend Mr H. Watson..[4] I have hitherto come to no plants with seeds particularly adapted for transmission; those that afford such facility most markedly, are the *Compositæ*, & those of the Galaps are the most widely dissimilar from those of any other country of all the Nat. Ords (as far I have seen)— It is I think high time throw overboard laying much stress on the subject of the *migration of seeds*, except in the cases of lands we know to have been recently formed, or, from devastating causes, to be recently clothed with vegetation— From what I have seen of the collection, I have no reason to suppose that more than one or two of the plants are introduced, even if they were. This leads me to another Question which I am obliged to you for directing my attention to.— "Whether the species of large mundane genera have as wide a range as those of small genera?"— Now the how to set about the solution of this question has puzzled me sorely. In the first place it is not fair to compare the large genus of one Nat. Ord. with its equivalent of small genera of another; & so, a Nat. Ord. must be taken which has a large genus & as many species as that genus contains scattered through a considerable number of smaller ones— Again the Nat: Ord. must be a Mundane one, or there will not be geogr. scope for the observations. This is not all;—*Ericeæ* is a large mundane Order, but its largest genus *Erica* is confined to Cape & Europe, & there are none of its species common to both, or any at all in all the New World. Lastly the Nat Ord must be worked out pretty well. There are 3 orders which answer these conditions well—*Ranunculaceæ*, *Cruciferæ* & *Caryophylleae*, then comes another difficulty. There are degrees of distribution. However I put the degrees aside & start with the question in its simplest & boldest form—thus, What species are common to the old & new World?—here two more bugbears occur, at the North where the world gets very small the countries become in a measure identical, & which am I to call Greenland whose Botany is Europæan? that is easily got over by asserting—lastly a Europæan plant being found in N W. America must not be included as a widely spread one, as the Botany of that corner is Identical with Siberia & Kansckatha, & many plants are found only in these two spots.. The following is then my proposition (theorem rather)— What proportion of the Ranunculaceæ of the Old World including Greenland & N W Am (W of the Rocky Mts) is common to the American continent?—& secondly Do the larger proportion of these plants belong to large genera or small.?— In answer to the latter

In *Ranunculus* proper the common are to the whole as 1: 10.6.
In as many species as R. contains together forming 10 small gen it is. 1: 56.3
Caryophylleæ—In *Arenaria* 140 species the proportion is ------------------ 1: 28.0
In 15 allied Genera together containing 140 sp. ------------------------------ 1: 46.6
In *Cerastium* a large genus of **Caryoph**. 69 sp -------------------------------- 1: 34.4
In *Stellaria* of 69 species (a N. Temperate genus) ---------------------------- 1: 23.0
but In *Silene* which is very N. Temperate & has 217 sp. ------------------ 1: 108.8
Cruciferæ—
In *Arabis* & *Cardamine* together, the two largest genera, 123 sp ----------- 1: 20.5
of 123 species scattered through about 30 allied genera --------------------- 1: 24.6
In *Draba* a genus of 77 species (very Arctic however) ------------------------- 1: 5.5

These results are any thing but satisfactory & yet the instances are the most strictly comparable I can adduce. In another form,

In 855 species comprising 8 genera the proportion of sp. common is to whole 1: 18.2
In 423------------------------55-- 1: 38.5.—

These results may I think be relied on as far as they go, but they would not have been attainable had we not the N. American flora of Torrey & Gray,[5] men of unerring sagacity & discrimination— The results, if De Candolle's[6] work alone had been taken, would be erroneous; because he makes species of N. Am plants since discovered to be forms of Europæan, & because the species of the genus have increased in a greater ration than small genera have, & plants common to the two have not turned up in the same proportion— Again D^c^ C had not the means of knowing whether some of those common were introduced into the New World or not—Torrey & Gray carefully discriminate these—

What a remarkable fact you mention that the Geog. distrib. of shells is proportional to their persistence in nature. This is a wrinkle to Botanists towards the detection of the orders of fossil plants.. But *Cycadeæ* are certainly not widely distributed. Pines are (*Coniferæ* I mean).—[7] Do you know any thing of a Mr (Count) Streletski who I hear is in Town & of whom we saw a good deal in V. D. Land?..[8]

I have a list of the principle peat earth plants with an attempt at arranging them according to the proportion each yields: I include more plants than you mention, but your Journal is not before me as I lent it a few days ago— Enclosed is a list of as far as I have gone with the Galapago Isld plants, whenever you return it I will add to it & send it again—I think I have about ⅓ done, the proportion of new sp. is terrible among Dicot. & I must perpetrate one or two genera.

It hardly appears, either that the genera are distributed equally through all the Isld.s,—or that seperate Islds have seperate genera; untill however I have gone through the collection I shall forbear any more remarks, as I am often woefully out when applying the numerical test to my preconceived Ideas— I hope to be at the Geological Soc. this next weeks meeting as I have not seen Mr Lyell or Dr Fitton yet,[9] & my Father will go if the weather is tolerable.— I think this letter will tire you out, I wish I could be as useful to you as you are to me in suggesting these most interesting questions from which when properly worked out we may begin to ascend to grand causes

Believe me to remain yours most truly | Jos D Hooker.

DAR 100: 10–11

CD ANNOTATIONS

1.11 *Euphorbiaceæ* . . . Tropical Islds;— 1.18] *scored pencil*; 'Large Genera' *added in pencil over scoring*
2.6 Malden . . . from] *scored pencil*
2.7 I was . . . Auckland, 2.9] *scored pencil*
2.9 Auckland, contains] '?representative species' *added pencil*
2.22 "Whether. . . small genera?"— 2.23] *scored pencil*
2.46 In *Ranunculus* . . . 77 species 2.56] 'Hence Species of small genera have narrower ranges than the species of large mundane genera' *added ink*
4.1 shells] *del pencil*; 'not forms'[10] *added pencil*
4.3 But . . . V. D. Land?.. 4.6] *scored pencil*
5.1 I have . . . Hooker. 7.1] *crossed pencil*

[1] Acotyledons, one of Antoine Laurent de Jussieu's three major plant divisions, equivalent to Linnaeus' Cryptogamia, i.e., fungi, algae, mosses, and ferns. At this time, Hooker included gymnosperms (cycads and conifers) in his definition of monocotyledons.

[2] J. D. Hooker 1846, p. 242, states that the number of grasses on the Galápagos is much less than on other tropical islands, like the Sandwich and Cape Verde groups.

[3] James Macrae collected plants on the Sandwich and Galápagos Islands for the Horticultural Society of London. He travelled with George Anson Byron in H.M.S. *Blonde*.

[4] Hewett Cottrell Watson was botanist to the H.M.S. *Styx* survey of the Azores in 1842.

[5] Torrey and Gray 1838–43.

[6] Augustin Pyramus de Candolle, whose *Prodromus systematis naturalis* was an authoritative botanical text, see A. P. de Candolle and A. de Candolle 1824–73.

[7] Both *Cycadeæ* and *Coniferæ* are very ancient groups of plants.

[8] Paul Edmund de Strzelecki, who had explored parts of the Australian interior and Tasmania (Van Diemen's Land) in 1839–40. He returned to Britain in 1843.

[9] Charles Lyell and William Henry Fitton were both on the council of the Geological Society in 1844.
[10] CD sometimes used 'form' to mean genera or higher groups. He is objecting to Hooker's extending his query on ranges of species to ranges of orders. See next letter.

To J. D. Hooker [6 March 1844][1]

Down Bromley Kent
Wednesday

My dear Hooker

I will not lose a Post in guarding you against what I am afraid is a geographical mistake, which I fear will have cost you some labour in vain.— Malden Isl[d] is not one of the Galapagos, but is 4°S. 154°W—it is a *coral* island. It was visited by Lord Byron in the Blonde & *I fancy* discovered by him; it is described at p 205 of his work, wh. I have not.—[2] I do not know who Macrae was.— & it is possible that some one of the Galapagos islands was so christened but is not now so called— you could make out by comparing dates of Byron's voyage &c &c.— I wish I c[d] help you.—

It will be curious if Malden isl[d] has any botanical affinity with the Galapagos, though one of the nearest Pacific isl[ds].—

A genus of birds, which I thought peculiar (Cactornis) to the Galapagos has quite lately been found in one of the Low Archipelago Islands.—[3]

Shall you study the Pacific Flora.— Lesson, I remember remarks on the uniformity of the Flora of the islands of the Pacific, but whether this uniformity was of species or merely of forms, I know not— He says, whole Flora is more Asiatic than Indian, but I presume he is no authority.—[4]

If you ever work the Pacific Flora, you will find the Appendix to my Coral Volume[5] useful geographically in just ascertaining whether the isl[d] is of coral or not.—

Thank you exceedingly for your long letter & I am in truth ashamed of the time & trouble you have taken for me; but I must some day write again to you on the subject of your letter.— I will only now observe that you have extended my remark on the range of *species* of shells into the range of *genera* or groups.— Analogy from shells would only go so far, that if two or three species of Cycas were found to range from America to India they would be found to extend through an unusual thickness of strata say from the upper Cretaceous to its lowest bed, or the Neocomian.—[6] Or you may reverse it & say those species which range throughout the whole Cretaceous, will have wide ranges; viz from America through Europe to India: (this is one actual case with shells in the Cretaceous period)— Yours most truly in Haste | C. Darwin

DAR 114.1: 7

[1] The Wednesday before the letter from J. D. Hooker, 9 March 1844.

[2] Byron 1826, pp. 204–6.

[3] *Cactornis* was one of the genera established by John Gould for CD's Galápagos ground finches. Gould regarded the fourteen species as 'strictly confined to the Galapagos' (Gould 1837, p. 6, and confirmed by CD at a meeting of the Zoological Society, 10 May 1837, see *Proceedings of the Zoological Society of London* 5 (1837): 49) until he described *C. inornatus* from Bow Island in the Low archipelago, about 3000 miles south-west of the Galápagos (Gould 1843, p. 104). See *Correspondence* vol. 2, letter from R. B. Hinds, 19 July [1843].

[4] Lesson and Garnot 1826–30, 1: 12, 14. René Lesson was primarily a geologist.

[5] *Coral reefs*, pp. 151–205, including the coloured frontispiece map.

[6] A series of lower Cretaceous rocks, named after Neuchâtel in Switzerland where they were first identified.

From J. D. Hooker 9 March 1844

West Park Kew
March 9. 1844.

My dear Darwin

I am very much obliged to you for your prompt information concerning Malden Isld., which has cost me much trouble, having looked in vain in all the Gazetteers & Atlasses we possess for it— The plants from that Isld being all mixed up with those of the Galapagos, I never doubted its being some obscure synonym of one of that group. As it has happened, I am exceedingly glad that I thought so, or I should probably have reserved them for a future examination to have proceeded faster with your plants, & thus have lost time in the end— I am very satisfied with what the results of Malden Isld. prove, the more especially as I now suddenly am called upon to look on it in a very unexpected relationship, which tests severely what I have *writ*. I find I have 6 Malden Isld plants. In the first place none of them are called by me common other of the Galapagos, so I cannot have confused either species or labels— Secondly two of them, the Fern (*Stenolobus*) & the *Amaranthaceous* plant (*Achyranthes*), were hitherto supposed peculiar to the S. Sea Isld^{s} till I put them into the Galap. Isld flora. Thirdly the *Poa juliflores* I called an Australian form, though I supposed it to come from Galapagos also— 4th. the *Phymatodes* is chiefly an E. Ind plant, & lastly the *Euphorbia* & *Ficus* the only remaining two are certainly not S. American & different from Galapago Isld species, at least the former is—the latter genus probably not Galapagean— If you do not object I think of introducing the Malden Isld plants either as foot notes, or appendix to the Galapagos Isl^{d}, & continue examining them together...[1]

I have long intended to pay particular attention to the Pacific Isld. flora, & to take the Galapagos as a starting point; though they are perhaps more S. American, I wish I could make the Botany proceed *pari passu* with the Geology, but I have so much on hand that it is at present impossible.— I am aware of Lessons remarks about the identity of the S. S. Isld floras, but my limited experience differs in the results it leads to. You ask whether the uniformity consists in species or forms. I am inclined to consider that uniformity of species is to a certain extent a sequitur to a uniformity of forms, & that it is a corollary to our Theorem.—[2] Thus, uniformity of Flora must depend upon the genera being widely diffused, genera being forms, I think that is evident, again we have (or suppose we have) proved that it is the largest genera which are most widely diffused, & that a larger proportion of their *species* have wide ranges than *those* of small genera, whence I think it follows that in all countries of uniform floras, certain single, species should to a certain extent, be widely distributed.— There is no occasion to suppose they are distributed to such an extent as to invalidate a hypothesis that "in each group of tropical S. S. Islds the several Islets have distinct floras;—".. I consider the S. S. Islds as a whole, or Oceania to have a most distinct & peculiar flora, not from possessing any one very large group peculiar to itself, (as America has her Cacti &c) but because she has a mixture of the peculiarities of N & S. America, Australia, India & perhaps N. Asia. Thus she has American *Vacciniæ*,.—E. Ind *Pandaneæ*, *Orchideæ*, & spices:—Australian *Casuarinae*, *Goodenoviæ*, & some *Leguminеæ* & *Proteaceæ*—S. American *Cordiaceæ* & *Amaranthaceæ* I am not however fully qualified to state any thing on this subject; these are my notions & so is this, that the Flora is more Indian than any thing else.

The very great interest the contents of your letters cause me to take in these subjects, has made me forget the principle object I have in troubling you so soon with this— The Admiralty or Treasury rather have, you will be pleased to hear, granted £1000 to be expended entirely on the Botanical plates of a work they have been pleased to entrust to me, to describe & figure the new & more interesting plants of the voyage. They stipulate that I am to provide about 500 plates for that, which will be required to fulfil my intention of giving complete floras of V. D Land, N. Zealand Fuegia & the so called Antarctic Islands.[3] And now I was going to ask a great favor of you, & that was if you could any way give me a hint of the amount that soft ground lithograph plates should come to, & how they had better be done. As my father keeps an artist[4] the drawings can be done comparatively cheaply, but £2 will not cover Drawing & copper plate engraving, which our artist could not perform: he can however draw a little on stone, so that by his doing some & the lithographer others, (copies of his

Incomplete
DAR 100: 8–9

CD ANNOTATIONS
1.1 I am . . . have *writ.* 1.10] *crossed pencil*
1.19 I think of . . . impossible.— 2.4] *crossed pencil*
2.10 again we . . . distributed.— 2.14] *scored pencil*
3.1 The very . . . copies of his 3.14] *crossed pencil*
Top of first page: 'Albemarl Id did not ascend' *added pencil*

[1] Eventually described in J. D. Hooker 1846, p. 253 n.
[2] See letter to J. D. Hooker, 23 February [1844].
[3] J. D. Hooker 1844–7, 1853–5, and 1860.
[4] Walter Fitch had entered the employ of William Jackson Hooker in Glasgow in 1834.

To J. D. Hooker 11 March [1844]

Down— Bromley Kent
March 11

My dear Hooker

I am truly pleased in every way at your grant; I hope & have no doubt, that it will give you more satisfaction than mine did me,[1] in as much as you will make a far worthier use of it— Pray never apologise for asking any question, which I can answer; it is a pleasure to me.— With all Sir William's experience, you will have a good start in publication-affairs— I fear I cannot help you much.— I enclose (which keep as long as you like, but sometime *return* to me) 1st. a paper (which read first) containing a scheme of my & Andrew Smith's system of publication[2] 2d rough copies of the account, sent in of 1s class expences for each number to the Treasury, (which I accompanied by a letter vouching for their accuracy) 3d a statement of sale & *all* expences of the Work, when it has more than half completed.— If you could get a sight of the Zoology of Beagle in the numbers, you could then see the cost of each Part.—[3] The Drawing is included in the engraving.— Colouring is the heaviest of the second-class expences. If the work had sold well (I think now between 150 & 200 copies are sold) money wd have been realised for further publication.— I am publishing my geology on the credit of the stock in hand.— If you want any other information, I could give you a note to Smith & Elder & I am *sure* they wd give you all information— I have found Smith & Elder a *most pleasant*, *fair*, *attentive*, & *obliging* firm to have any business with.— You will manage wonderfully well if you produce 500 Plates & any letter press for the grant. Let me caution you, that you will find expences increase beyond any estimate; always bear this in mind.

I sent on Saturday my volumcito to you by Deliverance Company; it is purely geological.—[4]

If you have any further doubts about Malden Isd. you had better consult Krusentern's Mem: on the Pacific,[5] which probably you know—it is in Geograph. Soc.— How capital & satisfactory your conclusions made unknown to yourself about the Flora of this island.

In the Flora of the Pacific, I shd think, judging from shells, that the great open space of water between the Low or Dangerous Archipelago & the American coast, was the dividing line; it is wonderfully so with the sea-shells.—[6]

The supposed Asiatic character of the Flora of Oceania, I had thought was connected with the heavy gales, or almost hurricanes, coming from that quarter & being opposed to the trade-winds; not that I suppose the actual species have been transported; but the *possibility* of communication seems to produce affinity in the organic beings of two regions. If you will look at the map in my Coral-volume, you will see that probably much more land existed within geologically recent times than now exists.—

To return, I know that an artist can almost immediately learn to draw on Lithographic stones.— I believe Hullmandell is a good Lithog. printer; but I found him rather troublesome.[7]

I forgot to tell you before that Dr. Boott has (I believe) all the Carex's which I collected, & I think one remarkable one from the Galapagos:[8] he no doubt wd inform you or lend the specimens, if he still has them.— I do not suppose I paid much attention to collecting the grasses at the Galapagos.— I think I told you there are no large Ferns or Palms or Palmettos. With respect to the different isld having different species, the main point appears to me, whether any two or three islands have close *representative* species of the same genus; the simple fact of one isld having a species & another isld not having it is far less wonderful. How curious is the distribution in this latter sense of the terrestrial Amblyrhynchus.— The tree Compositæ were, I think, all, certainly most, from the summits of the Islds: do not, pray, forget my question of the summits in these cases, having the most peculiar Flora.—

I suppose you will consider Juan Fernandez: Has not Bertero, in his list in Silliman's Journal,[9] published a list of plants of this isld? I think M. Gay has also written on this isld.—[10] I presume you have Endlicher's Flora of Norfolk Isd.;[11] otherwise I cd lend it you for any time.—

I must have some more thinking over your curious remarks on distribution of large genera; you have put the case rather differently from that which I had intended; but I will sometime trouble you with another letter. Excuse this untidy letter; as I am not well.

Yours ever | C. Darwin

DAR 114.1: 8

[1] The Treasury grant of £1000 for *Zoology* was not sufficient to cover the total cost of producing its 166 plates and 632 pages of letterpress, see *Correspondence* vol. 2, letter to A. Y. Spearman, 9 October 1843, n. 1.

[2] Andrew Smith published A. Smith 1838–49 with the aid of a Treasury grant. For CD's arrangements with Smith, Elder and Company and the Treasury concerning the publication of *Zoology*, see *Correspondence* vol. 2.

[3] The individual numbers cost 6, 8, or 10*s*., with the exception of number fifteen, which cost 15*s*. The total cost of all the unbound parts amounted to £8 15*s*. The bound work, in five volumes, was priced at £9 2*s*. (Freeman 1977, pp. 27–30).

[4] *Volcanic islands*, the second of CD's three volumes on the geology of the *Beagle* voyage, was published in March 1844 (*The Publishers' circular*).

[5] Krusenstern 1824–7b and 1835.

[6] CD discussed the conchology of this area in his essay of 1844 (*Foundations*, p. 179) and *Journal of researches* 2d ed., p. 391.

[7] Charles Joseph Hullmandel. For CD's difficulties with Hullmandel see *Correspondence* vol. 2, letter to Benjamin Waterhouse Hawkins, [*c*. 1 October 1843].

[8] Francis Boott included descriptions of two *Beagle* carices in Boott 1851. *Carex* does not occur in the Galápagos; however Hooker enumerated five species of the related *Cyperus* (J. D. Hooker 1845d, pp. 177–8).

[9] Bertero 1830. CD was thinking of Bertero 1831–3, which does not list Juan Fernandez plants.

[10] Gay 1833.

[11] Endlicher 1833. CD's copy is in the Darwin Library–Down.

To Ernst Dieffenbach 14 March 1844[1]

[Down]

'. . . I am very glad to hear that you are going to edit a German Geological Journal . . .'[2]
'If not there to be forwarded by favour of Prof. Liebig'[3]

J. A. Stargardt, Marburg (catalogue 574, 11–13 November 1965)

[1] Date taken from the Stargardt catalogue.
[2] Dieffenbach had corresponded with the German publisher Eduard Vieweg about the possibility of producing an encyclopaedia of science, but nothing came of the project (Bell 1976, p. 117).
[3] This note, 'If. . . Liebig', was written on the cover. Dieffenbach had studied with Justus von Liebig and served as his agent in England in connection with a fertiliser enterprise (Bell 1976, pp. 18, 107–11).

To J. D. Hooker 16 March [1844]

Down Bromley Kent
March 16th

My dear Hooker

I have had a note from J. Gray of Brit. Mus: asking for information about my system of publication.[1] As it wd be a long story to tell him, I have said, that you would in a few days send him, the bundle of papers, which I sent you, & I asked him to return them to you.— I hope this will not inconvenience you.— I received from London (as I had occasion to send there) your infusorial specimens for Ehrenberg—I am astonished that I have not yet heard from him.— Your specimens anyhow shall be carefully preserved—I fear they are rather too bulky to go to Berlin, & I understand, that merest scraps are sufficient.— Any part, which I find I cannot send, shall be returned to you.

Ever yours | C. Darwin

DAR 114.1: 9

[1] John Edward Gray, presumably inquiring for Richardson and Gray 1844–75.

To *Gardeners' Chronicle and Agricultural Gazette* [27 March 1844][1]

As you have noticed a communication made by me to the Geological Society in 1837, on the Formation of Mould,[2] I should be much obliged if you would correct an error into which I have fallen. In a postscript to that paper I state that marl was put on a pasture field, since ploughed, 80 years ago: I should have said 30 years, as I mistook the figures in the paper sent me. I found out this on visiting the place four years and a-half subsequently, and examining the old occupier of the farm.[3] Wishing to ascertain the accuracy of the stated depth at which the marl now lies buried, I had three long holes dug in different parts of the field, and in each I found the marl, together with some cinders and broken pottery, in a layer 13 inches beneath the bottom of the potato-furrows, which were about four inches beneath the general surface; so that these substances are now buried at a depth of no less than 17 inches. They will never, probably, be undermined by the worms, to any much greater depth, as they almost rest on the general substratum of pure white sand. I particularly examined the occupier, whether the field had ever been ploughed to a greater depth than six or eight inches, and he positively assured me that it never had. My original informant, therefore, rather underrated the depth at which the marl now lies; although probably in the interval of four and a-half years, between our observations, some soil may have been removed by the worms from beneath the marl. In the other fields, formerly examined, I found that the layers of lime and cinders were, in almost every case, about an inch lower than they previously were. It was curious to observe in some of the holes how distinct three layers were preserved; the uppermost of cinders being two inches beneath the surface (on the former occasion one inch below), the middle layer of lime at four inches, and the lowest of cinders and burnt marl, at from 10 to 12 inches. I found this lowest layer wherever I dug, and likewise the other layers, but less regular, owing to different parts of the field having been limed and cindered at different periods. When digging in this field, after a long drought, I noticed, that one single clod of earth, about as large as a man's two hands, was penetrated by eight upright, cylindrical worm-holes, nearly as large as swan-quills, so that I could see through them. Now this shows the quantity of earth in a small space, which is often probably removed by the worms and brought to the surface. The boggy field mentioned in the postscript to my Paper, on which two years and a half before a thick layer of bright red sand had been strewed, and which, I was informed, was then buried three-fourths of an inch beneath the surface, I found four years and a half subsequently (*i.e.* seven years from the sand being put on) was exactly two inches beneath the surface. In that field (also rather boggy) which I have described in my Paper, as first reclaimed 15 years before, the burnt marl was buried at a depth of four inches; so that in these two cases the rate of sinking, or more properly of being undermined, has been nearly the same, namely about two inches in seven years. In the fields, however, more particularly alluded to in this notice, in which the marl that was put on thirty-four years and a half before, then lay seventeen inches beneath the surface, the rate of being undermined has been much quicker, namely, three inches and four-tenths of an inch every seven years. This field is dry, and consists of black, poor, very light sandy soil. It has also been ploughed, which may make some difference; though it is clear, from the uniformity of the layer, that the marl must have sunk beneath the depth at which the plough could disturb it before the pasture had been broken up. I am surprised at the red sand on the most boggy field having been buried as much as two inches in the seven years, for I never saw a field on which there were so few worm-castings. One cannot, however, judge of the number of worms in a field from inspection at any one season.— *Charles Darwin, Down, Kent.*

Gardeners' Chronicle and Agricultural Gazette, no. 14, 6 April 1844, p. 218

[1] The draft of this letter (DAR 64.2: 7–9) is dated 27 March 1844.
[2] *Collected papers* 1: 49–53. This paper was discussed in the *Gardeners' Chronicle and Agricultural Gazette*, no. 11, 16 March 1844, p. 169.
[3] William Dabbs. CD's information originally came in a letter from Elizabeth Wedgwood, 10 November [1837] (*Correspondence* vol. 2).

To J. D. Hooker 31 March [1844]

Down Bromley Kent
March 31.

My dear Hooker

I have been a shameful time in returning your documents,[1] but I have been very busy scientifically & unscientifically in planting.— I have been exceedingly interested in the details about the Galapagos Isl^ds.—I need not say that I collected blindly & did not attempt to make complete series, but just took every thing in flower blindly.— The Flora of the summits & bases of the islands *appear* wholly different; it may aid you in observing, whether the different isl^ds have *representative species filling the same places in the œconomy of nature*, to know, that I collected plants from the *lower* & *dry region* in all the isl^ds. ie in Chatham, Charles, James & Albemarle (the least on the latter); & that I was able to ascend into the high & damp region only in James & Charles islands; & in the former I think I got every plant then in flower.— Please bear this in mind in comparing the representative species.—[2] (You know that Henslow has described a new Opuntia from the Galapagos).—[3]

Your observations on the distribution of *large* mundane genera, have interested me much; but that was not the precise point, which I was curious to ascertain;—it has no necessary relation to size of genus (though perhaps your statements will show that it has)—it was merely this; suppose a genus with ten or more species, inhabiting the ten main botanical regions, should you expect that all or most of these ten species would have wide ranges (ie were found in most parts of) in their respective countries. To give an example the genus Felis is found in every country except Australia, & the individual species generally range over thousands of miles in their respective countries: on the other hand no genus of monkey ranges over so large a part of the world & the individual species in their respective countries seldom range over wide spaces. I suspect, (but am not sure) that in the genus mus (the most mundane genus of all mammifers) the individual species have not wide ranges, which is opposed to my query.—[4]

I fancy from a paper by Don,[5] that some genera of grasses, (ie Juncus or Junceæ) are widely diffused over world, & certainly many of their species have very wide ranges— in short it seems, that my question is whether there is any relation between the ranges of genera & of individual species, without any relation to the size of the genera.— It is evident a genus might be widely diffused in two ways. 1st by many different species, each with restricted ranges, & 2^{d} by many or few species with wide ranges.— Any light, which you c^{d} throw on this I sh^{d} be very much obliged for. Thank you most kindly, also, for your offer in a former letter to consider any other points; & at some future day I shall be most grateful for a little assistance, but I will not be unmerciful.—

Swainson has remarked (& Westwood contradicted) that typical genera have wide ranges:[6] Waterhouse, (without knowing these previous remarkers) made to me this same observation:[7] I feel a laudable doubt & disinclination to believe any statement of Swainson's, but now Waterhouse remarks it, I am curious on the point. There is, however, so much vague in the meaning of "typical forms" & no little ambiguity in the mere assertion of "wide ranges", (for zoologist seldom go into strict & disagreeable arithmetic, like you Botanists so wisely do)[8] that I feel very doubtful, though some considerations tempt me to believe in this remark.—[9] Here again if you can throw any light, I shall be much obliged.— After your kind remarks, I will not apologise for boring you with my vague queries & remarks.—

Hamilton in his last Anniver. Geograph. Address refers to Leibmann's researches of the Alpine Flora of Mexico;[10] I mention this for the **bare chance** of your not having heard of him or his works, whatever they may be.—

I saw Smith & Elder the other day; & he told me he much regretted he could not make an agreement with you; but if you sh^{d} alter your plans, he sh^{d} be most happy and honoured by any fresh agreement with you—

Believe me | Very truly yours | C. Darwin

DAR 114.1: 10

[1] CD was returning Hooker's working list of Galápagos plants, see letter from J. D. Hooker, [23 February – 6 March 1844].

[2] D. M. Porter 1980b, p. 88, shows that CD did not mingle his plant collections from the separate islands, hence the materials were available for the requested comparison.

[3] Henslow 1837.

[4] See CD's essay of 1844 (*Foundations*, pp. 155–6).

[5] Don 1841.

[6] Westwood 1841, p. 417, referring to William Swainson, possibly Swainson 1832–3. CD copied out the relevant sentences of John Obadiah Westwood's work and added the comment: 'it rather proves converse ie. that some non-typical ['non-' *interl*] groups have wide ranges.—' (DAR 205.5: 97v.). Westwood had previously criticised Swainson in Westwood 1836, p. 563.

[7] Not published until Waterhouse 1845a, p. 19 n. The text of Waterhouse's footnote is given in the second letter from G. R. Waterhouse, [*c.* June 1845], p. 201. Waterhouse 1845a referred to earlier work on this subject (Waterhouse 1839), but this contains no discussion of typical genera and wide ranges.

[8] Following Alexander von Humboldt, plant distributions were often expressed in arithmetic terms describing the proportional representation of a taxonomic group in a given flora.

[9] CD kept the following note with his materials on divergence and classification (DAR 205.5: 97):

> March 31. *44* If Swainson's statement (& Waterhouse independently to me) that typical genera (which implies with respect to larger group) (for Ornithorhynchus ['O' *over* 'o'] can only be considered non-typical with respect to Mammifers) have wide ranges (?converse may still hold good? ['?' *added*]) is important; for the genera which are not typical are only rendered so by the extinction of allied genera, & that implies they are less adapted than other groups of genera to the [*over* 'their'] world ['& their co-inhabitants' *del*]— & therefore one might expect they w[d] be less widely distributed: they *(as genera) [*interl*] w[d] be rare, for they have or are decreasing in number—like individual species.— *(good) [*square brackets in MS*] Mem. Westwoods contradiction in Linn: Trans:.—

[10] Hamilton 1843, which refers to Liebmann 1843.

From J. D. Hooker 5 April 1844

West Park Kew
April 5. 1844.

My dear Darwin

Your queries & remarks have opened a wide field for research & investigation, for which I am truly obliged. These are all subjects which I ought to have attended to, without requiring to be reminded of them, by a more industrious Naturalist: truly I ought to have been able to answer you on the spot, if I had been half a Botanist, but I seem to know less every month than I did the month before as I find how much there is yet to learn—

I believe there is to a certain extent a "great relation between the ranges of Genera & individual species" Thus in the genus **Linum** (flax) several species are very common over all Europe, another (*monogynum*) abundant over the greater part of Australia & New Zealand.—a third (*saginoides*) has a wide range in Chili & Bonaria:[1]—one or two very common over the Cape district:—L rigidum has an immense range in U. States.—L angustifolium over Europe & Asia.—& I think another species over the Brazils.— Again in the genus Drosera.—D. *uniflora* has a wide range over Southern Chili Fuegia & Falklands D. *anglica* over all middle & N. Europe—several species over a great part of Australia & V. D L.—[2] D. cistiflora has (I believe) a very wide cape range, & two U.S. species an immense N. American. I think I could quote 50 instances similar to that of the two above large Genera, & plenty more striking— The contrary may however be as often true, but Negative evidence is always more difficult to find than positive; in looking for exceptions one is lost in a sea of doubt & does not know which way to look for what he wants, but when one is asked whether *such a thing* holds good the mind is immediately led to similar instances & corroborative evidence.— Now, though these large genera have each individual species with wide ranges, I believe they possess equally species which are remarkably local, (which however may mean no more than that the said local species have not been properly looked for). The Genus *Araucaria* is widely distributed, having representatives in Chili where the species cannot be said to have a very wide range, another in Brazil certainly of contracted limits—2 very sparingly indeed distributed in Australia, & a 4th. confined? to Norfolk Island.. Again some Genera of only single species have a mundane range, & others of but few are widely distributed, each species in its region; thus *Trientalis Europæa* is abundant all over North Europe & *T. Americana* over all Northern America. Few plants are however really so rare & so confined as many animals, from not having so many enemies.

On the whole I believe that many individual representative species of large genera have wide ranges, but I do not consider the fact as one of great value, because the proportion of such species having a wide range is not large compared with other representative species of the same genus whose limits are confined—& further because small genera have likewise individual widely extended representative species. The converse holds true in a certain degree & in the *Cacti* we have a parallel case to Monkeys, their geographical range is small being confined to warm & chiefly to tropical S. America & Mexico I have somewhere read that the species are remarkably peculiar to certain narrow limits. My Father (who I just asked) confirms this & adds that they are most remarkably local as species & equally so as genera, the Turks' Caps are chiefly confined to the W. Ind. Islands those with ribbed spines to Mexico &c & C. *triangularis* common to all tropical S. Am. is supposed to have been introduced.— The same holds good with the S. African Stapelias,—with the Nat. Ord. Rafflesiaceæ, with many peculiar Australian genera &c. I believe the rule is very good that "the Individual species of local genera are themselves local" as also that "In mundane genera many of the species have mundane ranges", but that it is not proved in Botany that "in mundane genera the representative species have a wide range each in its own country", to any remarkable extent. What you say of Junci is very true that they form a mundane Genus & many of the species have extended ranges but I do not know that it is the representative species that have wide ranges each in its own territory, even if true I would answer that water plants (as many are) are more widely diffused than dry land species— To conclude, I believe that most large mundane Genera contain both **1**st many different species each with restricted limits, & **2**nd also a large proportion of species with very wide ranges besides **3**rd many local species with very narrow ranges: **but** it is not apparent that the proportional number of species distributed under any one of these conditions is larger in general than those distributed under either of the others.— I shall however bear the matter in mind & hope for new lights in time..[3]

With regard to typical genera having wide ranges Swainson is an instance of the type of a certain class of Naturalists wandering very far indeed both mentally & bodily. I hardly know what is always meant by a typical form. The character of a group should be founded on the most important objects it contains in the œconomy of nature. The most important genus of a class is surely generally either the largest or the most widely diffused; if the largest genus is the type, we have already seen that large genera are generally most widely diffused. The type of a group often turns out (on extended knowledge of that group) to be the most aberrant form in it.— Perhaps Swainson has put the cart before the horse & should have said "a typical group or genus is that which is the most widely diffused"—[4] Some however I think define typical forms as those which are most fully developed or what they call most perfect, now though it may be very easy in any group to point out many which are *not* the most perfect or fully developed, a great many remain amongst which it is difficult to say which has the advantage of the other in organization.— I suppose you are acquainted with M^c^Leays writings, if his views are to be followed out all our theories must be[5] capsized as he leaves us nothing but the remnants of an Animal Kingdom to work upon.[6] Have you paid any attention to his circular & Quinary system?...[7] I cannot call this long prosy letter an answer to yours, I feel that I have given nothing but vague & unsatisfactory information; another time, ere long, I shall hope to have some naked facts sought out, bearing directly on the grand question, "Have the 10 representative individual species of widely distributed genera over 10 countries wide ranges each in its country."?— I think I state it as you wish it answered.

Smith & Elder gave the greatest & kindest attention to my book but I think the outlay required frightened them, as we wanted £1. on each plate to get them done in soft ground lithograph, which cannot be done at the Govt grant of £2 a plate, I also asked a share of the profits; an idea long abandoned; Every publisher was equally alarmed at the extent of the work &c except Mr Lovell Reeve of King W^m^. Street Strand, who, keeping his own lithog. press, works very cheaply. I give him all materials & plates on stone for a few copies of the work— it will be produced at 1/ a plate col^d^.; including letterpress, the same price as Belcher's Botany uncolored,[8] & 7^d^. a plate uncold.. I should have liked Smith & Elder better than any; but they did not seem to like so very extended a work, & further, to tell you the truth your estimates are so very much higher for printing &c that they do not seem to have gone so economically to work as they might have consistently with good execution—.— I have worked the Galapago Isld plants up to Compositæ & am now writing out clean the first part for printing

You must find *planting* a great recreation, I wish you would come to Kew & see the plans for planting 48 acres as an Arboretum. I wish I could send you a list of the depths at which we obtained live Corals—off New Zealand we dredged live *Hornera frondosa* from 400 fathoms..[9]

Believe me Yours most truly | Jos D Hooker

We have live Fagus Antarctica & Forsteri at Kew, old friends!—. also Berberis ilicifolia & Winters bark all from our voyage

DAR 100: 12–13

CD ANNOTATIONS

1.1 Your . . . learn—1.6] *crossed pencil*

2.18 which are remarkably local] *underl pencil*

3.28 I shall . . . time.. 3.29] 'You have' *added pencil*

3.29 time..] 'Rest on Types.' *added pencil*

4.3 I hardly . . . widely diffused; 4.6] *scored pencil and brown crayon*
4.4 important objects . . . widely diffused; 4.6] *underl pencil*
4.11 Some . . . theories must be 4.16] *scored pencil*
4.16 capsized . . . fathoms.. 6.4] *crossed pencil*
4.22 representative] *underl pencil*; '?' *added pencil*

[1] The Buenos Aires district of Argentina.
[2] Van Diemen's Land (Tasmania).
[3] See CD's essay of 1844 (*Foundations*, pp. 155–6), which incorporates some of Hooker's remarks and summarises CD's views on this question.
[4] William Swainson argued that typical or 'type' genera had a wide geographical distribution and, by implication, that 'aberrant' or 'osculant' genera had a limited distribution (Swainson 1832–3, 2 (*Insects*): unpaginated text accompanying plates 95 (*Papilio memnon*) and 133 (*Polyommatus cassius*)). CD kept the following note with his materials on divergence and classification (DAR 205.5: 97v.): 'v. Hooker's letter on what Typical means. I do not doubt it only refers to extinction ['or' *del*] rather fewness of forms.—' *ink*. Hooker's statement makes fewness of osculant groups a truism.
[5] The preceding part of this paragraph, was, at one time, excised along with its verso (5.1 to end). There is no record of where in his notes CD kept the excised fragment.
[6] William Sharp Macleay's principal theoretical writings are Macleay 1819–21 and Macleay 1830.
[7] CD had been familiar with quinarianism from his Cambridge days, since he commented knowingly on Macleay's system to John Stevens Henslow early in the *Beagle* voyage (see *Correspondence* vol. 1, letter to J. S. Henslow, [*c.* 26 October –] 24 November [1832]). During a brief visit to Cambridge in 1838, CD had an extract copied from Macleay 1819–21 by Syms Covington (DAR 71: 128–38). In CD's notebooks there are several remarks which show that he had given the system his serious critical attention (see S. Smith 1960). For CD's criticisms of quinarianism see *Correspondence* vol. 2, letter to G. R. Waterhouse, [3 or 17 December 1843].
[8] Bentham and Hooker 1844. George Bentham had been botanist to Sir Edward Belcher's expedition.
[9] J. D. Hooker 1845b. Hooker disputed Richard Taylor's claim that the greatest depth from which living animals had been dredged was 300 fathoms (see R. Taylor 1845).

To J. D. Hooker [17 April 1844][1]

London
Wednesday

My dear Hooker

I have waited in vain for Dr. Dieffenbach's answer to my queries to Ehrenberg for more particulars regarding what he wanted, & therefore I am going at once to send off a cargo of little packets to Berlin.— Those which I send are valueless, except to Ehrenberg, & therefore I am going to tell him *not* to return mine, & will you kindly send me a line by return of Post (to Down) telling me what I shall say to him about returning your more valuable cargo?—

Shall I tell him that the sea-weeds are undescribed & that you intend describing them, which will show that you do not wish him to describe them, or say nothing?—

Did you send my account-papers to Gray & has he returned them to you.[2] I hear poor Mr Gray's name was withdrawn last night from the ballot at the Athenæum. I wish you were in this Club;[3] we shd meet sometimes then, but I trust sometime you will pay me a little visit in the wilds of Down. I am going away for a 6 weeks in a few days time.—

Thank you *most sincerely* for your hint about the printing charges of my work—I am surprised at it, & sorry, but I am *truly* indebted to you for telling me what you think, & shall be more cautious in futuro.— I am going to call on C. Strzelecki of whose geological doings, I hear great things.

Your geographical-law-letters require being read and reread, & I have only read your last twice, & so will hazard no remarks on it— You seem, however, to have put the case of "typical forms", in a clearer point of view, than I ever saw it & stripped the word of half, if not all its mystery: I have long suspected that typical & abnormal forms consist only of those, in which a greater or less variety have been created or modifyed— with this *excellently*!! expressed sentence, I will conclude, | Yours most truly | C. Darwin

DAR 114.1: 18

[1] Dated on the basis of the following letter to C. G. Ehrenberg, 20 April [1844].
[2] John Edward Gray. See letter to J. D. Hooker, 16 March [1844].
[3] Hooker was elected to the Athenæum Club in 1851 (Waugh 1888). 'Mr Gray' is not John Edward Gray, who had been a member of the Athenæum since 1835, but possibly his brother, George Robert.

To Christian Gottfried Ehrenberg 20 April [1844]

Down near Bromley | Kent
April 20th.

Sir

I heard sometime since from Dr. Dieffenbach that you wished for any substances likely to contain infusoria from the Galapagos Islds & Tierra del Fuego.— I wrote to Dr. Dieffenbach for further particulars,[1] but not having heard in reply, I will forward immediately a packet containing such substances.— It would give me lively pleasure if anything I could send, should be of the slightest service to you.— Every little packet is labelled outside; & as they need not be returned, if they are useless to you, they will cost you only the trouble of reading the labels.— Should you want any further information about any of the substances sent, I shall be most happy to supply all I can.—

Two of the bottles & the majority of the parcels are from Dr. Hooker naturalist in the Antarctic expedition (the son of Sir W. Hooker), whom I applied to for earth from the roots of my plants which he is describing,[2] & besides such earth, he has sent earth, water & seaweeds from the Antarctic regions, in some of which when fresh, Infusoria were swarming— The seaweeds Dr. Hooker intends describing himself.—

I hope these Antarctic specimens will be useful to you.—

I can hardly venture to ask you to take the trouble, but shd any of these specimens prove of much interest, it would greatly gratify Dr. Hooker & myself to hear of their safe arrival to you.—

With the greatest respect | Sir | Your faithful & obliged servant | Charles Darwin

P.S. M. Baillière,[3] the Bookseller, has undertaken to send the Parcel by Steamboat, without charge to you, directed to the University, Berlin.

Postmark: 1844
Museum für Naturkunde der Humboldt-Universität zu Berlin

[1] See letter to Ernst Dieffenbach, 25 January 1844.
[2] See letters to J. D. Hooker, [27 January 1844] and [3–17 February 1844].
[3] Hippolyte Baillière, bookseller and publisher in London who specialised in French medical and scientific texts.

From George Robert Waterhouse 26 April 1844

10 Gloucester Grove West | Old Brompton
Thursday night 26 April 1844

My dear Darwin/

You wont let me off without definitions, you say!. I will try you with something between a definition and an explanation of what I mean by a typical species—*[1] The term "*typical species*" is used by Zoologists in two senses—it either refers to that species which possesses in the highest degree of developement *some* of the characters which distinguish the group xx[2] to which it belongs from other groups; or, it has reference to that species which is supposed to exhibit, in the best balanced condition, the greatest number of characters most common to the species forming the group of which it is a member— In the former case the type of the group would be that species which is most removed from other groups but in the latter such would not be the case—

By way of illustration I will suppose I am called upon to point out a type of the order *Carnivora.* According to the first definition I should select a Cat because in the Cat tribe some of the more striking characters of the *Carnivora* are most strongly developed; but, were I to adopt the second definition I should choose a Viverra because it may be said to possess most evenly developed the greatest number of characters which are found in the species of its order; and, in a Natural classification, such differences should (according to my views) be expressed by placing the Cats, among the Carnivora at the confines of the order which are most removed from the groups forming the orders which are most nearly related to the *Carnivora*, whilst the Viverra should be placed in the *middle* of the carnivorous order— I would distinguish the two, so called, types by terming the one *a type of a carnivorous Mammal*, and the other *a type of the Order Carnivora*— I will not however take upon myself to say that a type is a thing which exists in Nature—it may only be an abstract idea— It *may* be impossible to name any particular species which would be *generally* admitted as a type according to either of the two views above referred to, but it would not be difficult to show that some particular species approaches *very near* to the *idea* of the type in either case—

It may help to make my notions clear if I remind you of the general law of developement of parts in animals, viz. that when one organ is greatly developed it is at the expense, as it were, of some other organs—[3] thus the carnassial tooth, so characteristic of the order Carnivora, being a much developed tooth, other neighbouring teeth are robbed of their share of nutriment by it, and in the Cat, which has the carnassier most developed, the false and true molar teeth are *least* developed— It has in fact no true molar teeth in the lower jaw, and but a rudimentary one on either side of the upper jaw, but taking the order Carnivora as a whole the species could not be characterized as being almost destitute of true molar teeth for the greater portion of the species have two tolerably well developed true molars on each side of the upper jaw, and *all*, with the exception of the Cats have one at least on each side of the lower jaw— From this it will be perceived that the Cats furnish an exception among the Carnivora to the most common characters of the dentition, and so far would not would not serve as a good illustration of the order, though they might furnish the best illustration of a flesh-eating Mammal— On the other hand the Viverridæ have two true molars (let it always be understood I mean on *both sides* of the jaws) in the upper jaw and one in the lower (combined with a tolerably well developed carnassier) in which character they agree with the *Ursidæ* (but here there is no tooth which can functionally be called a carnassier) and are intermediate between the *Canidæ* and the *Mustelidæ*, the former of these two families having two molars to the upper jaw and ditto to the lower whilst the Mustelidæ have one to each jaw— So much for the dentition and were I to speak of other parts of structure in the Viverras I could show that they are intermediate between the Dogs, Bears, Cats and Weasels— I should therefore call them the typical family of the order Carnivora, and that species of the Viverridæ which was most removed from the Dogs, Bears, Cats and Weasels, the type of the family *Viverridæ*—*[4]

Believe me | Ever Yours | Geo. R. Waterhouse

*Species are the supposed descendants from a common parent. Animals are said by Zoologists *to be of the same species* when they perfectly resemble each other, or when they differ only in degree so far as from experience they have been found to differ in animals of the same parents—those parents being alike—

xxBy "group" I mean *any* assemblage of species—

*In a classification the family Viverridæ would be central,[5] the families of Dogs, Bears Cats & Weasels being arranged around it—and the type of the family Viverridæ would also be *central* being the most removed from the species of the other families, and according to the same system the type of a Carnivorous animal would, as I have before stated, be *external*. I know the question you are going to put to me now you have read this last paragraph! writing it has opened my eyes to a point for consideration from which something *may* spring.

Qy— What would be the type of a Central group, like the *Viverridæ*, according to my *first* definition? I dont know one—xx—but I am not answerable for that, for

I have never used the word *type* but in accordance with the *second* definition— this notion has just come into my head but I cant think it out for I am very sleepy— past two o'clock!!

DAR 181

CD ANNOTATIONS
2.1 By way of . . . its order; 2.6] 'When comparison with man excluded [*before del* 'exc'] typical = perfect.—' *added pencil*
3.3 thus . . . upper jaw, 3.11] 'Balancement' *added brown crayon*
Top of first page: 'According as one or other of these definitions typical form w^{d} the oldest or newest, & typical & perfection have some relation.—' *added pencil*
Margin of first page: 'On Types' *added pencil*
End of letter: 'I presume, no doubt, if the Viverridæ, had only one *genus, with few species [*above del* 'species'], yet it w^{d} be the typical family of the order Carnivora.—& if so the largeness of the genus has no relation to typicalness.— But I can hardly admit, that the one Viverra, w^{d} be called by any one typical' *added ink*
'When I ['discuss' *del*] allude to typical genera having wide ranges, I can bring all this in.—' *added pencil*

[1] The asterisk refers to a footnote added by Waterhouse to the bottom of the manuscript page. The note referred to is the first of three notes transcribed following the valediction.
[2] 'xx' refers to a second footnote added by Waterhouse. It is the second of the notes transcribed following the valediction.
[3] CD cited this remark in *Natural selection* (p. 305); however, in the *Origin* (p. 147) he dropped the reference to Waterhouse and attributed the 'law' to Johann Wolfgang von Goethe and Étienne Geoffroy Saint-Hilaire. CD had encountered the 'law of balancement' earlier. In his copy of E. Geoffroy Saint-Hilaire 1830 he has annotated passages in which it is discussed (pp. 215–19), and in 1837 he referred to Geoffroy Saint-Hilaire's work in *Notebook B*: 210–14.
[4] The asterisk refers to a footnote added by Waterhouse at the end of the manuscript. It is the last of the notes transcribed following the valediction.
[5] Waterhouse had attempted to represent taxonomical relationships as a group of adjacent circles (see Waterhouse 1843 and *Correspondence* vol. 2, letter to G. R. Waterhouse, [3 or 17 December 1843]).

From G. R. Waterhouse [after 26 April 1844]

My dear Darwin

Hope[1] is yet on the continent but I do not know his address, if however you send a note addressed to him under cover to me I will see that it is sent off through his man Mr Sibley who manages all these matters—

Your question about the *Viverridæ* puzzles me much—[2] *number* is certainly an element of some importance with me—especially in endeavouring to form an idea of the value of a character— I think were there only one species of *Viverridæ* known I should in all probability regard him as an aberrant form of some other group & should not select him as a type of the Carnivora—but I cannot for the life of me conceive what I should do for a type— I should be much in the same predicament with the Carnivora as I am with the *Edentata*— I cannot form any very distinct idea of the type of that group—

I will perhaps write again about this matter— I am at this moment so very 'queer' (with a cold) that I can hardly *think*— I am very sorry to hear you are sticking to your old bad practice of being unwell—

Believe me | Ever faithfully yours | Geo R Waterhouse

Tuesday in E

DAR 181

CD ANNOTATIONS
1.1 Hope . . . matters— 1.3] *crossed pencil*
3.1 I will . . . Tuesday in E 4.2] *crossed pencil*
Top of first page: 'Second note' *added pencil*

[1] Frederick William Hope.
[2] See CD's annotations to the previous letter.

From Erasmus Alvey Darwin [May 1844 – 1 October 1846][1]

$14\frac{1}{2}$ miles base with 200^{ft}
gives 0° 7′ 48″
35 miles with 674 feet
gives 0° 10′ 53″

Dear Charles.

Swale[2] sent here Lady Willogby's Diary[3] which I have transferred to Gower St[4] to take down to you. I hope Emma was none the worse for her journey & Granny[5] had a beautiful day for hers—

Yours. E D

Saturday

DAR 39.1: 29

[1] The date range is set by the publication of the first volume of *Lady Willoughby's diary* in May 1844 (*Literary Advertiser*) and the completion of *South America*. This letter is associated with two covers addressed to CD by E. A. Darwin, both with further calculations by E. A. Darwin:

$14\frac{1}{2}$ miles
188 feet, rise
gives 7′ 20″

486 feet rise in 35^{miles} (2025^{yards} to mile)
gives an angle 0° 7′ 52″
100 feet in $14\frac{1}{2}$ miles
gives 0° 3′ 5″

CD has annotated these covers:
7′ 20″] ‘p 182, p. 185’ *added pencil*
486 feet . . . 35^miles^] ‘((842 is the rise of the bottom of the lava in the 35 miles))’ *added ink*
0° 3′ 5″] ‘1′ [‘′’ *over* ‘°’]. 22″ (77 & 85 fathoms)’ *added ink*
The calculations relate to the inclination of lava flows in the valley of the Santa Cruz River described by CD in *South America*, pp. 116–17. According to CD's 'Journal' (Appendix II) he worked on the book from 27 July 1844 to 1 October 1846.

[2] Ralph and William H. Swale, Booksellers, 21 Great Russell Street, London.
[3] [Rathbone] 1844–8, a fictitious diary.
[4] The Hensleigh Wedgwoods lived at 16 Gower Street.
[5] Susan Darwin, CD's sister.

To Josiah Wedgwood III [May 1844][1]

Wednesday

Dear Jos.

Emma tells me that you have or will have next week 600 or 700£ to pay over.— My Father has kindly offered to take 600£ more from me, would you therefore please to send this money if now ready, with some which you have to send from Aunt Sarah.;[2] if not now ready, will you send it whenever it is ready.— If the sum is more than 600£, would you please to keep the difference till some future time.— My Father requests that you will be so kind, as to send him a copy of the same sort of receipt, which you had for the 1400£, & he will return it to you properly signed.— I forget, whether you require any document from Emma: if you do, w^d^ you please to send us the form, & I will keep a copy of it, so that we need not trouble you again, as I utterly forget the sort of document, you formerly required about the Railway Shares.—[3]

Yours affectionately | C. Darwin

Keele University Library (Wedgwood/Mosley 1028)

[1] Dated from an entry in Robert Waring Darwin's Investment Book, 1831–48 (Down House MS) in the account headed 'Trust Money—Emma & Charles Darwin', reading: '1844 Borrowed May 28 £600'. CD was at The Mount, his father's home in Shrewsbury, from 23 April to 30 May ('Journal'; Appendix II).
[2] Sarah Elizabeth (Sarah) Wedgwood, CD's aunt.
[3] A document, signed by Emma, requesting that railway shares be purchased from trust money left by her father is dated 16 October 1843 (Keele University Library, Mosley Collection). Josiah Wedgwood III was a trustee of CD and Emma's marriage settlement (see *Correspondence* vol. 2, letter from Josiah Wedgwood II and Emma Wedgwood to R. W. Darwin, 15 November 1838, n.1).

From Philip de Malpas Grey-Egerton 5 May [1844]

30 Eaton Place
May 5.

Dear Darwin.

I send you Enniskillen's's[1] account of the discovery of the Irish Yew.[2] "Old Hugh" (*not Yew*) "Willis of Ahaterourke under Ben Achlin found two upright

Yews in the mountain between the Cove and the Ben near Lugahurra hollow about 80 years ago. He brought one to his Landlord and planted the other in his own Garden where it now stands a fine tree. The remnants of the other are now in the Flower garden here. I have always heard that the first plants raised were from cuttings, and to judge from the appearance of the mother plant it must be true. I never heard of seed being sown till Mr Young our Gardener tried it and raised 3 plants which differ from the parent and are intermediate between it and the Common Yew

Florence Court[3]
April 26.

AL
DAR 163

CD ANNOTATIONS
1.10 Yew] ' " ' *added pencil*
Top of first page: 'Wild vars' *brown crayon*
'Ch IV' *circled brown crayon*
'Just allude to these & describe under wild vars.[4] *pencil*
End of letter: 'Ennuskillen. | P. G. Egerton' *pencil*

[1] William Willoughby Cole, 3d Earl of Enniskillen, Fermanaugh, Ireland.

[2] CD had heard of the discovery earlier from Egerton and had apparently asked him to obtain an account from Enniskillen. On 20 April 1844 CD made the following note:

> Sir. P. Egerton tells me that he [*altered from* 'has'] has seen original Irish Yew at [*over illeg*] Florence Court; it was found as young tree in open [*interl*] mountains & removed—. reproduced by cuttings & seeds. believes that former truest.— One seedling has come up different & Lord E. has promised one to Sir P. E.— I think this proves a natural seedling & weeping yew.— Species will turn out (N.B Leighton says there is difference in leaves of weeping yew??) made by jumps—curious both these yews coming true to seeds.— (DAR 163).

Leighton is William Allport Leighton and the reference is to Leighton 1841, pp. 497–8.

[3] Lord Enniskillen's seat.

[4] CD preserved the letter for possible use in chapter four of his 'big book' which was to be devoted to 'Variation under nature'. However, the yew is not mentioned in chapter four of either *Natural selection* or the *Origin*. The sudden appearance of useful or ornamental varieties of trees, including the weeping yew, is discussed in *Variation* 1: 361.

To Julian Jackson 23 May [1844]

Shrewsbury
May 23d

My dear Sir

I received from Down this morning the Paper on the R. Negro.—[1] I see by your note that it was sent from your office, on the very day, on which I left home.— I have given the best opinion I could on the Paper.— It appears to me pleasantly written, but most of the information might be found, scattered in other publications. I found it impossible to mark out whole passages, as you requested.

Pray believe me | Yours very truly | C. Darwin

Colonel Jackson

Royal Geographical Society

[1] Robinson 1844.

From George Brettingham Sowerby and Edward Forbes 28 May 1844

[Two lists of shells from James, Charles, and Chatham Islands of the Galápagos, one signed and dated by G. B. Sowerby, who summarises: '18 species in the two lists of which 8 identical and 2 identical with Tahitian species 1 [identical] with Hobart town species.' The other list, on the verso, is in Edward Forbes's hand, with additions by CD. On the Sowerby list CD has written: 'Dec. 20 /44/ | From this list of Galapagos shells nothing about Representatives can be made out']

A memorandum S
DAR 46.2 (ser. 2): 1–2

To J. D. Hooker 1 June [1844]

Down near Bromley Kent
June 1.

My dear Hooker

I write to ask you whether Gray returned you the estimates of the Zoology of Beagle's Voyage & if so, would you be so good, if you have quite finished with them, return them to me.— If Gray has not returned them, I must write to him myself at once, for I begged him to send them you soon back.— I am in no sort of hurry for them, only I sh^{d} be sorry they sh^{d} be lost. How busy you must be now: Your publication has commenced much earlier, than I expected—[1] I hope you will not work too hard; I well remember, when I thought it utter nonsense to talk of 8 hours being too much, but I now find 2 more than I can stand— You must be greatly tempted, I am sure, to overwork yourself, so do take care.— I heard from Henslow some month or two ago, saying he had found a lot of Galapagos plants, which he had omitted to forward to you.— this must put your calculations out.—

I have been away from home, the last six weeks, & have now on my return been reading over some of your geographico-botanical letters, with renewed interest.

Will you send me one line about my Papers & believe me, very truly your's | C. Darwin

DAR 114.1: 11

[1] The first number of J. D. Hooker 1844–7 was published 1 June 1844 (Wiltshear 1913).

To Emma Darwin [3 June 1844][1]

[Down]
Monday

My dear Em.

Thank goodness this is my last note & that in two more days, you will be here—

The horse cannot go for you, so Parslow[2] will go up & meet you at the Train, take you 16[3] & afterwards by ¼ past 3 to the Bolt-in-tun. You will come quicker this way, than by the Phaeton; I endeavoured to get the Coach call at 16, but they c^d^ not under some penalty— It is rather extravagant sending up Parslow, as John[4] might have done, but I thought you w^d^ like it best.— Marianne must walk from the Station to 16, & I fear one Coach will not hold you to the Bolt-in-Tun.—

You had better lie down & rest at 16 & be a good girl.— I do hope you will stand the Journey well.—

I have been wonderfully strong; on Saturday I sat up reading till ¼ before eleven, not dreaming it was so late & then went to bed & never awakened till I was called.—a thing which I can hardly remember having happened to me.— I long for you to be back—for I do so enjoy being at home again. Try & remember, just before you drive in at our gate, to rise from your seat & look over the wall, & see how nice the place looks.

Farewell, my dearest. | Yours, C. D.

I can fancy poor Annies scarlet face at Minnys kisses,[5] poor dear Boddy Bumpkins.—

DAR 210.19

[1] Endorsed by Emma Darwin: '1844 Down— I had been at Betley'. Betley Hall, Staffordshire, was the home of the Tollet family. Later, on the cover, Emma wrote 'On my return to Down after absence in Staff—1845?' The date given above is the most likely. CD and Emma had been in Staffordshire and Shropshire between 23 April and 30 May 1844 (see 'Journal'; Appendix II); CD had then returned to Down but Emma did not come home until 5 June, as evidenced by her diary and CD's Account Book (Down House MS).

[2] Joseph Parslow was the butler at Down House.

[3] Emma's brother and his wife, Hensleigh and Fanny Mackintosh Wedgwood, lived in London at 16 Gower Street.

[4] Probably John Jordan, a servant at Down House.

[5] Anne Elizabeth Darwin, Emma and CD's oldest daughter. Minny has not been identified.

To Henry Denny 3 June [1844]

Down Bromley | Kent
June 3d

Dear Sir

I am much obliged for your note.— You are at perfect liberty to mention Mr Martials story— I forget whether I said, he was a surgeon of a whaler, but a rather worthless, slightly educated man; perhaps, however, in some respects his story is less likely from this cause to have been invented.—[1]

I myself do not think our supposed knowledge of having come from one stock ought to enter into any scientific reasoning. Anyhow the inhabitants of eastern & western Europe have different species of intestinal worms.—

I fear I cannot at present offer to search for the specimens in Spirits, but if you will inform me, (supposing I do not send them before hand) at the latest period, when you absolutely require them for comparison, I will get them out of a chaos of specimens.—[2]

Believe me | dear Sir | Yours sincerely | C. Darwin

P.S. | I have been informed that the Pediculi generally, if not invariably, perish on wild animals in their passage to England, or in captivity. This, perhaps, may bear on their death in Mr Martial's story. A slight fever, or even a *broken limb with no fever* has been known to cause the evacuation of the intestinal worms in a person—facts which show by what slight changes in constitution parasites are affected.

American Philosophical Society

[1] While on Chiloé in 1834 CD made the following entry in his *Beagle* zoological diary:

> Mr Martial, a surgeon of an English Whaler assures me that the Lice of the Sandwich Islanders . . . if they strayed to the bodies of the English in 3 or 4 days died . . . If these facts were verified their interest would be great.— Man springing from one stock according his *varieties* having different **species* of [*added, pencil*] parasites.— It leads one into [*after del illeg*] many reflections.— (DAR 31.2: 315).

The last sentence was subsequently deleted in pencil and the passage beginning 'If these facts' to the end crossed in pencil. The pencil alterations appear to have been made after the voyage. The account was used later in *Descent* 1: 219.

[2] See letter to Henry Denny, 20 January [1844], in which CD offered to send Denny his sucking lice specimens. CD sorted his collections in July and provided Denny with specimens ('Journal'; Appendix II).

To *Gardeners' Chronicle and Agricultural Gazette* [before 8 June 1844]

Some of your readers may be amused at the style, as well as at the matter of the following quotations from "The Curiosities of Nature and Art in Husbandry and

Gardening," published in 1707.[1] They show that the value of the inorganic parts of manure, and the advantage of steeping seeds, were well known at that time. "The whole secret of multiplication consists in the right use of salts. Salt, says Palissy,[2] *is the principal substance and virtue of dung*. A field may be sown every year, if we restore to it by stercoration what we take from it in the harvest." . . . . "Seeing all multiplication depends on salts, the main business is to get together a great quantity at little expense, that the profit may be the greater." The author then describes a method of making liquid manure, in three old casks, into which objects are separately thrown, according to the ease with which they decompose. He further urges the importance of burning all wild plants, and of carefully dissolving the soluble parts of their ashes, and then proceeds—"Take as many pounds of saltpetre or nitre as you have acres of land to sow. For each acre dissolve a pound of saltpetre in twelve pints of the water that sanks from the dunghill. When the saltpetre is quite melted, throw in a little of those salts of plants (i.e. ashes) according to the quantity you have of them. This liquor is then called the 'Universal Matter,' because nitre is truly the universal spirit of the elementary world. This is the main point of the whole secret of multiplication. We will for the future call the water that is got ready in the casks, *Prepared Water*, and the water from which the salts are extracted from plants, and the nitre, *Universal Matter*. For one acre, take twelve pints of the prepared water, and mix with it immediately the universal matter, in which there ought to be a pound of dissolved nitre. The vessel into which you put these liquors must be large enough to contain the corn which you design for one acre. Then strew in your corn into these liquors; there must be two inches of water above the seed. Leave the corn to soak for twelve hours, and stir it up and down every two. If by that time it do not swell, let it lie longer till it begin to plump up considerably. One third less of seed than usual will serve for an acre; nay, you may safely use but half as much, and mingle among it some straw cut very small, that the sower may take it up by handfuls and sow it in the ordinary way, as I have said already." The explanation the author offers of the use of soaking seeds is whimsical. He says that the first action is to "cut the covers that infold the sprouts," and that the second action is "to serve each grain of corn, as it were, instead of a loadstone, to attract the nitre of the earth, which the subterranean fires have reduced and driven into steams and vapours in the low and middle region of the air, for the nourishment of vegetables and of animals. This is not a vain imagination, a chimera, or empty notion."— *C. Darwin.*

Gardeners' Chronicle and Agricultural Gazette, no. 23, 8 June 1844, p. 380

[1] Vallemont 1707.

[2] Palissy 1636, referred to in Vallemont 1707, pp. 172–4.

To G. R. Waterhouse 10 [June 1844 – March 1845][1]

Down Bromley Kent
10th

My dear Waterhouse

We shall be truly glad to see Mrs. Waterhouse, yourself & children on Wednesday. I am ashamed to say I forget what was the train, which we agreed you were start by: but please let it be, the train which leaves the Bricklayer's Arms at 2° " 20′ & which reaches Croydon at about 3. olock.— My Phaeton (one chesnut horse) shall be there (& shall wait for one later train)

I send the card again as it may be useful about the omnibuses: please bring it with you.—

Ever yours | C. Darwin

N.B. Have mercy on people, like myself, with bad memories & put your address to every letter,—I was in despair trying to remember & search your your address.—

Cleveland Health Sciences Library (Robert M. Stecher collection)

[1] Dated from the reference to the Bricklayers' Arms terminus of the Croydon railway, which was opened 1 May 1844; the passenger service to Croydon ended in March 1845 (Course 1962, p. 71). CD arrived back at Down from a visit to Maer, the Wedgwood family home, and Shrewsbury on 30 May ('Journal'; Appendix II).

To Ernst Dieffenbach 11 June [1844][1]

[Concerning the researches of C. G. Ehrenberg.] '. . . I have . . . sent him several packets of objects from my voyage & that of Dr. Hooker . . .'[2]

J. A. Stargardt, Marburg (catalogue 574, 11–13 November 1965)

[1] Date taken from the Stargardt catalogue.
[2] See letter to C. G. Ehrenberg, 20 April [1844].

From C. G. Ehrenberg 15 June 1844[1]

Berlin
d. 15 Juni 1844.

Hochzuverehrender Herr

Die mir von Ihnen unter dem 20sten. April freundlichst angekündigte Sendung von Erd-Arten habe ich bald darauf im May erhalten. Ich bin Ihnen aufs dankbarste verpflichtet durch diese Sendung und besonders auch durch die so zuvorkommende Weise mit welcher Sie meine durch Herrn Dr. Dieffenbach[2] ausgesprochnen Wünsche erfüllt haben. Ich fand alsbald diese Materialien vielfach so interessant, daß ich sofort Tag und Nacht mich ihen gewidmet habe.

In wenig Tagen hoffe ich Ihnen eine Ubersicht der Resultate meiner Untersuchungen übersenden zu können. Von den Gallopagos Inseln habe ich eine ansehnliche Zahl der mikroskopischen Formen nun zur Anschauung erhalten. Ganz besonders wichtig war mir aber der meteorische Staub oder die vulkanische Asche, welche aus der Gegend der Capverdischen Inseln stammt und die, etwa zu $\frac{1}{6}$ der Masse, aus Kieselschalen bestimmbarer Organismen besteht, deren einige bisher nur und allein (nicht in Africa) sondern in Cayenne von mir beobachtet worden sind, ungeachtet ich gerade vom Senegal sehr zahlreiche Formen kenne, die zum Theil ganz eigenthümlich sind deren keine aber jener Staub characterisirt. Ich nehme an daß das was Sie in Ihrer Reise von trüber Luft auf den Capverdischen Inseln sagen sich auf diesen Staub mit bezieht.[3] Wie viel Tage hielt wohl dieser Staubregen an? Es ist nun überaus interessant die Ihnen bekannten Umstände recht genau und detaillirt aufzuzeichnen.— Aus Neu Seeland habe ich noch nichts erhalten können. Was Sie gesendet haben ist unorganisch. Aus Neuholland besitze ich schon sehr viel Material, aber aus den vielen Inseln des Austral-Meeres[4] kenne ich noch wenige.

An Herrn W. Hooker[5] schreibe ich ebenfalls einige Zeilen des Dankes, da auch seine Sendung überaus interessant gewesen.

In etwa 14 Tagen schreibe ich Ihnen wieder, um die Resultate meiner Untersuchungen Ihnen zukommen zu lassen, bis dahin empfehle ich mich | Ihrer ferneren Gewogenheit und wiederhole nur meinen besten Dank.

Verharrend in groeßter Hochachtung | Ihr | dankbar ergebenster | Dr C G Ehrenberg | Professor und Secretaer der Akad d Wissensch.

DAR 163

[1] For a translation of this letter, see Appendix I. Ehrenberg's letter was delivered by hand, but not until late July or August. See letter to J. D. Hooker, [25 July – 29 August 1844], and letter to C. G. Ehrenberg, 5 September [1844].

[2] See letter to Ernst Dieffenbach, 25 January 1844.

[3] *Journal of researches*, p. 4.

[4] Ehrenberg probably used the term in a general sense to mean Oceania. See his letter of 11 July 1844 in which he refers to the 'Süd-australischen Archipel von Neu Guinea bis zu den Marquesas I.'

[5] Ehrenberg misread CD's letter of 20 April [1844]. The specimens were provided by Joseph Dalton Hooker.

To J. D. Hooker 29 [June 1844]

Down near Bromley | Kent
Saturday, 29th.

My dear Hooker

I think you will be pleased to hear about your infusoria in the enclosed letter,[1] which will you be so good as to return— Your data must have been grand ones for Ehrenberg. The statement, which surprises me most, is that about the dust, which is well known to fall on ships off N W. Africa, consisting entirely of Infusoria— I have a series of specimens & facts, which I must send to Ehrenberg.—[2] Shall I offer for you, *if he so wishes*, for some more Antarctic specimens, in which to hunt for Infusoria.—

I heard at the Athenæum, from some Botanists of you, about a fortnight since, & that your work is progressing well— I, also, heard that you were not looking well; do not trust to your own medical knowledge, & overwork yourself.—

You were so kind as sometime since to ask me to come to Kew; I should much enjoy seeing the Gardens & still more the pleasure of conversing with you.— W^{d} you when you return the enclosed, tell me at what hours the steam-boats leave London Bridge or, if there are none thence, elsewhere: how long they take to go to Kew, & at what hours, they return; I sh^{d} wish to get back here at night if I could.— If I can keep my steam & courage up for this great expedition, I will take advantage of your kindness, sometime in the course of next month.

Believe me | Your's very truly | C. Darwin

Have you heard that Stokes is going to publish an account of his voyage.—[3]

Is the steam-boat the best way? I get into London at London Bridge by the Croydon Railway.

DAR 114.1: 12

[1] A letter from Ernst Dieffenbach (see letter to C. G. Ehrenberg, 4 July [1844]). This letter has not been located.

[2] See letter to C. G. Ehrenberg, 4 July [1844].

[3] Stokes 1846. John Lort Stokes had returned to England in September 1843 after eighteen years' service in the *Beagle*.

To C. G. Ehrenberg 4 July [1844]

Down near Bromley | Kent.
July 4th.

Sir

I have been very much pleased to hear through D^{r}. Dieffenbach that you have found some of D^{r}. Hooker's & my specimens of service to you. Dr D. tells me that the dust off the N.W. coast of Africa consists of Infusoria: I am truly astonished at this, & I write now to inform you that I have specimens from several other stations in the same sea-region. There can, I think, be no doubt, that it is the finer matter blown from the dry country along that coast. From fusing under the blowpipe easily into a slag, it has generally been thought: to be fine *[illeg]* ashes from some active volcano. The dust falls during a great part, or all the year over very many hundreds of square miles of sea, & as no great rivers debouch on this coast, there must be a considerable deposit in progress of formation, which, it now seems, will consist of Infusoria. I have always intended drawing up a *brief* account of the area, over which, this dust falls. Should you like to have this little notice to communicate together with your account of the Infusoria contained in the dust, I shall be very proud to send it, with other specimens, if you choose, of the dust. If you sh^{d}

not care to have my notice, perhaps you would kindly take the trouble to let me know the names of the Infusoria, which you have found in the dust in order that I may refer to them, in the notice, which I will publish in some English Journal.[1]

Should you *wish* for any more specimens or substances in which to look for Infusoria, it will give me *great* & **undivided** pleasure to search my collection, although this will take me some time.— It is, however, probable that you are already overwhelmed with specimens from all quarters of the world. D^r. Hooker will be happy to send you any more antarctic specimens, which you would specify.— If you wish for any more specimens, perhaps you will be so good, as to direct me to what countries are most interesting to you, & what sort of substances will be most favourable— Would lignite or imperfect coal; or lumps of old coral from sea-beaches; or corallines & seaweeds from T. del Fuego, be likely.

In case you sh^d wish for any more specimens, could you direct me any cheap way to send them, as I found the carriage of the last Box expensive.—

Believe me Sir | With the highest respect | Your obliged & faithful sevt. | C. Darwin

P.S. | I believe there are many soundings (i.e. the mud & sand brought up on the lead) from great depths in the Antarctic & other seas, at the Admiralty: would portions of these be of service to you, for I daresay they could be obtained?

Postmark: 4 JY 4 1844
Museum für Naturkunde der Humboldt-Universität zu Berlin

[1] 'An account of the fine dust which often falls on vessels in the Atlantic ocean'. The paper was not read until 4 June 1845. See *Collected papers* 1: 199–203.

To Emma Darwin 5 July 1844

Down.
July 5th.—1844

My. Dear. Emma.

I have just finished my sketch of my species theory.[1] If, as I believe that my theory is true & if it be accepted even by one competent judge, it will be a considerable step in science.

I therefore write this, in case of my sudden death, as my most solemn & last request, which I am sure you will consider the same as if legally entered in my will, that you will devote 400£ to its publication & further will yourself, or through Hensleigh,[2] take trouble in promoting it.— I wish that my sketch be given to some competent person, with this sum to induce him to take trouble in its improvement. & enlargement.— I give to him all my Books on Natural History, which are either scored or have references at end to the pages, begging him carefully to look over & consider such passages, as actually bearing or by possibility bearing on this subject.—[3] I wish you to make a list of all such books, as some temptation to an Editor. I also request that you hand over him all those scraps roughly divided in eight or ten brown paper Portfolios:— The scraps with copied quotations from various works are those which may aid my Editor.—[4] I also request that you (or some amanuensis) will aid in deciphering any of the scraps which the Editor may think possibly of use.— I leave to the Editor's judgment whether to interpolate these facts in the text, or as notes, or under appendices. As the looking over the references & scraps will be a long labour, & as the **correcting** & enlarging & altering my sketch will also take considerable time, I leave this sum of 400£ as some remuneration & any profits from the work.— I consider that for this the Editor is bound to get the sketch published either at a Publishers or his own risk. Many of the scraps in the Portfolios contains mere rude suggestions & early views now useless, & many of the facts will probably turn out as having no bearing on my theory.

With respect to Editors.— Mr. Lyell would be the best if he would undertake it: I believe he wd find the work pleasant & he wd learn some facts new to him. As the Editor must be a geologist, as well as Naturalist. The next best Editor would be Professor Forbes of London.[5] The next best (& quite best in many respects) would be Professor *Henslow*??. Dr. Hooker would perhaps correct the Botanical Part probably=he would do as Editor=[6] Dr Hooker would be **very** good[7] The next, Mr Strickland.—[8] If no⟨ne⟩ of these would undertake it, I would request you to consult with Mr Lyell, or some other capable man, for some Editor, a geologist & naturalist.

Should one other hundred Pounds, make the difference of procuring a good Editor, I request earnestly that you will raise 500£.

My remaining collection in Natural History, may be given to anyone or any Museum, where it wd be accepted:—

My dear Wife | Yours affect | C. R. Darwin

If there shd be any difficulty in getting an editor who would go thoroughily into the subject & think of the bearing of the passages marked in the Books & copied out on scraps of Paper, then let my sketch be published as it is, stating that it was done several years ago[9] & from memory, without consulting any works & with no intention of publication in its present form—

PS | Lyell, especially with the aid of Hooker (& of any good zoological aid) would be best of all

Without an Editor will pledge himself to give up time to it, it would be of no use paying such a sum.—[10]

British Museum (Natural History) General Library

[1] Usually known as the essay of 1844, the name given to it by Francis Darwin, who published it with the pencil sketch of 1842 (*Foundations*; republished in de Beer ed. 1958).

[2] Hensleigh Wedgwood.

[3] The Cambridge University Library handlist entitled 'Darwin Library: List of books received in the University Library Cambridge March–May, 1961' records which works in the collection have CD annotations. A complete listing of CD's marginalia in these books is being prepared by Mario di

Gregorio. An earlier catalogue compiled at Francis Darwin's request by H. W. Rutherford lists the books from CD's library as they were in 1908 (Rutherford 1908).

4 After 1839, CD began the practice of filing his notes in separate classified portfolios. Many of the 'scraps' are still preserved together in various DAR volumes, e.g., DAR 46.1 has the notes assembled for writing 'Struggle for existence', chapter five of *Natural selection*, later chapter three of the *Origin*; other loose notes are in DAR 205.1 to DAR 205.11.

5 After this sentence CD wrote 'or Mr Lonsdale (if his health w^d^ permit).' These words were deleted by CD, possibly at the same time as other alterations made to the letter, see nn. 6, 7, and 8, below.

6 The passage 'Dr Hooker . . . Editor=' was written in the margin next to paragraph three; the intended position in the letter is unclear. CD at first wrote '=possibly he would do as Editor=' and then, perhaps at a later date, altered it to read 'probably=he'.

7 The sentence 'Dr Hooker would be **very** good' was added by CD, probably at the same time as he changed his previous remarks about Hooker (see n. 6, above) and queried Henslow's name (see Manuscript alterations and comments).

8 After this sentence CD wrote and later deleted 'Professor Owen w^d^ be very good, but I presume he w^d^ not undertake such a work.'

9 Francis Darwin (*LL* 2: 18) suggested that the words 'several years ago' were added later, but there is no evidence for this.

10 Ten years later, CD made the following note in pencil on the cover: 'Hooker by far best man to edit my Species volume | Aug. 1854' Another note in pencil on the cover, which has been crossed out, apparently by CD, reads: 'NB When new Will made make Trusts open.' CD also made pencil notes and markings inside the letter, presumably at the same time: the first page has been crossed out (ending with the sentence 'I wish you to make a list of all such books, as some temptation to an Editor.' 2.10) and at the top of the first page, the instruction 'Read Enclosure' was added. The enclosure has not been found. The pencil notes were undoubtedly connected with CD's intention to write his 'big book' (*Natural selection*). On 9 September 1854 he began sorting his species notes in preparation (see CD's 'Journal' (DAR 158); de Beer 1959, p. 13).

From C. G. Ehrenberg 11 July 1844[1]

Berlin
d. 11ten Juli | 1844.

Hochgeehrtester Herr

Vor nun 3 Wochen schrieb ich Ihnen durch die Gelegenheit des jungen Herrn Gladstone aus London, welcher sich erbot einen Brief zu besorgen. Ich hoffe daß seitdem dieser Brief in Ihren Händen ist.[2] Vorher hatte sich Dr. Dieffenbach erboten Ihnen die glückliche Ankunft der sehr interessanten Sendung zu melden. Meinen besten Dank für Ihre so große Freundlichkeit habe ich Ihnen bereits gemeldet und ich wiederhole ihn gern. Sie haben mich mit so reichem Material für meine Untersuchungen versehen daß ich noch lange daran zu zehren habe. Gestern sandte ich Ihnen durch die Gelegenheit des Herrn Gibsone aus Perth den gedruckten Auszug meines Berichtes über einen Theil Ihrer Sendung aus den Monatsberichten der berl. Akademie d. Wiss. Mai.[3] Ich hatte ein Exemplar für Sie und eins für Herrn Dr. Hooker an Herrn Francis unter der Adresse des Herrn Richard Taylor Fleet Street Red Lion Court London addressirt.[4] Ich hoffe daß Sie dieselben erhalten.

Auf Ihr heut angekommenes Schreiben vom 4ten Juli, melde ich Ihnen, daß alles was Sie für einer mikroskopischen Analyse für würdig halten mir höchst erwünscht ist. Alle Arten von See-Sand und Schlick oder thonartige Absätze des Meeres und großer Flüße sind auch in kleinen Mengen zuweilen reich an Formen. Was Sie mir über den Staubregen melden ist höchst interessant im Verhältniß zu meiner Analyse. Ich bitte daher um möglichst vollständige Details. Auch besonders was von Erd Proben aus großen Meeres-Tiefen erreichbar ist würde mir sehr willkommen seyn, zumal wenn auf der Admiralität vielleicht von den Rossschen kolossalen Grund-Messungen die Proben vorhanden wären.[5]

Herr Francis, Fleet Street Red Lion Court London bei Herrn R. Taylor, hat wohl die Güte was Sie ihm für mich übergeben an mich gelangen zu laßen, im Fall Sie es nicht direct schicken wollen was vielleicht am Besten ist. In Hamburg besorgt der Kaufmann Herr G. Morgenstern am alten Wandrahm die unter meiner Adresse an ihn gesendeten Dinge hierher sicher und wohlfeil für mich.

In ausgezeichneter Hochachtung und Dankbarkeit verharrend

Ihr | ganz ergebenster | Dr C G Ehrenberg.

Die Salz Ablagerungen in Patagonien sind nicht reich an Leben, obschon ich von andern Orten her sehr reiche concentrirte Soolwässer besitze.

Isolirte Inseln aller Erdgegenden interessiren mich rücksichtlich ihres kleinsten Lebens besonders auch. Von den Gallopagos und von Asoension habe ich durch Ihre Güte gute Ausbeute.

Der Sand aus Keeling Atoll enthält nicht bloß todte Theile, sondern zum Theil lebend eingesammelte trockene kleine Thiere worunter auch manche Kieselschalen von Polygastrischen Infusorien. Es ist nicht bloß ein verdautes Excrement sondern ein kleines Lebens Gewühl, worin viel Corallenfragmente von weichen Corallen, wie ich es auch im Rothen Meere fand.

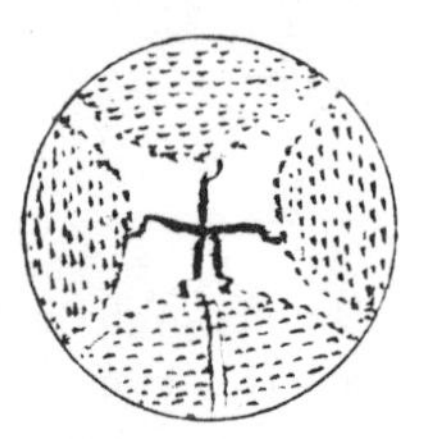

Asteromphalus
Darwinii.

Aus Terra del Fuego besitze ich jezt viel Material aber es fehlt mir dergleichen aus Californien, aus den Capverd. Inseln, von Sierra Leone, Congo, Angola, Ferner aus dem Süd-australischen Archipel von Neu Guinea bis zu den Marquesas I. Aus Neuholland besitze ich viel, aus Neu Seeland noch nichts, da die übersandten Stein Proben kein deutliches Resultat geben. Aus Guinea und Senegal besitze ich viel. Aus den Lakediven und Malediven noch nichts.

Prof. C. G. Ehrenberg. Secret. d. Akad. d. Wissensch. | Berlin.
Unter den Linden 21.

pr. adr. d. Herrn Gustav Morgenstern | Meissner Porzellan Niederlage | am alten Wandrahm | Hamburg.

DAR 163

[1] For a translation of this letter, see Appendix I.
[2] Ehrenberg's first letter to CD had not arrived. See letter from C. G. Ehrenberg, 15 June 1844, n. 1.
[3] Ehrenberg 1844a. CD's copy of the paper is in the Darwin Library–CUL.
[4] William Francis served as an assistant in Richard Taylor's printing firm, located in Red Lion Court, Fleet Street. The paper was being sent to Taylor for publication in English (Ehrenberg 1844b).
[5] CD had offered specimens collected by Joseph Dalton Hooker during James Clark Ross's Antarctic expedition, see letters to J. D. Hooker, [27 January 1844] and 23 February [1844], and letter to C. G. Ehrenberg, 20 April [1844].

To J. D. Hooker [14 July 1844]

Down near Bromley | Kent
Sunday

My dear Hooker

I propose to give myself the pleasure of paying you a morning visit on Thursday. I have to be in town on Wednesday & will come in my light tax cart,[1] about ten oclock in the morning & at midday return across country home. As I shall be in town, I will certainly come without the weather be *atrociously* bad.— W^{d} you send me **one single** line to say whether Thursday would suit you, if not I w^{d} come on Friday, but Thursday w^{d} suit me best.—

W^{d} you send your answer *as soon as you* receive this—then perhaps I shall receive it on Tuesday morning.

M^{rs}. Darwin is very much obliged for your kind invitation, but is afraid of the length of the drive.—[2]

Believe me Ever yours | C. Darwin

P.S. | I ought to apologise for coming in the morning & thus causing you to lose your best hours, but my visit will not be very long.—

N.B. As my health is always extremely uncertain, you must not be surprised if I fail: if I am not with you before eleven, you will understand that my health is to blame.—

Postmark: JY 14 1844
DAR 114.1: 13

[1] A light, two-wheeled farmer's cart on which a reduced tax had formerly been levied.
[2] Although CD declines the invitation on behalf of Emma, she did, according to her diary, accompany her husband on his trip to Kew. The diary records that they visited Hooker on Thursday, 18 July.

To J. D. Hooker 22 July [1844]

Down Bromley Kent
July 22^{d}

My dear Hooker

I enclose a letter from Ehrenberg, of which I wish you joy, if it costs you as much trouble in reading as it did me.[1] You will see that he begs for drawings & an account of the (as I suppose) dirty ice.— I told him that I believed, that the deep-sea soundings from your voyage were brought home & deposited at the Admiralty. Was I right? you will see that he is very anxious about them. I sh^{d} think Capt. Beaufort, would let you have portions.—

I mean to send a few things more, in a parcel through Mr Francis, as directed. If you like to send anything here (by the "Down Coach, from the Bolt-in-Tun Fleet St^{t}), I will enclose it & write & forward it to Mr Francis.—

Did you send anything from Kerguelen Land, you will see he is interested about isolated islands.— It is very provoking that his former letter has never come to hand— I will write to Mr Francis about the printed account & if your copy is sent to me, I will immediately transmit it to you.

On account of the directions, will you please sometime return my part of the letter.

I cannot tell you, how we enjoyed our day's excursion at Kew & pray thank Sir William in our names.— We got home by a little after seven, after a very pleasant drive.

Believe me | very truly yours | C. Darwin

P.S. | I find I have a Pyrus baccata— is it different from a common Siberian Crab.[2] & how is it possible to make Crabs into preserve? Do send me a gastronomic answer.—[3]

DAR 114.1: 14

[1] Presumably the letter from C. G. Ehrenberg, 11 July 1844.
[2] They are the same. The Siberian crab apple was then known as both *Malus baccata* and *Pyrus baccata*.
[3] For Hooker's answer see letter to J. D. Hooker, [25 July – 29 August 1844], n. 3.

To John Stevens Henslow [25 July 1844]

Down near Bromley | Kent
Thursday.

My dear Henslow

In looking through my collection, I stumbled upon some rusty wheat from the N. Bank of the Plata & it occurred to me, as you have been working at wheat,[1] that you might like to have the specimen, together with my rough notes, if not, the specimen & note can be thrown away together.

My more immediate object, however, in writing now, is to ask you to send me, if you can find **quite easily** the specimens, some bits of my Peat (with their country marked) for Ehrenberg. I am going to send another parcel to him. The specimens might be so small, that you could send them by Post.—

Did I give you a ball of white paint, with which the Fuegians colour themselves? if you can find it easily (*& it is* **quite unimportant** *if you cannot*) please to send me a little bit for the same end.[2]

I went to Hooker at Kew the other day[3] & admired the place much. Young Hooker seems to be working away & making capital progress.—

I saw a circular from you there, from which I perceive you are at work on the state of the People, & great need there seems to be for this in your side, & indeed on all sides, of the country.—[4] How I wish this house lay on the road to somewhere else, that we might have the chance of seeing you— we often regret on this one score having left London.

Ever yours | C. Darwin

Endorsement by Henslow: '26 July 1844'
Brown University Library (A. E. Lownes collection)

[1] See Henslow's papers on wheat diseases, 1841a and 1841b.
[2] See letter from C. G. Ehrenberg, 11 July 1844, in which he expressed interest in seeing additional specimens. CD had given the paint to Henslow for his museum, see *Correspondence* vol. 2, letter to J. S. Henslow, 16 September [1842].
[3] See letter to J. D. Hooker, [14 July 1844], n. 2.
[4] Possibly Henslow 1844.

To J. D. Hooker [25 July – 29 August 1844][1]

Down Bromley Kent
Thursday

My dear Hooker

I enclose one of Ehrenberg's letters, which had not arrived; I do not send you the one to me, as it contained nothing but *enquiries* about the dust and other such subjects.[2]

Very many thanks for all your and your cook's wisdom about Crabs, which has been duly copied out.—[3]

In looking over some coralls, I found the enclosed sea-weed from the Galapagos Isl^{ds} & I believe from 12 fathom's depth; I thought perhaps you would like to have it— I, also, send either for yourself or Mr Harvey[4] specimen. 390 & 391 of the little conferva in bundles described at p. 14 of my Journal: I have not, however, looked to see whether they are preserved. 392 is a minute attached conferva from 17 Fathoms off the Abrolhos Is^{d} coast of Brazil.[5] Please throw away these specimens if of no use.

Did you collect any pediculi fm your voyage; especially from the Penguin: H. Denny (to whom I have given all my specimens) has written to me expressing a great wish to have some from the antarctic regions.[6] If you are able to supply him: his address is "Philosophical Hall Leeds".

Whenever you have ready the specimens for Ehrenberg, they had better be sent here by the "Down Coach from Bolt-in-tun Fleet St!"

Ever yours | C. Darwin

DAR 114.1: 15

[1] The date range is based on Emma Darwin's recipe, see n. 3, below, and the receipt of the letter from Christian Gottfried Ehrenberg, 15 June 1844 (see letter to C. G. Ehrenberg, 5 September [1844]).

[2] Presumably CD refers to the letter from C. G. Ehrenberg, 15 June 1844, which arrived in late July or August. It seems likely that CD had also received a note addressed to Hooker, as mentioned by Ehrenberg in this letter.

[3] Emma Darwin's recipe book (DAR 214) contains the following recipe:

Compote of Apples

Cut any kind of apples in half, pare, core & put in cold water as you do them; have a pan on the fire with clarified sugar, half sugar half water; boil, skim, & put apples in; do them very gently; when done take them off & let them cool in the sugar, then set them in the ashes; & if the sugar is too thin set it again on the fire & give it the height required. (Dr Hooker). July 1844.

[4] William Henry Harvey.

[5] *Journal of researches*, pp. 14–15.

[6] See letters to Henry Denny, 3 June [1844] and 12 August [1844].

To Josiah Wedgwood III and E. A. Darwin[1] 25 July 1844

[Down]

We hereby request you to pay the sum of five thousand pounds now in your hands, being trust-money under our marriage-settlement, to Thos: Salt Esq:[2] for the disposal of Dr Darwin.[3]

Charles Robert Darwin

Emma Darwin

July 25th 1844

I hereby notify my assent to the above Arrangement

E. A. Darwin.

Keele University Library (Wedgwood/Mosley 1012)

[1] The trustees of the marriage settlement of CD and Emma.

[2] Of the Shrewsbury law firm of Dukes and Salt.

[3] The money was part of a £20,000 sum, made over to Robert Waring Darwin by CD and Emma, which formed the basis of a mortgage loan to Edward Herbert, 2d Earl of Powis, negotiated on 14 November 1844 (R. W. Darwin's Investment Book (Down House MS), pp. 57, 70).

To The Royal Geographical Society [30 July 1844 – 1 October 1846][1]

Down near Bromley | Kent
Tuesday Morning

My dear Sir

I sh^{d} feel *particularly* obliged to you, if you would take the trouble to search to see whether there is in the possession of the Society, a Spanish Map, by "La Cruz"[2] of the Cordillera of central Chile near St. Jago: It is one of the best, & I am in the *greatest* need of it.— If you can find it, w^{d} you send it directed to me, at the Athenæum, *not later than 12 oclock on Thursday* when I shall call there on my way home.

My dear Sir | Yours very truly | C. Darwin

P.S. Sh^{d} you have *any* map, of the Cordillera of *Chile*, I sh^{d} be very glad to see it.—

The rock-specimens have not been yet sent to the Geolog. Soc. for me.

Fitzwilliam Museum Cambridge

[1] The date range is that of the writing of *South America*. See 'Journal' (Appendix II): 'July 27th—/44/—/ Began S. America.' The first Tuesday after 27 July was 30 July.

[2] The map was drawn by William Desborough Cooley based on the diary of Luis de la Cruz (Cruz 1835), who, in 1806, travelled across the Andes to Buenos Aires. CD complained that maps of the area differed from each other and were 'all *exceedingly* imperfect' (*South America*, p. 176).

To Adolf von Morlot 9 August [1844]

Down near Bromley | Kent
Aug 9th

Dear Sir

I should have replied to your last obliging letter, had I not lately been much engaged, before this time. I regret exceedingly to say, that I cannot undertake to see your Journal published: my health during the last three years has been exceedingly weak, so that I am able to work only two or three hours in the 24: these are more than fully occupied & I have materials for several years' work, which is almost more than I dare undertake. Under these circumstances, I hope you will not think me either unkind or unreasonable in declining to add to my employments, & you must be well aware that everything going through the press costs time & trouble. The only channel of publication in England, that I can see, without great expence to yourself, *might possibly* be the Edinburgh New Philosophical Journal; but I cannot here aid you, as I am not acquainted with the Editor.—[1]

We in England are accustomed to believe, that publication is much less expensive with you, than with us: certainly Engraving is. The cheapness of German Books always astonishes me, & I wish scientific authors in England knew, how to follow so good an example. I may just mention to you, that "M. Baillière Bookseller Regent St" is one of the most spirited of scientific publishers, & it might, perhaps, be worth while to send your M.S. to him. Every publisher, however, in England looks at scientific books with a cold eye.— I hope sincerely that you will meet with success in whatever you determine on, though I cannot aid you.

You will think me a great sceptic, when I tell you that your letter has not convinced me. I daresay there may have been Glaciers on the mountains you describe, but my mind will require a long series of proofs to believe that they have come from Scandinavia.— One always puts too much stress on what one self has seen; but I cannot avoid suspecting that N. Wales offers an example of what has happened in many cases; viz, that the Boulders have been transported & rocks scored by floating ice, & that after & during elevation, glaciers have removed all these appearances, *except on the outskirts*, & have left in place their own marks.—[2] I do not believe the respective shares of work of these two agencies, will be clear, until the action of floating ice has been fuller studied in the north & some distinctive effect, *if any exist*, pointed out. The *piled* boulders, one on another, appears to me likely to be a distinctive character.—

Have you ever examined the bottoms of the pot-holes? I think it w^{d} be worth doing, for I found the shape of those, formed by **eddies** during floods in the Welch brooks, curious: they were formed like the bottom of the inside of a green glass bottle—a form evidently due to the centrifugal action of the revolving sand & pebbles.

Are you aware that Hopkins has lately published a paper in Cambridge Phil. Trans.[3] in which he disputes Forbe's semi-fluid theory,[4] & maintains that the movement cannot be compared with that of a viscid fluid:— he attributes all to gravity, with *the aid of the lower surface melting*.—

With my best wishes for your success & that your zeal may be rewarded by many discoveries; believe me, Yours sincerely | C. Darwin

Postmark: 9 AU 9 1844
Burgerbibliothek Bern

[1] Robert Jameson.

[2] See CD's 'Notes on the effects produced by the ancient glaciers of Caernarvonshire' (1842), *Collected papers* 1: 163–71. As opposed to CD's view, Morlot appears to have advanced the theory of erratic boulder transport by a great ice-sheet advancing from Scandinavia.

[3] Hopkins 1849, read 1 May 1843.

[4] James David Forbes believed that glacial ice flowed in the manner of a viscous fluid, whereas William Hopkins thought it simply slid downhill in its bed. Forbes's position (presented in numerous contributions to the *Edinburgh New Philosophical Journal* 1841–5, and summarised in J. D. Forbes 1843) involved him in a priority dispute with Louis Agassiz as well as in controversies with William Hopkins, John Tyndall, and others. See Rowlinson 1971.

To Henry Denny 12 August [1844]

Down Bromley Kent
Aug 12

Dear Sir

I took such especial pains, in myself always doing up every specimen, that I am astonished & can hardly believe there has been a mistake.— I have turned to my catalogoue, *made on the spot*, & I *there enter a memorandum to have these Pediculi, compared with those of the Domestic Guinea* Pig.—[1] It occurs to me that I may have transposed my numbers in copying them for you.— so I will recopy them.

658. from the Ctenomys Braziliensis (a *burrowing* rodent) or 〈 〉[2]
646 from the Aperea ie. the Cavaia Cobaya
638— from the Synallaxis[3]
1185. from Man[4]

It is possible I may have brought home the dead specimen in the same bag with birds, & the parasites from the latter have crawled on the former; but I feel no doubt that I with my own hands took the Lice off the Aperea & put them into spirits

Ever yours, &c | C. Darwi〈n〉

Postmark: AU 12 1844
American Philosophical Society

[1] The *Beagle* 'Catalogue for animals in spirits of wine' at Down House lists the specimens. No. 646 is 'Pediculi from the Aperea'. On the corresponding page of notes is CD's note: 'As this animal is supposed to be the wild Guinea-pig, it would be interesting to compare these parasites, with those inhabiting, an Europæan individual, to observe whether they have been altered by transportation & domestication: It would be curious to make analogous observation with respect to various tribes of men.—'

[2] The *Beagle* catalogue (Down House MS) reads 'Pediculi from Toco Toco' (the burrowing rodent).

[3] In both the *Beagle* catalogue and on a single sheet list of 'Insects in spirits of wine' (DAR 29.3: 44) number 638 is listed as from 'head of Certhia'. The *Beagle* catalogue (Down House MS) entry reads 'Pediculi, *very minute*, but curious from head of Certhia (1248)'. *Synallaxis major* was a later identification by John Gould, which was then corrected to '*Anumbius acuticaudatus*, G. R. Gray' in *Birds*, Corrigenda. CD's field descriptions in *Birds* (pp. 76–7) of *Synallaxis* correspond to those of *Certhia* made during the voyage (see *Ornithological notes*, p. 220).

[4] The *Beagle* catalogue refers to p. 315 of CD's Zoological Diary (DAR 31.2), which contains the account of the Sandwich Islanders' lice (see letter to Henry Denny, 3 June [1844], n. 1).

To J. D. Hooker 29 [August 1844]

Down. near Bromley Kent
29th

My dear Hooker

I send you a pamphlet, from Ehrenberg,[1] enclosed with one for me.— He asks anxiously after the deep-sea soundings. Have you any probability of getting them— My few additional specimens are all ready for him— Ever yours, in haste | C. Darwin

Postmark: AU 31 1844
DAR 114.1: 16

[1] Ehrenberg 1844a. The pamphlets had been forwarded via Richard Taylor, see letter from C. G. Ehrenberg, 11 July 1844.

To Leonard Horner 29 August [1844]

Down near Bromley | Kent
Aug 29th.

My dear Mr. Horner

I am greatly obliged for your kind note & much pleased with its contents. If one third of what you say, be really true & not the verdict of a partial judge (as from pleasant experience I much suspect), then should I be thoroughily well contented with my small volume,[1] which small as it is, cost me much time.— The pleasure of observation amply repays itself; not so that of composition, & it requires the hope of some small degree of utility in the end, to make up for the drudgery of altering bad English into sometimes a little better & sometimes worse.

With respect to Craters of Elevation, I had no sooner printed off the few pages on that subject, than I wished the whole erased.— I utterly disbelieve in Von Buch & de Beaumonts views, but on the other hand in the case of the Mauritius & St Jago, I cannot, perhaps, unphilosophically persuade myself, that they are merely the basal fragments of ordinary volcanoes, & therefore I thought I would suggest the notion of a slow circumferential elevation, the central part being left unelevated, owing to the force from below being spent & relieved in eruptions.[2] On this view, I do not consider these so-called craters-of Elevation, as formed by the ejection of ashes lava &c &c but by a peculiar kind of elevation, acting round & modifyed by a volcanic orifice.—

I wish I had left it all out; I trust that there is in other parts of the volume more facts & less theory.— The more I reflect on volcanoes, the more I appreciate the importance of E. de Beaumont's measurements[3] (even if one does not believe them implicitly) of the natural inclination of lava-streams & even more the importance of his view of the dikes or upfilled fissures in every volcanic mountain being the proofs & measures, of the stretching & consequent elevation which all such mountains must have undergone:[4] I believe he thus unintentionally explains most of his cases of lava-streams being inclined at a greater angle, than that at which they could have flowed.

But excuse this lengthy note & once more let me thank you for the pleasure & encouragement you have given me,—which together with Lyells never-failing kindness, will help me on with S. America, & as my Books will not sell, I sometimes want such aid.—

I have been lately reading with care A. d'Orbigny work on S. America,[5] & I cannot say how forcibly impressed I am with the infinite superiority of the Lyellian school of Geology over the Continental. I always feel as if my books came half out of Lyell's brains & that I never acknowledge this sufficiently, nor do I know how I can, without saying so in so many words—for I have always thought that the great merit of the Principles, was that it altered the whole tone of one's mind & therefore that when seeing a thing never seen by Lyell, one yet saw it partially through his eyes— it would have been in some respects better if I had done this less—but again excuse my long & perhaps you will think presumptuous discussion.

Enclosed is a note from Emma to Mrs Horner to beg you, if you can, to give us the great pleasure of seeing you here— we are necessarily dull here & can offer no amusements, but the weather is delightful & if you could see how brightly the sun now shines you would be tempted to come—

Pray remember me most kindly to all your family & beg of them to accept our proposal & give us the pleasure of seeing them.— Emma will tell how feasible the coming here is by a coach, which puts down by our door at dinner-time.

Believe me | dear Mr Horner | Yours truly obliged | Charles Darwin

American Philosophical Society

1 *Volcanic islands* was published in March 1844 (*The Publishers' Circular*).

2 CD discussed 'craters of elevation' in *Volcanic islands*, pp. 93–6. According to Christian Leopold von Buch (1836) and Jean Baptiste Armand Louis Léonce Élie de Beaumont (1838), volcanoes were caused by pressure from below which arched the strata into a dome-like formation until the centre collapsed and a vent was formed. Charles Lyell (1840a, 2: 238–50) argued that volcanoes were formed merely by the expulsion of underground material. CD attempted a compromise position according to which elevation took place only around the circumference of the volcano, requiring no collapse of the crater.

3 Élie de Beaumont had made measurements of the inclination of lava streams that convinced him that lava would cool into thick layers of volcanic rock only on nearly level ground (1838, pp. 173–8), a finding that seemed to exclude Lyell's notion that substantial rock layers would normally be formed by ejected lava on the growing slopes of the volcano. CD's copy of Élie de Beaumont 1838 is in the Darwin Library–CUL.

4 Élie de Beaumont argued that the growth and expansion of internal crevices meant Mount Etna, like other volcanoes, was constantly growing in volume and altitude (1838, pp. 116–21).

5 Probably *Géologie*, the third part of the third volume of Orbigny 1835–47, which was published in 1842. See letter to Charles Lyell, [1 September 1844]. CD borrowed the book from the Geological Society, see letter to the Geological Society of London, [3 January 1844].

To Charles Lyell [1 September 1844][1]

Down near Bromley | Kent
Sunday

My dear Lyell

I was glad to get your note & wanted to hear about your work—[2] I have been looking to see it advertised— it has been a long task— I had, before your return from Scotland, determined to come up & see you; but, as I had nothing else to do in town, my courage has gradually eased off, more especially as I have not been very well lately.— We get so many invitations here, that we are grown quite dissipated—but my stomach has stood it so ill, that we are going to have a month's holidays & go nowhere. The subject, which I was most anxious to talk over with you, I have settled, by having written 60 pages of my S. American geology— I am in pretty good heart & am determined to have very little theory & only short descriptions.— The two first chapters, I think will be pretty good, on the elevation & great gravel terraces & plains of Patagonia & Chile & Peru.— I am astounded & grieved over d'Orbigny's nonsense of sudden elevations; I must give you one of his cases.[3]

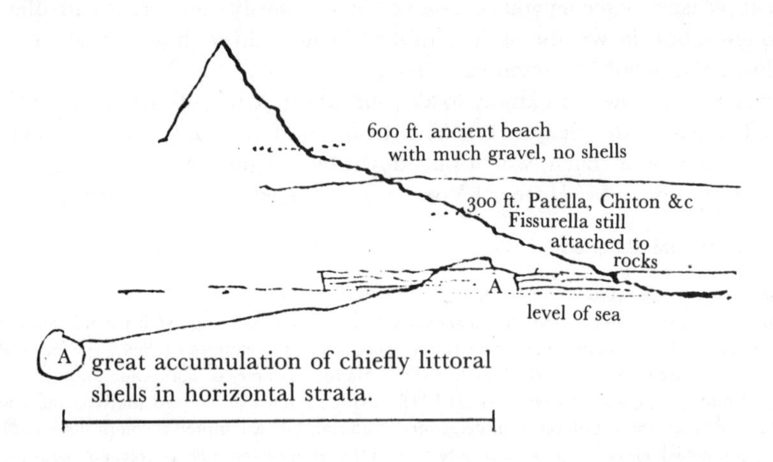

He finds an old beach 600 ft above Sea

He finds *still attached* to the rocks at 300 ft, six species of truly *littoral* shells

He finds at 20 to 30 ft above sea, an immense accumulation of chiefly littoral shells.

He argues the whole 600 ft uplifted at one blow, because the attached shells at 300 ft have not been displaced.

Therefore when the sea formed a beach at 600 ft, the present **littoral** shells, were attached to rocks at 300 ft depth, & these same shells were accumulating by thousands, at 600 ft. depth.— Hear this oh Forbes:[4] is it not monstrous for a professed conchologist?— This is a fair specimen of his reasoning. One of his arguments against the Pampas being a slow deposit, is that mammifers are very seldom washed by rivers into the sea![5] Because at 12,000 ft he finds the same kind of clay with that of the Pampas, he never doubts that it is contemporaneous with the Pampæan debacle, which accompanied the right Royal salute of every volcano in the Cordillera. What a pity these Frenchmen do not catch hold of a comet, & return to the good old geological dramas, of Burnett & Whiston—[6] I shall keep out of controversy, & just give my own facts. It is enough to disgust one with Geology;—though I have been much pleased with the frank, decided, though courteous manner, with which d'orbigny disputes my conclusions, given unfortunately without facts & sometimes rashly in my Journal.[7]

Enough of S. America: I wish you w^d: ask Mr Horner (for I forgot to do so, & am unwilling to trouble him again) whether he thinks there is too much detail, (quite independently of the merit of the book) in my volcanic volume; as to know this, w^d be of some real use to me; you could tell me, when we meet after York, when I will come to town: I had intended being at York, but my courage has failed; I sh^d much like to hear your lecture,[8] but still more to read it; as I think reading is always better than hearing.

I am very glad you talk of a visit to us in the Autumn; if you can spare the time, I shall be truly glad to see Mrs Lyell & yourself here; but I have scruples in asking anyone, for you know how dull we are here. Young Hooker talks of coming; I wish he might meet you— he appears to me a most engaging young-man.

I have been delighted with Prescott,[9] of which I have read Vol I, at your recommendation; I have just been a good deal interested with W. Taylor of Norwich Life & correspondence.[10]

Farewell—with our kind remembrances to Mrs Lyell | Ever yours | C. Darwin

We had, until this morning when we heard, much hoped to have seen M^r & Mrs Horner here at Down.

On your return from York I shall expect a great supply of geological gossip:

American Philosophical Society

[1] Dated by CD's reference to having read Prescott 1843 and Robberds 1843, see nn. 9 and 10, below. The only Sunday falling between these two dates was 1 September.

[2] Lyell was writing an account of his travels in North America, eventually published as C. Lyell 1845a.

[3] Orbigny 1835–47, vol. 3, pt 3: *Géologie*, pp. 93–8.

[4] Edward Forbes (1843) had recently demonstrated the existence of well-marked zones of water depth, each with its own distinctive fauna; he believed there were no animals living below about 300 fathoms (p. 170). See also Rehbock 1983, pp. 139–44, and Browne 1983, pp. 144–5.

[5] Orbigny 1835–47, vol. 3, pt 3: *Géologie*, pp. 85–6 n.

[6] Thomas Burnet and William Whiston. In Burnet's view the present state of the earth's surface was the result of catastrophic events during the deluge, while Whiston held that the earth had originally been a comet and that the deluge was due to the close approach of another comet. See Davies 1969 for accounts of these theories.

[7] Orbigny 1835–47, vol. 3, pt 3: *Géologie*, pp. 82–7. See *Journal of researches*, p. 171, and *South America*, pp. 93, 101.

[8] The British Association for the Advancement of Science met in York from 26 September to 2 October 1844. Lyell gave a special evening discourse on the geology of North America (*Report of the 14th meeting of the British Association for the Advancement of Science held at York in 1844*, p. xxx).

[9] Prescott 1843. CD's entry in his reading notebook (DAR 119) for 30 August 1844 reads: '1· Vol of Prescotts Hist of Mexico'. The two concluding volumes were recorded as read on 1 October (Vorzimmer 1977, p. 131).

[10] Robberds 1843, recorded in CD's reading notebook (DAR 119; Vorzimmer 1977, p. 131) on 5 September 1844. It was borrowed from the London Library on 7 August and returned on 23 September (London Library Archives).

From J. D. Hooker [*c.* 3 September 1844][1]

had all the work with them myself & think he ought to have given them up with the other colls'.— Have you seen Gray's 1st. number of Mammalia or Richardson's on fish?:[2] the latter is admirable as far as I can judge.

I have a sort of notion that the more varied the temperature is on a given surface the more species it produces—"ceteris paribus". The campos of central Brazil are, I believe, vastly richer in species than the woods, & every few miles adds new plants. The plains of the Andes are immensely rich in species of plants under the line, much more so than the sea coast.. A mountain immediately gives new vegetable forms, perhaps not so much because its temperature is lower but because its vicissitudes are more remarkable. I argue very much from the absence of new forms in proceeding from Chonos' Archip. down to Cape Horn. The temp. of the latter is certainly the coldest but it is quite *as*, if not *more* equable (I *suppose*). say that 12 degrees of Lat intervenes,—but go from Devonshire to John O'Groats & what a totally new vegetation is met with, not only have the old forms disappeared, but they are replaced by very numerous new ones. I do not suppose that any trees in the N. Temperate zone have the range in Latitude at the level of the sea that the Beech or Winters Bark have in S. America or the Dimon pine (Dacrydium cupressinum) in New Zealand.

This is a *young* subject to me, have you thought of it at all?. I shall be most happy to work out any suggestions on the subject.

I am trying very hard to get a month ahead with my book that I may go to Norfolk when my Mother returns.

With kind regards to Mrs Darwin Believe me | Ever yours most truly | Jos D Hooker.

Incomplete
DAR 104: 221

CD ANNOTATIONS
1.1 had . . . judge. 1.3] *crossed pencil*
Over first paragraph: 'Ranges determined by other Plants | Not on new Land | Isolation' *added pencil*
1.3 as I can judge.] 'Isolation' *added pencil, circled pencil*
2.10 but go . . . new ones 2.12] *scored pencil*

[1] Dated by CD's reply, see letter to J. D. Hooker, [8 September 1844].
[2] Richardson and Gray 1844–75.

To C. G. Ehrenberg 5 September [1844]

Down near Bromley Kent
Sept. 5th

Dear Sir

I waited until I had a packet ready, before writing to thank you much for the copies of your communications to the Berlin academy & for your obliging letters, for which I feel truly thankful: I have at last received the first letter you wrote.—[1] Dr Hooker has taken great trouble in getting the Admiralty specimens, but he tells me he has not succeeded very well: he encloses a letter in the parcel which I yesterday forwarded to Mr Francis, with a request to him to transmit it immediately to you: not living in London made it difficult for me, on account of clearance &c. to send it direct to you via Hamburgh.— I fear my specimens will not be of much interest to you; though I have picked out from as many isolated islands, as I could: anyhow they will not give you much trouble, as they are all labelled outside: if even one contains anything to interest you, I shall be amply repaid for the little trouble it has given me to get them together.

I have enclosed a copy of a little Paper, which I thought of sending to Taylor's Journal;[2] I think it will modify some of your opinions on the origin of the Dust; its *[illeg]* origin you will see, has been long out of the question: what great strata, abounding with Infusoria, must be now depositing at the bottom of the Atlantic; how curious, their aërial origin! I have sent specimens from four other stations: that numbered (3) is what you have already examined. Would you kindly take the trouble, the next time you send anything to England to return me my little Paper; & if you will send me any Abstract of the contents of packets 1.2.4.5. I wd allude to them in my Paper, before publishing it; if you will refer to the Latitudes & Longitudes, I shall know which packet is which.

Might I beg, as a great favour, that you would just look at the specimens of a white deposit which is of vast extent in Patagonia, between the fossiliferous tertiary strata & the overlying gravel: I have not seen any deposit quite like it & am very curious to know its nature, as I am now drawing up a short volume on the geology of S. America.—

Have you yet had time to look at any of the Pampas clay; M: Al: d'Orbigny considers it due to a debacle & of some considerable antiquity; I, on the other hand, cannot doubt that it is a slow, estuary deposit, contemporary with recent species of Mollusca, but with extinct species of Mammifers: would the Infusoria throw any light on this?[3] I shd. be *most* grateful for any such information; should you look at the specimens, which I before sent you, under this point of view, those from the central parts of the basin, would be the most instructive; namely from the tooth of the Mastodon of the Parana, or from any of the bones from the Uruguay or Banda Oriental.—

With my sincere thanks for the great honour you have done me, in the manner in which you have referred to me,[4] I have the pleasure to remain. Your's faithfully & obliged | Charles Darwin.

Might I take the liberty of asking you, the next time you see Dr. E. Dieffenbach, to tell him, that I *earnestly* beg him to be so good, as *immediately* to make enquiries, about the Copper-plate, &c &c & copy of the German Edit: of my Journal, which I have not received, & am very *anxious* about.—

I hope you will accept from me, a copy of my small volume on Coral Reefs, which I have directed my Book-seller to send to the care of Mr Francis, for you.—

Museum für Naturkunde der Humboldt-Universität zu Berlin

[1] Ehrenberg 1844a. See letter from C. G. Ehrenberg, 15 June 1844.

[2] Presumably a manuscript of CD's 'An account of the fine dust which often falls on vessels in the Atlantic Ocean' (*Collected papers* 1: 199–203). This did not appear in Richard Taylor's *Annals and Magazine of Natural History* but in the *Quarterly Journal of the Geological Society*. CD's notes and another manuscript copy of the paper are in DAR 188.

[3] Orbigny 1835–47, vol. 3, pt 3: *Géologie*, pp. 81–7, and *Journal of researches*, p. 52.

[4] Ehrenberg referred to CD as 'der geistvolle Beobachter der Corallen-Bildungen im Südocean' ('the ingenious observer of coral formations in the South Sea', Ehrenberg 1844a, p. 183) and as 'der bekannte verdienstvolle englische Reisende und Schriftsteller über die Korallenriffe' ('the well-known, meritorious English traveller and writer on coral reefs', *ibid.*,p. 194).

To J. D. Hooker [8 September 1844][1]

Down Bromley Kent
Sunday

My dear Hooker

I ought to have written sooner to have acknowledged your notes & the parcel, of such inestimable value, no doubt, in the eyes of Ehrenberg.— I sent it off the next day, with a few specimens of my own, but few, I fear of any probable interest. Really your collection must have been invaluable for Ehrenberg, now that he is summing up his labours in the distribution of Infusoria.— I have sent Eh. some more specimens of the Atlantic dust; I find that on one occasion, the dust when it began to fall on a ship, more than 300 miles from the land, was much coarser than on the succeeding days & many of the particles of stone are more than $\frac{1}{1000}$ of inch square: this shows how far the sporules of Cryptogams might be blown, & indeed common seeds, by winds of no great force.—

Have you seen Forbes Report;[2] it wd interest you in your speculation on ranging: he of all men, in the world, if he can spare time, ought to have the deep sea dredgings.

The subject of the greater number of species in certain areas, than in others, has long appeared to me a very curious subject: your example of East S. America, compared with Britain is very striking. Is not the case of New Zealand, with its varied stations, compared with the uniformly arid C. of Good Hope, opposed to your view, that the number of species bears a relation to the vicissitudes of climate?[3] When you speak of mountains, (as the plains of the Andes) being subject to vicissitudes, I am not sure, whether you means absolutely so on the same spot, or whether great differences, within short distances.— Is it not said, that the absolute changes of temperature are greatest on any one spot, in the extreme northern regions; & that equability is the characteristic of the tropics?

The conclusion, which I have come at is, that those areas, in which species are most numerous, have oftenest been divided & isolated from other areas,, united & again divided;—a process implying antiquity & some changes in the external conditions. This will justly sound very hypothetical.

I cannot give my reasons in detail: but the most general conclusion, which the geographical distribution of all organic beings, appears to me to indicate, is that isolation is the chief concomitant or cause of the appearance of *new* forms (I well know there are some staring exceptions).—[4]

Secondly from seeing how often plants & animals swarm in a country, when introduced into it, & from seeing what a vast number of plants will live, for instance in England, if kept *free from weeds & native plants*, I have been led to consider that the spreading & number of the organic beings of any country depend less on its external features, than on the number of forms, which have been there originally created or produced.— I much doubt whether you will find it possible to explain the number of forms by proportional differences of exposure; & I cannot doubt if half the species in any country were destroyed or had not been created, yet that country w^{d}: appear to us fully peopled. With respect to original creation or production of new forms, I have said, that isolation appears the chief element: Hence, with respect to terrestrial productions, a tract of country, which had oftenest within the later geological periods subsided & been converted into isl^{ds}, & reunited, I sh^{d} expect to contain most forms.—

But such speculations are amusing only to one self, & in this case useless as they do not show any direct line of observation: if I had foreseen how hypothetical, the little, which I have *unclearly* written, I w^{d} not have troubled you with the reading of it.

Believe me,—at last not hypothetically— | Yours very sincerely | C. Darwin

All your remarks are to me of real interest & value.

It used to strike me, that the great apparent change in the vegetation about Chiloe was fully as striking, as the apparent (to my non-botanical eyes) uniformity southward of Chonos Arch: There is no great or **sudden** change in climate till we reach near Concepcion, where less rain is the chief change.—

DAR 114.1: 17

[1] Dated on the basis that CD dispatched a parcel to Christian Gottfried Ehrenberg by the weekly carrier that left Down early on Thursday, 5 September. See preceding letter.

[2] E. Forbes 1843. CD's copy is extensively scored (Darwin Pamphlet Collection–CUL). See also CD's essay of 1844 (*Foundations*, p. 146).

[3] Essay of 1844 (*Foundations*, pp. 171–2).

[4] Essay of 1844 (*Foundations*, pp. 183–91).

To *Gardeners' Chronicle and Agricultural Gazette* [before 14 September 1844]

Mr. Groom[1] has stated in last Number that the leaves of some of his Pelargoniums have become regularly edged with white in consequence of his having watered the plants with sulphate of ammonia which had been exposed to the air for some time. Last autumn I planted many young Box-trees; and I have for some weeks observed that nearly all the young leaves in most of them are symmetrically tipped with white, giving the young branches a mottled appearance. I counted twelve trees thus affected. The older leaves are rarely tipped, with the exception of two bushes, in which they are regularly tipped, and the younger ones much less so. Mr. Groom states that in his Pelargoniums the older leaves are chiefly affected. The Box-trees are quite healthy, and growing well. I gave to some of them nitrate of soda, but it has made no difference in this variegation. Those growing in deep shade are not tipped, nor are some older trees. These facts may appear trivial; but I think the first appearance, even if not permanent, of any peculiarity which tends to become hereditary (as I fear is the case with the variegated Sycamore) deserves being recorded.— *C. Darwin.*

Gardeners' Chronicle and Agricultural Gazette, no. 37, 14 September 1844, p. 621

[1] Possibly Henry Groom, nurseryman in Clapham Rise, London. His remarks on variegated pelargoniums are in *Gardeners' Chronicle and Agricultural Gazette*, no. 36, 7 September 1844, p. 605.

To *Gardeners' Chronicle and Agricultural Gazette* [before 14 September 1844]

I should be extremely obliged if any of your chemical readers would inform me whether salt and carbonate of lime (under the form of sea-shells) would, if slightly moistened and left in great masses long together, act in any degree on each other?[1] It is, I believe, known that masses of the same substances will act on each other, of which smaller quantities will not. I do not ask this question for agricultural purposes (though possibly the answer might be of some interest in that point of view), but from having found in Peru a great bed of upraised recent shells, mixed with salt, which are decayed and corroded in a singular manner, so that the surfaces of the shells are scaling off and falling into powder.[2] I may mention, as explaining one element in the value of sea-shells as manure, that they are dissolved by water with greater facility than apparently any other form of carbonate of lime: one proof of this I observed in a curious rock, from Chili,[3] chiefly composed of small fragments of recent shells, which are all enveloped and cemented together by a pellucid calcareous deposit; but in some parts of this rock the little included fragments are in every stage of decay and disappearance; in other parts they are entirely dissolved, the little calcareous envelopes being left quite empty. Here we see that water, capable of dissolving shelly matter, has penetrated through their

thin films or envelopes of carbonate of lime, without having acted on them; these films, moreover, being a deposition from water within quite recent times.—[3]
C. Darwin.

Gardeners' Chronicle and Agricultural Gazette, no. 37, 14 September 1844, pp. 628–9

[1] Only one reader, who signed himself 'T. P.', responded to CD's query (see *Gardeners' Chronicle and Agricultural Gazette*, no. 40, 5 October 1844, p. 675), although there was considerable interest in the agricultural use of salt. In 1845 a short comment attributed to Christian Konrad Sprengel carried the discussion further (see *Gardeners' Chronicle and Agricultural Gazette*, no. 10, 8 March 1845, p. 157).
[2] See *Journal of researches*, p. 451, and *South America*, pp. 47–9, 52–3. CD was interested in the chemical decomposition of sea-shells as further evidence for the existence of elevated beaches on the west coast of South America. See also letters from Trenham Reeks, 8 February 1845 and 25 February 1845.
[3] Described in *Volcanic islands*, p. 144 n., and *South America*, pp. 36–7.

To E. A. Darwin [before 1 October 1844][1]

Down
Thursday

My dear Eras.

I have heard from Jos that he has been *obliged* (somehow owing to a transference being necessary, & it being impossible to transfer from his own name as executor, to his own as Trustee) to put the Trent & Mersey & Monmouths Canal shares, belonging to Emma, in your sole name. He sends a paper for you to sign, & date which he says I am to keep. He says the Trent & Mersey will send their receipt to you to sign, & when signed you can send it either to me [or] to Robarts for me, which latter w^{d} be best, as saving a letter.

For the Monmouths Canal shares you will have to write half yearly to a Newport Banker & he will pay the money direct to me.— I will half-yearly remind you & give you the Banker address, & number of the shares.

I am glad the Trent & Mersey shares, will give you no more trouble than your own ones do, & I sh^{d} think you c^{d} arrange for them to be sent together.—

I intend coming up on Tuesday & mean to dine at the Athenæum I shall sleep two nights in London

Ever yours | C. Darwin

DAR 210.15

[1] Dated by reference to the following letter, which deals with the same share transactions.

To Solicitor? 1 October 1844

Oct. 1st. 1844.

We hereby request you to have transferred into the Sole name of Erasmus A Darwin the two Monmouth canal shares & two Trent & Mersey Shares which we have agreed to take in part discharge of the legacy of Emma Darwin under the will of the late Josiah Wedgwood—[1]

Charles R. Darwin
Emma Darwin

Jan 9th. 1845

Keele University Library (Wedgwood/Mosley 977)

[1] The text is in Erasmus Alvey Darwin's hand. 'Jan 9th. 1845' has been added by CD and is presumably the date on which the request was signed.

To Adolf von Morlot 10 October [1844]

Down near Bromley Kent
Oct. 10th

Dear Sir

Your letters of the 3d of October & 30th of July arrived here *together* yesterday: how the delay of the latter was caused, I know not.— My answer will be a continued source of displeasure to you: I am heartily sorry for it, for I admire your zeal & wish you a plentiful harvest of discovery; & such zeal as yours, is a main element in discovery. I by no means underrate the importance of your observations on the ice-action over the many hills, through which you have lately travelled; but let me assure you, that no Editor of the respectable *English* scientific Journals would publish your letters in their present form, or with merely passages struck out, which I would have undertaken to have done. Your letters detail your belief, your theories & your conclusions, but they do not *detail* the facts. Would not your proper course (I can speak from experience when Secretary of the Geological Soc. that no other course would, or ought to according to the rules of the Society, be admitted) be to plainly & briefly describe every fact, which you observed on ice action, excluding all foreign facts; & after you have so described your facts, aid your reader drawing his conclusions, by pointing out, why a Scandinavian glacier explains your facts, better than local glacier, aided by icebergs (now the commonest view in England)—how your origin of the "till" explains your facts better than the iceberg-doctrine—

Pray observe I do not pretend to say your theories are not right, but a substratum of facts ought surely to be first given. Have you read Lyell's Paper on the "Till"[1] (you will see the *reference* in my American Boulder Paper)[2] & considered his curious case of disturbed beds *resting on* undisturbed? Again I am sure the publication of your Loëss views *in their present state* would injure your reputation: it is a most curious & difficult subject. I hope you may solve it.

I think Escholtz in Kotzebue's first Voyage describes a cliff of ice with earth on it[3] & in the Appendix to Beechey's Voyage to the Pacific there is some allusion to

Christian Gottfried Ehrenberg. By E. L. Radtke.
(Frontispiece to Laue, Max, *Christian Gottfried Ehrenberg . . . 1795–1876* (Heidelberg: 1895). By permission of the Syndics of the Cambridge University Library.)

James David Forbes. By W. H. Townsend.
(Courtesy of the Royal Society.)

the same class of facts: fossil bones were found in these frozen cliffs.[4] I suspect in these cases (according to Ermanns limit of perpetual congelation in depth) that the ice would be frozen to the undersoil.—[5] Granting, however, that these cliffs were moving glaciers covered with earth, & vegetation surely you ought to show that this is a common occurrence, before you apply it to your Scandinavio-German glacier; & then it w[d] not much signify whether you could or not explain how the earth was brought over your glacier. Surely you ought to give facts & explain (as you seem to admit that glaciers move by gravity, according to Forbes' beautiful views)[6] how your glacier was propelled across the Baltic.—

I could go on writing; but I well know that this comes very badly from me, who have dealt so largely in the sin of speculation, which I endeavour, though with little success, to check.— I have not the folly to oppose my opinion on the value of your observations, to that of the illustrious Germans, whom you mention; but at the same time every one must be individually guided by his own opinion, & my opinion is, that, though a good description of ice-action (which I daresay you are fully capable to give) on the mountains, which you have visited, would be very valuable, that your letters, in which facts are so mingled with speculation, are not fit to be published.— I know I shall appear to you unjust & unkind, & I am sorry for it.— I have, in accordance with your wish, expressed in your letter of the 2[d] July, communicated your observations to no one person.—

I am very glad to hear that Plutonic Geology is making progress; I have been prepared by Keilhau's clever papers (translated in the Eding: Phil: Journal)[7] for some great changes, but I am loth to give up the old views: the study of volcanic countries prejudices one in favour of heat-motions.— Evans Hopkins[8] & Hopkins the Mathematician,[9] are very different men.— I sh[d] have much enjoyed being instructed by you in the Alps in ice-action, but my health at present puts such an exertion quite out of the ⟨question. I fear the⟩ impression, which you will take from this letter is that I am a thoroughily selfish person, unwilling to take any trouble for anyone else.— I hope such an impression, though natural, is not altogether well founded.— I repeat, that I am sorry to disappoint you, & that I wish you success.

Believe me | dear Sir | Yours very faithfully | C. Darwin

Surely your better course would be to publish your facts & views in a Swiss or German Periodical; for they would naturally excite less attention here, than there, where other observers could test & profit by them.

Postmark: 10 OC 10 1844
Burgerbibliothek Bern

[1] C. Lyell 1840b.

[2] 'On the distribution of the erratic boulders and on the contemporaneous unstratified deposits of South America', *Collected papers* 1: 145–63.

[3] Kotzebue 1821, 1: 146–8. Johann Friedrich Eschscholtz was present when the glacier was discovered, but the description of the cliff was written by the naturalist of the expedition, Adelbert von Chamisso.

[4] William Buckland gives an account of the fossils listed in the journal of Alexander Collie, in Beechey 1831, 2: 331–56. A different edition of this work is in the Darwin Library–CUL. The reference is to a cliff face glazed over with ice and frozen mud. Alexander Collie was the surgeon on board H.M.S. *Blossom*, 1825–8.
[5] Georg Adolph Erman discovered that in eastern Siberia the ground might remain permanently frozen to depths of several hundred feet even though summer air temperatures were well above freezing. See Erman 1838.
[6] For James David Forbes's views see letter to Adolf von Morlot, 9 August [1844], n. 4.
[7] Keilhau 1838–40 and 1844, in which a theory of the origin of rocks by heat-induced crystallisation is proposed.
[8] Evan Hopkins.
[9] William Hopkins.

To James David Forbes 11 October [1844]

Down near Bromley Kent
Oct. 11.

Sir

I venture to take the liberty of addressing you, knowing how much you are interested on the subject of your discovery on the zoned structure of glacier-ice.[1] I have a specimen (from Mr. Stokes' Collection) of Mexican obsidian,[2] which judging from your description, must resemble to a considerable degree, the zoned ice. It is zoned with quite straight, parallel lines, like an agate, & these zones as far as I can see under the microscope, appear entirely due to the greater or lesser number of excessively minute flattened air-cavities. I cannot avoid suspecting that in this case, & in many others, in which lavas of the trachytic series (generally of very imperfect fluidity) are laminated in a very singular manner, that this structure is due to the stretching of the mass or stream during its movement, as in the ice-streams of glaciers.[3]

It has occurred to me, that you possibly might like to see the specimen of obsidian & some curious, most finely, laminated obsidians & trachytic rocks, which I collected at Ascension island. You would not of course, I presume, think it worth the expence of carriage to have the specimens sent to Edinburgh; but shd you at any time come to London, I should be proud, if you so like, to send them for your inspection. If the subject of the lamination of *volcanic* rocks should interest you, I would venture to ask you to refer to p. 65–72 of my small volume, "Geolog: Obser: on Volcanic Isld." which would be in the Public Library.[4] I there throw out the idea, that the structure in question may, perhaps, be explained by your views on the zoned structure of glacier-ice:[5] the layers of less tension, I may add, being in the case of the Ascension Obsidian-rocks, rendered apparent chiefly by the crystalline & concretionary action superinduced in them, instead of as in your ice, by the congelation of water.[6]

I hope you will excuse, should you feel no interest in this subject, the liberty I take in writing to you, & I beg to remain, your obedient servant | Charles Darwin

Postmark: 12 OC 12 1844
University of Saint Andrews Library

[1] Forbes argued that the zoned structure of glacier ice was produced by tension resulting from gravitational force acting on the viscous ice. See letter to Adolf von Morlot, 9 August [1844], n. 4.
[2] The specimen from Charles Stokes's collection is discussed in *Volcanic islands*, pp. 67, 69–70.
[3] See *Volcanic islands*, pp. 70–1.
[4] The Public Library of the University of Edinburgh, i.e., the collection belonging to the whole University and open to all its members.
[5] Forbes had already made a comparison between glacier ice and lava, although coming to different conclusions from CD (J. D. Forbes 1844). See letter to J. D. Forbes, 13 [November 1844].
[6] Forbes published parts of this letter with a paragraph from another letter from CD (see letter to J. D. Forbes, [November? 1844]) in the *Proceedings of the Royal Society of Edinburgh* 2 (1845): 17–18.

To Leonard Jenyns 12 October [1844][1]

Down Bromley Kent
Oct 12th

My dear Jenyns

Thanks for your note.— I am sorry to say that I have not even the tail end of a fact in English Zoology to communicate. I have found that even trifling observations require, in my case, some leisure & energy, both of which ingredients I have had none spare, as writing my geology thoroughily expends both. I had always thought, that I would keep a journal & record everything, but in the way I now live I find I observe nothing to record. Looking after my garden & trees & occasionally a very little walk, in an idle frame of my mind, fills up every afternoon in the same manner.—

I am surprised that with all your parish affairs that you have had time to do all, that which you have done. I shall be very glad to see your little work[2] (& proud shd I have been, if I could have added a single fact to it): my work on the species question has impressed me very forcibly with the importance of all such works, as your intended one, containing what people are pleased generally to call trifling facts.[3] These are the facts, which make one understand the working or œconomy of nature. There is one subject, on which I am very curious, & which perhaps you may throw some light on, if you have ever thought on it,—namely what are the checks & what the periods of life, by which the increase of any given species is limited. Just calculate the increase of any bird, if you assume that only half the young are reared & these breed: within the *natural* ie. if free from accidents life of the parents, the number of individuals will become enormous, & I have been much surprised to think, how great destruction *must* annually or occasionally be falling on every species, yet the means & period of such destruction scarcely perceived by us.

I have continued steadily reading & collecting facts on variation of domestic animals & plants & on the question of what are species; I have a grand body of facts & I think I can draw some sound conclusions. The general conclusion at which I have slowly been driven from a directly opposite conviction is that species are mutable & that allied species are co-descendants of common stocks. I know how much I open myself, to reproach, for such a conclusion, but I have at least honestly & deliberately come to it.

I shall not publish on this subject for several years— At present I am on the geology of S. America. I hope to pick up from your book, some facts on slight variations in structure or instincts in the animals of your acquaintance

Believe me Ever yours | C. Darwin

Bath Reference Library (Jenyns papers, 'Letters of naturalists 1826–78', Octavo vol. 1, 43(7))

[1] CD's activities in October 1844 make this a more likely date than 1845, as Francis Darwin dated the letter in *LL* 2: 31–2. He had recently written his 1844 essay on his species theory (*Foundations*) and was writing *South America.* In October 1845 he was not at Down and he suspended work on *South America* from April until 29 October 1845. See 'Journal' (Appendix II) and letter to Leonard Jenyns, 25 [November 1844].

[2] Jenyns ed. 1843.

[3] Jenyns 1846, which was to include notes on observing animals and plants and a calendar of periodic phenomena.

To Emma Darwin [20 or 27 October 1844][1]

[Shrewsbury]
Sunday

My visit is going off very pleasantly; and my father is in excellent spirits. I have had a deal of "parchment talk," as Catherine calls it, with my father, and shall have a good deal of wisdom to distil into you when I return, about Wills, &c. . . . My father says that Susan, the evening before she went, was enthusiastic in her admiration of you, in which you know how my father joins. I did not require to be reminded how well, my own dear wife, you have borne your dull life with your poor old sickly complaining husband. Your children will be a greater comfort to you than I ever can be, God bless them and you. Give my love and a very nice kiss to Willy and Annie and poor Budgy,[2] and tell them how much I liked their little notes, which I read aloud to grandpapa. I shall be very glad to see them again. I always fancy I see Budgy putting her tongue out and looking up to me. Good-bye, my dears. | C. Darwin.

Emma Darwin 2: 92

[1] Dated from the 'Journal' (Appendix II): '1844 . . . Oct 18th to 29th at Shrewsbury'.

[2] Henrietta Emma Darwin.

From J. D. Hooker 28 October 1844

West Park Kew
October 28. 1844.

My dear Darwin

It is a shameful time since I received your last valued letter, considering it is not answered yet—at the time I received it I had a good deal to say about the ranges of species, but have been so taken up with that dullest of all branches, "*specific Botany*", that I fear my ideas "tales quales" have flown or been absorbed.

Your communication was indeed most interesting & much food for me. I think I argued that there was a good deal of concomitancy in a uniformity of temperature & of vegetation, but did not mean you to infer that they were cause & effect: I suppose I have stated the thing too strongly

Still less do I suppose that an equable climate can account for an originally meagre vegetation: but I do think that it both favors the range of an existing one & excludes an increase from without, because many plants accustomed to change are impatient of so constant a drain on their constitutions, & because an equable temperature, causing a vigorous vegetation, tends to cover the surface with a growth of a few large things, which monopolize the *soil*, both *its* space & *its* nutritious qualities—I express myself very ill.

I did not suppose New Zealand with its limited flora opposed to my *notion*, (pray do not dignify it with the word *hypothesis*!) because I considered its climate as very equable—Grapes (I was assured) never ripened properly there in 35 & vegetation is luxuriant as far as 52 in the same longitude;[1] & I think its perennially verdant forests, arguing certainly *a priori*, tend to prove the climate tolerably equable. I should have thought that neither the diurnal or solstitial changes of N. Z. at all equalled those of N. Holl or V. D. L.—[2] It is also a blowy place & the breezes being all sea should be moist— Nor did I forget the *Cape*, with its sandy plains & abundant variety of vegetation, but had heard from Sir J. Herschell that the surface was heated to a surprizing degree during the day & I thought that nocturnal radiation was very considerable. My friend Mr Harvey tells me that they always require fires in winter, especially towards evening & in the morning. The barren plains of Patagonia & the Plate, district are my greater stumbling blocks, though the Geological formation has much to do with them. Again in judging of New Zealand we do so only of its lee shore, which is a mere modification (not a different one I think) of what prevails on its west which is horrible.

In talking of M^{ts}., I meant certainly absolute changes on the same spot, & grant that these are greatest in the xtreme North, but hardly thought it fair to cite a climate, where the duration & intensity of cold was beyond what all but a very small amount of plants could bear. To exist at all, Phænogamic plants must have a certain amount of heat during some part of the year & shelter during the remainder, or the very latent caloric of their reservoir of life will be absorbed.[3] I think therefore we must not compare a region which only affords 3 months in which it is possible for vegetation to progress, with the Tropics where 12 months allow in some places ½ as many vegetations. But compare the mean temp. of the variable, with that of a similar temperature where it is equable & I think if the Arctic & Antarctic regions be the regions we thus collated, the former will contain most forms for its amount of temperature. Again, I doubt extremely if any 23½ degrees in Lat of Tropics, will contain as many species of plants, as the 23½ to the Northward of it, in the old world at any rate. Except where vast table lands intersecting the tropic give the plants the benefit of a new climate, too changeable for the rank luxuriance of a tropical forest. I do not think any trees have so wide a range as many of the tropical ones. Nothing struck me so much as the few species of plants I was able to gather on the Corcovado at Rio.

Certainly there is not much land, after all, between the Tropics, especially to what there is to the Northward of them; to the Southward there is less still, but in leaving the South. Equinoctial line a new & varied flora is at once met with. The Botany of extra tropical N. H. is far more varied & peculiar than intratropical ditto. There is a vast similarity in the whole Flora of India, from the Chinese sea on the E. to the Zanzibar coast! to the W. & I am led to believe the Guinean flora may be added to it without much violence: but how totally different from any or all is the Flora of S. Africa. I am reasoning (if it be reason) on the broad principle of the tropical being an equable climate.

So much for *my side*.

I am not prepared to answer your remark concerning the sudden change of vegetation to N. of Chonos. Archip. without concomitant physical features. Allow me however to add, that the main feature of the vegetation to the South of that I suppose to be the Beech— Now the mean temp. of the year I suppose increases along that coast inversely with the Latitude (of course) & there is a temp. that trees cannot bear— Wherever that temp. is arrived at, the Beech will stop, perhaps suddenly, for though plants will "drag on a miserable existence" into climates too cold for their nature, I am not sure that they do so into warm ones. It is worth enquiring whether in the N. Hemisphere the Northern or Southern limits of any tree is most clearly defined?, I think the Southern. You dwell more on the facility of introduction of species into our or other climates, than I used to. After all, considering the hundreds of years our Island has been under cultivation, the 20,000 plants that have at one time or another been introduced into it, the times out of number that the same things have been imported with our foreign produce & which are immediately put into the most favorable situations for being naturalized. After all this, how many plants have we naturalized? Look at any garden neglected for 20 years & how few of the common continental hardy annuals or perennials survive—our equable climate (perhaps) leaves them no room, because it favors the undue increase of our own weeds. Fourteen points of the compass bring land-winds to us & yet how small our flora is compared to the continental one. There are few seasons of the year in which we cannot find some remains of nearly all our native plants: but abroad one vegetation replaces another.

All this however is a paltry subject in comparison to the question you propound. What I have aimed at is, to trace the connection between climate & the present state of vegetation; to account for the paucity of species *remaining* in an area from a supposition that certain states of climate are unfavorable to increase of species, either by importation or by modification of already existing forms: (if so be that many so called species are permanent alterations, due to climate or other physical cause). I fear no superstructure of inductive reasoning, built upon so narrow a base, even if a stable one, would lead to the solution of your question "The cause of the appearance of new forms."—

My great ignorance of Geology, or indeed of any thing but specific Botany, prevents my perceiving the truth of your hypothesis pray do not twist this confession into any other meaning, I do mean simply that not knowing what parts of the world have been most frequently divided & again united I cannot apply the test of proportionate number of species to it.

I was pondering the other day what materials we had after all for coming to a knowledge of the Geog. distrib. of plants, & was forced to confess it very small, from travellers & collectors invariably neglecting all but the most interesting things & from their passing by a thing of the utmost importance in the physical features of one area, because they gathered it in the last district passed through. We now do not know what 1000 square miles on the Earths surface produces most plants. Every Botanist has been crazy about the extraordinary richness of N. Zealand, because in the multitudes of new plants they overlook the great want of old ones—& because no *large* collector has ever been there to distribute his plants to all Botanists, who would thus find that every succeeding one brought home the same things.

Am I right in supposing from your hypothesis that Isld^{s}. produce most new forms: ceteris paribus. There are glaring examples on both sides. Kerg. Land, one of the most isolated spots in the universe has the Flora of Fuegia, with one two exceptions of which the *Cabbage* is an unanswerable one. Certainly Auck Id. & Camp. Isld. have a very extraordinary number of new plants for their vegetation, but I expect many of them to exist in South New Zealand. The Falklands have hardly a single peculiar plant. Iceland has none, with 400 sp. a most extensive flora for such a region.

However if you have reached this length of my letter you will be weary enough & so I shall drop the subject.

I have written to friends to be particular in the Bot. geog., especially with regard to the richness of areas & I expect very careful observations from N. India, Ceylon, & Brazil.[4] Nothing will give me so much pleasure as to get grounds for your reasonings & to carry out your theory of isolation.

I am quite agreed that Cryptogamic plants may have been & may be disseminated by winds &c, whether they will always grow where they may now be blown is another question: they may have occupied a small area on the original creation, & all or many may still be present over a great portion of the globe at this present time but only take root where they have done before. Most have such marvellous ranges that I think we do no violence to nature in supposing that of them only one individual was originally created, but that very few years sufficed to people the world with them. How phenogamic plants were circumstanced at the same time is a deep problem. One creation of Cryptogs would people the world as it now appears, whatever changes it may have gone through, short of *incineration of the whole sphere*!—but with animals & plants of other classes the conclusion is not so conformable to our notions.

Pray what writings on the subject of original creation will give me the best notions of the (mad) theories of some men from Lamarck's twaddle upwards. Species (or what we call species) may be muteable but I should not think they set about it themselves so systematically as he says.

And now all the previous is only a prelude to my main object of writing to you (don't laugh) which was to ask if you have seen a paper in the Tasmanian phil. Soc. Journal on "Atmospheric deposits of dust & ashes with remarks on drift pumice of the coasts of N.H. by Rev. W. B. Clarke, MA., FG.S., C.M.Z.S.,"?—[5] if not I will send it you—

With our kind regards to Mrs Darwin | Believe me ever yours | Jos D Hooker.

I have a paper on an Alga (allied to your ones on glass) it is a new genus which makes the red sea red!—according to somebody.— Montagne I see published in French institute[6] Have you seen it?

DAR 100: 16–21

CD ANNOTATIONS

1.1 It is a . . . was indeed 2.1] *crossed pencil*
5.1 In . . . certainly] *crossed pencil*
6.4 N. H.] 'New Holland' *added pencil*
8.8 too cold . . . tree is 8.10] *scored pencil*
8.11 After . . . continental one. 8.21] 'on Introduced Plants' *added brown crayon*
10.5 proportionate] 'onate' *added pencil*
12.1 that Isld^{s}. . . . a region. 12.8] 'Very important: Islands like Mountains— Isl^{ds} of Pacific most puzzling' *added pencil*
12.7 a most . . . reasonings 14.4] 'Look at my Coral Is^{d} Map & see whether most peculiar on Blue or Red. too many elements of confusion, viz neighbouring land, ∴ say cannot be tested, except similarity of S. Sea Is^{d}.'[7] *added pencil*
13.1 However . . . reasonings 14.4] *crossed pencil*
16.1 Pray . . . seen it? 19.3] *crossed pencil*

CD notes:[8]

Peculiar Floras
Small [*above del* '*Volcanic*'] Isl^{d} separate, *not 1 in [*added*] say 200 ['2' *over* '1'] miles. | measure | It must be volcanic for chance of rising
Galapagos | Juan Fernandez | St. Helena | Ascension. | Mauritius? | Campbell. Auckland | Sandwich | Norfolk Is^{d}
Against
Iceland | Azores | Kerguelen L^{d}.—
all Volcanic Is^{d} *have* or or are rising
This might be worth working out.— */I think not worth/ [*added*]
I have not means to know which are rising & wh not (*pencil*; DAR 100: 22).

Diffusion, when once introduced. | Coral-reefs. | time underrated; must not argue from stocked country. | cases like Scicily, force belief. =Pigeons from [*after del* 'from'] Holland= | land formerly joined; if Hooker is right, the upshot is, that plants cannot cross arms of the sea.— (*ink*; DAR 100: 23r.).

Invitation to here | doubt you have seen his account of fossil forms | 20th in London—*leave* send it to Athenæum or Geolog Soc. | Invitation | Cases of Variability in different countries. (*pencil, del pencil*; DAR 100: 23v.).

[1] New Zealand runs from *c.* 34° to 47°S, but Hooker is extending his definition southward to 52°S to include the Aucklands and Campbell Island.

[2] New Holland, now Australia, and Van Diemen's Land, now Tasmania.

[3] The caloric theory of heat, in which heat is held to be a material substance, was popular until the middle years of the century.

[4] Hooker is probably referring to Thomas Thomson in India, Robert Schomburgk in Brazil, and George Gardner in Ceylon, all three associated with Kew Gardens.

[5] Clarke 1842.

[6] Montagne 1844.

[7] *Coral reefs*, where the frontispiece map is coloured blue for atolls and lagoon islands, and red for fringing reefs. CD believed the blue-coloured areas indicated subsiding land. See also the first of the two CD notes transcribed with the annotations to this letter.

[8] Pin holes in the notes and the letter indicate that they were at one stage attached to each other. The notes are bound immediately following the letter, DAR 100: 22–3.

From Henry Denny 30 October 1844

Phill Hall Leeds
Oct 30th /44

Dr Sir

It was my intention to have written you a line long before this, but first the York Meeting[1]—at which I had 8 plates to get ready to exhibit to the Nat Hist Section took up nearly all my time, since which various domestic occurrences have intervened to prevent me, until now— I feared by your last,[2] that you felt I had doubted your statements as to the exact locality of the Pediculi last sent, or implied a want of care, on your part.— Now nothing could be further from my wishes than to suspect either. But the appearance of the specimens from the *Cavia* Cobaia, was so strikingly different, from the Louse of the Domestic Guinea Pig, that, I thought an interchange of Specimens might accidentally have taken place in the way you alluded that by mixing in your bag, when out *shooting* By this means a species might be actually taken from a Bird or Quadruped, & yet not belong to it.— The specimens on a cursory inspection appeared to possess the exact similitude to the Genus *Trinoton.* The occurrence of which on any other that *water Fowl* I believe has never been noticed, at all events, I never heard of an instance in which species of the *same Genus* were found, some on Birds others on Mammals.

I am about to institute a *rigid* examination of the specimens, for on a second glance at them previous to going to York, I was struck at the singular appearance of them, in some points they looked like *Trinoton*, & yet there was a something which said they *are not* of that Genus—as if they wore a sort of disguise, somewhat like Mr Kirbys *Heteromorpha.*— If I am confirmed in this I should like to name the genus *Pseudo-Morpha*—but have a suspicion, the term is occupied already—.[3] If they turn up true Lice of the *Aperea* will they not tell against the Aperea & Guinea Pig—being identical. Zoologists are not all decided about this, yet— I cannot see why an animal should be infested by *two* peculiar *parasites* in a Domestic State, in England, France Germany Prussia &c—& by a totally different Genus & species, in its wild state.— We find the same Louse on the Spoonbill in *Europe* & *Calcutta*, the Gannet in Europe & Cape of Good Hope,—the Curlew Europe & India &c &c. What is your opinion concerning the Aperea being the origin of the Guinea Pig?— They do not agree in all points as for instance in the wild state but *one* young at a Birth & that seldom. In the domesticate state they breed freely— The colour again to me appears rather strange. In domestic Rabbits we have the wild colour as well as the *varied* but the *Guinea* pig never is of the same colour as the wild *Aperea* I believe?—

Would it *not* be worth while if I could find any one living in the country of the Aperea to examine more specimens for me, & send the results in Ship letters, if all the *wild Aperea* had the *same* Louse & not the Genus *Gyropus* at all, it would be ground for separating them I think— besides if I could find an *Agent*, I should like very much to see the Louse of the Capybara & Agouti & Coypou all of which are common in particular localities, I believe.—

Believe me | Dr Sir | Yours respectfully | Henry Denny

Chas Darwin Esq | &c &c

DAR 205.3 (Letters)

CD ANNOTATIONS

1.1 It was . . . already—. 2.6] *crossed pencil*

2.9 a Domestic . . . Pig?— 2.14] 'Worth examining how this is with land Birds living in distant Countries?' *added pencil*

2.14 wild state . . . at all, 3.3] 'Land Birds of N America & Europe?' *added pencil*

[1] The British Association met in York in September 1844. Denny read a 'Report of the progress of the investigation of the exotic *Anoplura*', *Report of the 14th meeting of the British Association for the Advancement of Science held at York in 1844*, p. 392.

[2] See letter to Henry Denny, 12 August [1844].

[3] William Kirby (1825) described three species of insects which 'assumed the characters of another tribe or genus', one of which he named *Pseudomorpha excrucians* (pp. 98–101). Denny drew the plate which accompanies this description, but here the insect was called *Heteromorpha excrucians* (p. 109).

To J. D. Forbes [November? 1844]

"How singular it at first appears, that your discoveries in the structure of glacier ice should explain the structure, as I fully believe they will, of many volcanic masses. I, for one, have for years been quite confounded whenever I thought of the lamination of rocks which have flowed in a liquified state. Will your views throw any light on the primary laminated rocks? The laminæ certainly seem very generally parallel to the lines of disturbance and movement. Believe me, &c. C. DARWIN."[1]

To Professor FORBES.

Proceedings of the Royal Society of Edinburgh 2 (1845): 18

[1] In the *Proceedings of the Royal Society of Edinburgh*, this paragraph is preceded by two paragraphs taken from CD's letter to J. D. Forbes, 11 October [1844].

To Henry Denny 7 November [1844]

Down near Bromley | Kent
Nov. 7

Dear Sir

I am much obliged for your note & have been greatly interested by the facts you mention of the identical parasites on the same species of birds at immensely remote stations. I am sorry to say I cannot think of any possible means of procuring the parasites of the S. American Mammifers to which you refer. Some surgeon, or officer, interested in Nat. Hist w^{d} be the only means & I know none now there.

Are you aware whether the same parasites are found on any of our *land* birds in this country & in N. America. Some of the birds of Europe & N. America appear certainly identical; many form very closely related species or as some would think races: What an *interesting* investigation w^{d} be the comparison of the parasites of the closely allied & representative birds of the two countries.

Should you chance to know anything of the parasites of the *land*-birds of North America, perhaps, sometime you kindly w^{d} take the trouble to send me a line, as I am deeply interested in everything connected with geographical distribution, & the differences between species & varieties.— I hope you will turn in your mind the possibility of investigating closely the N. American land-bird-parasites.—

When the same bird in *immensely* remote countries, has the same parasite, do you never observe some slight difference in colour, size or proportions of such parasites? I have forgotten to answer your question, about the Aperea being identical with the guinea-pig, & I can only answer it by professing entire ignorance & doubt: I certainly sh^{d} disbelieve it, if you c^{d} show the parasites were different. How is the parasite of the wolf with the dog, if the latter has one?

I hope you will excuse this long note & believe me dear Sir | Yours very faithfully | C. Darwin

Cleveland Health Sciences Library (Robert M. Stecher collection)

From J. D. Hooker 8 November 1844

West Park Kew
Friday Nov.r 8. 1844

My dear Darwin

The accompanying are almost verbatim copies of the notes on Infusoriæ in my Journal: are they not luminous?!— I did think of dressing them a bit, but changed my mind; if they are worth any thing they are better as authentic— I have added an idea of my own of the formation of Pancake ice,[1] I never read an xplanation of it, but Ehrenberg seems to have been so puzzled about it in his former letters, that I thought I had better initiate him a little into the mysteries of the S. Pole.— I dare say he will think me very presumptuous for this & some other parts of these notes, but I give him them as I found them & he may make any use he can of them, providing he will promise to look on them as they really are—crude remarks made on the spot.

The figures referred to I will copy when the drawings are accessable, but they are now at Edinburgh in Goodsirs hands, who is doing the Crustacea, & have been there a long time..[2] The little story at the end about the Ice in columns I intend for you, & not for Ehrenberg's edification; though as I carelessly wrote it on the same sheet, he will have the benefit of it— I am presuming that you will read the ponderous mss. & forgetting that you are not a *Kopfnüsse* German.[3]

I have not forgot a question you once propounded to me as to whether the species of *large-groups-which-were-local* were local also, & am inclined to think that species of local large nat. ords. are local also, *Cacti* I think I mentioned before as an example & *Proteaceæ* & *Epacridæ* I take to be two others.

I find two fruitful sources of error into which specific Botanists fall, one is the describing two things as new, because they come from 2 far off countries; & the other of uniting 2 species because they come from the same. To a certain extent a safeguard will be the consideration of whether the group to which the individual belongs is local or widely diffused:—thus, I should hesitate to join two *Proteaceæ* because they both come from Australia, longer than I should to join two *Cardamines* one from England & the other from Australia— If this caution saves one single species, the Athenæum need no more cry "cui bono" to poor Watsons Geog. Distrib. of British plants.—[4]

I find *Lycopodiums* the same all over the world they do vary *miro modo*.. atmospheric & other causes have made (what have passed for) many species of them. How is this?, I am firmly convinced (but not sure enough to print it) that *L. Selago* varies in V. D L. into *L. varium*—two more different *species* (as they have hitherto been thought) per se can not be conceived, but no where else do they vary into one another nor does *Selago* vary at all in England.—[5] I am perfectly aware of the xtreme caution that should be used in this instance, & the propriety of keeping the two species distinct in V. D L., or finding some character between the V. D L. form of *varium* which appears there as *Selago*,—if in either case, any, however slight, a character can be found between them; but I can find none, the V. D L.

Selago is, to my eyes, identical with the English & V. D L. appears to me to possess *all* intermediate states between *selago* & *varium*: this is not the passage merely of one species into another, but of two groups of the genus differing originally *toto caelo*.. If I can only find a parallel case in the genus (& it is a weeks work to get the species together), I will be down upon *varium* very soon.

I do not know why I should bother you with these things xcept that you always take it so *kindly*:

My Father joins me in most kind remembrances to Mrs Darwin & believe me ever | Yours most truly | Jos D Hooker

P.S. The drawings referred to are not worth any thing but only shew what *form* of infusoria I refer to.—

If you go towards the *Ship* inn Charing X you can see by asking the waiter a very fine plant of Tussack belonging to Capt Ross.[6] I enclose you some *good* seeds, we have plenty young plants from them..

DAR 100: 24–5

CD ANNOTATIONS
1.1 The accompanying . . . German. 2.6] *crossed pencil*
5.4 *L. Selago*. . . one another 5.6] *scored pencil*
5.11 English] *comma added pencil after*
6.1 I do not . . . them.. 9.3] *crossed pencil*
Top of first page: 'Nov./10/44' *added pencil*

[1] Small, thin cakes of ice, which form on the surface when sea-water begins to freeze.
[2] Harry Goodsir was preparing a volume of 'crustacealogical researches', announced in Goodsir 1844, but never completed because of his death on the Franklin expedition in 1845.
[3] Hard-headed German.
[4] The *Athenæum* criticised Hewett Cottrell Watson for not drawing any scientific conclusions from the facts given in Watson 1835 (*Athenæum*, no. 458, 6 August 1836, p. 552); Watson rejected the criticism in Watson 1835–7. See also *Athenæum*, no. 529, 16 December 1837, pp. 909–10.
[5] J. D. Hooker 1844–7, pp. 115–7.
[6] Hooker was much impressed with the agricultural potential of the Falklands tussock grass with its six-foot blades, and hoped it would 'make the fortune of Orkney and the owners of Irish peat-bogs'. An account extracted from his *Erebus* letters was published in W. J. Hooker 1843.

To J. D. Hooker [10–11 November 1844]

Down Bromley Kent
Sunday

My dear Hooker

I had intended writing to you today, had I not received yesterday's enclosure to remind me. Many thanks for the seed & for sending me your curious account of the antarctic ice, which no doubt, together with your notes on Infusoria, Ehrenberg w^{d} be very glad to see.— I sent my last parcel through Mr Taylor of Fleet St. or rather, (I believe) through Mr Francis, who lives with him.— The homogeneous manner in which the brash[1] is coloured, certainly appears very curious; I presume you consider the quantity too great for snow; with respect to its rising from the bottom, I sh^{d} rather doubt it, though in fresh **running** water, it is well known that icy matter, I suspect like your brash, rises from the bottom, & brings with it stones. There has been much argument about cause of this, in which even Arago has joined:[2] the best explanation offered, as it appeared to me, was that the bottom of the stream lost its heat by radiation & and the water froze on it. Should you feel much interest on this subject, I could look you up, (I think) some references. One w^{d} doubt, whether the bottom of the sea would lose its heat by radiation through several hundred feet of thickness of water, & the whole body of water would have to be cooled to the freezing point of sea-water. On other hand Simpson & Deane (I think) found the bottom of *shallow* arctic sea hard frozen:[3] off Spitzbergen *masses* of ice suddenly rise from the bottom, I have fancied they were remnants of fixed & grounded icebergs.

What a curious, wonderful case is that of the Lycopodiums; I suppose you would hardly have expected them to be more varying than a phanerogamic plant. I trust you will work the case out & even if unsupported publish it, for you can surely do this with due caution. I have heard of some analogous facts, though on the *smallest scale*, in certain insects being more variable in one district than in another; & I think the same holds with some land-shells. By a strange chance, I had *noted* to ask you in this letter an analogous question, with respect to genera, in lieu of individual species[4]—that is, whether you know of any case of a genus with most of its species being variable (say Rubus) in one continent, having another set of species in another continent non-variable or not in so marked a manner.[5] Mr Herbert incidentally mentioned in a letter to me, that the Heaths at the C. of Good Hope were very variable, whilst in Europe they are (?) not so (?); but then the species here are few in comparison, so that the case, even if true, is not a good one.—[6] In some genera of insects the variability appears to be common in distant parts of the world: in shells, I hope hereafter, to get much light on this question through fossils. If you can help me, I sh^{d} be very much obliged: indeed all your letters are most useful to me.

Monday— Now for your first long letter & to me quite as interesting as long. Several things are quite new to me in it, viz for one, your belief that there are more extra-tropical than intratropical species. I see that my argument from the Arctic regions is false, & I sh^{d} not have tryed to argue against you, had I not fancied that you thought that equability of climate was the *direct* cause of the creation of a greater or lesser number of species: I see you call our climate equable, I sh^{d} have thought it was the contrary; anyhow the term is vague, & in England will depend upon whether a person compares it with the United States or T. del Fuego.— In my Journal, (p. 342) I see I state that in South Chiloe at height of about 1000 ft the forest had a Fuegian aspect: I distinctly recollect, that at sea-level in middle of Chiloe, the forest had almost a tropical aspect. I sh^{d} like much to hear, if you make out, whether the N. or S. boundaries of a plant are the most restricted; I sh^d^ have expected that the S. would be, in the temperate regions, from the number of antagonist species being greater. (N.B. Humboldt, when in London, told me of some river in N. E. Europe, on the opposite banks of which the Flora was, on the same soil & under same climate, *widely* different!)[7] I forget my last letter, but it must have been a very silly one, as it seems I gave my notion of the number of species being in great degree governed by the degree to which the area had been often isolated & divided;; I must have been cracked to have written it, for I have no evidence, without a person be willing to admit all my views, & then it does follow; but in my most sanguine moments, all I expect, is that I shall be able to show even to sound Naturalists, that there are two sides to the question of the immutability of species;—that facts can be viewed & grouped under the notion of allied species having descended from common stocks. With respect to Books on this subject, I do not know of any systematical ones, except Lamarck's, which is veritable rubbish; but there are plenty, as Lyell,[8] Pritchard[9] &c, on the view of the immutability. Agassiz lately has brought the strongest arguments in favour of immutability.[10] Isidore G. St. Hilaire has written some good Essays, tending towards the mutability-side, in the "Suites à Buffon", entitled "Zoolog: Generale".[11] Is it not strange that the author of such a book, as the "Animaux sans Vertebres",[12] sh^d^ have written that insects, which never see their eggs, should *will*, (& plants, their seeds) to be of particular forms so as to become attached to particular objects.[13] The other, common (specially Germanic) notion is hardly less absurd, viz that climate, food, &c sh^d^ make a Pediculus formed to climb Hair, or woodpecker, to climb trees.— I believe all these absurd views, arise, from no one having, as far as I know, approached the subject on the side of variation under domestication, & having studied all that is known about domestication.— I was very glad to have your criticisms on island-floras & on non-diffusion of plants: this subject is too long for a letter; I c^d^. defend myself to some considerable extent, but I doubt whether successfully in your eyes, or indeed in my own.

I sh^d^ be much obliged for a loan of the Tasmanian Journal with Rev^d^: Clarke's Paper: I suspect, however, it will turn out to be the same with a paper read by him before the Geolog: Soc: on the same subject.—[14] I shall be in town on the 20th.: could you leave it for me at the Athenæum (Have you your name down for this club? I think you w^d^ find it worth while, if you have not) or at the Geolog: Soc:—

Is there any chance of your being able to pay us a visit here soon; it would give M^rs^. Darwin & myself real pleasure. I sh^d^ have written sooner, had not my own plans, owing to a visit to my Fathers now paid, been rendered uncertain. We c^d^ send to meet you to Croydon; or you c^d^ come by Coach from the Bolt-in-tun Fleet St.— We are wholly disengaged, except on 19, 20, 21, when I must be in London on business.

Once again, thanks for your Botanical letters & believe me, my dear Hooker, | Very truly yours | C. Darwin

Will you tell Sir William, that the Deodar, which he gave me, is doing famously.

I am really ashamed how infamously this letter is written.—

DAR 114.1: 19

[1] Brash-ice, fragments of crushed ice.
[2] Arago 1833.
[3] Dease and Simpson 1838, pp. 218, 220.
[4] See CD's annotations to the letter from J. D. Hooker, 28 October 1844.
[5] CD was to repeat this question to Hooker and other naturalists. For example, on 28 December 1853, CD noted:

> Hooker went through the N. Zealand & Tasmanian Flora, & he thinks that all the genera which are variable in Europe are quite as variable in these localities. *Hence I must* ['clearly' *del*] *give up this kind of generic variability, as any aid in transmutation* (DAR 45: 5).

[6] See *Correspondence* vol. 2, letter from William Herbert, [*c.* 27 June 1839].
[7] CD met Alexander von Humboldt in January 1842 and made the following note:

> Jan. 29th /42/. Humboldt descanted on remarkable fact (as observed by Gmelin & Pallas) that the **banks* of the [*interl*] River Oby separates two Floras—on one side $\frac{8}{10}$ of plants same as in Germany with $\frac{2}{10}$ Asiatic—on other side reversed large proportion of Asiatic.— Remarked a similar case with respect to the distribution of oaks in some place, I did not catch up—with Astacus in all the brooks on one side & not on the other: says Bellis perennis extends to a certain limit & then ceases, but it is not the cold, for this plant will flower, ['far' *del*] within the limits of snow on some mountains— On the Oby there is no geological change=prepossession *of soil [*interl*] must here have done much.— ?Have two Floras marched from opposite sides & met here??—strange case.— (DAR 100: 167).

[8] Annotations on CD's copy of C. Lyell 1837 (Darwin Library–CUL) are discussed in S. Smith 1960.
[9] CD's copy of Prichard 1836–7 is annotated (Darwin Library–CUL).
[10] Agassiz 1842.
[11] CD's copy of I. Geoffroy Saint-Hilaire 1841 is annotated (Darwin Library–CUL).
[12] Lamarck 1815–22. CD's copies of this and the second edition (Lamarck 1835–45) are in the Darwin Library–CUL.
[13] CD's copy of Lamarck 1830 (volume one only) is in the Darwin Library–CUL. This particular opinion is not found there, but see pp. 223, 235, and 268, where CD has marked Lamarck's discussion of insect reproduction. See also *Notebook C*: 63.
[14] Clarke 1842, which was a longer version of a paper read to the Geological Society in 1839 (Clarke 1839).

To J. D. Forbes 13 [November 1844][1]

Down near Bromley Kent
Wednesday 13th—

Dear Sir

I had intended answering your very obliging note sooner, but have been prevented doing so. I am sorry I cannot break the specimen of zoned obsidian as it is not by own; but I should so much like you to see these specimens, that I have taken the liberty of paying the carriage & sending by steam-boat to you, directed to the University 14 fragments of rock, which even if not worth looking at, can not give you much trouble. One is apt to overrate the interest of anything one sees oneself, & I have no doubt I have done so, yet I think the series curious, as showing such perfect lamination or rather separation of minerals in parallel planes, in a pile of rock of undoubtedly volcanic origin.—

I think the Mexican obsidian, zoned with minute air-cells will interest you. Two of my specimens show the passage from obsidian into the intercallating feldspathic rocks, in which the laminar structure is chiefly rendered apparent by concretionary action; in the other specimens the laminar structure is redered apparent by crystalline action.—: one of these **like** gneiss appears to me interesting. All the specimens are labelled.— You can return them, whenever you like, directed to me to my Brothers House at

"7. Park St
Grosvenor Square
London."

I have forgotten to thank you for sending me your latest contributions on glaciers:[2] by a singular chance, the very day, after I had written to you, I received the Edin: New Phil. Journal & there read your remarks on the comparison of lava-streams & glaciers;[3] which if I had seen sooner, I sh^{d} not have troubled you.— I have heard lately from a young German who is very full of the ice-action-marks which he discovers in Germany & Hungary[4] how, I wish, someone could point out a clear line of separation, between glacier & iceberg action: on the outskirts of N. Wales, they appeared to me to blend together in the most puzzling manner. I observed one fact there, which I have always been curious to have explained; I have marked the passage in the enclosed: perhaps you can explain it—: is it due to your mobility of the ice.—[5] But I have much cause to apologise for troubling you with so long a note.

Believe me, My dear Sir. Yours very faithfully C. Darwin

University of Saint Andrews Library

[1] This date is the first Wednesday the 13th after the letter to J. D. Forbes, 11 October [1844].

[2] Three of J. D. Forbes's papers on glaciers from the *Edinburgh New Philosophical Journal* for 1844, bound together and signed by Forbes, are in the Darwin Pamphlet Collection–CUL. They include J. D. Forbes 1844, see n. 3, below.

[3] J. D. Forbes 1844.

[4] See letter to Adolf von Morlot, 10 October [1844].

[5] Presumably CD enclosed a copy of 'Notes on the effects produced by the ancient glaciers of Caernarvonshire' (*Collected papers* 1: 163–71). The passage CD refers to may be: 'how it comes that the glacier, in grinding down a boss [dome-shaped rock] to a smaller size, should ever leave a small portion apparently untouched, I do not understand' (p. 165).

From J. D. Hooker 14 November 1844

West Park Kew
Nov.[r] 14.. 1844

My dear Darwin

The only drawback I have to the pleasure your xcellent letters afford me is, the thoughts that you may be kind enough to feel yourself in some measure *obliged* to answer my interminable yarns—though it may be long before you will allow yourself to think so.

Very many thanks for all the information you give me, if everyone who had information to give, was equally ready to impart, I think that science would advance too fast for me at least

I cannot call to remembrance any *marked* case of a genus having a set of var[ble]. sp. in one area & better marked ones in another, but there are partial instances, & a good many—*Rubi* for instance in some countries as S. Am. & the E. Ind. are I believe more invariable than here, but I am not speaking on very good grounds as I have never xamined & there are few of them as you say— *Vaccinia* are certainly very bad in N. Am. but the *much fewer* Europ. sp. are good ones. Certainly Cape *Ericeæ* vary, but I am not prepared to say that the Europ. ones are very well behaved, but shall ask Mr Bentham. I have asked my father the question & as far as he can recollect he agrees with me, that the converse holds good to an alarming xtent. All the large mundane genera of Cruciferæ, Caryophylleæ, Ranunculus, Grasses, Cyperaceæ &c. are very bad every where. I am not however prepared to say that mundane genera vary in proportion to their size, as some solitary mundane genus of a single species will vary enough *for a whole Nat. ord.* On the other hand there a few mundane genera which are pretty good all over the world, thus Boott does not think *Carex* so bad.[1]

With regard to *Lycopodium* the more I examine them the more thoroughly I am convinced of the identity of a few species all over the globe, & my *Selago* xample is only one of several parallel cases in the same genus. All that I can at present say is, that if (as I believe) *Selago* is a variety of *varium*, then that it is not so variable a plant in the N. temp. zone as in tropics or South, for if once it is conceded that it is said var. there is no saying where the variation is to stop, without it turns into itself again as I hope it will, through whatever changes it may go I care not.— I am preparing to write (commit myself) on the subject.

I should very much like to hear about this river that Humboldt mentions, I cannot think of any analogous case.— There is a V. D L. bird *xcessively* common on the E. side of the Derwent which has never been seen on the W. south of Rosneath (about 10 miles above Hobart. I can vouch for this as far as my xperience of a few days goes & many most intelligent persons have told me the same & who pointed the fact out to me: I have a note of it somewhere.

The gum trees are an instance of a very large genus confined within very narrow limits, being *very* variable, yet one most distinct species covers about 2 acres only on one side of narrow valley opposite Hobart, its limits are as marked as possible, no one has seen it elsewhere, nor are there scattered young trees about it, it occupyes the ground for a little space or nearly so, it is distinguishable ½ mile off.— I believe parallel Australian cases are not rare in other constant genera, but not in so Protean a one as *Eucalyptus*

My name is down for the Athen. & has been I cannot tell how long, some 2 years I believe, but do not know, I xpect I have 2 more to wait at least xcept they will make a committee member of me as they did of Dr Graham & Richardson[2] & may of me if ever I attain their eminence, & am not black balled before. The expense of the Geolog. deters me, as I do not belong to the Royal yet, besides the almost impracticability of my attending the meetings: the Linnæan I am more in duty bound to meet with but can hardly ever go.[3]

I should like very much to steal a holiday about the beginning of Decr. & accept your & Mrs Darwin's most kind offer, but will you not (my father bids me say) come here one night of 19, 20, or 21st.; just one, I will meet you in Town; a ¼–5 Omnibus is plenty time for our 6 dinner we shall be alone & I will promise you quiet & may go as early as you like next morning, I want to be in town on one of those days, it is immaterial which. My mother is copying out the 1st. century of Gal. Isd. plants for L. Soc.[4] I did not tell you that I had to withhold it, from Henslowes sending me a supplement, which *my book* prevented my working up.[5]

Ever most truly yours, J D Hooker.

DAR 100: 26–7

CD ANNOTATIONS

1.1 The only . . . at least 2.3] *crossed pencil*

6.1 The gum . . . *Eucalyptus* 6.7] '(another species/of this Section very local)' *added ink; square brackets in MS*

7.1 My name . . . working up. 8.8] *crossed pencil*

Top of first page: 'Ch 4' *brown crayon*

'Boulders on Kerguelen Land' *ink, del pencil*

[1] Francis Boott was an authority on sedges.

[2] CD had been a member of the Athenæum Club since 1838. Thomas Graham was elected in 1842 and John Richardson in 1844; both were elected under 'Rule II', which permitted the committee to elect annually up to nine men eminent in science, literature, or the arts (Waugh 1888). Hooker underestimated the time his election would take for it was not until 1851 that he became a member under the same rule, a delay that annoyed him. On 7 April 1850 he wrote to his father from Calcutta (Royal Botanic Gardens, Kew (India letters 1847–51: 279)):

> As to the Athenæum I am rather disgusted at having to come on after my name has been down so many years—(1843) & after so many men have come on in that manner whose names were not so long, some not at all before the public.

[3] Hooker had been elected, without his knowledge, to the Linnean Society in 1842, while serving on James Clark Ross's Antarctic expedition. It was the first scientific society to which he belonged. (Huxley ed. 1918, 2: 429).

[4] J. D. Hooker 1845d.

[5] J. D. Hooker 1844–7. See letter to J. D. Hooker, 1 June [1844].

To J. D. Hooker [18 November 1844]

Down Bromley Kent.
Monday

My dear Hooker

I write one line to send my thanks to Sir William & yourself for your kind invitation, which I sh[d] have had real pleasure in accepting, had I been able.— But I assure you a morning's work in London totally unfits me for everything, even the quietest conversation, in the evening: I have long, to my sorrow, been compelled to relinquish the Geological evening meetings & attend only at the Council.[1]

Whenever, it would suit you to come here, it w[d]. give us very great pleasure & if you can spare the time do in December propose yourself. | In Haste | Ever yours | C. Darwin

Postmark: NO 18 1844
DAR 114.1: 20

[1] CD was an officer or council member of the Geological Society from 1837 to 1850. In 1844 he was a vice-president.

To Leonard Jenyns 25 [November 1844][1]

Down Bromley Kent
Monday 25th.

My dear Jenyns,

I am very much obliged to you for the trouble you have taken in having written me so long a note. The question of where, when, & how, the check to the increase of a given species falls appears to me particularly interesting; & our difficulty in answering it, shows how really ignorant we are of the lives & habits of our most familiar species. I was aware of the bare fact of old Birds driving away their young, but had never thought of the effect, you so clearly point out, of local gaps in number being thus immediately filled up. But the original difficulty remains, for if your farmers had not killed your sparrows & rooks, what would have become of those, which now immigrate into your Parish:[2] in the middle of England one is too far distant from the natural limits of the Rook & sparrow, to suppose that the young are thus far expelled from Cambridgeshire. The check must fall heavily at some time of each species's life, for if one calculates that only half the progeny are reared & breed,—how enormous is the increase! One has, however, no business to feel so much surprise at one's ignorance, when one knows how impossible, it is, without statistics, to conjecture the duration of life & percentage of deaths to births in mankind.

If it could be shown that apparently the birds of passage, *which breed here* & increase return in the succeeding years in about the same number, whereas those that come here for their winter—& non-breeding season, annually come here with the same numbers, but return with greatly decreased numbers, one would know (as indeed seems probable) that the check fell chiefly on full-grown birds in the winter season, & not on the eggs & *very young birds*, which has appeared to me often the most probable period. If at any time any remarks on this subject should occur to you, I sh^{d} be most grateful for the benefit of them.—

With respect to my far-distant work on species, I must have expressed myself with singular inaccuracy, if I led you to suppose that I meant to say that my conclusions were inevitable. They have become so, after years of weighing puzzles, to myself *alone*;; but in my wildest day-dream, I never expect more than to be able to show that there are two sides to the question of the immutability of species, ie whether species are *directly* created, or by intermediate laws, (as with the life & death of individuals). I did not approach the subject on the side of the difficulty in determining what are species & what are varieties, but (though, why I sh^{d} give you such a *history* of my doings, it w^{d} be hard to say) from such facts, as the relationship between the living & extinct mammifers in S. America, & between those living on the continent & on adjoining islands, such as the Galapagos— It occurred to me, that a collection of all such analogous facts would throw light either for or against the view of related species, being co-descendants from a common stock. A long searching amongst agricultural & horticultural books & people, makes me believe (I well know how absurdly presumptuous this must appear) that I see the way in which new varieties become exquisitely adapted to the external conditions of life, & to other surrounding beings.— I am a bold man to lay myself open to being thought a complete fool, & a most deliberate one.— From the nature of the grounds, which make me believe that species are mutable in form, these grounds cannot be restricted to the closest-allied species; but how far they extend, I cannot tell, as my reasons fall away by degrees, when applied to species more & more remote from each other.

Pray do not think, that I am so blind as not to see that there are numerous immense difficulties on my notions, but they appear to me less than on the common view.— I have drawn up a sketch & had it copied (in 200 pages)[3] of my conclusions; & if I thought at some future time, that you would think it worth reading, I sh^{d} of course be most thankful to have the criticism of so competent a critic.

Excuse this very long & egotistical & ill written letter, which by your remarks you have led me into, & believe me, Yours very truly | C. Darwin

Bath Reference Library (Jenyns papers, 'Letters of naturalists 1817–76', Quarto vol. 2, 51(8))

[1] This letter follows the letter to Leonard Jenyns, 12 October [1844]. November 25 was the only following 'Monday 25th'.

[2] In *Natural selection*, p. 185, CD referred to Jenyns on this point and made the same comment.

[3] The 'fair copy' of the essay of 1844 is in DAR 113; the original manuscript is in DAR 7. There is no record that Jenyns ever read the essay.

To Susan Darwin [27 November 1844?][1]

Wednesday

My dear Susan

I have to thank you for two business notes— I understand all about the money, & am much obliged for it.— it will just carry me through the half year.—

Thank, also, my Father for his medical advice— I have been very well since Friday, nearly as well, as during the first fortnight & am in heart again about the non-sugar plan.— I am trying the very bitter, weak, but thoroughly fermented Indian Ale, for luncheon & it suits me very well.—[2]

Our prize in the lottery, the China the Barberini vase, & wax releifs are all come & a very fine prize it is.—[3]

Poor Emma keeps very bad; I hope you will manage to stay more than one day.—[4]

Now for my main object in writing, viz to enclose Mr. Higgins[5] very clear & sensible note (& I, likewise, enclose his former one.). I doubt whether Mr. Higgins' information applies to the South of Kent, but, upon the whole I believe, I had better come into the Lincolnshire plan.— I keep quite of opinion, that it is very adviseable to have part of one's property in land. Sir John Lubbock was paying a long call here yesterday, & I consulted him a bit: he tells me, that in all *this* part of Kent, land is most absurdly dear; but he was quite of opinion, that it was very wise to invest something in land. If my Father still approves, I will write to Mr Higgins to thank him for his note; & shall I, in my Father's name, ask him to continue his look out & let my Father hear.— If the better one of the two estates, mentioned in his former note, remain unsold, perhaps it would do; & being within a few miles of Claythorpe, is an advantage, as, when you visit your estate you can rummage up my tenant.—[6] How very grand we shall be, when we go arm & arm & astonish our tenants.—

Please return Mr: Higgins' two notes.

Ever your's | C. Darwin

DAR 92: 9–10

[1] This letter was probably written between CD's visit to Shrewsbury 18–29 October 1844 (see Appendix II) and 26 March 1845, the date of an agreement (DAR 210.25) drawn up between Robert Waring Darwin and James Whiting Yorke for the purchase and sale of the estate at Beesby, Lincolnshire, which was eventually conveyed to CD (see letter to John Higgins, 15 March 1845). The precise date is conjectured from an entry dated 24 November 1844 in CD's Account Book (Down House MS) recording the receipt of £98 0*s.* 2*d.* from his father, which may be the money referred to in the opening sentence of this letter. See also nn. 2 and 4, below.

[2] For a discussion of CD's use of bitter ale and the 'non-sugar plan' see Colp 1977, p. 37. An entry in CD's Account Book (Down House MS) dated 21 November 1844 records the purchase of a quantity of ale from a merchant in Henrietta Street during a trip to London. Beer for normal household purposes was obtained from William Lewis, the Down brewer.

[3] This may refer to a lottery for some of the effects of Josiah Wedgwood II, or for items removed from Etruria prior to its auction in August 1844. However, the reference to the Barberini vase suggests

that it was for some of the belongings of R. W. Darwin, since CD is known to have acquired a Wedgwood copy of the Barberini vase through his father, and to have subsequently donated it to the Museum of Practical Geology (Meteyard 1875, p. 302). CD may have had a second vase, but there is no record of it and it was not in his possession in 1875 (Meteyard 1875, pp. 303–4). The wax reliefs are doubtless some of those prepared by John Flaxman and other artists as models for Wedgwood bas-reliefs; they may well have come into the Darwin family through Josiah Wedgwood II, who casually distributed them among his relations (J. C. Wedgwood 1908, p. 185). CD sold the wax reliefs in 1858 and probably gave away the Barberini vase at this time (*Calendar*, letters 2236 and 2323).

[4] Susan visited Down on Sunday, 8 December, see letter from J. D. Hooker, 12 December 1844.

[5] John Higgins was a land agent who managed farms owned by Susan and Robert Waring Darwin.

[6] If the conjectured date of the letter is correct then this estate cannot be the one that CD ultimately purchased (see letter from John Higgins, 15 March 1845) despite the similar locations.

To Adolf von Morlot 28 November [1844]

Down near Bromley | Kent
Nov. 28th.

Dear Sir

I received your letter of the 29th of October a week since. It would be presumptuous in me to give you credit for the spirit in which that letter is written; but I may say, that I think, that if you yourself were to see it in ten years hence, you would be pleased with it.— I am obliged for your German letter, which I must slowly spell out, for I am a miserable German scholar, & feel a difficulty in acquiring language, which I presume is unknown to yourself & many of your countrymen.—

Your news about the passage of trachyte is very interesting to me; & I shall be very curious to read, when published, Haidingers views on plutonic Geology:[1] seeing that the earth gets hotter with an increasing depth & seeing from how many points, liquefied *volcanic* rocks have or do come to the surface, I cannot, (not knowing the full reasons) give up yet the old plutonic view.— With respect to the Löess, have you read what Lyell says in his later Editions of the Principles,[2] & in his Paper in the Edin: New Phil: Journal, July 1834, & in the Proceedings of the Geolog: Soc: No. 36 & No 43.—[3]

With respect to the contemporaneousness of the now extinct great animals, with the ice Period, are you aware that in N. America, it can be clearly shown that they lived quite subsequently to the boulder-period. (V. Lyell in some late numbers of the Geolog. Proc:)[4] & this is the case with one of the monsters of S. America.[5] Dr. Falconer is, also, sure, that when the great quadrupeds of India were *alive*, the country was much cooler.[6]

As you attend to volcanic rocks, let me call your attention to the probability of the zoned-glacier-structure of Forbes (to whom I have lately sent some specimens) throwing light on *certain* streams of lava of the obsidian & trachytic streams: I have given some facts in my small volume on Volcanic Islds. on this subject.—[7]

Should you at any time come to England, I shall be happy to see you here, at my quiet country house; but if you expect any information from me, you will be disappointed, for I find myself falling far astern in the geological race of knowledge,—to an extent, which when I was strong, I had hoped, would not have happened for some years.—

With respect to passages in my letters, I feel almost certain that they contain nothing new or worth publishing; & they were written without care & with personal allusions to myself, which are not fit for any eye, but a correspondent's: Pray, therefore, if you do publish any passage, be cautious, for I repeat, I am almost sure, there can be nothing worthy of anyone's seeing.

Believe me | Yours very truly | C. Darwin

Burgerbibliothek Bern

[1] Wilhelm Karl Haidinger, who proposed that saline solutions at high pressures and temperatures percolating through rocks would lead to the chemical transformation of calcite to dolomite (Haidinger 1848). This was verified experimentally by Adolf von Morlot.

[2] C. Lyell 1840a, 1: 286.

[3] C. Lyell 1834. This paper was read to the Geological Society on 7 May 1834; an abstract of it appeared in the Society's *Proceedings*, but the full paper was printed in the *Edinburgh New Philosophical Journal.* Charles Lyell later communicated supplementary observations to the Geological Society (*Proceedings of the Geological Society of London* 2 (1833–8): 83–5, 221–3). Lyell's papers described loess as a recent alluvial deposit, laid down gradually.

[4] C. Lyell 1842.

[5] A reference to *Macrauchenia patachonica*. CD believed that his specimen was in a later formation than that containing the erratic boulders found in the valley of the Santa Cruz (*South America*, p. 97).

[6] See Falconer 1846 (read at meetings of the Royal Asiatic Society on 1 and 8 June 1844). Hugh Falconer's printed statements on the former climate of India are less definite than CD suggests, he actually stated that he considered the present climate of India to be as warm, if not warmer than it had been in the Tertiary period (p. 109). Falconer may have expressed his belief in a former temperate climate in India more strongly in private communications to CD and others. See also CD's manuscript note, dated 'Nov /44/', about the relationship of large mammals and cold climates: 'Falconer colder in India ['='*del*] Last period colder & yet more great quadrupeds' (DAR 205.9: 189).

[7] See letter to J. D. Forbes, 11 October [1844], and *Volcanic islands*, pp. 70–1.

From J. D. Hooker 29 November 1844

West Park Kew

Nov.r 29. 1844.

My dear Darwin

I am ashamed of not having written before, but have not been able to tell when I should be able to get down to Bromley, on account of Mr Bentham's coming here next week & Profr. Henslow the following. I find however that I have Saturday & Sunday 7th. & 8th. to spare: if perfectly convenient to Mrs Darwin & yourself I should much enjoy a run down & look at your habitat. Do you want any observations made in New Zeald: I have a friend now there, a fair naturalist & acute careful observer, who is going to the Middle & Southern Isld.s he is a Surgeon R.N. now acting Colonial Secretary to G.r Fitzroy.—[1] We had a long letter from Capt Sulivan the other day from the Plate—[2]

Brown (in Congo) says that the level of the sea in Tropics is not so rich in species as the temperate zone (or words to that effect), I never remarked that any one had said so before.[3]

I find, even down to the lower orders, the plants of Ascension are totally different from those of S^t. Helena. this is most remarkable as regards the Ferns.—of which there are 9 Ascension sp. only 2 (I think) of which are S^t. Helena's.. Even the Jungermanniæ are different!— I expect no parallel to this is on record.

Ever your's most truly | Jos D Hooker

DAR 100: 28

CD ANNOTATIONS
1.1 I am . . . Plate— 1.9] *crossed pencil*
3.1 I find . . . record. 3.4] 'Is not wind cross ways?? Surely not.' *added ink*

[1] Andrew Sinclair, who was colonial secretary in New Zealand, 1844–56. Robert FitzRoy was governor, 1843–5 (Mellersh 1968, pp. 207–35).
[2] Bartholomew James Sulivan, previously a lieutenant in the *Beagle*, had been appointed captain of the surveying vessel *Philomel* on active duty in the Falklands and on the coast of South America from 1842 to 1846.
[3] Brown 1818, p. 422.

To J. D. Hooker [2 December 1844]

Down Bromley | Kent
Monday

My dear Hooker

We shall be delighted to see you here on Saturday & I trust your engagement will allow you to stay more than Sunday.— If you come by Coach it is by the "*Down*" Coach which starts at 3¼ from the Bolt-in-tun, but I am sorry to say it is a very slow one, & you will not be at this house till past half past 6 oclock.— I have unfortunately made an arrangement to use my phaeton on that day, but I have just written to try to alter it, & if I so succeed, I will send you a line to inform you & will, on Saturday, send the phaeton to the inn *close* to the station at Sydenham on the *Croydon* Railway: Trains leave at 1° " 20′, 2°, 20′ 3°, 20′, both from London Bridge & from the Bricklayers arms; & if you would start by the 2° " 20′, you would arrive here at 4° " 30′, just time to get ready for our usual dinner hour. Sh^d you miss that train my phaeton shall stay one hour for the succeeding Train. The Bricklayers arms is ½ hours drive in a Cab from Charing Cross.— It is a pretty drive to here from Sydenham.

Ever yours | C. Darwin

If you do not hear, you will understand that I *cannot* liberate the phaeton.—

DAR 114.1: 21

To J. D. Hooker [4 December 1844]

[Down]
Wednesday

My dear Hooker

If I do not hear to the contrary my phaeton shall be at Sydenham, ready to return by the Train, which leaves both the Bricklayers Arms & London Bridge at 2° " 20′. Since I wrote there has been a change in the trains, & no train leaves the Bricklayers Arms at 3° " 20′.; so that if you chance to be too late for the 2° . 20′ at the B. Arms you must go to London Bridge for the 3° " 20′.

Ever yours | C. Darwin

DAR 114.1: 22

To William Benjamin Carpenter [11 or 18 December 1844][1]

Down. Bromley Kent
Wednesday

Dear Sir

I am exceedingly obliged to you for your kind note & very obliging offer of examining my specimens[2] I will send them tomorrow morning by our carrier to London & thence per coach to Ripley. The one without any apparent shells, is that which I want examining.— On looking round it, you will see the little, indistinct, embedded fragments. The specimens can be returned any time per coach, directed to me at "7 Park St. Grosvenor Sq^{re}"[3]

I am very much obliged by your kind offer of sending me some slices of shells, but I have not an achromatic microscope.

With very many thanks, believe me | dear Sir | Yours sincerely | C. Darwin

To | Dr W. Carpenter

University of Rochester Library

[1] Dated from Carpenter's reply, see letter from W. B. Carpenter, 21 December 1844.
[2] CD sent Carpenter concretions of marl from the Pampas and similar rocks from Chile for microscopic comparison (see *South America*, pp. 76–7).
[3] The address of Erasmus Alvey Darwin.

From J. D. Hooker 12 December 1844

West Park Kew
Dec 12 1844.

My dear Darwin

In the first place let me thank Mrs Darwin & yourself for the great kindness you shewed me during my most pleasant stay at Down, & which made me regret leaving you so soon very much indeed.[1] In Miss Darwin[2] I found a most pleasant companion to London; I hope that she did not suffer from the cold in the railway carriage.

I did not see Mr Brown at the Brit. Mus., but my Father went soon afterwards, & took the Agate to him: he is enchanted with it, & says it is very valuable & curious,—that it is either animal or vegetable!—but that is all; so I cannot give you much news about it: I am however to assure you of his gratitude— Curious I grant it looks, but I must confess I cannot see what analogies it has to make it so *very interesting* as he finds it.—

Neither Robertson nor McCormick in their accounts of the South allude to Boulders in Kerg. Land or elsewhere.[3] The collections, I believe at the Geogolog. Soc., will however throw some light on the subject of Kerg. Land ones I am sure; & I shall rout them out next week if I can. From Cockburn Isld.. (a conical volcano xtinct in 64 " 12′ S & 57 E.) I have rocks of 6 or 8 kinds, from Scoriæ[4] up to harder things, including Gneiss, & perhaps Granite, which must have been transported there: would you like a look at the Specimens?, if so I will leave them at the Geolog. Soc. & meet you there one day. I have laid out a piece of Kerg. wood to get cut & shall report to you when done. Also another very curious nodule, apparently containing fossils, from Cockburn Isld.s I enclose a little V. D. L. Fossil wood for your microscope to see the glandular tissue. Stretetski[5] will tell you all about it if you shd. see him—

I enclose a copy of the part of Gardners Ceylon letter,[6] mentioning the Europæan plants on the Mts.—*Gentiana prostrata* H.B.K. is the one with the xtraordinary range, which I have thus stated "In Europe it inhabits the Carinthian Alps, between 6000 & 9000 ft. In Asia Altai Mts in Lat 52. In America the tops of Rocky Mts Lat. 52° (where they reach 15–16000 ft.) & is also found on E. side of Andes of S. Am in 35 S. It descends to the level of the sea at Cape Negro in 53. & at Cape Good Hope in Behrings Straits in 68½N.—"[7] These are the authentic stations I have gathered together, there are doubtless lots more.

Anisotome is *not* the plant I was thinking of, in which the rudimentary female organ, though generally present, is sometimes wholly wanting. I cannot at present think what it is, but shall not forget the subject.

Brown certainly in Congo, seems to think the Banana indigenous only to East Asia,[8] but a Mr Ward;[9] a Mexican Consul, who dined here yesterday, assures me that it is wild there & ripens its fruit also—.— Lumley in Chancery Lane has all Flinders' voyage,[10] I shall be passing next week & ask the price. I cannot hear of Ann. Sc. Nat. to be picked up.— Sloane in Phil. Trans. mentions the seeds picked up on the coast of Ireland, which vegetated afterwards;[11] at least so says Gray of Brit. Mus. but I do not find it there. In Brown's Congo he says "I have no doubt that the nature of the integuments of the seeds of *Abrus precatorius* & *Guillandina Bonduc* would enable them to retain their vitality for a very long time in the currents of the ocean," & adds a note—Sir J. Banks received a drawing of *Guillandina Bonduc*, raised from seed found on W. coast of Ireland" (this is what I was thinking of.) also Linnæus is acquainted with similar instances of germination of seeds thrown on coast of Norway," vid. Coloniæ plant.—[12] We have seeds of *Entada Gigalobium* (old *Mimosa scandens*), thrown up on beach of Orkney & W. Ireland the seed is large flat 2 inches in diam. & comes from Carribees, it never germinated or was tried that I know of.

I send you for perusal Bot. Journal with Watson's Azores journal.[13] 3 Nos. of Tasm. Phil. Journal with Colenso on Caves & bones of N. Z. & other perhaps interesting papers.[14] A few numbers of Silliman[15] which are highly fossiliferous & contain other matters of interest.— Also one or two other little things I promised.

I am not yet prepared to give any further analogies between Juan Fernandez & St. Helena Floras', but shall remember the subject. *Anent* Mt. Floras I have nothing to say, but if I meet any account of those of the E Ind. Islds. shall remember it. all I can at present refer to are Gardner's Ceylon letter, (mentioned above) Wights[16] letters from Nylgherries alt. 5500 ft., who mentions incidentally Clematis, Circæa, Ranunculi, Geranium, Stellaria, Cerastium, *Docks*, Potentilla, a Rose, Galium, Rubia, Pedicularis, Osmunda Ophioglossum Vaccinium? Berberis.

Bishop Selwyn's Journal is at Hampstead,[17] I will send it you when returned.

The Quantity of Europæan Genera throughout the Andes, especially the Tropical one's, is quite amazing. I know of no other Mts more characteristically alpine in their Flora—Gentian's, Drabas & other Cruciferæ, Caryophylleæ are there as fully represented as in the Alps, & are characteristic of both.— I know of no materials for a comparison of similar Tropical heights with these, we know so little of the Alps of Java, even if they be high enough. A Botanist from the R. Gardens is now exploring the heights of Sta. Martha, which ought to bear on Humboldt's statement; when his collections come I will tell you the result.[18]

Schombugk[19] came here today P.P.C.[20] he meets Humboldt in Paris I wrote to the latter with some books my father had for him. I told him you were better wh. he will I am sure be glad to know. I asked Schombugk if he should see Dieffenbach to mention, *incidentally*, that your wood-cuts &c were not in England, am I right? It is no joke losing such xpensive things.—[21]

I send you some notes on Comparative Floras of N. Z. Terra Australis & S. Am West coast; it is very imperfect & I should be deeply obliged if would have the goodness to suggest how it could be clearer stated. I have reexamined many of the Lycopodium's & delivered a verdict in the Antarct. Flora of wh. I will send a proof sheet to Down when printed.[22]

I mention, in a letter home, that Robertson found a boulder of Syenitic granite in Kerg. Land.[23] I have come across a slight notice of the Stream of Stones[24] & am inclined to think I thought it more extraordinary at the time than I did a few nights ago, the lower part is very broad; I think I can hunt up some other account yet. What a curious thing if my ideas should change *per se*, like some species, in going from one country to another.

Ross wrote me yesterday that he had found some notes apparently mine!!![25] I have read Miss Martineau's mesmerism & would not engage that little Servant-maid with any character Miss M. would give her.—[26]

I have not yet compared the fossil leaves with the recent, but have laid out the latter to do so.

Galapago Flora is progressing

Henslowe spent parts of Tuesday & Wednesday here. Do you know of his finding 4 Whales ear-bones in the Crag.[27] Owen says 2 are types of N. Whales & 2 of South. Also Coprolites which are using as Guano from the abundance of Phosphate of Lime they contain.[28]

I hope M[rs] Darwin will do me the honor of accepting a copy of Backhouse's Cape & Mauritius book;[29] it, with the others, goes down by tomorrow's, Saturday's, coach. You need be in no hurry about returning them, The Sillimans' we send to V. D. Land in a few months. We always have two copies & if you care, we could regularly let you have a months reading of one of them.

With kind regards to Mrs Darwin Believe me Ever | Your's Jos D Hooker.

Tahiti & Owhyhee[30] comparisons are commenced.

The only Bourbon Flora is Bory's quatre-Isles de'Afrique[31] I will look at it.

DAR 100: 29–31

CD ANNOTATIONS

1.1 In . . . finds it.— 2.6] *crossed pencil*
3.5 I have . . . look 3.7] 'NB' *added and circled pencil*
3.7 at the . . . the M[ts]— 4.2] *crossed pencil*
5.1 *Anisotome* . . . subject. 5.3] *scored pencil*; 'My note wrong—'[32] *added pencil*
6.1 Brown . . . wild 6.3] 'Schomburgk' *added pencil*
7.1 I send . . . promised. 7.4] *crossed pencil*
9.1 Bishop . . . returned.] *crossed pencil*
10.3 in their Flora] 'in' *added pencil*; 'Flora' *crossed pencil and* 'Flora' *added in pencil above*
10.6 A Botanist . . . result. 10.8] *scored pencil*
11.1 Schombugk . . . printed. 12.5] *crossed pencil*
13.2 I have . . . progressing 16.1] *crossed pencil*
18.1 I hope . . . commenced. 20.1] *crossed pencil*

[1] For CD's notes on topics discussed during Hooker's visit to Down see Appendix III. Hooker refers to several of these topics in this letter.

[2] Susan Elizabeth Darwin, CD's sister.

[3] Robertson 1841, McCormick 1841 and 1842.

[4] Rough rocks formed at the cooling surface of molten lava.

[5] Paul Edmund de Strzelecki.

[6] George Gardner was superintendent of the botanic gardens, Ceylon, in 1844. A copy of part of Gardner's letter made by Hooker is in DAR 74: 141, accompanied by CD's notes on Gardner 1846. The copied excerpt describes European species on the mountains of Ceylon.

[7] J. D. Hooker 1844–7, p. 56.

[8] Robert Brown disputed Alexander von Humboldt's claim that there were species of banana in America (Brown 1818, pp. 469–71).

[9] William Robert Ward, chargé d'affaires in Mexico in 1843.

[10] Flinders 1814. The third appendix, on the botany of Australia, is by Robert Brown, naturalist to the voyage.

[11] Sloane 1696, p. 299 (misprinted in the volume as p. 399). Hans Sloane reported that one of these beans had been cast up on the coast of Kerry in Ireland, but did not mention their being grown in either Ireland or the Orkneys.
[12] Brown 1818, p. 481; Linnaeus 1768.
[13] Watson 1843–7.
[14] Hooker made William Colenso's acquaintance in 1841 while in New Zealand. The fossil bird *Dinornis* is discussed in Colenso 1843.
[15] Silliman's Journal: the *American Journal of Science and Arts* founded and edited by Benjamin Silliman.
[16] Robert Wight, superintendent of the botanic garden at Madras.
[17] Selwyn 1844, which had probably been lent to Hooker's aunt and uncle, Elizabeth and Francis Palgrave, who lived in Hampstead.
[18] Probably the claim, made in Humboldt 1814–29, 3: 494, that there are distinct species of the alpine genus *Bejaria* in the mountains near Caraccas, Bogota, and Santa Fé in Chile, which supported his general argument that the species present on alpine peaks are different from those of the surrounding plains.
[19] Robert Hermann Schomburgk.
[20] *Pour prendre congé.*
[21] Materials for the German edition of *Journal of researches* (1844), which Ernst Dieffenbach translated.
[22] J. D. Hooker 1844–7, pp. 112–17. See letter from J. D. Hooker, 8 November 1844.
[23] John Robertson was surgeon on board H.M.S. *Terror* during James Clark Ross's Antarctic voyage; Hooker was assistant-surgeon in the companion ship *Erebus*.
[24] In *Journal of researches*, p. 254, CD describes 'myriads of great angular fragments of . . . quartz rock' covering the bottom of the Falkland Islands' valleys as a 'stream of stones'. McCormick 1842, p. 27, mentions 'vast quantities of debris which have accumulated at the base of the hills'.
[25] James Clark Ross was provided with various materials by Hooker for his account of the Antarctic voyage, J. C. Ross 1847.
[26] Martineau 1845, p. 9. When her regular mesmerist could not attend the author, a maid successfully mesmerised her.
[27] The Crag is an East Anglian formation of shelly sand, traditionally placed in the Pliocene. The whale fossils are described in Henslow 1845a and R. Owen 1845. Henslow later retracted: the whale bones were from the much older deposits of Eocene London clay (Henslow 1847). Henslow had informed CD of his discovery (see *Correspondence* vol. 2, letter from J. S. Henslow, 17 October 1843).
[28] Henslow 1845b.
[29] Backhouse 1844 describes visits to colonial mission stations.
[30] Hawaii.
[31] Bory de Saint-Vincent 1804.
[32] See Appendix III, CD's notes of 8 December 1844, in which rudimentary and abortive organs of an unidentified umbelliferous plant are discussed.

To J. D. Hooker 16 [December 1844]

Down Bromley Kent
Monday 16

My dear Hooker

Really I do not know how to thank you half enough for all you have done for & sent to me: I might with truth do so for every single paragraph in your letter & every one volume. My wife begs to be most kindly remembered to you & she sends her best thanks for your valuable present.

I have not a quarter studied your botanico-geographical letter; & as I have to go to London tomorrow, I fear I must keep it for rather more than a week.— I will, then, also, commence first with the *lent* books, which will take me some little time, as I seldom am able to stand more than one hour's scientific reading.— I began the Tasmanian J: last night & was astonished at its interest.

To get all your geographical facts in one's head will be a hard task; I trust that your sketch[1] will not have caused you ultimately loss of time, as, judging by myself, preliminary sketches & resketches do much good. Your remarks are exactly the thing, which ever since being in Tierra del Fuego, I have felt a keen curiosity about, & have often complained to Henslow, how rarely I c^{d} find any such general remarks in Botanical works—I am far from a competent judge, but I cannot doubt, that your generalizations will be a most valuable & permanent gift to science. I cannot doubt that many others will be as much interested, as I am, in seeing all your results worked out.

Seriously I almost grieved, when I saw the length of your letter, that you sh^{d} have given up so much time to me,—Sir William will think me a bad friend to you—but anyhow, I trust, the sketch-part, of your geographical results, will not turn out lost time— When I return I shall have to learn, read & digest, & afterwards I will write my thanks again; for anything beyond my hearty thanks, I do not think I shall have to send.

Ever yours | C. Darwin

DAR 114.1: 23

[1] CD seems to be referring to some notes on the floras of New Zealand, Australia, and South America enclosed with the previous letter from Hooker, 12 December 1844.

To William Darwin Fox 20 December [1844]

Down near Bromley | Kent
Dec. 20th

My dear Fox

I was on the point of going to London when your note arrived, whence I returned yesterday, from visiting Erasmus with whom Susan staying. She begs me to say, (which I c^{d} have said) that my Father, she is sure will be very glad to see you at any time, whennever convenient to yourself, when you will write to propose yourself, & if my Father is not then very well, he will have no scruple in deferring your visit. My Father's health, I grieve to say, is now *very* uncertain: he has just lately been very well, but about 3 weeks since he had rather a severe attack upon his chest. He has been prevailed upon to sleep down stairs, but it was most painful to him, thus to give up one step in life. He was upon the whole very cheerful when I was there.— Illness with his figure & constitution is very dreadful.[1]

With respect to mesmerism, the whole country resounds with wonderful facts or tales: the subject is most curious, whether real or false; for in the latter case, what is human evidence worth? I am astonished at your zeal, with respect to Miss Martineau,[2] for to my mind, the girls case[3] bears more plainly, than I sh^{d} have expected, the mark of deception, possibly involuntary. You are no doubt aware that many doctors (for example D^{r}. Holland[4]) have long ago remarked how marvellous a diseased tendency to deception there is in disordered females. Sh^{d} your zeal still continue, I w^{d} write to Miss Martineau & propose your visiting her (my Brother wishes to avoid all communication with her on this subject)— When in London, I saw a letter from her (not to my Brother), in which she says **crowds** of people are coming to her from all parts of England; *she does not seem to dislike this*, but she says she is going very soon to leave Tynemouth for rest from visitors; so you w^{d}. have to go at once if you do go. I have just heard of a child 3 or 4 years old (whose parents & self I well know) mesmerized by his father, which is the first fact, which has staggered me.— I shall not believe, fully, till I see or hear from good evidence of animals (as has been stated is possible) not drugged, being put to stupor; of course the impossibility w^{d} not prove mesmerism false: but it is the only clear *experimentum crucis*, & I am astonished it has not been systematically tryed.

If mesmerism was investigated, like a science, this c^{d} not have been left till present day to be *done satisfactorily*, as it has been, I believe, left.— **Keep** some cats **yourself** & do get some mesmerizer to attempt it. One man told me he had succeeded, but his experiments were most vague, as was likely from a man, who said cats were more easily done, than other animals, because they were so electrical.!! (Miss Martineaus case was tumor, I presume, of the womb or Ovaria. the tumor is reduced greatly & her deafness decreasing: she is in a most excited state. My Father has often known Mania relieve incurable complaints.)—[5]

As long as you like to receive the Athenæums, I shall have *pleasure* in sending them. Sh^{d}. you happen to call this winter at the House, in which the "Rippons & Burtons (R. 389) Vesta Stove" is, w^{d} you be so kind as ask whether they continue to like it. 2^{d}. whether it often wants feeding. 3^{d} whether it gives out much heat, enough to warm a **cold** House. I must get a new one: & I am confounded with so many recommendations.—

Pray excuse this very untidy note, written in a Hurry & | believe me | ever yours | C. Darwin

Postmark: 20 DE 20 1844
Christ's College Library, Cambridge (Fox 70)

[1] On his 80th birthday (30 May 1846) Robert Waring Darwin weighed 22 stone 3½ pounds ('Weighing Account' book, Down House MS).

[2] Harriet Martineau's health had broken down in 1839. In 1844 she undertook a course of mesmerism, after which her health was dramatically restored. She published an account of her cure in the *Athenæum* (Martineau 1844) and subsequently published it in *Letters on mesmerism* (1845). See Arbuckle ed. 1983, p. 78.

[3] Jane Arrowsmith, a partially blind 19-year-old girl, thought by Harriet Martineau to be clairvoyant. According to Martineau, Jane, while in a mesmeric trance at a séance on 15 October 1844, predicted the safe return of the crew of a wrecked ship. See R. K. Webb 1960, p. 229.

John William Lubbock. By Thomas Phillips, 1843.
(Courtesy of the University of London.)

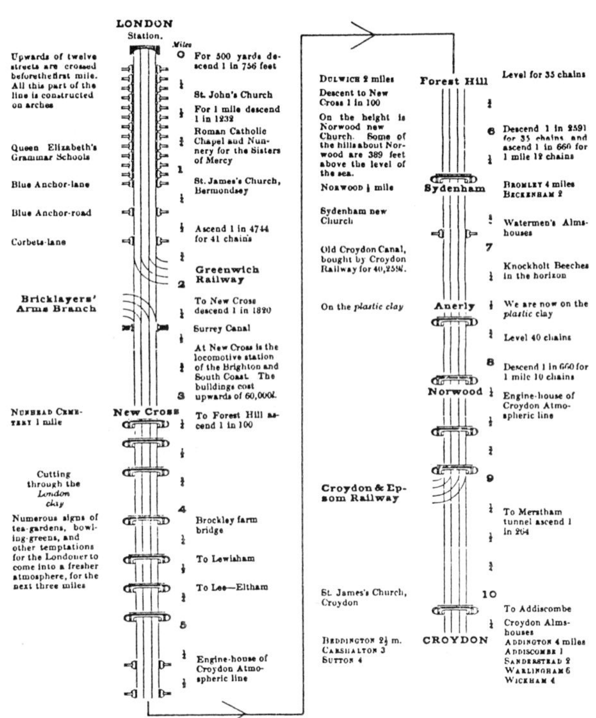

The Bricklayers' Arms Station. London terminus of the Croydon Railway, 1844–
(Courtesy of the Victoria and Albert Museum, London.)

Detail of a railway travelling chart, *c.* 1846. Issued by the *Railway Chronicle.*

[4] Henry Holland.

[5] Harriet Martineau suffered from 'prolapse of the uterus and polypous tumours' (R. K. Webb 1960, p. 193). She had been deaf from the age of 20 (p. 6).

From W. B. Carpenter 21 December 1844

Ripley—Surrey.
Dec.r 21—1844

Dear Sir

I have duly received your specimens, and have had several sections made of the fragment of the Pampas deposit, with one or two of the Chilian tufa for comparison— I fear that the result will not be very satisfactory to you, as not affording the evidence you desire.[1]

Notwithstanding the very close conformity in *aspect* between the Pampas specimen and the Chilian tuff,—or, rather, that portion of the latter which connects together the fragments of shell,—their minute structure is very different. The substance of the latter appears to be composed of very minute fragments of Shell, or of some other *organic structures*; it possesses a considerable degree of transparency in thin sections; and is destitute of any kind of *mineral structure*. The former, on the other hand, possesses more of the Oolitic structure; being composed of minute rounded grains, the spaces between which are filled up with amorphous matter. But it differs from the *true* Oolite in this,—that the sections of these granules do not present any of that concentric arrangement (precisely resembling that of Calculi) which the true Oolite exhibits; and that the spaces between them are not filled up with transparent crystalline matter, but with a peculiarly opake substance— Hence I should infer that the materials of the deposit had been subjected to long-continued attrition; and not by a process of alternate solution and deposit, such as appears to have taken place in the true Oolite— I cannot find in the Pampas deposit the least vestige of any organic structure, except some very minute fragments, which seem to be vestiges of Sponges or Alcyonia,—judging by the spicula they contain. If this shd be a point of any interest to you, I will send the sections to Mr Bowerbank for his examination,—that being a structure with which he is especially familiar.— I may add, that the Pampas deposit is of much closer texture than the Chilian tuff, and much more difficult to rub down. Its opacity is such, that it is only when brought to an extreme tenuity, that it begins to be translucent.

As I saw that the specimens were ticketed, I did not like to knock them to pieces for the purpose of trying different parts; but if you can allow me to do this, I will make further investigations

I am, dear Sir, | very sincerely yours | W B Carpenter

Chas Darwin Esq.

DAR 39.1: 33–5

CD ANNOTATIONS
1.1 I have . . . desire 1.4] *crossed pencil*

[1] CD reported Carpenter's observations in *South America*, pp. 76–7, having first asked for more detailed information (see letters from W. B. Carpenter, 2 January [1845] and 5 May 1845). For other microscopic examinations of CD's specimens, see the correspondence between CD and Christian Gottfried Ehrenberg.

To J. D. Hooker 25 December [1844]

Down Bromley Kent
Dec. 25th. Happy Christmas to you—

My dear Hooker

I must thank you once again for all your documents which have me interested me very *greatly* & surprised me. I found it very difficult to charge my head with all your tabulated results, but this I perfectly well know is in main part due to that head not being a Botanical one, aided by the tables being in M.S.[1] I think, however, to an ignoramus, they might be made clearer; but pray mind, that this is very different from saying that I think Botanists ought to arrange their highest results for *non*-botanists to understand *easily*. I will tell you, how for my *individual self*, I shd. like to see the results worked out, & then you can judge, whether this be adviseable for the Botanical world.

Looking at the *Globe*, the Auckland, Campbell I, New Zealand & Van Diemens so evidently are *geographically* related, that I shd wish, before any comparison was made with far more distant countries, to understand their floras, in relation to each other; & the southern ones to the northern temperate hemisphere, which I presume is to everyone an almost involuntary standard of Comparison. To understand the relations of the Floras of these islands, I shd like to see the group divided into a northern & southern half, & to know how many species exist in the latter

1 belonging to genera *confined* to Australia Van Diemens land & North N. Zealand
2 ---------------------- found only on the mountains of do.
(??) ---- 3. ———————— { of distribution in many parts of the world. (ie. which tell no particular story.)
4. ———————— { found in the northern hemisphere & *not* in the Tropics; *or only* on mountains in the Tropics

I daresay all this (as far as present materials serve) cd. be extracted from your tables, as they stand; but to anyone not familiar with the names of plants, this wd be difficult. I felt *particularly* the want of not knowing, which of the genera are found in the lowland Tropics, in understanding the relation of the antarctic with the artic Floras.

If the Fuegian Flora was treated in the analogous way, (& this would incidentally show how far the Cordillera are a high-road of genera.) I sh[d] then be prepared far more easily & satisfactorily to understand the relations of Fuegia with the Auckland Is[d] &c; & consequently with the mountains of Van Diemens Land. Moreover, the marvellous fact of their intimate Botanical relation between Fuegia & the Auckland Is[d] &c would stand out **more** *prominently*, after the Auckland Is[d] had been first treated of under the purely geographical relation of position. A triple division such as yours, w[d] lead me to suppose that the three places were somewhat equally distant, & not so greatly different in size: the relation of Van Diemen's land seems so comparatively small, & that relation being in its alpine plants, makes me feel that it ought only to be treated of as a subdivision of the large group, including Auckland, Campbell, New Zealand.

In Art VII—does the expression "more *remarkable* genera" (& sub-sections of genera) mean those *confined* to the stated countries?— Art VI does not appear to me clear, though I now understand it.

I think a list of the genera, common to Fuegia, on the one hand & on the other to Campbell &c & to the mountains of Van Diemens Land or New Zealand, (*but not found in the lowland temperate, & S. tropical parts of S. America & Australia, or New Zealand*), would prominently bring out, at the same time, the relation between these antarctic points one with another, & with the northern or arctic regions.

In Art III. Is it meant to be expressed, or might it not be understood by this article, that the similarity of the distant points in antarctic regions was as close as between distant points in the Arctic regions? I gather this is not so.— You speak of the southern points of America & Australia &c being "materially approximated" & this closer proximity being corelative with a greater similarity of their plants: I find on the globe, that Van Diemen's Land & Fuegia are only about $\frac{1}{5}$ nearer than the whole distance between Port Jackson & Concepcion in chile;—and again that Campbell Isl[d] & Fuegia are only $\frac{1}{5}$ nearer than the East point of North N. Zealand & Concepcion: now do you think in such immense distances both *over open oceans*, that one fifth, less distance say 4000 thousand miles, instead of 5000, can explain or throw much light on a material difference in the degree of similarity in the Floras of the two regions.—

I trust you will work out the New Zealand Flora, as you have commenced at end of letter: is it not quite an original plan?— & is it not very surprising that N. Zealand, so much nearer to Australia than to S. America, sh[d] have an intermediate flora; I had fancied that nearly all the species there, were peculiar to it.— I cannot but think you make one gratuitous difficulty in ascertaining whether New Zealand ought to be classed by itself, or with Australia or S. America,—namely when you *seem* (bottom of p. 7. of your letter) to say that genera in common, indicate only that the external circumstances for their life are suitable & similar. Surely can not an overwhelming mass of facts be brought against such a proposition: distant parts of Australia possess quite distinct species of Marsupials, but surely this fact of their having the same marsupial genera, is the *strongest* tie & plainest mark of an original, (so called) creative affinity over the whole of Australia; no one, now, will (or ought) to say that the different parts of Australia have something in the external conditions in common, causing them to be preeminently suitable to Marsupials; & so on in a thousand instances. Though each species, & consequently genus, must be adapted to its country, surely adaptation is manifestly not the governing law in geographical distribution.— Is this not so? & if I understand you rightly, you lessen your own means of comparison by attributing the presence of the same genera to similarity of conditions.

You will groan over my **very full** compliance with your request to write all I could on your tables, & I have done it with a vengeance: I can hardly say how valuable I must think your results will be, when worked out, as far as the present knowledge & collections serve.

Now for some miscellaneous remarks on your letter: thanks for the offer to let me see specimens of boulders from Cockburn island; but I care only for boulders, as an indication of former climate:[2] perhaps Ross will give some information: I hope you will write to N. Zealand on this subject. I see that there are cases in Van Diemen Land, which ought to be explored.—[3] Lieut: Britton speaks of a Fern above the coal of V. Diemen's Land, as being allied to recent Ferns of Tasmania:[4] did you collect any of the Coal-Plants there; I sh^{d} like to know something about them. **Many** thanks for your message to Dieffenbach.

I shall have done with the lent Books in about a week or 10 days & will return them by Deliverance company. They have interested me much, & I have ordered two of the numbers of the Journ of Bot, so that I shall return the new uncut copies. I was also very *glad* to see Silliman, as one or two Papers bore on my *present* geological writing. Thank you kindly for your offer of letting me see Silliman regularly; but I will not accept it; as I prefer reading each Journal in volumes, bound. I called on Bailliere[5] & he tells me the old series of Ann: des Sci. Nat: w^{d} cost 10£, & the new 20£: this is more than I can afford, though I sh^{d} much like to have them; so that sh^{d} you ever hear of a cheaper set, I sh^{d} be greatly obliged if you w^{d} inform me.— It will be extravagant to buy Flinders, without it be very cheap, as, upon reflection, I remember it is so easily borrowed from Public Librarys.—

Watson's Paper on Azores has surprised me much;[6] do you not think it odd, the fewness of peculiar species, & their rarity on the alpine heights: I wish he had tabulated his results: c^{d} you not suggest to him to draw up a paper of such results, comparing these isl^{d} with Madeira; surely does not Madeira abound with peculiar forms? A discussion on the relations of the Floras, especially the alpine ones, of Azores, Madeira & Canary Is^{d} would be, I sh^{d} think, of general interest:— How curious the several doubtful species, which are referred to by Watson, at the end of his Paper; just as happens with birds at the Galapagos.—[7] By the way, I see Pœppig states on the authority of Bertero, that Juan Fernandez has not (ie is not related to) a Chilian Flora, but a South Sea one:[8] surely this must be an error.—

What a proser Pœppig is!— I have half read through, with pleasure, the Norway tour;[9] how pleasantly it is written.

Any time that you can put me in the way of reading about Alpine Flora, I shall feel as the *greatest kindness*: I grieve there is no better authority for Bourbon, than that stupid Bory: I presume his remark that plants, on isolated Volcanic isl[ds] are polymorphous (ie, I suppose, variable?) is quite gratuitous.[10]

Farewell, my dear Hooker. This letter is infamously unclear, & I fear can be of no use, except giving you the impression of a Botanical ignoramus.— Ever yours | C. Darwin

DAR 114.1: 24

[1] CD refers to Hooker's notes on southern floras, enclosed in the letter from J. D. Hooker, 12 December 1844. The notes have not been found.

[2] That is, their possible distribution by icebergs during previous cool conditions.

[3] CD had just read volumes one and two of the *Tasmanian Journal of Natural Science*, lent to him by Hooker, see letter from J. D. Hooker, 12 December 1844, and 'Books Read' (DAR 119, entry for 25 December 1844; Vorzimmer 1977, p. 132). CD's notes on volume two, commenting on Colenso 1843 and Breton 1843, are in DAR 205.3: 108. CD may be referring to McCormick 1842 or Colenso 1843.

[4] Breton 1843, p. 135.

[5] Hippolyte Baillière, London dealer in French medical and scientific books.

[6] Watson 1843–7.

[7] Watson 1843–7, 2 (1843): 407–8.

[8] Pöppig 1835, 1: 288.

[9] In CD's list of 'Books Read' (DAR 119; Vorzimmer 1977, p. 133) there is an entry for 30 January 1845: 'Laings Tour in Norway', probably Laing 1836.

[10] Bory de Saint-Vincent 1804, 3: 161–6.

From J. D. Hooker 30 December 1844

West Park Kew
Decr. 30. 1844.

My dear Darwin

When I sent you my crude notes I had no idea of imposing, by my importunity, the task you have so kindly performed, I do not know whether I was more ashamed at my own conduct, or pleased with its consequence, when I read your kind letter. I do indeed thank you very much for your hints, most of which must I plainly see be acted up to & I do not doubt all, when I have time to do something more to the subject. Botanizing in narrow Geog. limits does give one very narrow views, which I shall hope to get over as I proceed to the other floras' of the South. The material approximation of S. New Zeald & Ant. Amer. amounts of course to a nonentity, but when once carried away with the discovery of their being a certain Botanical similarity between the two, I fancied what did not exist. Nothing was further from my intentions than to have written any thing which would lead one to suppose that genera common to two places indicate a similarity in the external circumstances under which they are developed, though I see I have given you excellent grounds for supposing that such were my opinions: it will puzzle you to see how, but that identical sentence was more intended to express the very contrary, I will set it right before I do any thing more to that subject, which cannot be before I get the cryptogamic part of Ld Aucklands & Campbells Isld. finished.

I am not in the least in want of the books you have, nor shall be for a month, so pray do not hurry with them. I do hope you did not order the numbers of L.J.B.[1] because I sent you uncut ones, as the cutting can make no difference in the world, we have not a wholly uncut copy in the house of any duplicate part. Bailliere is a shark, I am going to Paris if possible in Feb^{y}, & will enquire about the Annales. I asked Brown about Flinders' & he has put me on the scent of a copy of his part for 10/ or 15/ which I shall buy at any rate & send to Colenso if you do not think it worth that to you. I w^{d} certainly not give over 20/ for the whole work or 30/ at the outside.

The paucity of peculiar Azorean species is very strange & more particularly the want of W. Ind. or N. Am. forms, though the current washes up canoes (if all be true) on their shores. I have written to ask Watson.[2] I doubt if Madeira *abounds* with peculiarities; Plants have such wide ranges, especially over some Islds. that we are forced to look on a few peculiar species (proportionally to the whole) as constituting a peculiar flora, in many instances. I have been greivously at a loss to get any thing about the J. Fernandez flora, without *going through* the Herb. Hook.!—[3] I am not inclined to believe Pœppig & Bertero. As to Bory I should think on looking over my list of St Helena plants that they were remarkably *non* polymorphous.—?did he know any other flora's at all or even that he argues from.

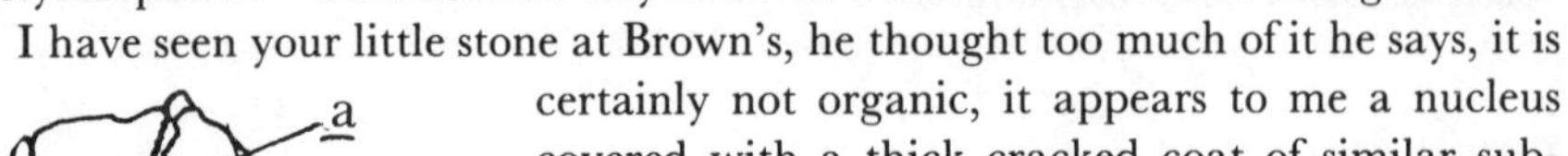

I have seen your little stone at Brown's, he thought too much of it he says, it is certainly not organic, it appears to me a nucleus covered with a thick cracked coat of similar substance, & the cracks are filled up with also similar substance, which projects at the lips of the cracks, forming the anastomozing ridges: cavities, as at (*a*), occur between the nucleus & its coat, not filled up with the deposition which filled the cracks & *there* are small chrystals, it is nearly homogenous & all chalcedony.[4]

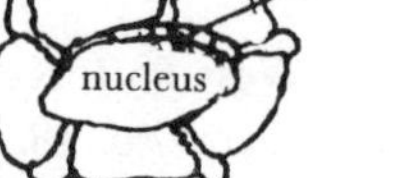

Brown seems to haul out of the V. D L. fossils being Eucalypti, but will give no opinion. I have taken one lump to the polishers to cut the apparent petiole of a leaf, which sticks out of the specimen. My Kerg. wood is coniferous, not of Arau*caroid* structure however. The V. D L. wood is imbedded in solid basalt & Streletski says the trees were imbedded in fluid lava in that & other cases. He believes that similar geological formations, of the same era, have similar floras, & instances Illawarra,[5] two other very remote spots of Australia & a spot in V. D. L. all isolated, all of greenstone, (which he knows for many reasons to have been erupted on the same day) as containing precisely the same plants. The said rocks are the same by chemical analysis, but though other rocks are also the same in composition they have not the same flora because not erupted at the same time. If true this would argue that each geol. era was peopled per se & of course that the same species was planted in two places at once: but I doubt Streletskis means of judging of the floras.— I think I quote him properly, his book appears in February—[6]

I have been delighted with *Vestiges*,[7] from the multiplicity of facts he brings together, though I do [not] agree with his conclusions at all, he must be a funny fellow: somehow the books looks more like a 9 days wonder than a lasting work: it certainly is "filling at the price".— I mean the price its reading costs, for it is dear enough otherwise; he has lots of errors. Have you read "MaCullochs proofs & attributes"?[8] After all what is the great difference between *Vestiges* & Lamarck, whom he laughs at.[9] In one places he implies that species are made by the *will of the mother*,[10] under which I wonder he does not quote a subject I have lately been struck with, & that is, the real or apparent effect that a mental emotion of the mother may have on her unborn offspring.— I thought till lately that all these nursery stories were laid upon the shelf; but have lately heard some staggering circumstances related. All Sealers have told me that the young, taken out of the clubbed mother, bear similar club marks on their heads & this they swear to. If you care to hear any thing on the subject I will go on at some future time. Do not think I am arguing this for the developement of species!—

I have drawn up the Sandwich Isld & Society Isld lists,[11] they are woefully imperfect, the representative species are very few, there is a certain similarity between them, from both containing, in common with other Pacific Islds, many plants apparently derived from the E. Indies i.e. from Asia. The Sandwich Isld Flora is by far the most peculiar & the least tropical, perhaps the most allied to the American. Its situation under the influence, I suppose) of the cold winds of the two continents, combined with its own high snow clad Mts. must make its climate different from that of Tahiti. As far as materials go, the several Islds of the Sandwich present numerous instances of representative species, nothing in this respect is known of the Society. Shall I send you the lists?— The Sandwich is certainly a very peculiar flora. I am reading up *the Pacific*, to the end of investigating its flora & that of the coral & other Islds.: more particularly to know what plants are best fitted for transportation, whose names I am getting together. I often come across bits I think you might want to know of, if you would tell me any subjects you particularly want information on I can just as easily as not note them, premising I shall do it highly ignorantly, such as occurrence of coral blocks high above water mark.[12] These books do make me earnestly long to go to those Islds. Colenso sent some more *Dinornis* bones, I took them to Owen who found they were proofs of two of his species (which were dubious) being good, there was no head.[13] I have written to my "plains of India" friend[14] about angular (sharp), boulders, can I ask him any thing else: he is son of Dr Thomson prof of Chemistry & a good Naturalist. I am getting confused in considering the multiplicity of ways that have been proposed for peopling our globe with plants, as it now appears. McCulloch argues for a developement continued up to the present moment! & double creation of the same species ad libitum, which is taking a very sharp knife to the Gordian knot.

Galapago flora goes on well, I have stuck at a highly curious new genus, amongst the supplements. I had occasion to grub up some Cape de Verd's & thought, when at it, I would name one Nat. Ord. took Malvaceae, one of the largest, found 10 species almost all common not only to Africa but also to W. Indies & so gave up in disgust, feeling I had made a fool of myself in ever supposing them any way peculiar: what difference there is between the Islds must depend on local causes.

I shall not forget your Alpine Floras but have nothing to add at present

Many many happy returns of the season, I hope that each may find you more fully restored to perfect health. | Ever your's most truly | Jos D Hooker.

DAR 100: 32–4

CD ANNOTATIONS

1.1 When . . . outside. 2.9] *crossed pencil*
4.1 I have . . . related. 6.12] *crossed pencil*
7.11 I am . . . other Islds.: 7.12] *pencil cross in margin*
7.11 I am . . . supplements. 8.2] *crossed pencil*
7.24 McCulloch . . . knot. 7.26] *scored pencil*
9.1 I shall . . . Hooker. 10.2] *crossed pencil*
On cover: '10s for Brown | Annales | pleasure to write | Streletski with one of the fossil Kangaroos has told him, how can he pretend to know; *I think it [*interl*] has it not been shown that geolological Composition of the under rocks, & indeed of the soil has a far less effect on distribution. his whole assertion sounds very rash.
I was less pleased with the Vestiges.: *though admir written [*interl*] his geology is [*over* '&'] bad & zoology far worse
Effect of imagination, thanks for | cow | =Sealers oath= | (delighted to hear you are working out Pacific Isd) | I shd like to see the lists of Sandwich at any time= | Henslow Keeling Isd.— Lesson— | ((Coral-facts))' *pencil, crossed pencil*

[1] The *London Journal of Botany*.
[2] Hewett Cottrell Watson.
[3] The combined collections of William Jackson Hooker and Joseph Dalton Hooker, the most extensive herbarium in Britain at that time.
[4] This was referred to in the letter from J. D. Hooker, 12 December 1844; agate is a form of chalcedony.
[5] New South Wales.
[6] Strzelecki 1845.
[7] [Chambers] 1844.
[8] MacCulloch 1837. For CD's notes on MacCulloch see Gruber and Barrett 1974, pp. 414–22.
[9] [Chambers] 1844, p. 231.
[10] [Chambers] 1844, pp. 218–9.
[11] The Sandwich Islands, now called the Hawaiian Islands; the Society Islands include Tahiti.
[12] *Coral reefs*, pp. 131–4, where CD explained exposed coral rocks as the result of elevation.
[13] Colenso 1843. *Dinornis*, an extinct New Zealand bird, was first described by Richard Owen in 1839 from a single bone. A more complete description based on additional specimens was given in

R. Owen 1843b. William Colenso's specimens may be those listed in Flower 1879–91, 3: 430, as metatarsal bones having been presented by 'J. Colenso Esq.' The skull of *Dinornis*, an important classificatory feature, was not described until R. Owen 1846a.
[14] Thomas Thomson.

From Woodbine Parish [1845?]

[Sends names of species found in banks of marine shells near Buenos Aires. Shells identified by G. B. Sowerby. (See *South America*, pp. 2–3.)]

A memorandum 2pp
DAR 43.1: 56a–57

To G. B. Sowerby [1845?][1]

Down near Bromley Kent
Saturday

My dear Sir

I hope to bring the shells (if it be convenient to you) *about* eleven oclock, on next Thursday morning.— If you will not be at home, would you kindly send me a line by return of Post: if I do *not* hear, I shall understand, that my visit will be convenient to you.—

I hope I shall not fail this time, but my engagements *must* always be uncertain, owing to the great uncertainty of my health.

Believe me | dear Sir | Yours very faithfully | C. Darwin

Muséum National d'Histoire Naturelle (Cryptogamie), Paris

[1] The conjectured date is based on several lists of shells in the Darwin Archive dated 1845 by George Brettingham Sowerby. During that year CD was at work on both *South America* and *Journal of researches* 2d ed., and for both he consulted Sowerby on shell identifications and descriptions.

To David Thomas Ansted, assistant secretary, Geological Society of London[1] [*c.* January 1845][2]

Down n^r^ Bromley | Kent
Sunday

My dear Sir

I have written to Mr Lonsdale[3] about the descriptive catalogue of Fuegian specimens.—

Mr. Lonsdale says he remembers well the *large* boxes being exposed, & that they were taken below in the crpt, & are labelled outside either "Tierra del Fuego" or "Patagonia" or "S. America" presented by either "Capt. King" or "Capt.

FitzRoy". He thinks probably that the Catalogues are inside one of these boxes & on the top of the specimens.

Would you be so kind as to direct Charlton[4] to search for these boxes, which are large & heavy, for I well remember them.— Mr. Lonsdale says that if not there, the catalogue is probably in some table-drawer or cupboard *in the upper museum*.— & if not there, must be *together* with several other catalogues, belonging to the specimens in the upper museum.—

As it could not take long to have these places searched, would you oblige me by having this done by Wednesday.—[5] If I cannot consult the catalogue soon, it will be useless to me.—[6]

I hope you have finished the labour of the first number of the Journal; I shall be exceedingly anxious to see it, & hope I may able to get one at the Society.

Pray believe me | My dear Sir | Your's very sincerely | C. Darwin

Smithsonian Institution (Archives, George P. Merrill photograph collection, R.U. 7177)

[1] Ansted served as assistant secretary of the Geological Society from 1844 to 1847 and edited the *Quarterly Journal of the Geological Society* for many years.

[2] The date is based on the reference to the appearance of the first issue of the *Quarterly Journal of the Geological Society* in February 1845.

[3] William Lonsdale, assistant secretary and librarian of the Geological Society from 1838 to 1842.

[4] Isaac Charlton, the house steward of the Geological Society (Woodward 1907, pp. 243, 309).

[5] CD attended council meetings of the Geological Society on 8 January and 5 February, either of which may have been the Wednesday referred to (see Appendix II).

[6] CD completed the manuscript of *South America* on 25 April 1845 (see 'Journal'; Appendix II).

From W. B. Carpenter 2 January [1845]

Ripley—Surrey—
Jan 2—1844[1]

Dear Sir

I ought to have written to you, ere this, in answer to your last note; but was unwilling to do so, until I should have subjected the specimens to another careful investigation, under a higher power. This I have been prevented from doing until last night; and the result is such, as to make me wish to make a further examination of the Pampas tufa, before expressing a decided opinion respecting it.— As to its non-identity with the Chilian tufa, I am equally clear as before; but I have detected in it some additional traces of distinct organic forms, which do not correspond with anything I have elsewhere seen. My impression is, that they are remains of Spongeous bodies. As you have been good enough to allow me to "work my wicked will" with the specimens, I shall make a thorough examination of them, and I hope to be able to arrive at some definite conclusions.[2]

I can quite relieve your scruples about the expense of preparing the sections, by telling you that they are made by a boy who acts as a servant in my small establishment, and the share of whose time devoted to these objects is amply covered by the grant of the Association;[3] under the purposes of which I consider that the examination of your specimens most legitimately comes. I am hoping to induce the Government to take up the subject in connection with the Geological department of the Ordnance Survey; and shall be very glad of your good word, if you have an opportunity of speaking it.

I am, Dr Sir | yours very truly | W B Carpenter

DAR 39.1: 31–2

[1] An error for 1845. This letter clearly follows the letter from W. B. Carpenter, 21 December 1844.
[2] See letter from W. B. Carpenter, 5 May 1845, for Carpenter's more detailed report on the comparison.
[3] Carpenter had received a grant from the British Association for research on the microscopic structures of fossil and recent shells (*Report of the 14th meeting of the British Association for the Advancement of Science held at York in 1844*, p. xxiv).

To J. D. Hooker [7 January 1845]

Down Bromley Kent
Tuesday

My dear Hooker

I will send back the books, which have much interested me, on Thursday by our carrier & so by the Deliv: Comp: to you: I did not buy the numbers of Bot: Journ. because I had cut the leaves, but because, I was interested by several of the papers: I sent for the 15 & 20th No^s, but Bailliere has sent me the 19th instead of 20th, & as it w^d have cost almost the price of another number to have sent it back & got it changed, I have thought you would excuse me sending two numbers of XIX, & as you *must* often have communication with Bailliere you c^d without trouble change it for the XX;—so that, like a wise man, I have saved myself trouble,—by giving it to another.

Have you in your Library Capt: Porter's Voyage in the Essex (in the Pacific) I have long wished, but never been able, to see it?[1] Would you oblige me, by looking at the two or three specimens, with the books, of a tertiary modern sandstone of T. del Fuego, in which there are leaves, & which I thought, when collecting them, were of the Beech: w^d you give me your opinion on this? You can keep the specimens in perpetuity, or send them back again any time, you may have any other parcel.

Thirdly (& lastly, you will say, gracias a dios) w^d you make sense for me out of the following note:

"Mr Brown tells me, that the section (of silic: wood) in the direction of the medullary rays, has the discs in a double row, placed alternately & not opposite as in the common (?)*Araucaria*, & therefore in this respect would, according to *Nicol*(?) be said to have the (?)*Araucarian* structure."

I presume the word Araucarian is miswritten.[2]

I do not know, whether you will think it worth while to refer to it, but Pœppig in his Reisen Band 1. p. 367. has a list of genera, from Cordillera by Concepcion & remarks on their Europæan-Alpine character, with relations to T. del Fuego, the Tropics, & Australia.—[3]

I am glad to hear that you are going to Paris & hope that you may enjoy it; thank you much for your offer of enquiring about the price of the Annales.— With respect to Brown in Flinders, I sh^d^ not like to give more than 10^s^, as, though in itself so valuable, it is easily procurable.— I shall be curious to see Streletski's book, though how he is to tell that ancient eruptions happened on the same day, I cannot see or believe: your account of his views seems wild enough.—[4]

I have, also, read the Vestiges, but have been somewhat less amused at it, than you appear to have been: the writing & arrangement are certainly admirable, but his geology strikes me as bad, & his zoology far worse.[5] I sh^d^ be very much obliged, if at any future or leisure time, you w^d^ tell me on what you ground your doubtful belief in imagination of a mother affecting her offspring. I have attended to the several statements scattered about, but do not believe in more than accidental coincidences. W. Hunter[6] told my Father, then in a lying-in-Hospital, in many thousand cases, he had asked the mother, *before her confinement* whether anything had affected her imagination & recorded the answers, & absolutely not one case came right, though, when the child was anything remarkable, they afterwards made the cap to fit. Reproduction seems governed by such similar laws in the whole animal kingdom, that I am most loth (& sh^d^ require much better evidence, than the oaths of all sealers) to believe that in *mammifers*, there is so intimate a connection between the embryo & mother, that the latter's mind can affect the former, whereas in *Fish* this is impossible, & in *Birds* not probable.

I am delighted to hear that you are attacking the Pacific Flora; I am unwilling to trouble you, but I sh^d^ like just to glance over the lists of Society & Sandwich Isl^d^., though I do not care much about it, as hereafter, I shall be able, thanks to all the Saints, to study the whole subject in your works.— Take care of yourself, though you have been a Doctor; it strikes me that you must be doing too much.— With respect to coral plants, I suppose you know that Henslow has described my Keeling Plants in the Annals.—[7] Lesson in the Voyage of the Coquille, has some lists; as, also, I *think* has Lutke (his book if you do not know it, is in the Geograph: Soc:)[8]

Thanks for your offer of collecting facts about coral-reefs, but I will not trouble you, as I shall never publish a second Edition.;[9] *but* sh^d^ you meet anything about *subsidence* of the land in the Pacific, or about Elevation in out-of-the-way-Books, ⟨I⟩ sh^d^ be very much obliged if you w^d^ note it.

Why do you speak, as if writing to you was a '*task*' to me; it is a great pleasure, & I only sh^d^ be better pleased, if I thought my random observations could possibly be quarter of the interest to you, which you are pleased to say they are.— I am often frightened for you, when I think how hard you must be working: the time was, when I thought that speaking about too much work was a chimera—

Farewell. Ever yours | C. D.

N.B. I have enclosed my rough note from Pœppig, (if you can read it) which please to return: the translation may not be very accurate.

Postmark: JA 7 1845
DAR 114.1: 25

[1] D. Porter 1823.

[2] CD presumably copied this information from notes taken following conversations with Robert Brown in 1837 (see *Correspondence* vol. 2, letter to J. S. Henslow, 18 [May 1837]), some of which still survive (DAR 42: 45). It refers to an anatomical test for identifying coniferous wood, proposed by William Nicol (1834). According to Nicol *Araucaria* is characterised by having pits or 'discs' in its cell-walls arranged alternately in double rows. CD's confusion results from the bad punctuation of the note; a comma after 'opposite' would make the meaning unambiguous. Brown's identification of the silicified wood was used in *Journal of researches*, p. 406, and *South America*, p. 202.

[3] Pöppig 1835, 1: 367–8.

[4] Strzelecki 1845. Paul Edmund de Strzelecki's claims were not as extreme as CD suggested: he asserts there were a series of distinct eruptions, each clearly distinguishable from the next (pp. 120–2, 150).

[5] CD's immediate reaction to the *Vestiges* ([Chambers] 1844) is reflected in the following note, which he kept with his material on divergence and classification (DAR 205.5: 108):

> Nov/:—/44/. After the "Vestiges of *Nat Hist [*interl*] Creation", I see it will be necessary to advert to Quinary System, because he brings it to show that Lamarck's willing (& consequently my selection) must be erroneous— I had better rest my defence on few English, *sound* anatomical naturalists assenting & hardly any foreign.— Advert to this subject, after Chapter on classification, & then show, from our ignorance of comparative value of groups, source of error—

[Chambers] 1844, pp. 231–2, favoured the quinarian system of classification because, among other things, he felt it showed Lamarck's theories were untenable. Regularities in animal structure, as revealed by the quinarian arrangement, were 'totally irreconcilable with the idea of form going on to form merely as needs and wishes in the animals themselves dictated' (p. 232). CD apparently believed that the quinarian system might be used by other naturalists to refute his theory of natural selection.

[6] William Hunter.

[7] Henslow 1838.

[8] Lesson and Garnot 1826–30. Lütke 1835–6.

[9] CD eventually revised *Coral reefs* for a second edition in 1874.

From Bartholomew James Sulivan 13 January – 12 February 1845

HMS Philomel Rio Gallegos | Patagonia
Monday, Jany. 13. 1844[1]

My dear Darwin

As I have some information to give you that I think will interest you, I will write at once while it is fresh in my memory. About a week since we left the Falklands to get Soundings, but did not at the time intend going far from the land, and consequently had only a few days water on board, but the weather being very fine, as we had to get soundings across to the main I ran across, and having unusually fine weather and light winds, I found when we were off Santa Cruz, that it would be necessary to go into some place for water to take us back, and made for the Straits of Magellan wishing to go there, but having very light winds from SE in sight of Cape Fairweather, and being anxious to get back as soon as I could, I ran into this river. (and a precious intricate place it is, we were in a ticklish situation once or twice coming in) three days since in the Evening and the next morning started for the head of the Harbor where Capt Stokes[2] says fresh water can be got about 25 miles off. it is a second Santa Cruz but more intricate and you can only go up and down with the tide. You know the kind of Cliffs at Cape Fairweather, having the appearance of Chalk. Cliffs of the same kind commence at about 12 miles up this River on the North shore (the South being low and flat) and are about 200 or 250 feet high. Tho pushed by tide & time I thought I would land for a moment and expected to see some of the oyster shells &c like St Julian, but could not find a single fossil shell of any kind, and therefore I think no part of the cliff can consist of your "Fossiliferous Strata"—(refering to sketch at page 202 of your book) with this you will recieve a drawing of the cliff,[3] but I will now roughly discribe it The whole height appeared to be composed of what you have called "white sediment"—that is a yellowish white friable stone stratified horizontally about two hundred feet high and in numerous beds of different shades of colour, but three beds of from 12 to 20 feet wide were of a more compact nature, one near the level of the sea a hard conglomerate (I suppose) of gravel the next about one third up the Cliff a grey soft sand stone, a third nearer the top of a blueish grey sand stone more compact and looking at a distance like limestone all the rest was the light colour friable clay earth none of which was hard enough to fall in solid pieces but the three harder rocks were in large masses at the foot of the cliff, some, 15 or 20 feet square. The whole formation being covered by a bed of shingle, covering every slope of the Hollows in the cliffs and, therefore, I think must not be the regular top bed of shingle you describe. It was from 6 to 12 feet deep forming a border to all the cliffs like this

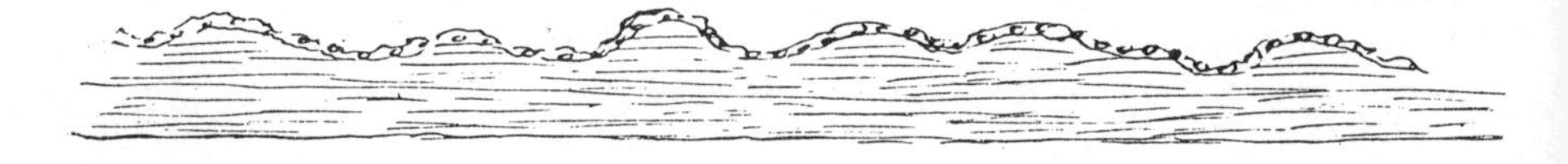

while I was looking at the masses of rock that had fallen I saw what I thought a concretion sticking out of one of the blocks of grey sand stone but on breaking it with a hammer saw that it was a piece of *fossil bone* after digging it out I saw another piece of bone close to it with part of the same sand stone adhering to it. it was the joint of an annimal as large as a horse at least. This made me search more and I soon saw what appeared to be the thin bones of the nose of a small head, and I succeeded in getting it out whole but the stone was so soft and the bones so very old that it crumbled to pieces after I put it in paper and proved to be the head of a small Animal about the size of an Armadillo and as the *Skull* and some Teeth remain good I hope it will be still worth something The skull is about this size

and the teeth about this

but all so old that even the teeth are quite black and rotten. near the same place I also found a leg bone of an Animal about the size of a *deer*, but only the joint and about half the bone could I get out perfect. I was then obliged tho reluctantly to leave, on account of the tide; and as it was, our boats got aground about twelve miles further up, and we were away all night, and did not get back till yesterday, (Sunday) and I knew if possible I must sail today as the work must be finished at the Falklands this season; and as I really have no business here, I did not like to remain, but much to my delight this morning it blew right in; and it was impossible to sail, so I started early and remained at the cliffs as long as the tide would allow searching for more ('*big bones*)' and I am glad to say it was well worth going, for I got a great many (not very large) which I hope will be valuable. some I cannot tell what they are as rather than risk breaking them I brought stone and all cutting them out in a large piece of stone, but I am sorry to say we were in such a hurry that I was obliged to have my boats crew working away also and many are broken and many only partly got out however, the following I can make out. *Parts* of the head and teeth of two Animals one large, teeth this size

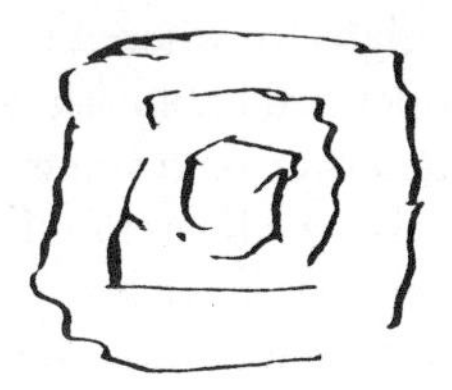

the other smaller,

Pieces of bone from 1½ inches to ½ an inch in diameter. The whole back bone and ribs of an Animal about the size of a small deer, this is in one large piece of stone so I did not attempt to clear out the bones but cut all the stone away but what contained the bones There may be more in the stone than what show externally.— One large piece of stone has I think a small head and other bones in it, and I think there are parts of one or two smaller heads, but am not certain and what I am much pleased at is getting a *piece* of the *shell* of an *Armadillo* quite perfect, about the size of those now existing. the piece is very small but there are many perfect scales about this size [] There are six or seven bands about 4 inches long in all and the breadth of three or four scales perfect like this

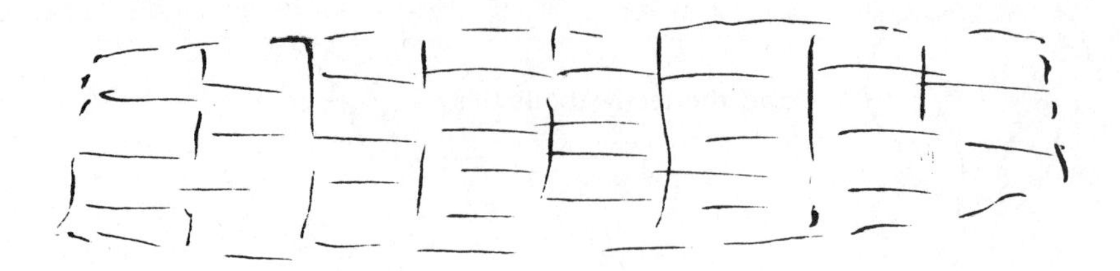

There were also some very small black things in the rock which may be some minute fossils and some things which may not be fossils, but I have taken one in case they are, these are like this

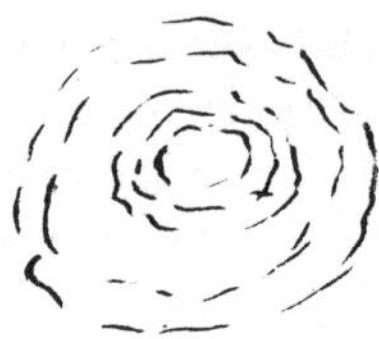

the lines are dark and very perfect, perfectly circular, about 9 inches in diameter. I looked carefully for shells but could not find a single one of any kind. Now comes the important part, of the bed in which they are: and I have not the slightest doubt but that the blocks came from the lowest of the two beds of sand stone I have described. unfortunately I could not get up the cliff to examine the bed, but I could see that the color was the same. There were only two beds of *stone* as I have said before in the cliff except the conglomerate at the bottom, and in one place the cliff having fallen enabled me to get up to the Summit of a high cliff where I could examine closely all the upper beds of clay earth and the upper bed of blueish sand stone as well as some part of the Clay between that and the lower sand stone and certainly they were not at all like the rock in which the fossils are Therefore I am as convinced in my own mind that they came from the lower sand stone bed as if I took them out. The color and appearance being exact and there were no traces of fossils, in the upper sand stone bed the only other one it could possibly have been but now comes another singular thing while examining the upper part of the cliffs I saw near me a piece of the upper white clay earth that had fallen it was about the size of my head and sticking out of it were two beautifully perfect teeth about this size

They seemed less altered than the other ones and more perfect I was within two yards of them and with some difficulty reached the piece but unfortunately directly I touched it it crumbled in pieces and roll'd down the cliff two hundred feet I searched every part I could get to where the Fragments rolled but could not find the teeth and the pieces that fell to the bottom had not even the marks of the teeth in them, but I have brought a bit of the piece to show you that it is totally different from the rock the other fossils are in. Therefore even the upper white part of the "white sedimend" must contain organic remains but I could not see any more in that part of it I could reach. among the shingle on the surface there were some things which I think are marine fossils but I forget the name They look something like this

I have now told you all that I can think of, till you see the fossils; I think you will complain how mutilated they are but if you had seen the time we had and the strong tide running, and the rough way the men were working at it in spite of my urging them to take care, you would not wonder at it. Then sometimes they were on the side of a piece of stone as big as my Cabin and it is so easily broken that it was almost impossible to get a piece out solid, where we got them at a corner or end of a piece we were more successful. had I a few days to spare, I would get a boat load for I did not examine more than half a mile of the Cliffs. I think I should be well pleased if the wind would blow right in tomorrow but it has come round this evening and is quite fair so I must not lose any more time, for I have much to do to finish this season, and it might be thought at head quarters that I had no business remaining here. As I hope to go home this year myself I shall not send them home but keep them till we return that I may ensure their going safely. of course they are for you, but I suppose you will prefer their going with yours to the College of Surgeons.[4] I may after all be thinking them of more importance than they are, but I think they must be much older than any of the Bahia Blanca ones.[5] they are nearly black and some quite rotten. I was amused by hearing one man calling out that he had got marrow in his bone, thinking he was joking yet I found that the piece of stone had broken so as to divide the bone length ways and really it was full of a pith or marrow *still soft* but something like the pith of a piece of Elder tree— in some of the bones little sharp spines project from the inside and tho fine are very perfect. I think there are *leg*, *shoulder* and a *blade* bone (or I should say pieces of them) among them.

January 14th: Another foul wind today so I started early and had another tide at the cliffs and added considerably to the collection, some the same kind as before, but others certainly different, and very puzzling to me. I think there are pieces of several small heads, and some look like large insects more than animals, & others are I think fish or reptiles, but I am sorry to say few are perfect. even in the rocks they are not perfect being in bits here and there but we seldom succeeded in getting a piece out whole where the rock was large, but I thought it better to get as many as possible even if broken than lose time to get out a few more perfect. I got today also some smaller bones that are small enough to match the small heads one so small that a man said he had found a "little birds leg". They are all much scattered about in the rocks as if they have been separated and spread about before being deposited. I also got *one or two* in different kinds of strata, showing that the remains are not confined to one bed. in one block of very friable *rubble* or clay there was a piece of a Jaw larger than a horses, with several teeth, but all quite loose and separated in small pieces exactly like the clay they were in so that they all separated as they were taken out. but now I must confess that I was wrong in my Idea about the bed the fossils are in. I found today, that the sketch I intended sending you of the Cliffs only applied to one part and that the two upper beds of darker sand stone I mentioned were not the same in other places and sometimes instead of two darker and harder beds, there were three and four, in other places only one, and all within a mile in length. The lower bed of gravel conglomerate only appeared here and there. The best description I can give you of them, is, that beds of light coloured clay alternate with thiner beds of blueish and grey sand stone, the clay varying in colour from nearly white to a Yellowish brown The upper beds of clay are much the thickest— The strata most marked and general is a bed about 20 feet thick of the light coloured clay, *immediately above* the lower dark sand stone that I thought contained the fossils. The upper part of this bed is nearly white but it is darker below and at the bottom is a yellowish brown. it is perfectly Horizontal and as far as the Eye can reach along the Cliffs it can be seen plainly, while the other beds occasionally change both in colour and size. I will try and give you a rough Idea of it

to begin with (e) the low part of the cliff up to about 60 feet is composed darker colour clays, on which here and there is a bed of Conglomerate (d) is the lower grey sand stone in which I thought the fossils were it is about 20 feet thick but occasionally is not seen or is changed in color and compactness. Then comes the regular straight bed (c.) about 20 or 25 feet thick the lower part darker than the upper which is nearly white, then comes (b) which may be about 50 or 60 feet sometimes one two or even three beds of narrow darker stone showing plainly in it, but often changing— above it is (a) the upper white and yellow beds that extend to the top where they are covered by the shingle, these beds must be above a hundred feet thick as the level of the plain is some height above the top of the cliffs except in one or two places but in those places these white beds extend to the top, so that they are evidently directly below the shingle. There are many beds in this portion all thicknesses and shades sometimes darker sand stones, some narrow ones; red, and white, & blue, but the whole appearance, is the light colour which with the bed (c) give the cliffs the appearance that used to be considered chalk. I fear I have explained this sadly and you may think me too minute but I am anxious to explain *where* the fossil bed is which I will now do. in one place I found them today I saw that the Grey sand stone I thought they were in was not to be seen on that cliff while large pieces of the bed (c) were to be seen on a slope below it evidently resembling in size & appearance the stones in which the fossils are but not in colour This I saw was accounted for by, the large pieces at the foot of the cliffs being covered at high tides which had made them so much darker— after some difficulty I managed to get to a corner of the bed in the cliff, and saw that the lower part which is of a brownish yellow color, tho looking friable clay, had in it large masses of a more compact nature which fell to the foot of the cliff, where the high tides constantly covering them, and a rapid tide—washed off the outer & softer parts of the blocks, leaving the hard stones in which the fossils are, and from the wearing away of the stone the bones are seen projecting; I could find no bones in the part of the bed I reached but by cutting away the softer part of a block saw the sand stone inside exactly the same as on the beach except a lighter colour. in the blocks that had caught on different ledges of the cliff lower down I could not find any fossils either, but when they have these softer surfaces worn away by the sea they will I suppose appear. I hardly like to say I am certain they are in this bed after saying last night that they were in the one below. however there is one strong proof, which is, that wherever the valleys are so deep as to remove this bed, as at (x) there are no stones of the same kind at the foot, nor any fossils; and the same when a *Slip* of the cliff had taken place in which the upper beds covered the bed (c) there were no fossil stones below while in every part where the bed was showing in the cliff, there were the fossil blocks underneath. in many places the cliff showed something like this section from a landslip—

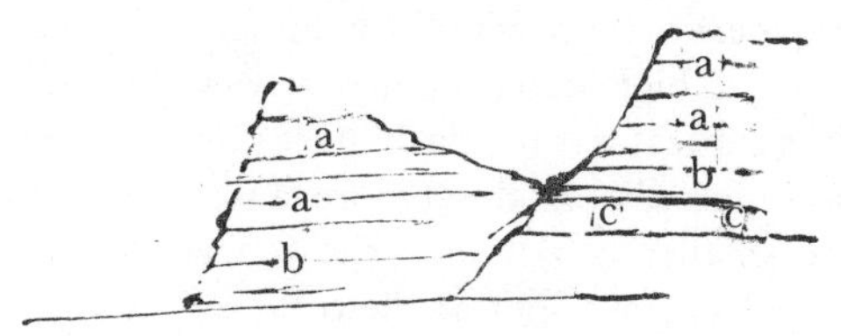

and wherever that was the case there were no fossil stones below. of course the bed (c) was in these cases hidden, and unfortunately I could not find any slip where (c) had slipped also and so come low down to be easily examined it was generally only the upper mass of white (light coloured) beds that slipped— if you cannot make out my long winded explanation, you must live in hope of my explaining it better verbally when we look over the fossils, and when you abuse me for not having got them more perfect— I am anxious to ask you how it is that the valleys and hollows should be covered with the shingle bed, unless after the land had risen and the valleys been worn out, it had again sunk and been covered by the shingle yet I know that this has not been the case from your account— Perhaps a sea that flowed over the level plain might after washing away the strata in the hollows wash some shingle off the level, and cover the valleys, but then it must have washed it away again so I cannot understand it. The most extraordinary thing to me is the fact that such numerous and deep hollows should be worn away at all, and what is more, a few Isolated hills rising out of the low land to the southward about 15 miles off, show that the plain once extended round them, for they are about the same height (to the Eye) but where is all that enormous mass of clay and sand stone gone? and what a time must it have taken; one cannot fancy any time sufficient and yet all this is the work of a day (Geological) compared to the time that these same beds were gradually forming from the parent rocks and from what rocks were these beds formed? and what was the nature of the country when the little Armadillo lived whose remains are in these beds? They surely could not have had such different habits as to have lived in a mountainous country, and yet what other kind of country could have been the parent of the plains of Patagonia. I may be mistaken as to the scales being armadillo but they are *exactly like them*; if not, they must be a lizard, or extraordinary fish scales, but I have written enough to tire you without troubling you with any more of my fancies but I must tell you that *here* in *Patagonia* we have had rain every day, and no sun to be seen. Yet it cannot be the rainy season, for the land looks dried up, and there are marks of the Animals feet when the ground was wet now quite hard and dry. we have even rain with the wind *West*, showing that there must be an approach to Southern weather. I suppose the Andes lower allow some clouds to pass them. There are also several little springs and tuns[6] of good water in the cliffs and hollows, and a little inland a *fresh water pond* I do not think we ever saw such a thing further North; Further I must add that it is a horrid place to get into, coming in we were nearly on shore, and as Beagle's Chart of it had 4 and 6 fathoms I thought we were out of the channel, and trusting to that I weighed today on my return to run out and ran right on shore, there being where the channel is marked only 4 *feet* water and as we cannot get out till high water tomorrow we were obliged when the tide floated us to run back again as it was too dark to see our way. we had just got back when I began writing. I must say one thing more I was very anxious to get you a good specimen of the smaller ostrich but tho a few were seen none of our sportsmen have managed to get a shot at them.[7]

West Falkland Jany. 29. I fear you will be sure to say before reading all this "How much more is he going to write"—but I must risk tiring you as I think you may like to hear the result of our soundings between the Islands and the main. To explain this I must give you a sketch

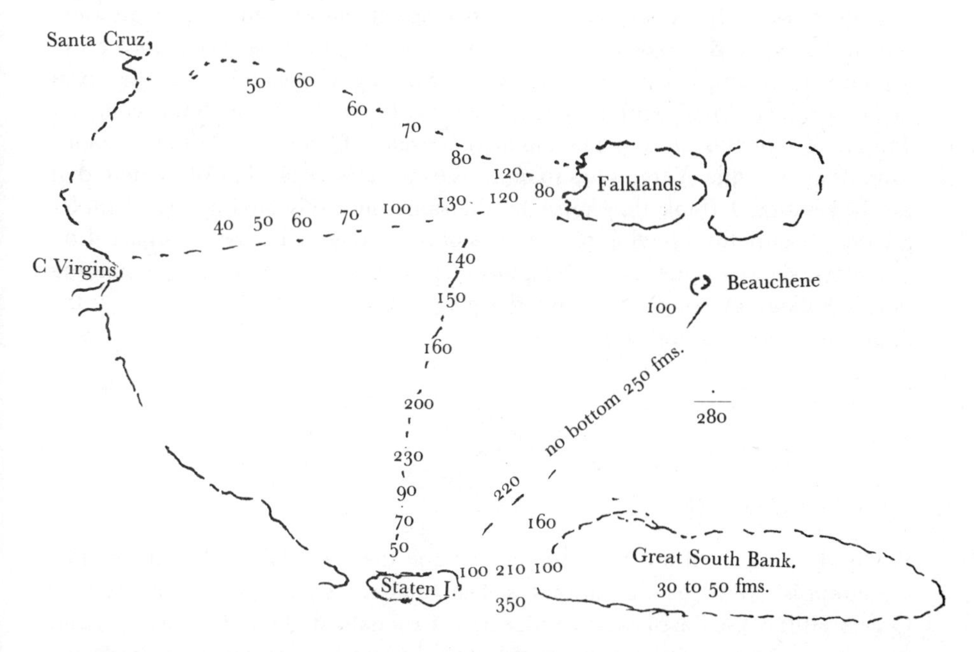

mind that this is only a rough sketch The soundings are just written by eye but it will do to explain You will see that from the west Isd of the Falklands to the main on three lines we got bottom right across, the deepest being towards Staten Island while the main bank further North reaches nearly to the Island. I shall yet get more soundings round the NW end of the Falklands to see if it is shallower there. we also got bottom between the great South bank and Staten Island— I know you are interested about the Patagonian bank and that is why I mention this. we have also got the Edge of it in several Places. between the Falklands & Cape Corientes there is little variation in the depth (50 & 60 foms) till near the Edge where it deepens faster, and from 80, very fast, so that a mile or two more you have 100 and a few more miles, *no bottom with 1400 fathoms.* is it possible that the bank can have a thickness equal to that at its outer edge or does it rest on a foundation of older rocks which slope off in that manner? Yet if that were the case the bank would not be so level. The only thing more I have got to bother you about is my discovery that the nearly Horizontal beds of Quartose sand stone in the Swan & New Islands group of West Falkland have veins of basalt (or some trap rock) traversing them. I have thought so for some time, but from not knowing enough of the rocks did not like to be sure about it, particularly as I had only found in in Numerous

Dykes running perpendicular in a North and South direction in what appeared to be regular openings in the rock and therefore I thought they might possibly have been filled (as mineral veins are) by some compact sedimentary rock, but in all the dykes the rock was of a very compact dark coloured crystalised structure and I thought must be Igneous, & have long had specimens of it to show you on my return but yesterday passing a small Island half of which has been worn away leaving a high cliff, I saw a dark black vein running diagonally across the strata and on sending a boat to bring some of the rock I found it was a still more compact Trap rock very difficult to break and full of crystals of Felspar and others which I suppose are Olivine Yet it is has so much the appearance of what I have found in the Dykes that I think they must be the same only this having cooled under greater pressure from having such a mass of rock above it is more compact. I do not know whether there is any thing unusual in these dykes but being something new in Falkland Geology as I am writing I think you would like to know it: the appearance of the Iland is like this

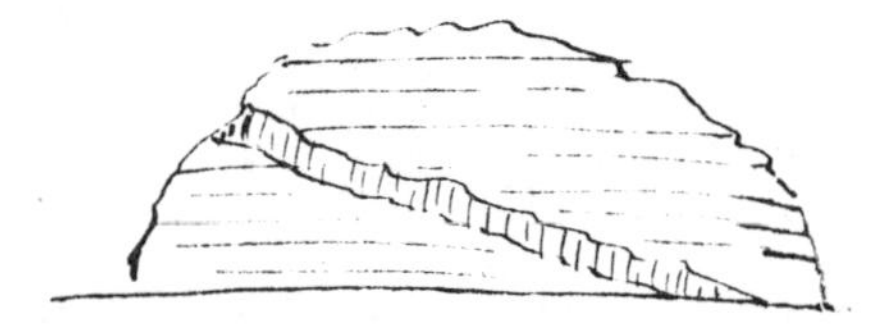

the height of the cliff being about 100 feet and the vein about 6 feet thick The sand stone is I think the same as that (so bent, and altered by heat,) composing the Quartz Hills of the East Island, as here we occasionally find it partly changed and sometimes almost as much as on the East Island; one reason that made me doubtful before about the dykes being Igneous was, that the rock on each side was so little altered (tho for some *inches* it was decidedly so, and where the vein had fallen out the two sides or walls of the crevice stuck up in this manner

and were very much harder than the rest of the sand stone but only for two or three inches,) there are hundreds of these dykes running miles in length, as you see them on opposite sides of channels and Islands and some are twenty feet wide, and all sizes from that down to two inches, yet they run nearly in the same direction and till I saw this Island I never saw any with the least inclination.—

I must not finish this without trying to give you a better opinion of the Falklands as to weather than our visit in Beagle would lead you to adopt. from the middle of November (when we arrived) the weather has been so fine that we have not had more than two *rainy* days and not more than three or four more showery (one day of which there were hail showers.) besides these we have had perhaps four or five days thick weather with northerly winds and all the rest have been nearly as cloudless as at Santa Cruz the most rain and cloudy weather we have had was during the week we were in and near the Gallegos. So dry is the ground for want of rain, that even in the lowest hollows we cannot find water enough to *wet* our *feet* in walking *miles*, and it is only in the ravines coming from the highest hills were there are good rivulets that we can get water, and I have need all the season to carry water with me in my boat from the ship whenever I go away: during the past month we had plenty of strong breezes but since Christmas we have hardly had a gale of wind and few even strong breezes. I was surprised at the dry weather both summers before as we often had difficulty in getting water; all but the large streams being dry; but this year it is much drier—. I must also tell you one more thing which you will think singular after our bad luck with Gales in Beagle. we were three weeks last year on the bank to the Eastward of Staten Land—and this year ten days off Staten Land, and eight days at sea going to and coming from Gallegos and we have only had one breeze that could be called a gale which lasted two or three hours and besides that, had not once in either last year or this to close reef the topsails I might almost say, "*double reef* them"; and from leaving England to this time, including five passages to Monte Video and back we have only had one heavy Gale (and that a fair one) *at sea* so as to be obliged to batten our hatches down: this will seem almost incredible to you after the weather we had in Beagle— *occasionally*, we have had gales enough and hard enough down here, but have either been at Anchor, or in sheltered bays and sounds, so that we had no ocean sea. But enough of this, as I am sure you will think there is quite enough said, but you must thank the *fossils* for it, as I should not have written at all had it not been for them. I hope you will think them worth the infliction of such a long Yarn

Feb[y]. 3 As we shall I hope be back at Stanley (Port William) in a few days I must have this ready to leave there in case a vessel calls. I have been looking over your old letters that I might recollect all the things you mention, so as to give you what information I can on these points. You will recollect that I was told, that one reason why the horses did not increase so fast as the cattle was, the studs driving the mares on before the Colts could walk and so leaving them to die.[8] since that I have myself seen several remains of very young Colts tho I have only been a few walks in the part where the horses are, while I never yet found a dead calf—so that the reason given me, may be correct, but cannot I should think be the only cause. There is no lack of pasture for the number of horses as they have some very good vallies among the hills and have always had a supply of Tussac besides but they do not apparently like leaving the Hills for the low land to the Southward. There are numerous spots where they collect for shelter under the high Quartz Rocks, which are called their stables, and round them the grass is very good the peaty land having been trodden into a rich soil and the grass being eaten close grows quite green in the spring Perhaps the shelter the rocks afford them may make them prefer the high ground. You will recollect my mentioning how easily the rabbits are tamed. I think that is also the case with the horses. I saw a little filly about 18 months old, that was run down and brought in, and in four days she was so tame that you could play with her and the soldier (sapper) who had her, in less than a week had taught her to give her *paw* like a dog, but of course this being a single instance may not be of much value— The wild cattle cannot have lost their original habits entire or they could not be driven about a week after they are caught as they easily are. in fact I am convinced that they need not be *caught* but that the wild herds may be *driven* on to Peninsulas where by rounding them with a few tame ones for a few days, they may be driven to any part of the Islands. I have driven a herd back onto a point with a party on foot and there killed and caught ten, and therefore several men on Horseback might easily do it— of course they must let the old bulls pass as they cannot always drive them, but cows calves and young bulls I think they can. one day on my way to Mount Usborne we found a small herd of cows and young ones half way up among the stones in running away they got into a lane of grass running into a stream of stones where there was no outlet as they are afraid to cross the stones I was pushed for time as it was clear on the top for the observations but thought that if I could keep them there till I had finished I might get a good supply of beef, so I left two of my boats crew to guard them and they kept them rounded up in a mass for *three hours* till I returned from the mountain when we killed out of the herd five cows and caught three young ones. this I think shows that they cannot be the same as wild animals. I do not recollect any thing more that I can tell you on these points, except that I have a rat in spirits for you that I caught two miles inland and he is very unlike a *domestic* rat yet there are so many about the Islands that I think they must have come from vessels. It is quite incorrect what we were told respecting the difference in the Foxes of the two Islands. they are the same both in size and color we have never been able to detect any difference they are very numerous in the west Island that is on every Harbor or bay we have been in there have been two or three couple seen. I had two last year quite tame having been eight months on board, and I sent them to Rio in hopes of getting them sent home, but the Packet would not take them and I have never heard whether they were sent by a merchant vessel. The Packet could not well take them as their smell was any thing but pleasant and she had numerous Lady Passengers. I wish they could have gone home safe, particularly as the Female was with Young. I have seen several more basalt dykes during the last two days one twenty feet wide running across a point and forming the Hill on the summit about 200 feet high it was composed of numerous concentric masses which when moved against each other, rank like metal one small dyke looked exactly like a pile of shot (where its side was uncovered) when seen at a little distance so small and regular were the *balls*. there must be a very great mass of it contained in all the dykes as they extend over at least 400 or 500 square miles and sometimes there are three or four in a hundred yards and they are at least five feet wide (the mean of the whole). it is singular that they should be found where the sandstone is nearly horizontal while in the other Island where it has been so twisted they are not seen

I have found a few more Fossil shells in the sandstone on the South side of Saunders Island but, I think they must be similar to those you have tho I had previously supposed the sand stone to be a different formation.—

I fear you have mistaken the position of the bed of soft sand stone containing the fragments of granites &c. You say it is the "Ice formation", and suppose it to be on the South side of the Islands.[9] it is quite the contrary. if you look at the chart you will see in the SE corner of Byron Sound a small Island called Skip Is. The bed lines the *South* Shore of the sound, in the depth of the bay to the SE of that Island, and is situated at the *Northern* base of the Highest Mountains in the whole Islands rising directly from the sea to 2300 feet high this chain runs down to West Point tho in some places much lower (600 or 700 feet) The Fragments appeared to have been well worn before being deposited tho not round like pebbles on a beach. the cliffs where it lined the shore were about 30 feet high and I saw it for about two miles in length, but could not see how for it ran inland tho the hills come so close that it cannot be more than two or three hundred yards but in one place it projected nearly a mile and formed a low point You may suppose how glad I am that I am on the point of finishing this survey. The actual chart work will be complete in three days but we have plenty of "*off shore*" soundings to get and lots of work at the *paper* still to do, but I find I have had quite enough of it, and long to be settled quietly on shore again tho every thing has gone so happily for me having my wife and chicks with me (tho I see little of them while down here) and every thing comfortable in the vessel besides which I am blessed with health such as I never knew for years and never expected to know again in that my getting this employment was in every way most providentially ordered for me, but I trust now it is over *here* I shall be allowed to go home as I do not wish to stay out to work in the River Plate— All our little ones have benefitted much by the Falklands and are pictures of health and our little *Falkland* now nearly a year old does his birth place credit.[10] what is more singular is that just before leaving the River they all caught the hooping cough and were getting *worse* up to the day they arrived here and one was *very ill*. Yet from that day they began rapidly to get better and tho they were daily with the Doctors children in fact almost living together, and there were numerous other children round, not one child got the Cough as if it could not exist at the Islands. my wife tried a singular remedy, for she *bathed them in the sea* while they had the Cough, but it certainly did them good—and when I went back after being away two months I found them all well and looking better than I ever saw them. it certainly is a most healthy climate but still I shall be glad to leave it and hope I shall never again have to leave home, in fact I do not think any thing would get me to do it again—

I forgot to mention that on my last trip to the Cliffs I made a sketch which I think gives a pretty fair Idea of them and therefore I will send it to you. I have endeavoured to color it as nearly as I can.[11] it does not exactly represent the cliff at any spot but shows nearly all the different beds that were within a range of a mile tho several were very broken and the darker ones were often replaced by others of a lighter color still the general appearance of the cliffs was very like the sketch. the light colored gravels at the bottom were generally the rubble from the other beds above but the bottom bed of hardened gravel (or fine shingle, for the pebbles were quite round) stuck out boldly, often in overhanging masses. I am very curious to know whether you consider all these beds included in the "white sediment", that is whether they all come between the fossiliferous strata and the shingle. I cannot think they were all formed by the action of the sea on a beach as you suppose the white sediment to have been; because many of them are compact and regular sand stone; and the bed of shingle must have been formed previously; as the lighter particles according to your theory must of course always be underneath. the bed of sand stone immediately below the fossil bed must also I should think have been deposited regularly, had there been no compact beds of sand stone above the fossils, I could have supposed that the carcasses had floated down some river or estuary and had sunk by degrees among the "white sediment" as it was forming, but on the whole I cannot help thinking that all these beds except the upper white ones must have been regularly deposited by running water. I do not know whether it is possible for bones to cause the sediment in which they are contained to harden round them, but certainly the bed appeared to be very friable, except at the lower part where the Concretionary Hard lumps are that I believe contain the fossils. If it is possible for bones to impregnate the stone with any cementing ingredients, I think they must have do so in this case; they must I should think from there appearance have parted with all their phosphate of lime— do not laugh at my conjectures if they are sheer nonsense, which I have no doubt some of them are. I wish I could have found shells with them so as to give some Idea of their age, but I hope some of the things which I cannot make out will turn out to be marine animals of some kind— I recollect your mentioning somewhere in your book that ""Volutes" have been found some way South (I think 40 or 42) and may yet be found further South"[12] I have been searching the book but cannot find it; perhaps I mentioned in a former letter that they are numerous at the Falklands but in case I did not I will now. we often found *Numbers* of dead ones on the beaches, but for nearly two years never found a live one till one day a "*red bell*"[13] was seen with one he had just brought up to eat. and since then in one place myself and the doctor each got one at a very low tide. we then saw that the reason our dredge had never brought them up was that they are buried just under the surface of the sand the shells are about 6 inches long and 2½ in breadth. I could not myself have vouched for their being Volutes but my late First Lieut. and the Doctor, who are old Pacific shell collectors say that they are and as it confirms what you said I thought you would like to know it. from the numbers on all the beaches they must be abundant—

Speaking of shells, I must not forget to tell you than in the River Plate a few miles from the entrance of the Uruguay I found a bed of apparently a more compact nature than the sandstone Cliffs near it, and in it there were a great number of *casts* of shells the whole shell being replaced by a highly crystaline substance I am pretty certain that they are marine shells, & I do not recollect whether you found any similar ones or not I will try to give you an Idea of their shape

I think the crystaline substance is lime but the rock appears to be a hard clay— in case I should not come home this year if you want any more particulars about them I will try and go to the place again but at the time I thought you would be sure to have seen others like them and I still suppose you have the bed they were in was not more than 15 feet above the river I also picked up a few miles from the same place two teeth on the beach at the foot of some cliffs which from having some of their Crevices filled with a hard substance I think may be Fossil I will give you the shape of the end of the tooth

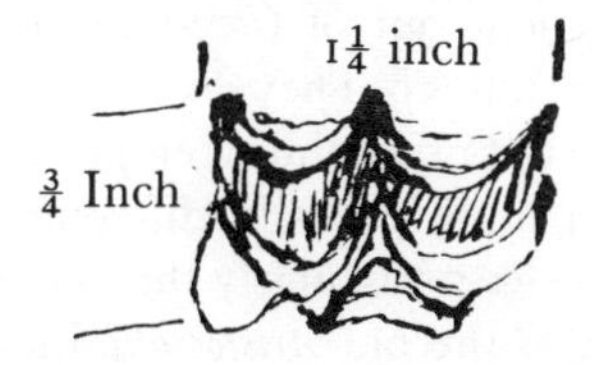

This is miserably done and not like it, but it has five cutting edges two lower line the two top rows and a fifth between the two lower ones. I could find no bones near, and these may be perhaps teeth of some existing Animal, but they appear to me rather large. The thickness one way being an inch & quarter and the other way three quarters of an inch

While I am giving you specimens of my *talent* for *drawing* I may as well give you an Idea of the fossil shells I picked up at the West Falkland since I sent you the others for I fancy the sand stone is a different formation. They are only casts of shells but are very numerous being 15 in number on one bit of stone about 15 square inches I do not think there were any like them in those I sent you before. that is not so small

and now I really will leave off scribbling for I am sure you will be well tired of me long before you have made out all my hurried scrawl, but you must know it has been written at intervals of a few minutes when being tired of bending over the chart board I have given myself a spell.

Stanley (Port William) Feby 12. 45

we arrived here two days since and the day after a vessel arrived going home direct from New Zealand so I must send this. I have been much interested by the accounts of FitzRoys career. two of the *Radical* Members of Council are in this vessel and they have told me all and every thing that has happened since his arrival. it is Beagle over again, *temper violent* saying *any thing* to any body, doing most hasty and extraordinary things, but on the whole much liked and he has (they say) done some things which will save the Colony tho so contrary to Ideas at home that he himself and many others think they will recall him.[14] but only fancy FitzRoy the High Tory adopting the most ultra Radical views, free trade in *every thing*, no customs duties, and still more given up the *principle* of the Wakefield System of buying land which the Government so insist on and given up all right of the Crown to the land so far as allowing all to purchase direct from the natives, paying a fee of *1 penny* an acre to government instead of *10 Shillings* an acre as formerly.[15] this is the point that all are in such doubt about tho they agree it must safe the Colony— The old system of *caring nothing* for expense. Surveying a Harbor now at his *own* expense, repairing Government House at his own expense on which said house *20.000£* had been spent out of *Government money*, previously while nothing had been done to make roads &c. They say he has done more with the few Hundred Pounds he could spare out of an almost empty Treasury than had been done before with Thousands opening paths in different directions &c. in fact doing much to please every one—& particularly the natives who go to him with all their troubles and complaints. the old *Aristocratic* Ideas still showing themselves, and occasionally so violent as even to cause a *deputation* to take up their hats and walk out of his room in the middle of his *attack* on them, leaving him storming at them; in fact, being the Captn. of a ship over again as for the Gov^{t}. Officers under him they appear to have no opinion but his; he deciding on things without consulting them; and adopting the popular and Radical views that they were opposed to so much so that his measures are obstructed by them in working out. unfortunately he is inclined to be *High Church* and the Bishop being a *Puseyite*, they will do harm in that way.[16] already the Bishop has caused strife and dispute between the *Church Mission* converts and the Methodist forbids Methodist communicants from receiving the sacrament in the church, while before both Church and Methodist Missionaries received each others members at their sacrament he has also refused to allow them to be *buried* together He tells the Wesleyan Converts himself that they have never been baptized because it was not done by a properly ordained successor of the Apostles—and not only this but has so worried and interfered with the missionaries that they say themselves all their influence is gone. Mr Williams[17] has lost the influence he possesed. and for this the Church Missionary Society have assisted to send a Bishop out— FitzRoy tried to get an Addition to the Bishops salary voted from the *Public Money* but it was successfully opposed—

They say that Mrs. FitzRoy is "the greatest blessing to the Colony"—that her influence and example was much wanted and is doing much good in fact she seems almost to be worshiped—but I fear she is delicate the old General a great Favorite and keeps clear of all parties and all public affairs. I know Mrs. Darwin and yourself will be glad to hear all about them so I have ventured to add to my long Yarn

Mrs. Sulivan joins me in kind remembrances to Mrs. Darwin who I trust with your little ones is quite well. if it please God I hope to see you before eight months more have passed—[18]

Believe me dear Darwin | Yours very sincerely | B J Sulivan

Do not attempt to wade through this long yarn except when you have little to do of any consequence

B J Sulivan[19]

DAR 46.1: 76–86

CD ANNOTATIONS

1.1 As I . . . sand stone more 1.28] *crossed pencil*
1.15 the appearance of Chalk.] *underl pencil*
4.1 I fear you . . . sketch 4.4] *crossed pencil*
4.10 I know . . . *fathoms.* 4.15] 'Patagonian Bank' *added pencil, circled pencil*
4.18 The only . . . perpendicular 4.23] 'Basalt veins Falkland Isls' *added pencil, circled pencil*
4.46 there are hundreds of these dykes] *underl pencil*
5.1 I must not . . . Santa Cruz 5.7] 'Weather Falkland Isds' *added pencil, circled pencil*
5.11 I have . . . points. 6.4] *crossed pencil*
6.4 You will recollect . . . cause. 6.9] 'Death of Horses' *added pencil*
6.24 in fact . . . animals 6.38] *crossed pencil*
6.38 I do not . . . vessels. 6.42] 'Rat Falkland' *added pencil*
6.42 It is quite . . . difference 6.44] 'Foxes same on the two islands' *added pencil*
6.51 I have seen . . . *balls.* 6.56] 'Dikes' *added pencil, circled pencil*
7.3 formation.—] *followed by a pencil line drawn across the page*
8.2 granites] 'granites?' *added pencil*
8.2 You say . . . contrary. 8.3] 'Boulder Formation' *added pencil*
8.4 Byron Sound] 'Byron Sound is NW part of Western island' *added pencil*
8.8 have been well worn 8.9] *underl pencil*
8.9 tho not round like] *underl brown crayon*
8.10 30 feet] *underl pencil*
8.10 two miles 8.11] *underl brown crayon*
8.15 but we have . . . Sulivan 18.1] *crossed pencil*
On cover: 'Death of Horses only Species fact.' *pencil*
'Sounding to Falkland Isd' *brown crayon*
'Boulder [formation]' *brown crayon*
'Pebble island rabbit' *pencil, del pencil*
'Ch V [*below del* '1']' *brown crayon, circled brown crayon*

[1] 1844 is an error for 1845. The correct date occurs later in the letter.
[2] John Lort Stokes.
[3] *Journal of researches*, p. 202, where CD gave a diagram of the strata of coastal Patagonia. He showed the cliffs as being composed of 'Fossiliferous Strata'. See also n. 11, below.
[4] At least one of Sulivan's fossils was presented to the Royal College of Surgeons, see Flower 1879–91, 2: 436. Richard Owen described this and some further fossils sent by Sulivan at the British Association meeting in Southampton in 1846 (see *Report of the 16th meeting of the British Association for the Advancement of Science held at Southampton in 1846* Transactions of the sections, p. 66). He named them *Nesodon imbricatus* and *Nesodon sulivani*.
[5] CD's fossil quadrupeds from Bahia Blanca were described by Richard Owen in *Fossil Mammalia*.
[6] Literally, barrels or casks.
[7] For CD's remarks on a second, smaller species of South American *Rhea* see *Journal of researches*, pp. 108–9, and 'Notes upon the Rhea Americana', *Collected papers* 1: 38–40.
[8] See *Correspondence* vol. 2, letter from B. J. Sulivan, [10 May 1843].
[9] Presumably mentioned in one of CD's letters to Sulivan.
[10] James Young Falkland Sulivan, Sulivan's oldest son, whose name reflected the family's belief that he was the first British subject born on the Falkland Islands (*DNB*).
[11] Sulivan's watercolour sketch of the Rio Gallegos cliffs is preserved in DAR 46.1: 75.
[12] *Journal of researches*, pp. 611–12.
[13] *Haematopus leucopus*, the Falkland Islands oyster-catcher or red-bill.
[14] Robert FitzRoy was recalled 30 April 1845. The news reached him on 1 October 1845. For the circumstances see Mellersh 1968, pp. 224–35.
[15] According to the Treaty of Waitangi, land purchases were outlawed, and only the government could buy land from the Maoris. The 'system', however, was not Edward Gibbon Wakefield's but that of the first governor, William Hobson. (See Mellersh 1968, pp. 203–4).
[16] George Augustus Selwyn was bishop of New Zealand, 1841–67.
[17] Henry Williams, archdeacon of Waimate.
[18] Sulivan's family returned to England in August 1845. He returned in June 1846, see letter to Richard Owen, 21 [June 1846].
[19] 'Do not attempt . . . Sulivan' was written on the cover.

To Charles Hamilton Smith 14 January [1845]

Down near Bromley | Kent
Jan 14th

Dear Sir

I venture, on the remembrance of the kindness, which you showed me several years since, when starting as naturalist on board the Beagle,[1] to ask you to do me a favour.— I have been reading your interesting Paper on the original Population of America, in the last New Eding. Phil. Journal,[2] & at p. 8 you refer to a "ruined city or cities", in the Caroline Group, of vast size, & to other appearances indicating that the land has subsided or is again rising; I am particularly anxious to know, where I could find any account of these facts, & if you would be so kind as to take the trouble of sending me the briefest reference, I should esteem it a great favour. From the structure of the coral-reefs in the Caroline Group, I have been led to believe it has subsided, & in a small work, published two or three years since, (at p. 127)[3] I have given an extract from an Australian newspaper describing the ruins of houses on Pouynipète[4] (or Seniavine), which are "*now only accessible by boats*", but I did not like trusting too implicitly to such authority &

therefore laid no stress on the statements.— Admiral Lutké[5] who has so well explored this group & whom I saw some time since in London, does not appear to have seen anything of these ruins.

Hoping that you will excuse the liberty, which I have taken in troubling you, & will kindly favour me with a reply; I beg to remain | dear Sir | Yours faithfully & obliged | Charles Darwin

American Philosophical Society

[1] See *Correspondence* vol. 1, letter to Caroline Darwin, 12 November [1831].
[2] C. H. Smith 1845.
[3] *Coral reefs*, p. 127 n.
[4] An island located at 6° 55′ N, 158° 15′ E, known for its ruins. It is now usually known as Ponape.
[5] Fedor Petrovich Lütke.

To J. D. Hooker 22 [January 1845]

Down Bromley Kent
22^d

My dear Hooker

I sit down to go through your letter categorically— 1st. There is no chance of any other fossil leaves from T. del F. being in England: sh^d you make out anything more about them hereafter, please to let me know: pray keep them for ever.—

I am delighted to hear of the Galapagos flora being done: would you, when it is printed in the Transactions give me a copy (& thanks for your offer of copy of Paper on Southern Coniferæ);[1] as you must have so many Botanical friends, to whom your papers w^d be more valuable than to me, any old proof-sheet copy would do perfectly for me, & any such copy of *any* paper of yours, I sh^d be **truly** obliged for.— Thanks for Araucarian information;—but pray do not trouble yourself to look for any more Papers of Nicols or send them to me, as I wish only just to state the character of the wood which I found.[2] (I intend going through the Eding. New Phil Journal myself some time; I have gone through all late Parts.) I will take care of your Pacific Isl^d notes, till your return.—

I am very **much** obliged for the loan of Wilkes[3] & will send by our weekly Thursday carrier next week for it: how magnificent a soul you have about books: I presume the 20£ d'Urville is the second & last Voyage, which I had not heard was out: hereafter I shall get you to look & see whether the geology of T. del Fuego is treated of in it.[4] Pray do not waste any time about the Annales in Paris, though if in Book-shops, I certainly sh^d like to know the cost. Also I shall be very glad if you could urge Dieffenbach for the *copper-plate, wood cuts & M.S. notes of mine*: I am the more anxious about them now, as I am in a sort of negotiation with Murray, who wishes to get the power from Colburn & publish a 2^d Edit:[5] I have no doubt that you will work on him; Lyell recommended me to write to the great Humboldt & set him to worry the little Devilbach.

I heartily wish you a pleasant journey: by the way you will of course see Ehrenberg, & I will write to him to ⟨ask⟩ him to give you for me, a little MS. Paper which I sent for him to read on Atlantic dust.[6]

Farewell, with thanks for the long catalogue of things, which you have done & intend doing for me. Ever Yours | C. Darwin

Postmark: JA 22 1845
DAR 114.1: 26

[1] J. D. Hooker 1845a. CD's copy is in the Darwin Pamphlet Collection–CUL.
[2] *Journal of researches* 2d ed., p. 332.
[3] Wilkes 1845.
[4] Jules Sébastien César Dumont d'Urville participated in three exploratory voyages. CD refers to the second voyage of the *Astrolabe*, 1837–40. By January 1845, parts 1–8 of Dumont d'Urville [1841–54] were published (Stafleu and Cowan 1976). The volumes containing the geology of the voyage were not published until 1848–54 (2 vols.). The geology of Tierra del Fuego is described at length in volume one (1848).
[5] *Journal of researches* 2d ed. See letter to Charles Lyell, [8 February 1845].
[6] 'An account of the fine dust which often falls on vessels in the Atlantic Ocean' (*Collected papers* 1: 199–203). See letter to C. G. Ehrenberg, 5 September [1844].

From J. D. Hooker [22–30 January 1845][1]

is versed enough in Nat. Hist. even for that. I send the comp. Flora of Sandwich & Society: it is very imperfect, I could not send it before, as I had some things to add to it. as it is I think it shews positively no relation between the two groups. Even the *canaille* are not the same in the two. I have had to drop the Pacific Flora, from want of time ⟨ ⟩

⟨ ⟩ from the context he[2] cannot mean, without he compares the M^{ts}. to the plains below: Certainly the genera remind one of Europe, especially Wicken (vetches *Vicia.*) &c. I can however hardly see how so *very* few Proteaceae as there are, can much remind him of New Holland a place I suppose he never saw & singular fruited Legum. grow all over the world & ⟨ ⟩ Holld. that I remember.. Gnapha⟨lium⟩

Incomplete
DAR 104: 247

CD ANNOTATIONS
1.1 is versed . . . that.] *crossed pencil*
1.3 it shews . . . time 1.5] *scored pencil*
2.1 from . . . Gnapha⟨lium⟩ 2.6] *crossed pencil*

[1] This fragment of a letter was pinned to the following note, in ink, by CD:

> Feb. 1845. Hookers list of Society Is^{d} includes about 250 Phænerog excluding grasses, & 270 from Sandwich group. out of these there are only 32 species in common, including no doubt (one by name Convolvulus) Scævola &c) some coral-plants.— At the Soc Is^{d} there 25 gramineæ & at Sandwich 24. & 6 [*over* '5'] in common.— Of Lycopodiæ Soc & Sandwich have each 12 with 5 in common.— (DAR 104: 248).

On the verso of CD's note is a table of numbers as summarised above. The date of Hooker's letter must follow that of the preceding letter and is probably before Hooker's departure on his European tour (see Huxley ed. 1918, 1: 177–8).

2 Eduard Pöppig in Pöppig 1835, 1: 367.

From C. H. Smith 22 January 1845

Plymouth
22 Jany 1845.

My dear Sir

In reply to yours, received this morning, I beg to say that on referring to my vast collection of notes I find the following entry.

"Ascensio an Island of the Caroline group in 11 North Lat in the S. Seas, lately discovered by H.M. Sloop Raven. A gentleman who went to reside on this spot, found a place called Tainen covered with extensive ruins of an Antique town; but settled so low that it is only accessible in boats; the water coming to the steps of the houses. the stones are laid artificially, but without cement; some being 20 feet long."— "the walls have doors & windows. the present inhabitants are different in habits & manners from the other natives ⟨ ⟩ South Sea Islands: the social system is ⟨ ⟩ [more] advanced; ⟨ ⟩ on a par with the men & their customs approximate ⟨ ⟩ Europeans ⟨ ⟩. Lotzky who communicated ⟨ ⟩ to society March 2^{d} 1839 had ⟨ ⟩ letter from ⟨ ⟩ stating that the ⟨ ⟩ had since visited ⟨ ⟩ & a survey.[1] he stat⟨ed⟩ ⟨ ⟩ since ⟨ ⟩ monuments ⟨ ⟩ I have not noted whence I extracted the above but ⟨ ⟩ may have been from the ⟨ ⟩ intelligence very similar to the above has likewise reached me from New Holland, where I ⟨ ⟩ & other correspondents. As a mere fabrication I could not accept it, although ⟨ ⟩ what I have expressed on the nature of ⟨ ⟩ of the materials brought forward in the consideration of the population of America, you will perceive that I had left the absolute authenticity open as regards perhaps several of ⟨ ⟩

I shall now be exceedingly glad to receive from you a few lines in order to have your opinion whether the above Tainen & your Pouynipète or Seneavine are the same. If I understand it rightly, they must be distinct for Ascension cannot well be the same; it has been long [known]. I may remark that the [Capped] parallelitha[2] of Tinian seem to belong to the same people; and it is a question whether the pillar Idols of Easter Island & the stone sculptures of Pitcairns may not likewise be referred to them.

Now let me add a few words respecting yourself: It gave me very sincere pleasure to hear from you, as in some measure it was an evidence of your health ⟨ ⟩ in that condition at least which I call "working order" for that is my [opinion] of my own, when I can get to my table to write or draw as if I were strong. Sometimes I have wished you were resident in this place. the usual mildness of the climate being favourable ⟨ ⟩ upper part of the town is in a sufficient ⟨ ⟩ I own it would be a sacrifice for you ⟨ ⟩ kindred [minds]. ⟨ ⟩ three persons of extensive information to be ⟨ ⟩ there is no public library richly stored. I find that the classes of books that I want in my several pursuits are all, or with rare exceptions missing 〈 〉 I have to purchase them myself unless the private library of a very kind hearted friend help me on.

My eldest daughter[3] who had the pleasure of seeing you in town about a twelve month ago, desires to be kindly remembered.

I am | My dear Sir | Most sincerely yours | Charles Ham^{n} Smith

Charles Darwin Esq | &c &c &c

P.S. would you have the goodness to favour me with the title of the little work wherein you have spoken of the coral formations of the Caroline Islands?

Water-damaged and torn
DAR 177

[1] Johann Lhotsky presumably communicated information from a third party, but no record of any publication has been found. It appears that Smith's information on subsiding ancient ruins in the Caroline Archipelago was derived from the same source as CD's, that is, information from a Mr Cameron in Sydney, transmitted through Lhotsky. See *Coral reefs*, p. 127 n., and letter to C. H. Smith, 26 January [1845].

[2] Parallelith is an arrangement of parallel ranks of stones in the form of a lane or avenue (*Funk and Wagnall's New 'Standard' Dictionary of the English Language*, New York 1952).

[3] Probably Emma Smith, who was her father's companion and assistant until his death.

To C. G. Ehrenberg 23 January [1845]

Down near Bromley | Kent
Jan 23^{d}

Sir

Would you be so good, if convenient, to return to me, by Dr J. Hooker, who will be at Berlin in a fortnight, the short sketch, which I sent you, of the facts regarding the Atlantic dust, & which I thought, perhaps, worth inserting in some English Journal: if you have examined the other specimens of dust, would you be so good as to inform me, that I might refer to your descriptions in this sketch.[1]

Amongst the little packets, which I sent you, there were some of the earth of the Pampas, in which so many extinct mammifers are embedded, should you feel any interest on this subject, I should feel *particularly thankful* to hear the result of your examination. Should the subject not interest you, I should be sorry to think even of asking you to waste any of your time on it. There are, also, specimens of a singular white bed, (I now believe of a very fine tufaceous nature) which extends for hundreds of miles on the Patagonian coast, about which I am curious.

I beg to apologise for having thus troubled you, but might I further request you, should you see Dr Dieffenbach, kindly to take the trouble to ask him, to return me my copper-plate, woodcuts, & M.S. corrections for his German edition of my Journal, which I cannot get him to return to me, though I shall be thus put to considerable loss of time & money, in preparing a second English Edition.—[2]

Believe me dear Sir | With great respect | Your's faithfully & obliged | C. Darwin

To | Professor C. Ehrenberg | &c &c &c

Museum für Naturkunde der Humboldt-Universität zu Berlin

[1] See letter to C. G. Ehrenberg, 5 September [1844].
[2] See *Correspondence* vol. 2, letter to Ernst Dieffenbach, 19 July [1843].

To C. H. Smith 26 January [1845]

Down near Bromley. | Kent
Jan 26th

My dear Sir

I am exceedingly obliged for your note & your most kind expressions regarding my health, which is still very indifferent, but allows me to do my light work. The extract you send me is rather fuller, than anything which I have seen; it refers certainly to the same island, to which I refer in my volume on "the Structure & Distribution of Coral Reefs" (p. 127. & p. 168) & I believe without doubt the same with Pouynipète or Seniavane; I came to this conclusion after accurate comparison of various charts at the Geographical Soc. Out of the about 43 islands (or atolls) of the Caroline Arch:, about 40 are *low* coral-islands, & only 3 high islands, encircled at a distance by barrier coral-reefs; & it appeared that the description referred to a high island. Every one knows how greedily a theorist pounces on a fact, highly favourable to his views, so that I was very unwilling not to make much of this fact, but there was something in the account, which made me rather sceptical: moreover the writer spoke of granite-blocks & I have reason to believe the isld is volcanic, though, to be sure trachyte & granite are easily confounded. Lastly I heard (perhaps very unjustly) very indifferent accounts of Dr Lloghtsky's moral character.[1] I perceived by the manner in which you referred to the case, that you did not think it fully established; & I agree with you, upon the whole, that it probably is not a mere fabrication. I have very little doubt that hereafter, the existence of former wide tracts of land, since buried in the ocean by subsidence, will turn out the chief means of the migrations & passage of animals, plants & man, from one part of the world to another. How rapidly do facts in geology accumulate, showing that most of the sedimentary strata of Europe & N. America, have during their accumulation (as shown by upright buried trees, footsteps once on the surface, presence of beings, which cannot live in deep-water &c &c) slowly subsided.

In the Pacific & Indian oceans, if there be any truth in my explanation of the singular structure of coral-reefs, wide tracts have subsided, since the existence of the present reefs.—

But I must apologise to you, for having run on at this length: pray accept my sincere thanks for your prompt & most obliging answer, & believe me, with my best compliments to Miss Smith,

My dear Sir | Yours sincerely obliged | Charles Darwin

American Philosophical Society

[1] Doubts about Johann Lhotsky's character probably arose from his being 'careless with money and too outspoken in criticism of those in high places' (*Aust. Dict. Biog.*).

To Emma Darwin [3–4 February 1845][1]

[Down]
Monday night

My dear Wife

Now for my day's annals— In the morning I was baddish, & did hardly any work & was as much overcome by my children, as ever Bishop Coplestone[2] was with Duck.[3] But the children have been very good all day, & I have grown a good deal better this afternoon, & had a good romp with Baby—[4] I see, however, very little of the Blesseds— The day was so thick & wet a fog, that none of them went out, though a thaw & not very cold; I had a long pace in the Kitchen Garden: Lewis[5] came up to mend the pipe & paper the W.C. in which apartment there was a considerable crowd for about an hour, when Mr Lewis & his son William, Willy Annie, Baby & Bessy[6] were there. Baby insisted on going in, I daresay, greatly to the disturbance of Bessy's delecacy— Lewis from first dinner to second dinner was a first-rate dispensary, as they never left him— They, also, dined in the Kitchen, and I believe have had a particularly pleasant day.—

I was playing with Baby in the window of the drawing-room this morning, & she was blowing a feeble fly (fry) & blew it on its back, when it kicked so hard, that to my great amusement Baby grew red in the face, looked frightened & pushed away from the window.— The children are growing so quite out of all rule in the drawing-room, jumping on everything & butting like young bulls at every chair & sofa, that I am going to have the dining-room fire lighted tomorrow & keep them out of the drawing-room. I declare a months such wear, w^d^ spoil every thing in the whole drawing-room.—

I read Whately's Shakspeare[7] & very ingenious & interesting it is—and what do you think Mitford's Greece[8] has made me begin, the Iliad by Cowper,[9] which we were talking of; & have read 3 books with much more pleasure, than I anticipated.— I have given up acids & gone to puddings again.—

Tuesday morning— I am impatient for your letter this morning to hear how you got on.— I asked Willy how Baby has slept & he answered "she did not cry not one mouthful". My stomach is baddish again this morning & I almost doubt, whether I will go to London, tomorrow; if I do you won't hear. Poor Annie has had a baddish knock by Willie's ball in her eye.—it is swelled a bit, but not otherwise bad.

C. D.

Your cap cannot ⟨be⟩ found anywhere: Jane says you took one. $\frac{9}{10}$ of the snow is gone & the children are going out. Very many thanks for your letter[10]

Sotheby (28 Mar 1983)

[1] Date based on nn. 7 and 8, below, and on Henrietta Litchfield's statement, before her transcription of parts of this letter, that Emma went to Maer in February 1845 (*Emma Darwin* 2: 92). Emma's diary records that she was away between 31 January and 11 February.
[2] Edward Copleston.
[3] Henrietta Litchfield notes, 'This must be some family joke. Bishop Copleston had been a friend of Sir James Mackintosh.' (*Emma Darwin* 2: 93).
[4] Henrietta Emma Darwin, born 25 September 1843.
[5] John Lewis was a carpenter in Down village (*Post Office directory of the six home counties 1845*.)
[6] Elizabeth Harding, nursery maid at Down House (see *Emma Darwin* 2: 80–1).
[7] T. Whately 1785. The London Library borrowing list records that CD borrowed Thomas Whately's book on 30 January and returned it on 27 March 1845 (London Library Archives).
[8] Mitford 1784–1818. Volumes two and three of William Mitford's *History of Greece* were borrowed from the London Library on 9 January and returned on 27 March 1845 (London Library Archives).
[9] Cowper 1791.
[10] The final paragraph was written in pencil.

To Edward William Brayley[1] 7 February 1845

Down, Kent
Feb. 7th. 1845.

My dear Sir

You have my best wishes for your success in your present application. You are aware that I have never had an opportunity of hearing you lecture, & therefore cannot speak of your qualifications in that line; but I have great pleasure in adding, that I have always been struck by your remarkable powers in acquiring scientific knowledge of varied kinds, & by your extensive reading. I think it will be generally acknowledged, that a capacity of this nature, must be eminently serviceable in teaching a subject of so divesified a nature as Geology.—

Believe me dear Sir | Yours very faithfully | C. Darwin

E. W. Brayley Esqe

American Philosophical Society

[1] Brayley was applying for the professorship of geology at University College London. His application was not successful and the post went to Andrew Crombie Ramsay in 1847, T. Joyce having taught geology in the interim. See Bellot 1929. This letter was published in a volume of testimonials for Brayley, see Freeman 1977, pp. 63–4.

To Emma Darwin [7–8 February 1845][1]

[Down]
Friday night

My dear Emma

I shall write my Babbiana tonight, instead of before breakfast. It is really wonderful how good & quiet the children have been; sitting quite still during two or three visits conversing about everything & much about you & your return— When I said I shall jump for joy, when I hear the dinner bell Willy said, "I know when you will jump much more—when Mamma comes home" & so shall I responded many times Annie. It is evident to me, that *you must* be the cause of all the children's fidgets & naughtinesses.— Annie told me that Willie had never been *quite* round the world, but that he had been a long way, beyond Leave's Green— The Baby has neglected me much today & would not play; she c^{d} not eat any jam, because she had eat so much at tea; but not like Annie of old she did not care. She was rather fidgety, going in & out of the room & Brodie declares she was looking for you— I did not believe it, but when she was sitting on my knee afterward & was looking eagerly at pictures, I said "where is poor Mamma" she *instantaneously* pushed herself off, trotted straight to the door, & then to the green door,[2] saying Kitch & Brodie let her through, when she trotted in, looked all round her & began to cry; but some coffee-grains quite comforted her— Was not this very pretty? Willy told me to tell you that he had been very good & had given Annie only one tiny knock, & I was to tell you that he had pricked his finger.—

My own annals are of the briefest, I paced half-a dozen times along Kitchen Garden in the horrid cold wind, & came in & read Monsters & co,[3] till tired, had some visits from children, had very good dinner & very good negus[4]—played with children till 6 oclock read again & now have nothing to do, but most heartily wish you back again.—

My dear old wife, take care of yourself & be a good girl. C. D.—

Sat. Morn.

All right—Willy said to me "poor Poor laying all by himself & no company in the drawing room."

Farewell to our Slip of Land[5]

Is not poor Eliza's letter wonderful, pray beg Harry to give some kind message from us—[6]

DAR 210.19

[1] Dated on the basis of the previous letter to Emma, [3–4 February 1845].

[2] The door separating the kitchen quarters from the rest of the house.

[3] I. Geoffroy Saint-Hilaire 1832–7. CD's annotated copy is in the Darwin Library–CUL. He recorded that he finished reading this work in March 1845 (DAR 119; Vorzimmer 1977, p. 133).

[4] 'A mixture of wine (esp. port or sherry) and hot water, sweetened with sugar and flavoured [with nutmeg]. Invented by Col. Francis Negus, British soldier (d. 1732).' (*OED*).

[5] CD had difficulties acquiring some strips of land over which access to Down House was gained and other small plots adjacent to his property. This note may refer to the land under discussion in

Correspondence vol. 2, letter to Susan Darwin, [8 December 1843]. See also *Correspondence* vol. 4, letter to John Higgins, 10 September [1847].

6 Sarah Elizabeth (Eliza) Wedgwood was Emma's cousin; her sister, Jessie, married Emma's brother, Henry Allen (Harry) Wedgwood, in 1830.

To Trenham Reeks [before 8 February 1845][1]

☞ Besides numbers on the labels, they are scratched on the glass.
(Please to keep this paper for me)

1312. A bed of recent upraised shells from Peru falling into fragments & decaying. The shells are often filled with common salt, & have their surfaces, *deeply & peculiarly corroded*; therefore I *suspect* double decomposition has slowly taken place: *if so there sh^{d} be in this mixture some carbonate of soda & traces of muriate of lime*:[2] is this so?

I should esteem it a great favour, to be informed, or referred to any work whether muriate of soda & carbonate of lime, when mixed in large masses & damp, will partially decompose each other. I see some Agriculturists recommend mixing salt & chalk for this purpose, but they do not state positively the result.[3]

1313 The same *continuous* bed at a greater height, in which the shells have absolutely decayed— *does this contain carbonate of soda, or muriate of soda, or carbonate of lime or muriate of lime, or some of all four?*

763 is this sulphate of Soda? it was originally in long pure crystals:

1633 (paper packet) superficial incrustation; *what is this*? is it sulphate & muriate of Soda? or both?

1227. Nitrate of soda as quarried: *does it contain any* **muriate of lime**? this is the only point, which I want to know.

3052. *is this soil damp from muriate of lime*? the surrounding soil under the Peruvian climate, was impregnated with muriate of soda, but was *not* damp.

3048. *is this anhydrite or gypsum*?: does it contain any carbonate of Soda or lime? a superficial crust over the country.—

1264. *What is the opake, white, saline crust* on the common salt?

759. Salt from great salt-lake of the R. Negro: it does not serve well for curing meat. *What does it contain, or not contain, that makes it different from sea-salt?*

762. Salt from salt-lake of Chiquitos: how does it differ from sea-salt?

954. Saline matter, abundant. In veins with gypsum in a tertiary sediment; *what is it?*

25 Bone of head of Glossotherium, from the Pampas;[4] it had an extraordinarily fresh appearance & even emitted a flame:; *what percentage of animal matter does it contain?*

Mr. Trenham Reeks, | of the Museum of Economic Geology.[5]

AL
DAR 39.1: 51–2

CD ANNOTATIONS
1.1 Besides . . . so? 2.6] *crossed pencil*
3.5 The same . . . *four?* 3.7] *crossed pencil*; 'Whether any Carbonate of lime?' *added pencil*
3.24 Bone . . . *contain?* 3.26] *crossed pencil*

[1] Apparently enclosed with a lost covering letter and returned with Reeks's letter of 8 February. The questions were not necessarily addressed to Reeks and may have been forwarded to Reeks by a third party, see n. 5, below.
[2] See CD's second letter to the *Gardeners' Chronicle and Agricultural Gazette*, [before 14 September 1844].
[3] The passage 'I should esteem . . . result.' was added on the verso of the first page with the symbol '*a)' used to indicate its position in the text.
[4] In *South America* CD identified the bone as from a *Mylodon*, stating in a footnote that Richard Owen had at first considered the head to be that of *Glossotherium*, a closely related but distinct genus (*South America*, p. 92).
[5] The last sentence is not in CD's hand. Reeks was a mineralogist employed in the laboratory of the Museum of Economic Geology.

To Charles Lyell [8 February 1845]

Down Bromley Kent
Saturday

My dear Lyell

I find that d'orbigny (Voyage dans l'Amer. Merid. Partie Gèolo. p. 226)[1] describes ten Silurian fossils from the eastern line of the Bolivian (or Upper Peru) Cordillera; he states that they are all distinct, names them, but says they present the strongest *general* resemblances to those of Europe.— p. 230 describes seven Devonian fossils with same remarks.— p 239 describes 23 Carboniferous fossils with same remarks;[2] two, however of these 23 perhaps are not new, for I see that they are not described by him, viz. *Natica antisinensis* & *Spirifer Roissyi*— Amongst his Cretaceous fossils he states that 5 are common to the Paris Basin.— There is a copy in the Athenæum, if you wish further to refer to it.

I forgot, when with you on Thursday, to ask you, if you have any opportunity, to have another talk with Murray about my Journal.—[3]

After being with you, I went & had a long talk with Mr Cuming[4] about S. American shells, & especially their range with respect to my Tertiary species; & I find the only series, which he has not examined & which he wishes to examine, are the shells from Patagonia & T. del. Fuego. For *several* reasons, I am *exceedingly* anxious to get this done, & hear his report; but which way can this be done with least trouble to yourselves? I fear it must be very troublesome, & I send my most humble apologies to Mrs. Lyell.— I see there are from the list, about 90 specimens, which I shd like Cuming to look at. I could send the list marked; I at first thought it w^d^ be perhaps least trouble to you to get them out, & let me take them away to Cuming's, & bring them back; but as there are so many, perhaps it w^d^ be least to get Cuming to meet me at Hart St.[5] some morning: he is an awful proser & very difficult to make to stick to his work. About half a dozen from Bahia Blanca, I must beg leave to get out, for Sowerby to compare with those embedded with the extinct mammifers—[6] I fear you will think this so much trouble, that you will wish I had never given you my collection. Would Mrs Lyell be so kind, as to send me a line to tell me, what she thinks w^d^ be the least troublesome plan?

Believe me | my dear Lyell, | C. Darwin

Postmark: FE 8 1845
American Philosophical Society

[1] Orbigny 1835–47, vol. 3, pt 3: *Géologie*.
[2] The correct reference is p. 233.
[3] CD refers to plans for a second edition of *Journal of researches* to be published by John Murray.
[4] Hugh Cuming provided CD with information about the distribution of South American shells (*Journal of researches* 2d ed., pp. 390–1, and *South America*, p. 133). CD considered the distribution of Tertiary shells an important indicator of climate during that period (*South America*, pp. 134–5).
[5] Lyell's London address.
[6] George Brettingham Sowerby had previously examined the fossil shells associated with the extinct mammals of Bahia Blanca (*Fossil Mammalia*, p. 9). CD required a more precise account for *Journal of researches* 2d ed., see letter to G. B. Sowerby, [May 1845].

From Trenham Reeks 8 February 1845

Museum of Economic Geology
8th February 1845.

Dear Sir,

I have great pleasure in forwarding the answers to your questions,[1] together with the extract upon bones from Liebig.[2]

You will observe in N° 1313 that there is but a very little Carbonate of Lime, which appears to me strange, as, in N° 1312, a continuation of the same bed, there is an abundance.

I have tested the salts N°s 759, 762 for Iodine but cannot find a trace of that body or any other which would account for its inefficacy in curing meat.[3]

If you will direct me how to send the remainder of the Specimens, they shall be forwarded.

I am | Dear Sir | Your obedient Servant | Trenham Reeks

C. Darwin Esqre | &c &c &c

[Enclosures]

N° 1312. Contains no Carbonate of Soda, but Muriates and Sulphates of Lime & Soda.

* I believe Carbonate of Lime and Muriate of Soda to have no action upon each other.
Nº 1313. Contains Sulphates and Muriates of Lime and Soda, but very little Carbonate of Lime.
Nº 763. Is Sulphate of Soda.
Nº 1633. Contains besides Earthy Matter and Carbonate of lime, Sulphates and Muriates of Soda.
1227. Contains plenty of Muriate of Lime
3052. Contains plenty of Muriate of Lime which may account for its dampness.
3048. Is neither Anhydrite nor Sulphate of Lime but consists of about 64 per cent Earthy Matter, with water & Carbonate of Lime, but no Carbe. of Soda.
1264. The Crust is Gypsum.
759. Contains a little Sulphate of lime & Magnesia with Mechanical Impurities. I can find nothing in it to account for inefficacy in curing meat.
762. Same as 759.
954. Is chiefly Sulphate of Magnesia.
25. Contains about 7 per cent Animal matter and 8 water.

T. Reeks.

One hundred parts of dry bones contain from 32 to 33 per cent of dry gelatine; now supposing this to contain the same quantity of nitrogen as animal glue, viz: 5.28 per cent then 100 parts of bones must be considered as equivalent to 250 parts of human urine.

Bones may be preserved unchanged for thousands of years, in dry or even moist Soils, provided the access of rain is prevented; as is exemplified by the bones of antediluvian animals found in loam or gypsum, the interior parts being protected by the exterior from the action of water. But they become warm when reduced to a fine powder, and moistened bones generate heat and enter into putrefaction; the gelatine which they contain is decomposed, and its nitrogen converted into carbonate of ammonia and other ammoniacal salts, which are retained in a great [measure] by the powder itself. Liebig's Chemistry of Agriculture and Physiology. Page 194. 2nd. Editn.[4]

DAR 39.1: 43–4, 49–50

CD ANNOTATIONS
2.1 You . . . abundance. 2.3] *crossed ink*
4.1 If you will . . . Reeks 5.1] *crossed ink*
Top of first page of letter: '1845 Gardeners Chronicle p. 93—on salt very [good] for cheese'[5] *pencil, circled pencil*
At end of letter: 'Gardeners' Chron. 45 p. 93. on salt for cheese' *ink*
Verso last page of letter: 'Mr Phillips[6] (about Patagonian tuff.) | Dela Bech about Muriatic acid'[7] *pencil, del pencil*

Enclosures:
1.1 Nº . . . as 759. 2.15] *crossed ink*
2.5 Nº 763 . . . Gypsum. 2.12] *crossed pencil*
2.5 Nº 763 . . . Soda.] 'Test this for Carbonates' *added pencil, circled pencil*
2.6 1633.] 'Port Desire' *added pencil*
2.8 1227.] 'Nitrate of Soda as quarried' *added pencil*
2.12 1264.] 'Arica' *added pencil*
2.15 as 759.] '—Chiquitos Salina' *added pencil*

1 See letter to Trenham Reeks, [before 8 February 1845]. For the use CD made of these replies see *Journal of researches* 2d ed., pp. 66, 155, 370, later elaborated in *South America*, pp. 52, 69, 72, 74.
2 Liebig 1842, p. 194. CD possessed an earlier edition of this work (Liebig 1840) in which the cited passage occurs on p. 202. His copy is in the Darwin Library–CUL.
3 Specimen 759 was a salt, from a salina near the town of Patagones (El Carmen), which CD recorded as being particularly inadequate for preserving meat. For Reeks's later analysis see letter from Trenham Reeks, 25 February 1845.
4 CD had apparently requested this reference in connection with his question concerning No. 25 in his letter to Reeks, [before 8 February 1845]. CD cites Reeks and Liebig in his description of the fresh appearance of the fossil bones he found at Banda Oriental (*South America*, p. 92). The word 'measure', which Reeks failed to copy, has been supplied by the editors from Liebig 1842.
5 *Gardeners' Chronicle and Agricultural Gazette*, no. 6, 8 February 1845, p. 93. Cited by CD in *Journal of researches* 2d ed., p. 66: 'those salts answer best for preserving cheese which contain most of the deliquescent chlorides'; i.e., they are not *pure* sodium chloride.
6 Richard Phillips (see letter from Trenham Reeks, 14 March 1845).
7 Henry Thomas De la Beche, director of the Geological Survey, at whose request Reeks was answering CD's queries (see *South America*, p. 52).

To J. D. Hooker [10 February 1845]

Down near Bromley | Kent
Monday

My dear Hooker

I am much obliged for your very agreeable letter; it was very goodnatured, in the midst of your scientific & theatrical dissipation, to think of writing so long a letter to me. I am astonished at your news & I must condole with you in your *present* view of the Professorship,[1] & most heartily deplore it on my own account. There is something so chilling in a separation of so many hundred miles, though we did not see much of each other when nearer.— You will hardly believe how deeply I regret for *myself* your present prospects— I had looked forward to seeing much of each other during our lives. It is a heavy disappointment; & in a mere selfish point of view, as aiding me in my work, your loss is indeed irreparable.— But on the other hand, I cannot doubt that you take at present a desponding, instead of bright view of your prospects: Surely there are great advantages, as well as disadvantages. The place is one of eminence; & really it appears to me there are so many indifferent workers & so few readers, that it is a high advantage, in a purely scientific point of view, for a good worker to hold a position, which leads others to attend to his work.— I forget whether you attended Edinburgh, as a student,[2] but in my time, there was a knot of men who were far from being the indifferent & dull listeners which you expect for your audience.[3] Reflect what a satisfaction & honour it would be to *make* a good Botanist—with your disposition you will be to many, what Henslow was at Cambridge to me & others, a most kind friend & guide.

Then what a fine garden, & how good a Public Library; why Forbes[4] always regrets the advantages of Edinburgh for work; think of the inestimable advantage of getting, within a short walk, of those noble rocks, & hills & sandy-shores near Edinburgh. Indeed I cannot pity you much, though I pity myself exceedingly in your loss.— Surely lecturing will in a year or too, with your *great* capacity for work (whatever you may be pleased to say to the contrary) become easy & you will have a fair time for your Antarctic Flora & general views of distribution. If I thought your Professorship would stop your work, I sh[d] wish it & all the good worldly consequences at el Diavolo: I know I shall live to see you the first authority in Europe on that grand subject, that almost key-stone of the laws of creation, Geographical Distribution.— Well there is one comfort, you will be at Kew, no doubt every year.—so I shall finish, by forcing down your throat my sincere congratulations.

Thanks for all your news— I grieve to hear Humboldt is failing;[5] one cannot help feeling, though unrightly, that such an end is humiliating: even when I saw him he talked beyond all reason.— If you see him again, pray give him my most respectful & kind compliments, & say that I never forget that my whole course of life is due to having read & reread as a Youth his Personal Narrative.[6] How true & pleasing are all your remarks on his kindness: think how many opportunities you will have, in your new place, of being a Humboldt to others. Ask him about the river in NE Europe, with the Flora very different on its opposite banks.—[7] I have got & read your Wilkes.—[8] what a feeble book in matter & style, & how splendidly got up. Shall I return it, (with your Sandwich Lists, which have interested me much; ah what labour) to Sir William.—

Do write me a line from Berlin—also thanks for the proof sheets; I did, not, however, mean proof-Plates: I value them, as saving me copying extracts.—

Farewell, my dear Hooker, with a heavy heart, I wish you joy of your prospects. | Your sincere friend | C. Darwin

Postmark: 10 FE 10 1845
DAR 114.1: 27

1 Hooker's 'agreeable letter' has not been found. Hooker had been invited to teach at Edinburgh University during the spring of 1845 as substitute for the ailing professor of botany, Robert Graham, and was plainly Graham's choice as successor (see letter from J. D. Hooker, [23] March 1845).

2 Hooker had studied for his medical degree at Glasgow, where his father, William Jackson Hooker, had been professor of botany.

3 For CD's time at Edinburgh see *Autobiography*, pp. 46–53, and *Correspondence* vol. 1.

4 Edward Forbes, previously at Edinburgh University, but at this time living and working in London.

[5] Hooker met Alexander von Humboldt in Paris shortly after his arrival there on 30 January (Huxley ed. 1918, 1: 179).
[6] Humboldt 1814–29. See *Autobiography*, pp. 67–8.
[7] The Obi in Siberia, see letter to J. D. Hooker, [10–11 November 1844], n. 7, and letter from J. D. Hooker, [late February 1845].
[8] Wilkes 1845.

To W. D. Fox [13 February 1845][1]

Down Bromley Kent
Thursday.

My dear Fox

I am very glad to be able to tell you that my Father is at present very well; he has, however, till lately been suffering a good deal from a cough.— When do you intend to pay your Shrewsbury visit? Thanks for your letter & all your news of your Noah's Ark;[2] it w^d^ really be a curiosity to see the place, besides other pleasures. Emma would have as much pleasure I think, as myself, in accepting your very cordial invitation, but I fear there is a reason against it, which even you will admit to be paramount, viz her confinement in the middle of July[3] & as soon as the Baby can travel viz Sept 1st we must visit Maer[4] & Shrewsbury.— It really is one of my heaviest grievances from my stomach, the incapability & dread I have of going anywhere: I literally have not slept, I believe, out of inns, my own, Erasmus' Shrewsbury & Maer houses, since I married.— My stomach continues daily badly, but I think I am decidedly better than one or two years ago, as I am able with rare exceptions to do my three hour's morning work, which at present is on the Geology of S. America, & very dull it will be.—

You ask about Down, the house is now very comfortable; & the garden will be tolerably so, when the evergreens are grown up; I continue to like it very well; & its thorough rurality is invaluable.— By the way, do not ask about the Vesta stove (& thanks for your remembering it) for after hearing it from other quarters, also, highly praised, I *saw lastly* an Arnotts stove in action, & that struck the balance, & I have got one, & it answers admirably, requiring feeding only twice in 24 hours, & once will do.—

I am very glad to hear you are coming to London, & do, if you possibly can, come on here; we shall be heartily & cordially glad to see you, & you now know, how quietly & in the eyes of most, dully we go on.—

I have not heard anything lately about Mesmerism; except Sidney Smiths dream, (who is said to be dying)[5] that he dreamt he was in a madhouse & that he was confined in the same cell with Miss Martineau & the Bishop of Exeter.[6] It is said that the remarks on Miss. M. in the Athenæum, were by Brodie;[7] I thought them much superior to the general writing in that paper.— I hear Miss. M. is in so excited a state about Mesmerism, that she can hardly keep on peace with her old friends, who are unbelievers; I wonder how she & Erasmus will get on.—[8] I simply feel, that I *cannot* believe, in the same spirit, as it is said, that ladies do believe on all & every subject.

I read aloud Arnolds Life[9] to Emma & liked it very much; I wish he had had rather a more lightsome & humorous spirit: as Carlyle would say, he was no "sham".[10]

Farewell my dear Fox. | I hope we shall soon meet here. Ever yours | C. Darwin

P.S. I sh^d. like very much sometime a few of my potatoes,[11] & chiefly to get true seed from them, & see whether they will sport or not readily.—

Postmark: FE 〈 〉 1845 A
Christ's College Library, Cambridge (Fox 69a)

[1] The date is based on a note on the manuscript, in an unidentified hand, reading, 'Feb 14 1845' and on the death of Sydney Smith, see n. 5, below.

[2] Fox kept a large number of different varieties of domesticated animals for observation and breeding purposes. The 'Ark' enabled him to respond to numerous requests for information from CD.

[3] George Howard Darwin, CD's second son, was born 9 July 1845.

[4] Emma's mother, Elizabeth (Bessy) Wedgwood, and sister, Sarah Elizabeth (Elizabeth) Wedgwood, still resided at Maer.

[5] Sydney Smith died on 22 February 1845.

[6] Harriet Martineau and Henry Phillpotts. Martineau was an outspoken Radical, while the bishop was an equally vehement Tory.

[7] Benjamin Collins Brodie. The anonymous article (*Athenæum*, no. 896, 28 December 1844, pp. 1198–9) was a reply to a series of articles by Harriet Martineau in the *Athenæum* testifying to her own recovery from illness, apparently as a result of mesmerism (Martineau 1844). See R. K. Webb 1960, pp. 230–3.

[8] Erasmus Alvey Darwin was a close friend of Harriet Martineau, but for his reaction to her involvement with mesmerism see letter to William Darwin Fox, 20 December [1844].

[9] Thomas Arnold. The 'Life' is Stanley 1844, recorded in CD's list of 'Books Read' on 30 November 1844 (DAR 119; Vorzimmer 1977, p. 133).

[10] Thomas Carlyle decried Britain for having 'quitted the laws of Fact, which are also called the laws of God, and [mistaken] for them the laws of Sham and Semblance, which are called the Devil's laws' and sought 'a world reformed of sham-worship' (Carlyle 1843, pp. 42, 45).

[11] The potato tuber had been sent to Fox by John Stevens Henslow from plants he had raised in 1836 from CD's seeds (see letter to W. D. Fox, [before 3 October 1846]).

To Leonard Jenyns 14 February [1845]

Down Bromley Kent
Feb. 14th.

Dear Jenyns.

I have taken my leisure in thanking you for your last letter, & discussion, to me very interesting, on the increase of species. Since your letter, I have met with a very similar view in Richardson, who states that the young are driven away by the old into unfavourable districts, & then mostly perish.—[1] When one meets with such unexpected statistical returns on the increase & decrease & proportions of deaths & births amongst mankind & in this well-known country of ours, one ought not to be in the least surprised at ones ignorance, when, where & how, the endless increase of our robins & sparrows is checked.—

Thanks for your hints about terms of "mutation" &c; I had had some suspicions, that it was not quite correct, & yet I do not yet see my way to arrive at any better terms: it will be years before I publish, so that I shall have plenty of time to think of better words— Development w^{d}. perhaps do, only it is applied to the changes of an individual during its growth. I am, however, *very* glad of your remark, & will ponder over it.

We are all well, wife & children three, & as flourishing as this horrid, house-confining, temper-souring weather permits.—

With thanks, believe me | Your's very sincerely | C. Darwin

Bath Reference Library (Jenyns papers 'Letters of naturalists 1826–78', Octavo vol. 1: 43(8))

[1] J. Richardson 1829–37, 2: xix–xx. CD recorded this work as having been read on 29 January 1845 (DAR 119; Vorzimmer 1977, p. 133). It is in the Darwin Library–CUL. John Richardson referred to the practice of adult birds driving the young into less favourable breeding grounds but said nothing about the young perishing. For Jenyns' comments on this subject see Jenyns 1846, pp. 113–17.

From Alcide Charles Victor Dessalines d'Orbigny 14 February 1845[1]

Monsieur,

Je venais enfin de terminer le travail relatif à vos fossiles lorsque votre dernière lettre m'est parvenue. Je me suis empressé de rédiger mes observations générales et je vous adresse la caisse par l'entremise de M. Ballière, comme vous me l'avez demandé.

Vous remarquerez, comme je le pensais bien que votre magnifique collection de Fossile, ne montre aucune contradiction avec ce que j'ai vu en Amérique. Les différences qui pourraient exister tenaient toutes à de fausses déterminations de M^{r}. Sowerby. Les fossiles sont des élemens d'une valeur immense en Géologie, mais ils doivent être pris avec toutes les précautions désirables. Sans cela ils peuvent donner des resultats tout a fait opposés avec les faits. Si je me suis fait beaucoup attendre pour vos déterminations, c'est que je voulais ne pas vous donner un travail provisoire. cela m'a demandé des recherches trés longues, que je ne voulais pas néanmoins négliger, pour ne pas Vous donner des renseignemens fautifs. J'espère que vous me pardonnerez le retard, en faveur du motif qui me guidait.

Votre intéressante collection est sans contredit magnifique, et demanderait une publication spéciale faites avec tous le soin désirable—car elle peut servir de base à l'étude de l'Amérique Méridionale. Je vous engage bien à faire facon des Planches de toutes les espèces Nouvelles, afin que votre travail et le mien puisse completer la Géologie de l'amerique du sud.[2] Il serait bon surtout que les fossiles fuissent représentés sur plusieurs faces et avec beaucoup de series. Des figures imparfaites ne font qu'encombrer la science sans lui servir aucunement.

Je dois à votre collection de pouvoir rectifier une erreur que j'ai commise sur l'age des grès de Concepcion,[3] que j'ai placée dans les terrains tertiaires et qui d'après votre *Baculite* nouvelle est une dépendance des *terrains Crétacés*.

Pour le reste je me suis souvent trouvé embarrassé pour les noms des espèces. Lorsque Mr Sowerby avait, par exemple donné un nom connu à une espèce que je crois nouvelle, j'ai cru devoir lui donner une *dénomination*, mais je n'y attache aucune importance, si vous ne la conservez pas changez la ou faites ce que vous voudrez, je n'ai voulu que vous fixer, et pouvoir m'entendre avec vous s'il devenait nécessaire de vous donner de nouvelles indications sur les détails et sur l'ensemble.

Voici du reste, le resumé suivant l'age des terrains que j'ai pu reconnaitre d'après vos fossiles. En commencant par les plus inferieurs.

Terrains Jurassiques?[4]

une partie de la cordillère de copiapo, l'alto de guasco, (chile) et la cordillere de coquimbo

Terrain Néocomien ou *crétacé inferieur*.

La cordillère centrale du chile. La chaine centrale de la Tierra del Fuego. La Cordillère du Perou près de Pasco.

Terrain turonien, ou craie chloritée

Concepcion du chile. Cordillera de Copiapo, Cordillera de Coquimbo.

Terrain patagonieen ou Tertiaire inferieur

Sur l'ocean atlantique. La Bajada de Santa fé. Baie de San Joseph. Port desire. Port St Julian. Rivière de Santa Cruz.

Sur le grand océan Chiloé; (Côte est) Huafo. Navidad. Coquimbo (couches inferieures et moyennes) Payta, la Mocha.

Tous ces fossiles ne se trouvent plus vivans dans les mers actuelles et appartenaient à une periode passée.[5]

Terrain Diluvien ou de L'epoque actuelle

Buenos Ayres. Maldonado. Coquimbo, (couches superieures de la Plaine.).

Vous voyez, par cet apercu rapide, dont vous avez les details avec mon travail, que de cette manière tout se classe sans aucune indécision et sans mélange.

Vous avez dans vos fossiles le plus beau fait que je connaisse pour prouver les mouvemens brusques de la période qui a determiné leur extinction. C'est un groupe de Crépidules. Ces coquilles sont encore réunies, telles qu'on les rencontre vivante au Callao. Le moindre mouvement, même la mort des individus si elle était survenue avant d'être enveloppée, aurait séparé les differentes coquilles qui composent le groupe. C'est un fait très curieux.[6]

Je serais heureux, Monsieur, si le travail que j'ai fait peut vous être utile, j'insiste encore sur l'importance de vos collections pour qu'elles soient publiées.

Veuillez, je vous prie, disposer d⟨e⟩ moi en toutes circonstances et me croire | Votre très humble et très | Obéissant Serviteur | Alcide d'Orbigny

Paris ce 14 fevrier 1845

Votre caisse est chez M Ballière.

DAR 43.1: 62–5

CD ANNOTATIONS

Top of first page: 'M. Hombron | astonished at amount of my labour' *ink*
'(B. Blanca | Silurian Falkland)'[7] *pencil*

[1] For a translation of this letter, see Appendix I.

[2] CD eventually included four plates of shells in *South America*.

[3] Orbigny (1835–47, vol. 3, pt 3: *Géologie*, p. 90) had earlier described the Concepción sandstone as Tertiary. Though CD was inclined to accept the new Cretaceous dating, he noted the possibility that the baculite may simply have been a survivor from an earlier epoch (*South America*, pp. 126–31).

[4] According to CD, the shells distinguished by Orbigny as Jurassic and Turonian (Upper Cretaceous) had all come from the same deposit. CD, who preferred the term Oolitic to Jurassic, compromised on the expression 'Cretaceo-Oolitic' (*South America*, p. 216).

[5] Orbigny (1835–47, vol. 3, pt 3: *Géologie*, p. 191) had criticised CD's opinion (*Journal of researches*, p. 423) that a few living species were to be found in the lower beds at Coquimbo. Following George Brettingham Sowerby's species determinations, CD qualified his position in *South America*, pp. 128–9, and in *Journal of researches* 2d ed., p. 344.

[6] Orbigny believed that the shells were joined together in a way that would have been easily disturbed, and that the molluscs were therefore still living when the sea-bed was elevated. According to CD, the molluscs appeared to have died long before the shells were lifted up (*South America*, pp. 46–8).

[7] Roderick Impey Murchison and James de Carle Sowerby considered CD's Falklands specimens to be Silurian (*Journal of researches*, p. 253). The specimens were later described and dated by John Morris and Daniel Sharpe (Morris and Sharpe 1846). Three were described as Silurian and three as resembling Devonian forms. CD summarises the results in 'On the geology of the Falkland Islands', *Collected papers* 1: 203–4.

From A. C. V. D. d'Orbigny [14 February 1845?]

[Identifies CD's fossil shells and marks new species. Many species have been described previously by Orbigny. The list is annotated by CD and George Brettingham Sowerby.]

A memorandum 21pp
DAR 43.1: 68–89

From Edward Forbes [after 14 February 1845][1]

Thursday

Dear Darwin—

Any letters directed to the Survey—at 6 Craigs Court Charing X—will always find me—[2]

Gryphaea orientalis[3] is from the Southern India* beds & is Cretaceous—Probably upper GreenSand.

* Verdachellum[4] not Pondicherry in this case.

I shall look out for the Copiapo Avicula which had escaped me.[5]

D'Orbigny & Von Buch no doubt mean the same thing.[6] Exogyra or Gryphæa couloni is a characteristic Neocomian fossil. Von Buch's conclusion of the beds forming a passage from the oolites to the Chalk is exactly what I believe to be true. Von Buch wrote before the "Neocomian" was investigated.[7]

The relative positions would be as follows:

England	Switzerld [France]	S America.
Lower Green Sand	Neocomian	Your beds.
Wealden	oolites	
ool. beds		

D'Orbigny probably looks on the Amer beds to be purely "Neocomian" since he holds the doctrine of definite divisions or formations everywhere.

Ever, most sincerely | Edward Forbes

DAR 43.1: 47–8

[1] Date based on Alcide d'Orbigny's naming of CD's South American shells (see letter from A. C. V. D. d'Orbigny, 14 February 1845).

[2] Forbes had begun work at the Geological Survey on 1 November 1844 (Wilson and Geikie 1861, p. 378).

[3] Forbes regarded one of CD's *Gryphæa* specimens as identical to *Gryphæa orientalis* (*South America*, p. 212).

[4] Viruddhachalam or Vriddhachalam, south-west of Pondicherry.

[5] See *South America*, p. 223, where *Avicula* was listed as a new species by Orbigny. CD stated that George Brettingham Sowerby considered it to have been *A. echinata* and noted that the specimen had been lost.

[6] According to Orbigny, CD's *Gryphæa* was similar to *G. couloni* from the Neocomian formations of France and Neuchâtel. Christian Leopold von Buch, who had also examined a *Gryphæa* (*Exogyra*) from a South American deposit, declared it to be identical to specimens of *G. couloni* from Southern

France and the Jura (Buch 1836, p. 471). The passage is scored in CD's copy in the Darwin Library–CUL. CD recorded their views and Forbes's clarification in *South America*, pp. 181, 193.

[7] The geographical extent and chronological position of the Neocomian beds were hotly debated throughout the 1840s, following their identification by Jules Thermann in 1835 (*EB*).

From Trenham Reeks 25 February 1845

Museum of Economic Geology
25th. Feby. 1845.

Dear Sir,

I have much pleasure in enclosing you a copy of the analysis of the Salt (No 759.)[1] I can find nothing in it to account for its inferiority and my former suspicion of its being Potassium proves groundless.

I am, Dear Sir | Your obedient Servt. | Trenham Reeks

C Darwin Esqre.

Analysis of Salt (No. 759)
from Patagonia.

Chloride of Sodium—	99.52
Sulphate of Lime	0.26
Earthy matter	0.22
	100.00

DAR 39.1: 45–6

CD ANNOTATIONS

Top of first page: 'Tuff specimens' *pencil, del pencil*
'Second d'Orbigny about value of Salt'[2] *pencil, del pencil*
'Not in Geology' *pencil*

[1] From a salt lake near the Rio Negro (see *Journal of researches* 2d ed., p. 66, and *South America*, p. 74, where the analysis is quoted).

[2] A reference to Orbigny 1835–47, 2: 130, in which the mining of salt from the pans of Andres Pas is discussed.

From J. D. Hooker [late February 1845]

Ghent
Saturday Night.

My dear Darwin

I feel that I should have answered your letter some time ago, if it were only to thank you for the most kind, but far too flattering sentiments towards me which it contains. My last to you must have been, I feel sure, a most grumbling & ill tempered communication, for it was composed under the irritation produced by news which did not please me, though if I am to believe me friends, who do surely know better than myself, I ought to be very thankful that such a situation as Edinbro' affords is open, if only to be tried for. My Father is so anxious for me to be home soon, that he desired me not to go to Germany, but allowed of my return by Belgium & Holland, though he did not grant me so much time as I have staid; but then I have taken the liberty to allow of a little paternal over-anxiety, & have also a strong wish to see all in these countries, that Germany may be taken by itself another time, should I ever be so fortunate as to compass a second continental tour. Your messages to Ehrenberg & to Dieffenbach I sent in my own name to my good friend Klotsczh, of the Herb. Reg. Berol.,[1] & they will I am sure be duly attended to.

Most heartily did I wish for your company over & over again in Paris, for Humboldt had lots of time there, & such quantities of things to ask about, that demanded much better answers than I could give: the more I saw of him, & he either came to my room or sent for me almost every morning, the more I liked him, to be sure his amazing volubility & the constant practice of quoting himself, his travels & his works, for every subject, savours somewhat of old age, but those who know him best, say that he must not be so judged, as he was ever the same in his younger days. Further, his habit of always asking questions & seldom proposing a subject for mutual discussion, or giving his own information except when asked, leads one to suppose, that he is collecting more materials than he has time to arrange or dispose of: but as I saw more of him, it became more & more evident, that his mind was still vigorous, that he was still a most extraordinary man.

Nothing proved this more than my proposing the question, (which I am **truly** *ashamed* to have forgotten till you so kindly reminded me of it), concerning the N.E. Europe river dividing two Bot. regions. I do not suppose that he drew breath for 20 minutes, during which he was engaged in telling all he knew on the distrib. of Siberian plants &c. The river is the Obi, to the E. of the Oural, to its W. bank Rhododendrons some Coniferæ & other marked plants proceed, & occupy the plains on both sides of the Irtych, but though these inhabit the W. bank itself of the Obi they do not cross it: some other facts connected with this river & subject, are to be found in Gmelin's Botany of Siberia,[2] a work we have, but which I have almost neglected. Another most singular fact in the Botany of these regions Humboldt also told me of, & that was, that all the rivers to the W. of the Oural are covered (their banks) with Oaks: none of them to East are, nor are these trees met with in any part of Siberia, until reaching the waters of the Arnour & other Chinese rivers, given off from the Yablonoi & Stanavoi ranges, what is still more remarkable is, that the said rivers both of W. Oural & Arnour have fresh-water lobsters, equally foreign to all the Siberian waters. The absence of Lobsters & Oaks in all the countries watered by the Siberian rivers is a wonderful fact & to Humb. quite inexplicable. The only analogous facts I know off are those connected with the difference of the Floras of Greenland & W. Baffins bay, which are in every respect trifling in comparison. Such are Humboldts strong arguments against the *migration of species*, a doctrine he has most studiously & repeatedly warned me against, as wholly untenable, ever quoting the to him unaccountable fact, that the Befarias of the Caraccas & Andes should be the same, without a double creation;[3] (there is no smothering the truth that he is garrulous upon his own observations).

Fancy my amazement, on being shewn by him one day, 15 sheets of *Kosmos*, all printed, all just arrived from Germany & all to be corrected together!—[4] In common with many other Paris men, I had given Kosmos up, especially as he himself had told me that it would not be finished for two years—the two first parts are to appear this year, no 1 is ready, 3 will conclude it. As an instance of the man's extraordinary memory, I may tell you that he gave me the heads of all the subjects of the two first parts, without once stopping, I do think he was almost ¼ hour incessantly going on, from one head to another: the general nature of the work is, a review of the present state of our knowledge of Astronomy physics & natural History

The Paris Botanists have little or nothing to say about Geographical distribution, they are far more occupied with Anatomical & Physiological questions. One thing they all appear to agree in is, that we want facts to generalize upon & all incline to the migrating side, as however they do so without examining for themselves, their opinion does not carry much weight. You know what my own sentiments were, that I considered migration to a great extent, as, at any rate a precipitate conclusion, & this after having paid some little attention to the subject. I now, chiefly from the results of working out your suggestions, incline to consider migration as the only cause of the dispersion or diffusion of a so called species. Nothing that Humboldt has said hitherto alters my opinion, though I can no more account for Rhododendrons not crossing the Obi, than for Eryngium campestris & many other Continental weeds not crossing the channel, though they run wild when brought across. Nothing will be more likely to settle my own mind satisfactorily, than a comparison of my Southern plants with those of the N. Hemisp. of those there are a proportion common to both temperate regions, should I be able to trace the majority of them from one zone to the other, I shall declare myself a good migrationist, if not I must hold the question still unsettled; in this work I am at present interrupted.

The change of Wheat into Rye is here wholly disbelieved; Lindley, who you would suppose from his English writings, puts some faith in it,[5] strongly denies all belief in it to Decaisne:[6] the latter told me that a person had produced accidentally, by sowing wild Strawberry seeds one plant with *simple*, not *ternate*, leaves, & that it again reproduced *simple*-leaved plants, this is something like Dorking fowl.[7] Some make use of such *exceptions* to deny the existence of species at all, I have a tract on the subject for your perusal.[8] Perottet declares the vegetation of the tops of the Nylgherries to be quite European.[9] Have you seen Hind's paper on Sandwich Isld Botany?[10] from what I read of it he seems *wholly*, **entirely** ignorant of the subject, I saw it hurriedly at Delesserts[11] & it appeared that he takes no notice of the most striking facts.

Decaisne is certainly the most promising Botanist of the day, he has lately established as good sexes in Algæ as are to be found in other cryptogamic plants,[12] which however is nothing to a new light[13]

I will send you D'Urville when I return, I have got letterpress only.[14] You could not have pleased Humboldt better than by your message, he was delighted, he certainly is failing fast, he never talks of his Asia voyage, all is of S. America.

DAR 100: 165–6

CD ANNOTATIONS
1.1 I feel . . . Siberian plants &c. 3.5] *crossed pencil*
3.5 The river is . . . trifling in comparison. 3.20] 'Can this be like Flora on opposite sides of Australia, two great islands joined' *added pencil*
3.9 Gmelin's Botany of Siberia,] *underl brown crayon*
3.18 I know . . . comparison. 3.20] *scored pencil*
3.22 ever quoting . . . creation; 3.24] *scored brown crayon*; 'When formerly hotter.' *added pencil*
4.1 Fancy . . . without examining 5.4] *crossed pencil*
6.1 The change . . . Lindley,] *crossed pencil*
6.3 Decaisne] *square bracket added before in brown crayon*
6.3 the latter . . . leaves, 6.5] *scored pencil*
6.7 I have . . . subject, 6.10] *scored pencil*
7.2 good sexes in Algæ as are to be found] *underl pencil*
8.1 I will . . . S. America. 8.3] *crossed pencil*
Top of first page: 'Ehrenberg Humboldt Hinds Pamphlet' *pencil*

1 See letters to J. D. Hooker, 22 [January 1845], and to C. G. Ehrenberg, 23 January [1845]. Johann Friedrich Klotzsch was keeper of the Royal Herbarium in Berlin.

2 Gmelin 1747–69. See also letter to J. D. Hooker, [10–11 November 1844], n. 7.

3 *Bejaria*, a member of the heather family. According to Alexander von Humboldt there were distinct species of this alpine plant near Caracas, Bogota, and Santa Fé (Humboldt 1814–29, 3: 497).

4 Humboldt 1845–62.

5 John Lindley described plant breeding experiments performed by Lord Arthur Hervey in 1843 (Lindley 1844). This article was used by Robert Chambers to defend the transformist position, see [Chambers] 1844, p. 221, and [Chambers] 1845, p. 111.

6 Joseph Decaisne.

7 Antoine Nicolas Duchesne found a race of strawberry with simple leaves in 1763, documented in Duchesne 1766. White Dorking fowl is an ancient British breed with five toes, thought to have been introduced with the Romans. The several coloured forms of Dorkings owe their origin to spontaneous variations (Tegetmeir 1866–7).

8 Gérard 1844. There are two copies in the Darwin Pamphlet Collection–CUL; one is marked 'À Monsieur Hooker père, hommage de l'auteur' and 'C Darwin Esq, from JDH'; the other is marked 'Given me by Dr Lindley'.

9 Perrottet 1838.

10 Probably Hinds 1845. This paper contains a sketch of flora of the Sandwich Islands (pp. 92–5).

11 Benjamin Jules Paul Delessert.

12 Decaisne 1845. The process of sexual reproduction in cryptogamic plants was first demonstrated by Johannes Japetus Smith Steenstrup in 1842 (Steenstrup 1842).

13 This letter is unsigned and may be incomplete. The final paragraph was written above the salutation.
14 Dumont d'Urville [1841–54].

From Edward Forbes [March? 1845]

[Comments on George Brettingham Sowerby's identifications of South American fossil shells. Notes from more than one memorandum.]

A memoranda 5pp
DAR 43.1: 53–5

From William Hopkins 3 March 1845

Cambridge
Mar. 3 1845

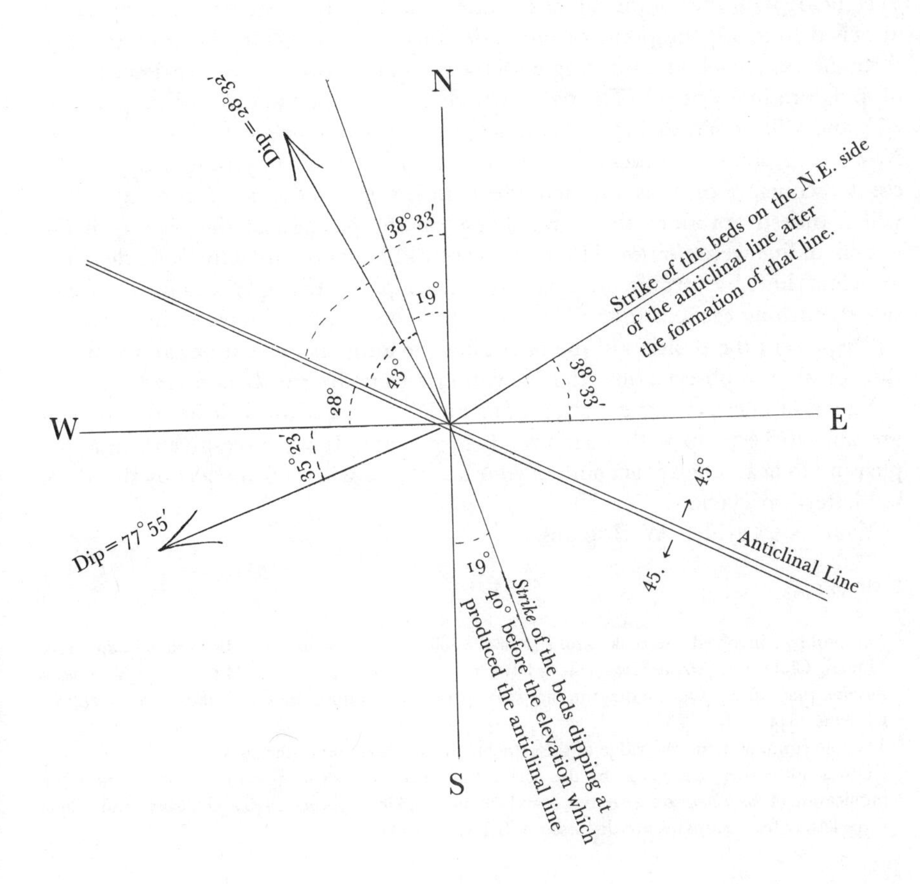

My dear Sir

The above diagram contains the solution of the problem which I received from you some time ago.[1] The directions of the lines are not given with accuracy in the figure, but the proper measure of each angle is given, so that the diagram might be made as accurate as you please.

All problems of this kind have a certain degree of complexity, but no real difficulty. I would have sent you an answer immediately but since I received your letter till within a few days I have been so entirely occupied during my leisure time in writing on glaciers for the Phil: Magazine[2] and in making experiments on the same subject, that I have really not had a moment till just now to work out the numerical results of your problem. But let me assure you that so far from thinking it a trouble to solve any problem of the kind, I consider the proposal of any geological problem, when put in a definite form, as your's is, as a positive favor. When ever one suggests itself to you, therefore, send it to me.

You might represent the problem pretty well to yourself by fixing a circular piece of stiff paper on an axis passing thro' the centre of the paper, and making with it an angle of about 40°. Place the axis horizontal and in a direction supposed to coincide with that of the anticlinal line;[3] and then turn the axis and the paper attached to it, till the plane of the paper meets an imaginary horizontal plane thro' the axis, in a line coinciding with the *strike* of the beds or laminæ described as dipping originally at 40°. The paper will then be inclined to the horizon at about 40°. and will be parallel to the laminæ just mentioned before their disturbance. Now in forming the anticlinal line we may (instead of supposing a *ridge* to be *elevated*) concieve the mass on each side of the line to *turn downwards* thro' 45°. This will manifestly produce the same change in the position of the beds as if the anticlinal line were *elevated*. This movement will be represented for one side of the anticlinal line, by turning your axis and the paper to the *right*, and for the other side by turning the axis to the *left*, thro' 45°. The plane of the paper in each case will represent the position of the beds after the anticlinal elevation, and you will thus be able to observe how that elevation affects their *strike* and *dip*.[4]

I do most sincerely regret both the fact and the cause of the little intercourse you are able to keep up with your friends in general. It w^{d} have given me much pleasure to hear a better account of your health— I will send you a copy shortly of my letters on glaciers.

Your's very truly | W Hopkins.

DAR 39.1: 53

[1] The problem involved the strike and dip of the foliated mica-schist CD observed at Cape Tres Montes, Chile (see *Correspondence* vol. 1, letter to J. S. Henslow, [10–]13 March 1835, and *South America*, pp. 158–9). On the diagram are some pencil calculations, by CD, relating to the angles.

[2] Hopkins 1845.

[3] The line running along the ridge of the anticline in the direction of the strike.

[4] CD was not entirely convinced by this account and evidently wrote again to Hopkins before the publication of *South America*, see letters from William Hopkins, 27 April 1846 and 5 May 1846. These three letters from Hopkins are discussed in Schwartz 1980.

From William Hallowes Miller 8 March [1845]

8 Park Terrace Cambridge
8 March

My dear Darwin

Pray excuse me for not having sooner given an account of the crystals you sent me—I was busy with lectures and other matters for some time after they came— The cleavage planes are very dull and uneven. I could make nothing of them except by using sun light for the upper signal. The angles usually fell within 15′ of 90°, but different parts of—apparently the same cleavage plane—tho' in fact cleavage planes of different crystals aggregated together in positions nearly but not quite parallel—differed in position full 15′ not unfrequently. So that I have little doubt of the identity of the crystals with Potash Feldspar[1]

I regret very much to hear that your health is indifferent. I am well. I have at last been able to escape from College and marry—[2] I live in Cambridge where I hope some time to have the pleasure of seeing you

I remain | Dear Darwin | Yours very truly | W H Miller

DAR 39.1: 42

CD ANNOTATIONS
1.1 Pray . . . in fact 1.5] *crossed pencil*
1.4 within 15′] *underl pencil*
2.1 I regret . . . Miller 3.1] *crossed pencil*

[1] Miller assisted CD by identifying a number of minerals in his South American collections (see *South America*, pp. 140–1, 153, 154). The description of the Portillo chain in the Chilean Cordilleras refers to pebbles containing crystals of 'reddish orthitic or potash feldspar (as determined by Professor Miller)' (*South America*, p. 182).

[2] Miller married Harriet Susan Minty on 5 November 1844 (*Alum. Cantab.*).

From C. G. Ehrenberg 13 March 1845[1]

Berlin
d. 13 Maerz 1845.

Mein hochzuverehrender Herr

Leider ist Herr Dr. Hooker nicht nach Berlin gekommen und somit haben sich die Erfüllungen Ihrer Wünsche noch mehr verspätet.[2] Da sich heut eine Gelegenheit biethet Ihnen das Manuscript über den Meteorstaub wieder zuzusenden, so ergreiffe ich dieselbe obschon ich dadurch gehindert bin Ihnen ausführlicher zu schreiben. Zuerst spreche ich einige Worte über Dr. Dieffenbach, der leider schon bei Ankunft Ihres ersten Briefes nicht mehr in Berlin war. Der arme Mann hat, wie Ihnen wohl bekannt ist, sein Vaterland verlaßen müßen, weil er in politischen Beziehungen gewesen war, die nicht geduldet wurden. Um dieß für sein weiteres Fortkommen weniger schädlich zu machen hatte er eben eine Reise angetreten mit der er sich zu gerichtlicher Vernehmung und wo möglich Reinigung in Gießen stellen wollte, in der Hoffnung bald los zu kommen. So sizt er aber nun bereits $\frac{1}{2}$ Jahr in Gießen, ohne das gewünschte Resultat zu erreichen.[3] Im Herbste war ich selbst auf mehrere Monate verreist und dann sehr beschäftigt. Ihren zweiten Brief habe ich stets bei mir zur Erinnerung gehabt, ich habe alsbald an Dr Dieffenbach geschrieben und von ihm beiliegende Antwort erhalten. Hoffentlich haben Sie durch den Buchhändler die gewünschten Sachen nun schon erhalten.

Was Ihre vortrefflichen Nachrichten und besonders Materialien über den atlantischen Meteorstaub anlangt, so kann ich Ihnen denn nun auch das Resultat der Untersuchung mittheilen.[4]

Außer den schon früher verzeichneten 37 mikroskopischen Organismen,[5] welche den Meteorstaub erfüllen und bilden helfen haben sich nun noch 29 andere erkennen laßen

Cocconema Lunula
Eunotia Argus
—— granulata
—— longicornis
—— Pileus
—— quaternaria
—— tridentula
—— Triodon
Gallionella decussata
Grammatophora oceanica

Gomphonema gracile
Navicula Bacillum
Pinnularia viridula
——
Amphidiscus armatus
—— obtusus
Lithodontium rostratum
—— platyodon
Lithostylidium biconcavum
—— crenulatum
—— Clepsammidium
—— Emblema
—— obliquum

Lithostylidium Rhombus
—— rostratum
—— Rajarum
—— unidentatum
Spongolithis cenocephala
—— Fustis
——
Textilaria globulosa.

Diese sind nun dadurch sehr merkwürdig, daß auch unter ihnen keine einzige der eigenthümlichen mir schon mannichfach bekan⟨n⟩ten Küstenformen von Senegambien und aus der Sahara überhaupt aus Africa ist. Nur Lith. Rajarum ist mir auch aus Isle de France[6] bekannt.

Unter allen 66 Formen ist keine neue, indem Eunotia longicornis sehr ähnlich in Ungarn fossil vorkommt.

Dagegen wie unter den ersten 37 Arten 2 characteristisch für America waren, so sind unter den lezten 29 wieder 4 welche Guiana und Senegambien gemeinsam sind, nämlich Eunotia quaternaria Eunotia tridentula [Eunotia] Pileus Amphidiscus obtusus.

Ferner verhalten sich alle 6 von Ihnen mir übersandten Staub-Arten ganz gleich. Sie zeigen alle eine sehr reiche Mischung von Kieselschaligen Infusorien und auch gleicher Arten.

Man kann nun mit Sicherheit außprechen, daß aller Staub der 6 Proben aus einer und derselben Quelle kommen muß. Dieser Quelle läßt sich noch näher kommen.— Vorher waren nur Süßwasser Thierchen vorgekommen, diese konnten aus der Mitte eines Festlandes seyn. Jezt sind auch 2 reine Seethierchen dabey gefunden, Grammatophora oceanica und Textilaria globulosa. So muß nun die Quelle des Staubes eine Küsten-Gegend seyn. Am nächsten liegt freylich die afrikanische Küste, aber es sind keine afrikanischen Formen darunter, obschon viele davon Weltbürger sind. Dagegen sind nun 6 Südamerikanische dabei.

Ferner habe ich ermittelt, daß der Wind mit welchem der Staub niederfällt, nach Angabe der Schiffer nie ausdrücklich der Harmattan ist, sondern daß es der Passat-Wind ausdrücklich ist. Sie haben ihre eigne Erfahrung mit scharfer Unterscheidung nicht hervorgehoben, was aber sehr wünschenswerth ist. Wahrscheinlich stehen Ihnen noch Schiffs-Journale zu Gebote um dieß scharf zu ermitteln. Lebende Dinge habe ich nicht dabey gefunden, solche nemlich die schnell eingetrocknd noch Organe zeigten.— Es wäre gut auf San Jago und an der afrikanischen Küste directe vergleichende längere Beobachtungen über diesen wunderbaren Staub zu veranlaßen. Sie scheinen alle Mittel nahe zu haben um dieß zu erreichen und werden nun gewiß es versuchen.

Recht interessant war mir auch Ihre Schminke aus dem Feuerland die ebenfalls aus Infusorien besteht, das erste dort vorgekommene fossile Lager. Ich habe bis jezt 18 Species ermittelt, wovon keine neu ist aber lauter Süßwasser-Formen.[7]

Die weiße Erde in Patagonien würde ich für ein vulkanisches Product halten.[8] Sie erscheint bimsteinartig, nämlich wie verwitterter Bimstein und ich habe auch darinn, aber nur wenig, fast unkentliche, doch mir noch deutlich werdende Infusorien Schalen erkannt. Es ist vorläufig eine Species der Gattung Fragilaria, wie es scheint, nennbar. Ich habe leider jezt zu viel andere Beschäftigung um recht intensiv die Sache zu verfolgen, doch halte ich das Verhältniß für sehr interessant wenn es nicht allzu lokal ist. Ist die Erde nur staubig, nicht festes Gestein? Wie mächtig schätzen Sie die Lagerung? wie weit die Ausdehnung? Stunden? Meilen? Sollte es wirklich Hunderte von Meilen weit ausgedehnt seyn, wenn ich recht verstehe, so müßen deutliche vulkanische Gebirgs Formen es so weit begleiten.

Weiter kann ich Ihnen für heut keine Nachricht geben. Nur noch großen Dank für die übersandten reichen Materialien, Ihnen und Herrn Dr. Hooker der leider verschmäht hat einige Zeit unter uns, die wir ihn recht freundlich aufgenommen haben würden, zu verweilen.

Aus großen Meeres Tiefen von beiläufig 10000 bis 20000 Fuß, man will ja 24000 F. gemessen haben wären mir die kleinsten Proben sehr erwünscht, mit oder ohne Talg des Senkloths.

In herzlicher Hochachtung verharrend | Ihr | ganz ergebenster | Dr C G Ehrenberg[9]

DAR 39.1: 59a–b

CD ANNOTATIONS

1.1 Leider . . . erhalten 1.17] *crossed pencil*

3.11 Textilaria globulosa] *added pencil cross*; 'marine' *added ink*

3.13 Grammatophora oceanica] *added pencil cross*; 'marine' *added ink*

4.1 Diese . . . bekannt. 4.4] '?' *added pencil*
9.7 Es wäre . . . versuchen. 9.10] *scored pencil*
9.9 Mittel] *underl pencil*
10.2 das erste dort vorgekommene] *underl pencil*; '?' *added pencil*
11.10 Gebirgs] *underl pencil*; '?' *added pencil over* '?'
12.1 Weiter . . . verharrend 14.1] *crossed pencil*
Top of second page: 'Patagonian tuff—& Fuegian paint' *ink*
Bottom of last page: 'Pampas— | Cordillera Honestone | my infusoria' *pencil*
'I believe nothing particular' *pencil*

[1] For a translation of this letter, see Appendix I.
[2] Since Joseph Dalton Hooker intended to visit Berlin, CD had asked Ehrenberg to return by him the manuscript of 'An account of the fine dust which often falls on vessels in the Atlantic Ocean' (*Collected papers* 1: 199–203) and also to urge Ernst Dieffenbach to return the materials CD had sent to him for the German edition of *Journal of researches*. See letter to C. G. Ehrenberg, 23 January [1845], and letter from J. D. Hooker, [late February 1845].
[3] Ernst Dieffenbach had been implicated in the April 1833 storming of the Frankfurt guardhouse by a militant liberal group and had had to flee Germany. In 1843 he was informed that proceedings against him had been dropped in Prussia; however, he was not able to return permanently to his home in Giessen until he had permission from authorities in Hesse-Darmstadt (Bell 1976, pp. 19–21, 87–9).
[4] Ehrenberg (1845a, pp. 64–6) had reported on these findings before the Berlin Academy on 27 February 1845.
[5] Described in Ehrenberg 1844a, pp. 194–207.
[6] Mauritius.
[7] Described in Ehrenberg 1845a, pp. 63–4.
[8] Described in Ehrenberg 1845b, pp. 143–8.
[9] CD's abstract of this letter is in DAR 39.1: 59c.

From Trenham Reeks 14 March 1845

Museum of Economic Geology
14th. March 1845.

Dear Sir,

I had some conversation with Dr. Kane of Dublin,[1] last Evening upon the decomposition of Carbonate of Lime and Common Salt[2] and he mentioned that M D'Arcet[3] had found upon mixing these two substances after a certain time an efflorescence of Carbonate of Soda made its appearance. Upon this ground a manufactory for Carbonate of Soda was established[4] which very soon failed from the small quantities produced. So that it appears there must be some amount of action take place upon these bodies *en masse*.

Mr Phillips[5] desires me to say his time is so very much occupied he has hardly had time to write but that he finds the rock to contain only traces of Sulphate of Lime.

I am, Dear Sir | Your's sincerely | Trenham Reeks

C Darwin Esqre. | &c &c &c

DAR 39.1: 47–8

CD ANNOTATIONS
1.1 I had . . . *en masse.* 1.7] *crossed ink*
Bottom of last page: 'Tuff from Patagonia' *pencil*

[1] Robert John Kane.
[2] CD had written for information on this subject to the *Gardeners' Chronicle and Agricultural Gazette*, 14 September 1844 (*Collected papers* 1: 198–9), see second letter to *Gardeners' Chronicle and Agricultural Gazette*, [before 14 September 1844].
[3] Jean d'Arcet, French chemist, much of whose research had industrial applications.
[4] Possibly a reference to a French government report dealing with the LeBlanc process and other similar procedures (Arcet *et al.* 1794). See *South America*, p. 52 n.
[5] Richard Phillips, curator of the Museum of Economic Geology.

From John Higgins 15 March 1845

Alford
15th March 1845

Dear Sir,

I have obtained the offer of another Estate in this Neighbourhood; which I think very likely to suit you; and I enclose a Plan & Particular of it.[1]

The Estate consists of a small Farm House, but moderate out buildings; (for which a suitable reduction is made in the price) and about 325 Acres of Land of average quality, in the occupation of an industrious good Tenant.—

It is situate about 2 Miles from Claythorpe, where your Father has Property; 2½ Miles from Alford (my residence) 11 Miles from Louth; and 11 Miles from Spilsby.—[2] The Tithes are commuted at about 4s/3d Pr. acre; parochial charges are very moderate; and the Church has been lately nearly rebuilt; and the Money borrowed has been paid off.—

The Price will be about *£12'500*; and it will yield a steady clear Rental of 3¼ Prcent, free from deductions.— Upon the whole, I consider it a *good* and *safe* Investment; and I think the price asked does not exceed the value; on the contrary I know a Person who will give that sum if it is declined by you.[3]

I am pressed for as early an answer as possible; and

Incomplete
DAR 210.25

CD ANNOTATIONS
4.1 The Price . . . and 5.1] *crossed pencil*

[1] There is a plan of an estate at Beesby, Lincolnshire (see nn. 2 and 3, below) in DAR 210.25. The plan is dated 'March 1845'. For CD's intention to invest in farmland and his earlier dealings with Higgins see letter to Susan Darwin, [27 November 1844?].
[2] This location corresponds to that of the farm at Beesby CD subsequently purchased, see n. 3, below.
[3] Robert Waring Darwin paid a deposit of £620 on 7 April 1845 for the purchase of the farm at Beesby

(R. W. Darwin's Investment Book (Down House MS), p. 74). The same amount and date are recorded by CD in his Investment Book (Down House MS).

To John Murray 17 March [1845]

Down near Bromley Kent
March 17th.

Sir

Having heard from Mr. Lyell that you feel inclined to publish a second edition of my Journal, I take the liberty of addressing you on the subject.[1] Mr Lyell has shown you a copy of the agreement by which I am legally free to publish with anyone, and I am bound to Mr Colburn by no other claim: I could explain by word of mouth a transaction of such a nature on his part, that nothing should induce me to have any further business-transaction with him.[2] It was entirely through Capt. FitzRoy, that our work was published by Mr Colburn.— I should propose to shorten a little, in a second edition, the geology & natural History & add something to my notices on the Fuegian savages &c &c; though I cannot add much, as I did not take many notes, leaving the subject to Capt. FitzRoy: I apprehend the corrected volume would be nearly its present size without the appendix perhaps a little smaller: I hope to derive advantage from the notes to a late German Edition.[3] Some few woodcuts might be added if you thought it adviseable.

The only difficulty which I see in the way of a second edition, is my ignorance how many copies remain unsold,[4] & how to ascertain this or how their purchase, if as I believe, they are few, could be negotiated:[5] I should be most unwilling to enter into any communication with Mr Colburn, & I am sure, from past experience, I shd not manage it successfully.

Mr Colburn has not for a long time noticed my volume in his advertising lists; & even if ⟨he⟩ had done so, the channel is an indifferent one: many people, to this day, believe that my work is united to Captain King's & FitzRoy's

If I may at all judge from the opinion of scientific acquaintances, I should hope for a considerable sale: I may mention, that Capt. Basil Hall, long ago, urged me to publish a cheaper edition.— I shd be perfectly satisfied to see my Journal in your excellent Colonial Library, or published in any other manner by you, which you might think preferable.[6] I beg to apologise for the length of th⟨is⟩ ⟨ ⟩ & I hope, whenever convenient to yourse⟨lf⟩ that I may hear your views on this subject.

I beg to remain | Sir | Yours faithfully | C. Darwin

Postmark: MR 17 1845
John Murray Archive (Darwin 354–5)

[1] Charles Lyell had talked with Murray about a possible second edition of CD's *Journal of researches* (see letter to Charles Lyell, [8 February 1845]).

[2] This refers to the incident described in CD's letter to John Murray, 12 April [1845].
[3] The notes were made for Ernst Dieffenbach's German translation (1844). CD had asked that they be returned (see letter to C. G. Ehrenberg, 23 January [1845]) but they have not been found and may not have been recovered by CD.
[4] According to Freeman (1977, p. 33) Henry Colburn printed 1500 copies of *Journal of researches*. In 1842 CD reported that 1337 copies had been sold, see *Correspondence* vol. 2, letter to Susan Darwin, [22 February 1842].
[5] Apparently this arrangement was not made, since Colburn advertised the *Journal of researches* in 1849, four years after the second edition published by Murray (Freeman 1977, p. 33).
[6] John Murray's Colonial and Home Library was a popular series. Works were issued in monthly parts at 2*s*. 6*d*. before being published in bound volumes.

To J. D. Hooker 19 March [1845]

Down near Bromley | Kent
March 19th.

My dear Hooker.

I presume that you are either returned home, or will soon return.— I have not sent back Wilkes,[1] & your Pacific MS. as I did not like doing so till I could hear that it should arrive safely, so as to be able to make enquiries should it not arrive. When I hear from you I will send it by the first weekly opportunity: I have had the other volumes from the London Library, so need not trouble you. I seldom read a book with so little observation on anything except savage mankind, & as I do not much care about this, I find the work very tedious. I am reading Forbes Alps, which I find a wonderful contrast in style and spirit.[2]

I shall be *very much* obliged for a loan of d'Urville,[3] whenever you can spare it, in the Autumn or summer.—& even more so, for the pamplet, you refer to, on, variation.—[4] I am particularly anxious to collect all such stray facts as the hereditary simple-leaved strawberry. I thought Hinds Regions of Vegetation, which I presume you refer to, pompous & very poor.[5]

Your last letter interested me exceedingly; how high you appear to have been in favour with Humboldt; I wish you had had time to have gone to Berlin & seen the great men there: have you seen Ehrenberg's statement of infusoria in certain pumice![6] How capitally it upsets the metallic-nucleus-oxidation theory,[7] not to mention all other theories, or notions.—it is so extraordinary, that it is almost mesmeric, though I beg Ehrenberg's pardon for comparing him in the remotest degree with the mildest mesmerist.

How I am to get my paper back from Ehrenberg now, I do not see.— I was very glad to hear Humboldts views on migrations & double creations: it is very presumptuous but I feel sure, that though one cannot prove extensive migration, the leading considerations, proper to the subject, are omitted, & I will venture to say, even by Humboldt.— I shd like sometime to put the case, like a lawyer, for your consideration, in the point of view, under which, I think it ought to be viewed: the conclusion, which I come to, is, that we cannot pretend, with our present knowledge, to put any limit to the possible & even probable migration of plants. If you can show that many of the Fuegian plants, common to Europe, are found in intermediate points, it will be grand argument in favour of the actuality of migration; but not finding them, will not in my eyes much diminish the probability of their having thus migrated.— My pen always runs away, when writing to you; & a most unsteady, vilely bad pace it goes.— What would I not give to write simple English, without having to rewrite & rewrite every sentence.

Ever yours | C. Darwin

When will you set off for the north?

I presume you are aware that Henslow has described a Galapagos Cactus.[8]

Would you tell me, what would be my best book in French or English on the morphology of plants; *especially* any book giving details on vegetable monsters & curious races of plants.

DAR 114.1: 28

[1] Wilkes 1845.
[2] J. D. Forbes 1843. CD had finished reading this work by 25 March 1845 (DAR 119; Vorzimmer 1977, p. 133).
[3] Dumont d'Urville [1841–54].
[4] Gérard 1844.
[5] Hinds 1843. There is a separately printed version in the Darwin Library–CUL. But Hooker was probably referring to Hinds 1845, see letter from J. D. Hooker, [late February 1845], n. 10.
[6] Ehrenberg 1844c. Christian Gottfried Ehrenberg claimed that pumice, tuff, and some other volcanic rocks were organic in origin and were formed by beds of peat, turf, or mud surrounding volcanoes being sucked into the craters and ejected during the next eruption. Infusoria contained in the original deposits fused into a variety of shapes and substances, all listed by Ehrenberg. Ehrenberg's investigations in this field were further reported in Ehrenberg 1845b.
[7] The theory that the earth contained a solid metallic core and that internal heat came from chemical processes, such as oxidation. Under this view, volcanoes were thought to be the result of intense chemical activity which would exclude the possibility of finding organic material in erupted rocks (Brush 1979).
[8] Henslow 1837.

To John Murray 20 March [1845]

Down Bromley Kent
March 20th.

Dear Sir

I am much obliged for your note, & have the pleasure of enclosing a copy (on the other side) of the agreement with Mr Colburn.

Believe me | Yours faithfully | C. Darwin

London Sept. 22d. 1837[1]

This is to certify that after the completion of the sale of the first Edition of the Voyages of Captns. King, & FitzRoy and Mr. Darwin, Mr Darwin is to be at liberty to publish his own Journal in any manner he pleases, separately from the general narrative, but this permission is not to prevent Capt^s King & FitzRoy from continuing to unite it with their own Journals in subsequent editions of the general work | Henry Colburn.

John Murray Archive (Darwin 8–9)

[1] See *Correspondence* vol. 2, letter to William Shoberl, [22 or 23 September 1837], for CD's acknowledgment of his copy of this agreement.

To C. G. Ehrenberg 23 March [1845][1]

Down near Bromley Kent
March 23

Dear Sir

I am exceedingly obliged to you for all the trouble, which you have so very kindly taken for me; I hope soon to receive my plates &c from Dr Dieffenbach's publisher.

Your account of the Atlantic dust is most interesting; I will add every particular, which I can about the direction of the wind, but I must say, that as the months, during which it falls on the African side of the Atlantic, are the same with those of the Harmattan, & as the first beginning of the falling of the dust has been observed in several cases & has always began with the wind between NE and SE, I cannot doubt that it comes from Africa, though it may have originally travelled from S. America. Could the Gulf-stream, which is said to sweep round as far as the Bay of Biscay, formerly (or still now) have brought S. American minute organisms, & thrown them up on the African continent?

I know well how fully your time is occupied, but if could afford time to send me one line, telling me whether the little packet of dust, *which I myself collected* contains infusoria I sh^d be much obliged. Also, whether you have, or intend publishing the substance of the letter to me, that I may refer to it & copy correctly in my notice (which shall be read before the Geological Society) the names of the Infusoria?

I am grateful to you for your remarks on the white Patagonian stone; I had come, from several reasons to the same conclusion, with you, on its primarily volcanic origin. Unfortunately you do not tell me which of the specimens of the white stone, contains infusoria;[2] I believe I sent several with their localities. The formation is a great one; it is associated with much sulphate of lime; it is of the consistence of our chalk, perhaps a *little* softer, & has an immense extension. At Port St Julian it cannot be much less than 800 feet in thickness; it extends continuously for 200 geographical miles (& probably is of great breadth) & I believe is of much greater extension, for I have specimens from the northern parts of Patagonia & layers having exactly the same *external* characters at the Rio Negro, which gives an extension in a N. & S. line, of at least 550 miles. Should you be led *from your own curiosity* to make any further examination, would you kindly inform me of the *result*.

I have specimens of great beds, from the upper parts of the late-Secondary, or Cretaceous-epoch, formations of the Cordillera of Chile, which from their appearance I suspect abound or are formed of infusoria; & sh^d. you wish for such, I shall be delighted to send them.

Believe me | dear Sir, Your's sincerely obliged | C. Darwin

P.S. If you could spare time this Spring to examine the mud of the Pampas, to see, whether it contains fresh-water or salt-water infusoria, it would be the most important assistance to my work and kindness;[3] I pledge myself I would ask you to look at nothing else.

I forget whether I told you that the Fuegian white paint *is collected* in *fresh-water* brooks: how beautifully do your microscopical researches reveal the origin of things!

Museum für Naturkunde der Humboldt-Universität zu Berlin

[1] The postmarks are illegible, but the letter is clearly a reply to the letter from C. G. Ehrenberg, 13 March 1845.

[2] Ehrenberg reported on CD's specimens in April 1845 (Ehrenberg 1845b). In his paper he printed extracts from this letter and in a footnote said, 'I found them in all of the specimens' (p. 143).

[3] See letter to C. G. Ehrenberg, 21 May [1845], and *Journal of researches* 2d ed., pp. 129–30. CD was anxious to establish that the mud had been deposited in fresh or brackish water. Alcide d'Orbigny had maintained that it had been formed by a sudden inundation (*débâcle*) of sea-water. See *South America*, pp. 88, 248, and *Correspondence* vol. 2, letter to Charles Lyell, [16 December 1843].

From J. D. Hooker [23] March 1845

West Park Kew.
Sunday Evening | March 1845.

My dear Darwin

I had the pleasure of receiving your last welcome letter the morning after my arrival & thank you much for it. Wilkes you may return at your convenience, either sending it to Hiscock's Kew boat Hungerford Stairs, or by Parcel's delivery Co^y., the former is, if any thing, the most convenient, but it makes little matter. I quite agree with you in your opinion of (what I have read of) Forbes' Alps, it is written in a vigorous & manly style, he appears at once the naturalist & traveller & though personally unacquainted either with him or his friends, I look forward with much pleasure to making his acquaintance in Edinbro'. His glacier theory[1] is I believe much disputed, especially by Sedgewick & Hopkins of Cambridge,[2] at least I judge so from what a young Cambridge friend told me some time back; true or false it has the merit of great originality & certainly it is "not a bad idea".— D'Urville you shall have the moment it arrives, the more willingly because I do not like a new bought book to be idle, & I have no time to look into it myself; how foolish I must look to you for having bought books I have no time to read, but my present circumstances must be plead as some excuse.

With regard to Morphology (Vegetable) the best work I know is St. Hilaire's "Lecons de Botanique comprenant principalement la Morph. Veg. &c" Paris 2 Parts.—P. J. Loss. 10 Rue Hautefeuille—1841.[3] I shall send you a little pamphlet on the Doctrine, written by the man "Murray", which will I hope have a salutary effect on your course of study..[4] I see it will go by post & so it will prepare you for the subject. By the bye, Decaisne of Paris wrote to Lindley asking him about the transmutation of Cerealia, & the latter wrote back word that he did not believe in the change at all.. this rather surprized me I must confess, for though Lindley has never actually avowed himself a believer, the tenor of his communications apparently shewed him to be one, & this full recantation to a foreigner makes me think him not very candid to us his country men.

From all that I heard at Leyden, the Indian Islands seem not only to be peculiarly rich in species, but also to present many curious facts regarding the distribution of the individuals & species in the different localities. I talked much with Schlegel,[5] who appears a very nice fellow, he is strongly in favor of a multiple creation & against migration, & as he drew most of his arguments from Zoological grounds, I could not follow him well, he says he has long studied the subject & has come to that conclusion after a full consideration of the number of cases, in which a species is common to two narrow areas seperated by large tracts equally capable to all appearance of supporting the said species: from what I know of the Botany of these regions I incline decidedly to the migration principle, the number of dispersed species being very great & belonging to very transportable orders. Blume[6] told me that the Bos (bubalus?) of Java is decidedly the same as that of India, but that the species is nowhere found (not even fossil) in Sumatra, the high road to Java if it migrated: this is to me startling but Blume may be mistaken, or Bos may have been imported by the Javanese, a very different & more energetic people I suppose than the Sumatrans: I did not think of this latter explanation when with Blume, but Horsfield[7] would doubtless solve the difficulty.

The Holland Botanists are Miquel of Rotterdam, a most agreeable person & accomplished Botanist,[8] Blume of Leyden who has published a most beautiful work & knows the plants of Java well,[9] & de Vriese of Amsterdam,[10] all these, & to the first I attach some importance, are strong anti-migrationists. I do not think however that the subject has engrossed much of their attention— I have set Miquel to collect facts for you, which will probably lead to the modification of his own opinions, as a similar course did of mine. Certainly the further I trace a diffused species, the more natural its voyages seem & there are remarkably few plants that inhabit all countries, they are the exceptions. Schouw was one of the first, I think, who proposed a double creation amongst plants, & attributed the reappearance of a species in a remote spot to similar *momenta cosmica* (or some such name) influencing both spots, or rather producing the same form in both.[11] Unfortunately for his theory, besides these plants being the exceptions in the flora of a country, we further have dispersed species abounding under circumstances where all the momenta that we can appreciate are opposite. Except St. Helena & the Galapagos there are few spots not largely indebted to neighbouring lands for their vegetable productions. I do not know Owens opinion upon these subjects, nor do I like to ask him, for I never propose such subjects to these master minds without being soon convinced of the feebleness of my own reasoning in such matters: indeed my own opinion is wholly formed upon the arguments of others & I shall be quite content to be a gatherer of facts for you to work with. I advanced the subject with great trepidation to Brown the other day, & found him a strong migrationist, he quoted Schouw & shewed the folly of his reasoning by adducing the opposite characters of the vegetation of the temperate regions of the 2 hemispheres, though the *momenta cosmica* were similar, whilst at the same time there were so many species common to the two as to render it probable that had their appearance been owing to such a cause the whole vegetation would naturally have followed the impulse. he further alluded to the borowed nature of the flora of the S. Sea Islands & the plants of the coral reefs being such as are the most easily transported by birds winds & waves. I certainly cannot account for plants which put themselves to the trouble of going from the American Continent to New Zeald & L^d. Auckland group, not also visiting Tasmania; but there is no accounting for tastes!—

I shall not go N.[12] till the beginning of May I hope, & have quite a month full of work before that, Graham & Brown are trying to get me appointed as Assist. & Successor.[13]

S^t. Hilaire is perhaps too much upon pure Morphology for you, I will look out for a book on *Monsters* which is I suppose what you want, such things are generally the subject of detached monographs, of which there are several in the Ann. Sc. Nat. of considerable interest. if I was going to stay at home I would hunt them up by degrees. At present I do not see how we are to get your mss from Ehrenberg & the blocks from Devilbach but must wait a little, I expect to hear from Klotssczch soon & perhaps Schomburgk will take charge of them.. Reinwardt[14] at Leyden asked most particularly after you, as did Schlegel, but I could tell neither of them how you are.—Why?

Ever your's most truly | Jos D Hooker.

My pamphlet on variations of species has not yet arrived from Paris where I left it..[15]

DAR 100: 41–2

CD ANNOTATIONS

1.1 I had . . . country men. 2.11] *crossed pencil*

4.22 strong migrationist 4.23] *underl pencil*

4.30 I certainly . . . tastes!— 4.33] *heavily scored pencil*; 'How are currents in Antarctic regions Does the west wind go all round.' *added pencil*

5.1 I shall . . . left it.. 8.2] *crossed pencil*

[1] James David Forbes attributed the movement of glaciers to the viscosity of glacial ice.

[2] Adam Sedgwick and William Hopkins. Hopkins' views are set out in Hopkins 1845.

[3] Saint-Hilaire 1841.

[4] Murray 1845.
[5] Herman Schlegel, director of the Natural History Museum at Leiden, with its great collection of specimens from the Dutch East Indies.
[6] Carl Ludwig Blume.
[7] Thomas Horsfield, keeper of the East India Company Museum in London and previously stationed in Java.
[8] Friederich Anton Wilhelm Miquel.
[9] Blume and Fischer 1828[–51].
[10] Willem Hendrik de Vriese.
[11] Joakim Frederik Schouw. Schouw's 'momenta cosmica' were rather more like environmental factors than the vague influences that Hooker implies (Schouw 1816). CD's notes for further questions to ask Hooker on these subjects are in DAR 206: 17.
[12] To Edinburgh.
[13] Robert Brown and Robert Graham were trying to secure Hooker as assistant and successor to Graham, professor of botany at Edinburgh.
[14] Caspar Carl Reinwardt.
[15] Gérard 1844. See letter from J. D. Hooker, [late February 1845].

To J. D. Hooker [26 March 1845]

[Down]
Wednesday

My dear Hooker

I received your letter & the edifying pamphlet yesterday & as our carrier goes tonight I will return it & Wilkes,[1] & your Pacific M.S. by him.

I see the Un. St. Ex. Ex. make the Sandwich Flora eminently peculiar to itself.[2]

Your books will go by the Boat on Thursday or Friday.— I will write again in a few days

Ever Yours | C. D

DAR 114.1: 29

[1] Murray 1845 and Wilkes 1845.
[2] Wilkes 1845, 4: 282–3. The United States Exploring Expedition, carried out during 1838–42 under the command of Captain Charles Wilkes, surveyed the Pacific and southern oceans. The botanist was William Rich. See letter from J. D. Hooker, 30 December 1844, in which Hooker arrived at the same conclusion concerning the Sandwich Island Flora.

To J. D. Hooker 31 March [1845]

Down Bromley Kent
March 31st

My dear Hooker

I hope your Book has arrived safely with your M.S.. Have you noticed in the fourth vol. of Wilkes, there is a short discussion on the Flora of the Sandwich Ids., & he considers it as of a **very** *peculiar* & confined character: I shall be curious to see the real scientific reports, if they turn out as trust worthy.—[1] What a capital tour you have had, & how many great men, you have become acquainted with: by the way I have heard from Ehrenberg, who grieves much at your not having come, &

says you would have been hospitably received at Berlin. He has returned me my M.S. & most goodnaturedly has written to Dieffenbach, from whom also I heard seven weeks ago assuring me that my &c &c sh^{d} arrive in a few days & laying all the blame on the Publishers; but nothing has arrived! I have many things to write about, but am determined to write nothing, which will require any answer from you, as I am sure your time must be now fully occupied & more than occupied; but I shall keep some memoranda hereafter to screw knowledge out of you. Nothing would do you so much good as a little vanity, & then you would not talk of collecting facts for others, when, say just what you please, I am sure no one could put them to better use than yourself.—

I hope & trust you will find Edinburgh far pleasanter than you expect, though the lecturing must be a direful break in your Antarctic flora (of which there is a *little* recommendatory notice in L'Institut of last week):[2] I sh^{d} think that Forbes[3] was one of the cleverest men there; I have found him very civil in correspondence, but I am told he is as frigid as one of his own glaciers: & a capital theory I fully believe his to be.—

You are very kind in your enquiries about my health; I have nothing to say about it, being always much the same, some days better & some worse.— I believe I have not had one whole day or rather night, without my stomach having been greatly disordered, during the last three years, & most days great prostration of strength: thank you for your kindness, many of my friends, I believe, think me a hypocondriac. How late shall you be in Edinburgh: I ask because I think I shall probably take a tour, for my unlucky stomach's sake, to the Eildon hills near Melrose, in September to see some appearances like the 'parallel roads of Glen Roy'.—[4] & perhaps I might go further on & see you in Edinburgh if there.— I see I have kept to my determination, in a highly praiseworthy manner, & asked you nothing which requires an answer— one of the subject I am curious to discuss hereafter with you—is the position, as a method of induction, in which morphology stands; it seems to me a very curious point.— I will order St Hilaires book;[5] I have just finished three huge volumes by Is. St Hilaire on animal monsters, and a nasty curious subject it is.—[6]

Farewell with all good wishes Ever yours | C. Darwin

N.B. You may see that I have dubbed you a D^{r} again, as you are to be a Professor: have I not done right?—

DAR 114.1: 30

[1] Wilkes 1845, 4: 282–3.

[2] *L'Institut* 1ère sect. 13 (1845): 120.

[3] James David Forbes.

[4] Instead of making this tour, CD visited Shrewsbury in September and from there went on to see his new property in Lincolnshire and to visit William Herbert, the plant hybridiser, and Charles Waterton in Yorkshire (see 'Journal'; Appendix II).

[5] An annotated copy of Saint-Hilaire 1841 is in the Darwin Library–CUL.

[6] I. Geoffroy Saint-Hilaire 1832–7, recorded in CD's list of 'Books Read' (DAR 119; Vorzimmer 1977, p. 133) in an entry dated March 1845.

From J. D. Hooker [2–6 April 1845][1]

Gunn sent it, & Berkeley pronounces it as the same species with your's,[2] where then did the species originate? on the Beeches of Fuegia or of Tasmania? I suppose Vestiges would call it a case of parallel developement & arrestation in each country.[3] McCulloch & others a double creation[4]—Schouw a similar momentum cosmicum exerted in each; Lamarck would be nonplussed as I am amongst them all.

I fear I shall not be in Edinbro, in September but cannot in the least tell. I lecture D.V.[5] (if I do not stick) from the first week of May till the end of July, whether I shall have to Examine for degrees or no it is impossible to say, even if so it is uncertain whether that would keep me till September. most heartily I should enjoy to have a ramble with you, but as I enter upon the probability of giving also a winter course at Edinbro—from Jany–March, there would be little time left over to be in England during the year. Were you going in any other direction but Scotland I should be tempted to make a tour & meet you somewhere, but I am as much at sea regarding my prospects for the next 6 months as at any time during the Ant Exped. I have not a word of lectures prepared & Graham writes me that he preaches from *notes* which are at my service: not having *modest assurance* enough to be able to speak two words impromptu they will be very useless.

The principles of morphology appear to me to rest upon two modes of reasoning: one consists in following the modification of any given organ through the vegetable kingdom, this I suppose is the Inductive method, & the other is following the said modifications through the different states of a single species, tracing monsters in short; applying the rules for such instances to all Botany would be deductive reasoning if I remember my logic aright.[6]

I know some of the Officers of this N. Pole Exped.,[7] if you want any Questions solved during the voyage pray send them to me, in black & white, & I will give them to a careful officer.

I do not doubt the Flora of the Sandwich Islds being very peculiar, but the difficulty is to settle what amount of new species or of new genera produces peculiarity. One species will sometimes render a whole vegetation peculiar in the eyes of some. In some instances, which I mentioned to you before, & which Hinds has wholly overlooked,[8] the Flora of the Sandwich Group is quite singular, in the preponderance chiefly of *Lobeliaceæ & Scævoleæ* (if I remember); they are not however likely to strike a casual observer or to give a feature to the vegetation. Wilkes is probably indebted to his Botanist for the observation, which is just: no missionary book, nor does Cook (I think) or any other unpractised observer particularize the group as having any peculiarities of vegetation but the very contrary. I have not read Wilkes yet.. Our ideas of peculiarity are most loose, we have no standard, in the first instance we must know the absolute numerical amount of peculiar species, this must ever be the primary point, the leading fact, all other causes of peculiarity, as preponderance of a species, genus or higher group, or insulation of individuals &c &c must be secondary considerations. Except Brown & Humboldt, no one has attempted this, all seem to dread the making Bot. Geog. too exact a science, they find it far easier to speculate than to employ the inductive process. The first steps to tracing the progress of the creation of vegetation is to know the proportions in which the groups appear in different localities, & more particularly the relation which exists between the floras of the localities, a relation which must be expressed in numbers to be at all tangible.

I think you would like to read Kingdon's translation of Decandolle's Vegetable Organography,[9] do not buy it however. I asked Brown since I saw you & he said St Hilaire is the best of its kind for Morphology.[10] You should get Jussieus introduction to Botany,[11] it is only 6/, & worth in every respect 16/— admirably done & inconceivably cheap.

Ever with best compliments to Mrs Darwin | most truly yours | Jos D Hooker.

Incomplete
DAR 104: 220, 219

CD ANNOTATIONS
1.1 Gunn . . . your's,] 'a Cyttaria from a Beech of Van Dieman's land.' *added ink*
2.1 I fear . . . useless. 2.12] *crossed pencil*
3.6 aright.] '11' *added pencil*
4.1 I know . . . them to a 4.3] *crossed pencil*
5.1 I do not doubt] 'Geograph. Distribut' *added brown crayon*
5.8 Wilkes . . . species, this 5.13] *crossed pencil*
6.1 I think . . . Hooker. 7.1] *crossed pencil*

1 Dated by a letter from M. J. Berkeley to J. D. Hooker of 1 April 1845: 'The Cyttaria is I think the same with the Fuegian species' (Archives, Royal Botanic Gardens, Kew). By 8 April, Hooker had apparently informed CD that Ernst Dieffenbach was in London, see letter to Ernst Dieffenbach, 8 April [1845], and letter to J. D. Hooker, [16 April 1845].
2 See n. 1, above. Ronald Campbell Gunn collected plants, including the fungus *Cyttaria darwinii* in Tasmania, for the Hooker family.
3 A reference to *Vestiges of the natural history of creation* ([Chambers] 1844).
4 John MacCulloch put forward a progressionist interpretation of the fossil record and assumed successive creations and extinctions (MacCulloch 1837, 1: 128–46).
5 Deo volente.
6 Paragraph three was at one time excised and kept in CD's portfolio on classification and divergence (DAR 205.5) as evidenced by CD's pencilled annotation '11'.
7 Sir John Franklin's ill-fated expedition to search for a north-west passage left Britain on 19 May 1845. The vessels were the *Erebus* and *Terror*, the same ships that James Clark Ross commanded on the Antarctic voyage of 1839–43; on board were Francis Rawdon Moira Crozier and Robert McCormick, known to Hooker from the Ross expedition, and Harry S. Goodsir from Edinburgh.
8 Hinds 1845.
9 A. P. de Candolle 1839–40.
10 Saint-Hilaire 1841.
11 Jussieu 1842.

To John Murray [5 April 1845][1]

Down near Bromley | Kent
Saturday 6th

Dear Sir

I am very much obliged for your note. I have always understood the agreement with Mr Colburn[2] to signify, that he was not to have the power of publishing a second edition, except in conjunction with a second edition of Captś King & FitzRoy, (which is an extremely improbable event); & I am almost certain, from several circumstances that this was the sense in which Mr Colburn really meant it to be understood. But of course I must abide by the letter of the agreement, & of this you must be a far better judge than myself.—

My only objection to the Colonial Library is the great requisite reduction; I presume you have calculated that not more than $\frac{1}{5}$ must be reduced; though I am doubtful, whether when every fifth page is struck out, it will be possible to keep its present character of a mixture of science & Journal & I presume you would wish the scientific part to be chiefly sacrificed: however, I am a full believer, that it is hardly possible to reduce too much & perhaps I may find it easier than I expect.— I will consult on this, my excellent & kind friend & adviser Mr Lyell.—

I write now to ask, whether you would so far greatly oblige me, as to negotiate with Mr Colburn,[3] which I particularly dislike & shd do badly.— if you have any great objection I must write myself & I wd take the liberty of showing you my letter: it occurs that, as my Journal wd be reduced, & as my volume is required by Mr Colburn to complete the sets of the three volumes of the General Work he would perhaps be willing to allow the second edition, for some not large sum of ready money from me, without purchasing those of mine unsold.— if purchased I shd not know what to do with them.—

As I have never received one penny from Mr Colburn, I have some claim on him, & I think mere shame wd prevent him being rigid with me; though that is a weak hold on such a man.— Would you kindly let me hear from you on this latter subject pretty soon, & sincerely apologising for the trouble I give you.

I remain Dear Sir | Yours very faithfully | C. Darwin

P.S. | In case you should be unwilling to enter into the negotiation for me with Mr Colburn, I enclose a draft of the letter, I shd then propose to send to him: I daresay it is a very impolitic one, & I trust you will be so kind, as to give me your opinion on it with entire freedom,—criticising the whole or any part.—

To | J. Murray Esqe—

[Enclosure]

Dear Sir

I should be much obliged if you would inform me, how many copies of my Journal remain unsold.— My object in asking is, that in accordance with the

Agreement with you in my possession, I have entered into a communication with Mr. Murray respecting a second Edition in a considerably reduced form in the Colonial Library. I wish to know for what sum of money, paid at once, you would dispose of the remaining copies to me;—or for what not large sum, you would be willing to allow me to publish soon a reduced Edition, retaining the copies you yet have in order to dispose of them, with the volumes of Captains King & FitzRoy.—

As I have not yet received any sum of money for my volume, which you have sold to a considerable extent in the *separate form*,—a circumstance which all authors & others to whom, I have mentioned it, have thought hard on me,—I trust you will enter on the present negotiation in a favourable spirit.

Believe me | yours faithfully | C. Darwin

H. Colburn Esqe

P.S. An early answer would oblige me.—

John Murray Archive (Darwin 3–7)

[1] Although CD dated this letter 'Saturday 6th', the actual date of the Saturday was 5 April.
[2] See letter to John Murray, 20 March [1845].
[3] Murray evidently agreed to do this, see letters to John Murray, [10 April 1845], 12 April [1845], and 17 [April 1845], n. 1.

To Ernst Dieffenbach 6 April 1845[1]

[Down]

[With thanks for Dieffenbach's publication.][2] 'I consider your having made my work known in Germany, a full & ample recompense to such exertions, as I made during our Voyage'.

J. A. Stargardt, Marburg (catalogue 574, 11–13 November 1965)

[1] Date taken from the Stargardt catalogue.
[2] The German translation of *Journal of researches* (Dieffenbach trans. 1844).

To Ernst Dieffenbach 8 April [1845]

Down Bromley Kent
April 8th

My dear Sir

I have just heard that you are in London.— I wrote a few days ago to Giesen[1] to thank you very much for your copy of my Journal & to say how exceedingly pleased I am with its appearance & how much gratified I am in my work being thus rendered in some degree known in Germany.—

I also wrote to explain the circumstances, which makes me so anxious to have the plates & M.S. back again, & unfortunately I have not yet received them!

Have you written to Viewig?[2] Pray write again— I am sincerely sorry thus to give you so much trouble; but I am sure you will not think me unjustifiable in urging you to write again to Viewig to send the goods to Smith & Elders 65 Cornhill.

I hope you will write to me & tell me your plans & how long you are likely to be in England.

I sh[d] much like to see you: but unfortunately, I was in London last week & have been unwell ever since.

Believe me dear Sir | Yours very faithfully | C. Darwin

P.S. | Pray tell Viewig not to wait to alter the Plates. I hope, however, there is no cause for writing & that you have taken care to bring the Plates &c with you.

Hans Rhyn

[1] Dieffenbach had been in Giessen to deal with personal legal problems (see letter from C. G. Ehrenberg, 13 March 1845).

[2] The firm of Friedrich Vieweg und Sohn was the publisher of Dieffenbach's translation of *Journal of researches*.

From C. G. Ehrenberg 8 April 1845[1]

Berlin
d. 8 April 1845.

Hochgeehrtester Herr,

Ihr Schreiben vom 23 Maerz habe ich erhalten und freue mich daß Ihre Wünsche einigermaßen befriedigt sind. Beigehend sende ich Ihnen den angezeigten Auszug aus meinem lezten Vortrage in der berliner Akademie der Wissenschaften, welcher auch auf die von Ihnen mir gesandten Materialien Rücksicht nimmt, so viel es nämlich bisher möglich gewesen ist dieselben durchzusehen.[2] Möge das Resultat mit Ihren übrigen Forschungen übereinstimmen. Was den Staub anlangt so wäre es wohl sehr interressant wenn Sie die Windrichtungen noch genau ermitteln könnten. Daß er zuweilen bis nahe an America hin vorkommt scheint mir nicht entscheidend für den afrikanischen Ursprung zu sprechen, da der Passat Wind von Africa her horizontal nach America hinweht, der Harmattan aber wohl schwerlich so tief ins Meer reicht. Was dem Passat anheim fällt muß mit ihm von Africa her am Meer hin wieder nach America oder bis nahe daran gehen, wenn es auch erst aus America oberhalb nach Africa getragen worden wäre. *Doch ist diese Vorstellung nur ein vorläufiger Versuch zur Erklärung des auffallenden Mangels ächt afrikanischer Formen.* Der von Ihnen selbst gesammelte Staub ist wohl Nr. IA meiner Tabelle,[3] alle übrigen sind von Mr.

James.[4] Eine besondere Bezeichnung haben Sie dem Päckchen nicht gegeben. Es war das der ersten Sendung. Sie können von meinen Mittheilungen beliebigen Gebrauch machen.

Was den weißen Patagonischen Fels bei St. Julian anlangt so habe ich folgendes daraus ermittelt. Es ist ein ganz entschiedener vulkanischer bimsteinartiger Tuff, dessen Gyps-Beimischung in den übersandten Proben nicht zu erkennen war. Wenn die Ausdehnung der Masse so groß ist so gehört sie zu den wichtigsten mir bekannten Oberflächen Verhältnißen der Erde. Die Masse ist so reich an mikroskopischen Kiesel-Schalen Organismen, daß auch das bimsteinartige scheinbar unorganische nur durch Fritten des Organischen entstanden zu seyn scheint. Das noch jezt erkennbare organische Element beträgt doch wohl sicher $\frac{1}{10}$ der Masse. Das meiste ist aber in sehr kleinen und etwas veränderten Fragmenten. Diese organische Beimengung ist bei Port St. Julian und Port Desire (Device??) ganz gleichartig, bei New Bay aber geringer, doch noch deutlich. Ich habe bis heut 28 species aus Ihren Proben herausgefunden. Darunter sind 21 bekannte, 7 neue mithin das Land characterisirende. Ihre Mittheilungen haben mich so sehr interessirt daß ich mit aller Kraft und Anstrengung die Analyse von Neuem vorgenommen habe.

Das Merkwürdigste bei diesem Verhältniß ist, daß während bisher die vulkanischen Massen überall nur gefrittete *Süßwasser- Formen* erkennen ließen, in dem Patagonischen Tuff vorherrschend *Meeres-Formen* sind, ein Verhältniß welches ich bisher umsonst gesucht hatte, das aber zu erwarten war.[5] Folgende Formen kann ich Ihnen nennen:

Die mit * bezeichneten sind neue Arten.

Kieselschalige Polygastrica:

	Port St. Julian	Port Desire	New Bay
Actinocyclus Venus?	–	+	–
Coscinodiscus marginatus	+	+	–
radiatus	–	+	–
* spinulosus	–	+	–
Diploneïs didyma	+	–	–
*Discoplea Mam̄illa	+	+	–
Fragilaria rhabdosoma	–	+	–
vulgaris?	–	–	+
*Gallionella ?coronata	+	+	–
* ?plana	+	+	–
sulcata	+	+	–
Goniothecium hispidum	+	–	–
*Hyalodiscus patagonicus	+	+	–
*Mastogonia ?Discoplea	+	+	–
Pinnularia borealis?	+	–	–

Kieselerdige Phytolitharia:

Lithasteriscus tuberculatus	+	—	—
Lithosphaera stellata	—	+	—
Lithostylidium amphiodon	+	—	—
articulatum	—	+	—
rostratum	—	—	+
Spongolithis acicularis	+	+	—
appendiculata	+	+	—
aspera	+	—	—
Caput serpentis	+	—	—
Clavus	—	+	—
Fustis	+	+	—
porosa	—	+	—
*Thylaecium hispidum	+	—	—

Kalkformen sind gar nicht vorgekommen und es scheinen daher diese beim Fritten mit den angrenzenden Kieselschalen das Glas gebildet zu haben was eben jezt den Bimstein des Tuffes ausmacht, und den organischen Character verloren hat. Über das Verhältniß des Gypses (Sulphate of lime) in Quantität und Vorkommen hätte ich mir gern etwas nähere Auskunft erbeten.

Es wäre sehr wichtig die Probe von Rio negro auch genau zu vergleichen, da das äußere Ansehn leicht täuscht. Überhaupt hätte ich bei dem Interesse des Gegenstandes gern *ein etwas groeßeres* wenigstens 1 Zoll großes *wirklich festes Steinstück* zur Prüfung gehabt. Sie besitzen wohl selbst nicht viel? Was Sie mir sandten waren pulverige kleine Mengen, nichts Festes. Ich habe aber, voraußetzend daß diese kleinen Proben von Ihnen selbst gewählt und als characteristisch für das Ganze erkannt sind, sie mühsam untersucht, besonders als ich Resultate kommen sah.

Hierzu füge ich auch noch die von Ihnen gewünschte Notiz über die Pampas. Sie haben mir zwar keine Proben mit dem Ausdruck Pampas bezeichnet zugesendet, allein ich glaube daß Sie damit die Proben aus Patagonien und den La Plata Ebenen meinen, welche Mastodonten einschließen. Da ich einiges davon bereits untersucht habe, so kann ich Ihnen wohl ein vorläufiges leitendes Resultat mittheilen.

Allerdings enthalten beide Lager auch mikroskopische Organismen. Das patagonische mit der Aufschrift: Earth attached to fossil Bones M. Hermoso in Patagonia hat mir 3 Formen polygastrischer Thiere und 6 Formen von Phytolitharien erkennen laßen, welche bis auf eine der lezteren sämtlich Süßwasserbildungen sind.[6]

Eine andere vorläufig untersuchte Probe hat die Aufschrift: Mastodon tooth with earth from the Plata Aestuary Mud.—Banks of the Parana. In dieser habe ich schon 7 Polygastrica und 13 Phytolitharia erkannt. Es sind fast zu gleichen

Theilen entschiedene Seethiere und Süßwasserformen leztere überwiegend. Daher ist die Ablagerung eine halbsalzige, brakische gewesen.

Das patagonische Lager ist vielleicht eine Süßwasserbildung, doch wird auch diese wohl vorsichtiger zur brakischen gerechnet, da die einzelne beobachtete Seeform deutlich ist.[7]

Ich besitze durch Dr. Tschudy zwar einige Kreideproben von Peru,[8] würde aber Ihre Proben von der Cordillera von Chile gern vergleichen.

Haben Sie nicht unvollständig verglaste Obsidiane von verschiedenen Orten mitgebracht? Diese könnten besonders interessant werden. Entglaste Obsidiane sind weniger interessant.

In ausgezeichneter Hochachtung verharrend | Ihr | freundlich | ganz ergebenster | Dr C G Ehrenberg

Der Staub welcher zwischen dem Senegal und Cayenne im hohen Ocean beobachtet worden, ist leider nicht gesammelt und es läßt sich daher vorläufig nicht entscheiden ob er nicht eine ganz andere Art von Staub gewesen. In Dr. Meyens Reise (der war mit Princeß Luise) steht, daß die Erscheinung näher an der afrikanischen Küste war.[9] Er hielt die Röthung der Segel für Generatio Spontanea einer kryptogamischen Pflanze, die er Ærophytum atlanticum nennt. Wahrscheinlich hat er die Thau-Perlen als Pflanze gemeint. Er war ein oft sehr flüchtiger Beobachter, der mit der Phantasie beliebig nachhalf.[10]

DAR 39.1: 60–1b

CD ANNOTATIONS

End of final paragraph: 'Princip Source dust not collected | remark character possibly [*interl*] quite different' *pencil*

[1] For a translation of this letter, see Appendix I.

[2] Ehrenberg 1845a. CD's copy is in the Darwin Library–CUL.

[3] Ehrenberg enclosed with this letter a chart listing species of Infusoria found in Atlantic dust. The chart, in somewhat different form, was printed in Ehrenberg 1845a, pp. 85–7. On the MS version CD noted totals of species, which he cited in 'An account of the fine dust which often falls on vessels in the Atlantic ocean' (*Collected papers* 1: 201).

[4] Robert Bastard James, who had collected dust that had fallen on his vessel in the Atlantic Ocean (see *Correpondence* vol. 2, letter from R. B. James to Charles Lyell, [*c.* 10 March 1838]).

[5] Ehrenberg's results were read to the Berlin Academy on 24 April 1845 and published in Ehrenberg 1845b, pp. 143–8. CD's annotated copy is in the Darwin Library–CUL. For CD's use of Ehrenberg's information see *South America*, pp. 111, 118–19, 248.

[6] Ehrenberg 1845b, p. 147. CD used these results in *South America*, p. 81.

[7] If the Pampas mud had been deposited in fresh or brackish water, it would explain the absence of shells and the presence of large numbers of fossil mammal remains (see *South America*, pp. 98–9, and letter to C. G. Ehrenberg, 23 March [1845]).

[8] Johann Jacob von Tschudi, who had travelled in Peru in 1838.

[9] Meyen 1834–5, 1: 54–5.

[10] The final paragraph was written on the verso of Ehrenberg's chart, see n. 3, above.

To John Murray [10 April 1845][1]

Down Bromley Kent

Dear Sir

I feel as certain as I can about anything, which has passed some years since, that I never signed any paper of any *kind whatever* with Mr Colburn: I am almost sure it w^d^ have impressed itself on my memory, had I done so. Indeed, I sh^d^ **assuredly** have made a copy, as I did of the Agreement between Capt F & Mr Colburn, which Capt F. made without any consultation with me.

I enclose my rough copy, *which please to preserve*;[2] I w^d^ have made a better copy, but I am compelled to leave home for the day immediately.

I hope your negotiation will be successful & believe me dear Sir | Yours very faithfully | C. Darwin

P.S. | In case you sh^d^ come into any terms with Mr C. he will perhaps have forgotten, that he let the Editor of the German Edition[3] have the loan of the Copper plate & one or two of the woodcuts— these have not been yet returned owing to the carelessness of Viewig of Brunswick, but I have no reason to doubt that they are safe, & that I shall soon have them.—

John Murray Archive (Darwin 10–11)

[1] Dated by CD's reference to a trip away from Down and by an entry in CD's Account Book (Down House MS) recording that he paid for a trip to London on 10 April 1845. This was the only trip recorded between the dates of letters to John Murray, [5 April 1845] and 12 April [1845].

[2] It seems that CD now sent his original copy of the agreement between Robert FitzRoy and Henry Colburn, after having first sent a transcript to Murray in his letter of 20 March [1845]. For the text of the agreement see letter to John Murray, 20 March [1845].

[3] Ernst Dieffenbach.

To John Murray 12 April [1845]

Down near Bromley Kent
April 12th

Dear Sir

I am exceedingly vexed to find by your letter just received, that I have unintentionally deceived you. I did not perceive the drift of your questions, with respect to the signing of an agreement, & thought they referred only to a second Edition. I certainly have never signed *any agreement* of *any kind*; but the separate sale of my Journal in the first edition was with my *verbal consent* & *approval*. Mr Colburn urged me strongly, directly after the date of the first publication, to agree with him for a second edition & I most unwillingly assented on the same terms with the first edition with Capt FitzRoy, of which you have seen a copy. He accordingly sent me a paper to sign, but on reading it over, I found inserted, that I sh^d^ share risks: on this I indignantly sent it back by his clerk, with a message, that if not immediately returned in its proper form, I would have nothing more to do with it: it was never returned & I am assuredly free both legally & morally. I had heard a short time before of a similar scandalous proceeding by which an author was entrapped; & this was the subject, to which I alluded in my first letter: I inform you of this transaction in strict confidence, though I am ready to assert every word on oath. Further considering that Mr. C. printed 1500 copies of my Journal, & has obtained so high a price as 18s for the sale it appears of 1400, I have just cause to complain that he has never paid me anything; not even repaid me, what I was compelled to pay him for my presentation copies. I have no scruples, on any ground whatever, with respect to a second edition, but in point of honour, whatever it may be law, I presume he has a claim on the sale of the first, as by my ignorance I managed the affair so ill.— I think it would be better if you were to inform him, that you were mistaken in supposing that I meant to say, that I had not consented verbally to the separation of my Journal in the first edition; but you can add, or not as you think fit, that I am so far from thinking that he has any claim on me, that if treated illiberally, I would use every legal means to free myself from him.— If he writes, I must say that the separation was with my consent; but I will not write to him, without showing you my letter, as you have so obligingly interested yourself in this matter.

I assure you in truth, that I am more annoyed at having led you into this mistake, that I can well express: I fear you will repent of having had anything to do with my work. Of course, you will permit me to pay for the legal opinion, if our negotiation fails, at which I shall be most sincerely grieved.

I again apologise for the trouble, I have given you & believe me | dear Sir | Yours faithfully | C. Darwin

J. Murray Esqre.
Private

John Murray Archive (Box 37, loose letters)

To J. D. Hooker [16 April 1845][1]

Down Bromley Kent
Wednesday

My dear Hooker

I am particularly obliged to you for having informed me of Dieffenbach's arrival: other business compelled me to go to town & I called at the Geograph. Soc to enquire after him & by good luck there met him: he seems really sorry for the trouble his publisher has caused me, & he declares he will return through Brunswick on purpose.—

Many thanks for your information about books &c: how can you ask whether it bores me: I assure you deliberately that I consider all the assistance, which you have given me, is more than that I have received from anyone else, & is beyond valuing in my eyes.

When you tell me not to buy Kingdons trans. of Decand: veg: organ: do you mean you can *sometime* lend it me.[2] I will order Jussieu:[3] I hope you will not forget the French pamphlet on variation. & I sh^{d} like to see the iceberg-paper in Boston Journal.[4] I have got Couthouy's paper on Coral-Reefs.[5]

Your ideas on the formation of the flat icebergs appears to me quite new & very probable; I have occasionally wondered at their flatness, but never thought about their origin: on your view there ought to be, I should think, horizontal stratification. Goodsir, I have been very glad to hear, is going on the Arctic Expedition, as Naturalist.[6] At Forbes request,[7] I have sent him some suggestions (so will not trouble your friend) on icebergs, boulders &c &c

By the way, let me say, whilst I think of it, I saw C. Streletski the other day & found that Morris is describing for him the plants, in the tuff from Van Dieman land, & I said, that I had no doubt that you would put the few specimens, which I gave you, into his hands if he so liked.[8]

What a very curious fact is the discovery of the Cyttaria in Van Diemens land—the means of distribution in such cases, is indeed a non-plusser; & if one was forced to believe that the land & water had held even nearly the same relative positions, since the first existence of our present species, it would be a complete poser: it is indeed a most perplexing subject. I shall be very curious to know at some future day, the result of your examination of the species of T. del Fuego & the Cordillera: I have been reading parts of d'Urvilles sketch of the Flora of the Falkland Is^{ds}.[9] & I see, he makes a good many species identical with Europæan.

I fear your Edinburgh lectures must interfere for a time with your Antarctic Flora; though you seem, as yet, to have taken your lectures very coolly. I sh^{d} think the "modest assurance" to lecture from notes, instead of from a fully-written paper, would soon come to any one well understanding his subject: do not you think it is much pleasanter to hear a speaking than a reading lecturer. I heartily wish you well through them.

I must go to breakfast so farewell my dear Hooker. | Ever yours | C. Darwin

Whenever an Abstract of your paper on the Galapagos plants appears, I hope you will try & get me a copy.—

The enclosed little lichens, came from near summit of **most** barren isl^{d} of San Lorenzo off Lima: what on earth made me think them worth collecting I know not—please throw them away.

I have by this mornings post just heard from Murray, that he agrees to publish my Journal in the Colonial Library.[10] Did you make any comments criticisms or corrections on the margin of your copy of my Journal: if so, will you kindly lend me your copy & never mind if any of your criticisms are severe or short & few. I shall have to shorten my Journal a little.

I am determined not to give you much trouble or ask many questions now you are busy, but I must beg *sometime* for a single sentence about the Galapagos plants. viz what per-centage are (as far as is known) peculiar to the Archipelago? you have already told me that the plants have a S. American physionomy. And how far the collections bear out or contradict the notion of the different islands, having in some instances representative & different species.

Will you tell me, may I not leave out, without any loss, the little & imperfect account (p. 14–16) in my Journal of the oceanic confervæ? But I am breaking my vow of not giving you trouble, now that you must be so very busy.—

DAR 114.1: 31

[1] Dated from CD's reference to having received a letter from John Murray and his reply to Murray, 17 [April 1845].
[2] A. P. de Candolle 1839–40.
[3] Jussieu 1842.
[4] Hayes 1844, in which several of CD's observations on icebergs in South America are cited.
[5] Couthouy 1844. CD's annotated copy is in the Darwin Pamphlet Collection–CUL. Joseph Pitty Couthouy endorsed CD's coral reef theory.
[6] Harry Goodsir.
[7] Presumably Edward Forbes.
[8] See Strzelecki 1845, pp. 245–54, for John Morris's description of plant fossils. Tuff is any light porous rock, but CD may be referring to 'the yellowish compact limestone' near Hobart that Morris described (p. 254).
[9] Dumont d'Urville 1826.
[10] *Journal of researches* 2d ed. See the following letter to John Murray, 17 [April 1845].

To John Murray 17 [April 1845]

Down Bromley Kent
Thursday 17th

Dear Sir

I was very much pleased to receive your letter yesterday. Mr Colburn has behaved more liberally than I anticipated.[1] I find three Nos of the Colonial Library, each of 171 pages, will exactly hold my present vol: & index (without the appendix); now though I shd here & there like to add a sentence, some parts I am sure wd be improved by being shortened, & therefore I want to know how many pages you would prefer the three volumes together to be; & I will try as accurately as I can, to meet your wishes.[2] I also wish to know (if you would let me hear soon either from yourself, or a clerk) what are your opinions & wishes with respect to woodcuts: I do not much care about them myself, but here & there, I think there is a subject which would be well illustrated by one.[3]

Thirdly, when you would like to publish? & how quickly the numbers follow each other? I have engagements for about 14 days, & then I wd, without any halt, complete the whole copy,[4] & superintend the correcting the press (the quicker the better) whenever you like: I shall be however a rather slow worker in getting ready my few alterations in the first (or already printed) copy.

You will see, I have gone on the assumption of taking advantage of your offer of three numbers: I could not have made up my mind to have wasted so much time, as it would have cost to have shortened my work $\frac{1}{3}$; & I firmly believe it w^d have lessened whatever merit it possesses;—in this opinion I found Mr Lyell even stronger than myself.—

I shall gladly accept your offer of 100£ for the copyright, to be paid 3 months after date of publication, & shall be happy to sign an agreement to that effect, whenever you like to send one.[5]

I hope not to give you much more trouble & remain dear Sir with thanks, | Your's very faithfully | C. Darwin

In my calculation I have allowed one of your pages to hold 451 words & mine to hold 377 words.

Would the map appear in the First number?[6]

John Murray Archive (Darwin 23–4)

[1] Henry Colburn's letter of agreement is in the John Murray Archive and reads:

> Dear Sir:
> I hope you understand me clearly with respect to Mr Darwin's Volume, that although according to his agreement he is not entitled to republish until all copies are sold, yet so long a period has passed since he first published and he now wishes *you* to publish the new edition, I will offer no impediment in the way & you can proceed immediately
> I am Dear Sir | Yours very truly | H. Colburn

The letter is undated, but is endorsed 'April 1845'.

[2] In a letter to Leonard Horner dated [January 1847] (*Correspondence* vol. 4) CD recalled that:

> With respect to my *Journal*, I think the sketches in the second edition are pretty accurate; but in the first they are not so, for I foolishly trusted to my memory, & was much annoyed to find how hasty & inaccurate many of my remarks were, when I went over my huge pile of descriptions of each locality.—

[3] The number of woodcut illustrations was increased in the second edition from four to fourteen.

[4] CD worked on the second edition from 25 April to 25 August ('Journal'; Appendix II). His annotated copy of the first edition is in the Cambridge University Library. Emma Darwin also read and commented on this volume.

[5] No agreement has been found at John Murray's.

[6] *Journal of researches* 2d ed. had no map. See letter to John Murray, [4 June 1845].

To John Murray [23 April 1845][1]

Down Bromley Kent
Wednesday

Dear Sir

I am much obliged by your note.— I shall not be ready by the first of May.[2] My health is very uncertain & I have now lost a week & therefore I prefer not sending a line to the Printer till I have the corrected copy of the whole 1st No ready.

I presume you are aware that I have the copper plate of the map: according to my own likings & most people, it is a great advantage having a map bound up with any & every work of travels.

About woodcuts. I am not very eager I think, however two or three subjects w^d be improved by them.

With respect to the title, I am greatly puzzled, but will think it over; I am of opinion it ought to be called a Journal,—but I will not trouble you at present.

Yours very faithfully | C. Darwin

John Murray Archive (Darwin 12)

[1] The Wednesday following letter to John Murray, 17 [April 1845].
[2] CD recorded in his 'Journal' (Appendix II) that he began work on the second edition of *Journal of researches* on 25 April.

To W. D. Fox [24 April 1845][1]

Down Bromley Kent
Thursday

My dear Fox

It is some time since we have had any communication. I write now chiefly to say, that I heard some little time ago from Shrewsbury, in which they said they had wished to have asked you to have come to the Shrewsbury Agricult. Meeting,[2] but some invited & more self-offerers have filled the house, more in my opinion than ought to have been allowed. I shall keep out of the way. M^r & M^rs. Wilmot of Nott:[3] will be there & M^r & M^rs. Miss Gifford that was G. Holland & E. Holland.[4] By the way, was it not an odd & friendly thing, M^rs. Darwin & young Mr D. of Elston[5] called on my Brother a few weeks ago; & it seems the young man, whose appearance my Brother liked, has called several times formerly at Grt. Marlborough St. They would not let my Brother return the call.— I have forgotten to add that they desire me to say that they shall be particularly glad to see you at Shrewsbury, if you are inclined to go there any other time either before or after July. My Father has been pretty well lately; but yesterday we heard that his leg has suddenly inflamed & was very painful I hope it will not last; I intend going there for a week very shortly.[6]

We have had Ellen Tollet staying with us, & heard indirectly much of you: the Tolletts & you seem to have many acquaintances in common.

Our children are very well, notwithstanding this most cold-catching weather: poor Emma is as bad as she always is, when she is, as she is. (this last sentence is quite Shakespearian)[7]

As for myself, my most important news is that I have agreed with Murray for a second Edition of my Journal in the Colonial Library in three numbers; & thanks to the Geological fates, I have written my S. American volume the first time over.[8]

The only other piece of news about myself is, that I am turned into a Lincolnshire squire! my Father having invested for me in a Farm of 324 acres of good land near Alford.[9]

Have you read that strange unphilosophical, but capitally-written book, the Vestiges,[10] it has made more talk than any work of late, & has been by some attributed to me.—at which I ought to be much flattered & unflattered.

Ever yours My dear Fox.— | C. Darwin

Christ's College Library, Cambridge (Fox 69)

[1] The date is based on CD's completion of the first draft of *South America* (see n. 8, below).
[2] In 1845 the annual meeting of the Royal Agricultural Society took place in Shrewsbury. The principal day of the show was 17 July.
[3] Possibly Edward Woollet Wilmot of Worksop Manor, Nottinghamshire, a governor of the Society.
[4] The George Henry Hollands and Edward Holland.
[5] Elizabeth de St Croix Darwin and her son Robert Alvey Darwin.
[6] According to his 'Journal' (Appendix II), CD left for Shrewsbury on 29 April, returning on 10 May.
[7] Emma Darwin was pregnant with her fifth child.
[8] CD completed the first draft of *South America* on 24 April ('Journal'; Appendix II). Regarding the agreement with John Murray, see letter to John Murray, 17 [April 1845].
[9] The Beesby farm. CD's Investment Book (Down House MS) shows that he paid £213 13*s*. 8*d*. interest to his father on 10 August 1846. The total amount advanced by his father was £13,592 0*s*. 7½*d*. See also Keith 1955, p. 222.
[10] [Chambers] 1844.

To Thomas Bell [26 April – August 1845][1]

Down
Saturday

My dear Bell

Would you kindly take the trouble to send me a line (to 7. Park St. Grosvenor Square) to tell me Bibron's[2] address, without indeed, which w^d^ be much better, you could tell me the two points, I am anxious about.—viz, the name of the Trigonocephalus[3] from B. Blanca and secondly whether there be more than one species of snake at the Galapagos, & whether such have a S. American physiognomy.[4]

I am right, in thinking that you sent all my snakes to Bibron?

My reason for wanting to know is that I am preparing a second Edit. of my Journal for Murray's Colonial Library.

I hope you are well; does not this rain gladden your heart, when you think of the meadows of Selbourne?[5]

Pray give my kind compliments to M^rs^. Bell, and believe me | Ever yours truly | C. Darwin

W^d^ you let me have a line soon.—

New York Botanical Garden Library (Charles Finney Cox collection)

[1] The date range is that of the composition of *Journal of researches* 2d ed. ('Journal'; Appendix II).

[2] Gabriel Bibron. Bell had sent CD's *Beagle* collection of snakes to him (see *Reptiles*, p. vi) for use in Duméril and Bibron 1834–54, vols. 6 and 7: in vol. 6, p. xii, the authors acknowledge having received specimens from Bell.

[3] *Trigonocephalus* was the then accepted term for a genus which included pit vipers. The snake in question interested CD because, though it lacks a rattle, it displays the rattlesnake's habit of rapidly vibrating the end of its tail when irritated or surprised: 'showing how every character, even though it may be in some degree independent of structure, has a tendency to vary by slow degrees' (*Journal of researches* 2d ed., p. 96). In a note at the end of the volume (p. 506) CD reports that Bibron thought it to be a new species, which he proposed to call *T. crepitans*.

[4] 'There is one snake which is numerous; it is identical, as I am informed by M. Bibron, with the Psammophis Temminckii from Chile' (*Journal of researches* 2d ed., p. 381).

[5] Bell, a great admirer of Gilbert White, had purchased White's house, 'The Wakes', in Selborne, Hampshire.

To J. D. Hooker [28 April 1845][1]

Down Bromley Kent
Monday

My dear Hooker

I return your specimens & elegant drawings with many thanks; I truly hope you did not think of making those drawings for my sake; for the information is more copious than I required & the name, which Berkeley has given will do instead of my poor description: I wish I knew the name of the double conical confervoid bodies, which I have figured.—[2]

I enclose Berkeley's note, as I have copied all I want: w^{d} you direct my note of thanks to him.

I have been vexed at my stupidity in having mentioned the Galapagos Plants to you at this time, for my Journal comes out in 3 numbers & the Galapagos will be in the last.[3]

If you have a scored copy of my Journal, with any corrections, I sh^{d} be much obliged for the loan of it; but I beg you not to give yourself any trouble; if you do, (as you have done about the confervæ) I must lose the advantage of asking you a few questions.

I am going tomorrow to see my Father at Shrewsbury for 10 days[4] & on my return shall set to & finish my Journal.

Ever my dear Hooker | Yours most truly | C. Darwin

DAR 114.1: 32

[1] Dated on the basis of n. 4, below.

[2] An illustration of a double *Conferva* in *Journal of researches*, p. 16, which remained unnamed in the second edition (*Journal of researches* 2d ed., p. 15). Miles Joseph Berkeley, however, named the other species as *Trichodesmium erythræum*.

[3] The three parts of *Journal of researches* 2d ed. appeared on 28 June, *c.* 2 August, and *c.* 30 August (Freeman 1977, p. 35).

[4] CD set out for Shrewsbury on 29 April 1845 ('Journal'; Appendix II).

From J. D. Hooker [28 April 1845]

should shew him during my absence if you liked.

Either you have misunderstood me, or I have ill expressed myself about the Galapago plants, it was only the *first part* that I had got ready, I however devoted 10 days uninterruptedly lately & finished the species of all the rest, except the *Leguminosæ*, some 18, which Bentham will do. Said *first-part* has been 3 months before the L. Soc. but is not printed, when it is I shall not forget you.[1] The proportion of new species is prodigious & even the old ones have been most difficult to name, as there are no floras of Mexico Peru & Chili, There are 185 species in all, a goodly number, (of flowering plants & 42 Cryptogamic)

Of the flowering, 100 I am sure are new & most probably confined to the group. thus distributed

Jas Isld has	71 species	38 of which are confined to the Archipelago of these there are peculiar to the *individual Isld*—	30
Chas Isld —	68 —	29 ———	21
Albem ——	46 —	26 ———	22
Chatham —	32 —	16 ———	12

These results are most delightful & as soon as the *Legum.* are done, which will alter very slightly, the above, (they are included roughly determined by myself) I shall set to work with the essay on the distrib. of the species.[2] Thus, though Albemarle is one of the largest, it is the most sterile & the most peculiar, it is not only most different from the coast, but its new species are all but 4 peculiar to itself. I shall work the whole thing out very carefully at Edinbro. There are 21 *compositæ* of 12 genera. all but one species peculiar to the Archip. & 10 of the genera ditto. The most remarkable genus of these is *Scalesia* (arborescent) it has 6 species, & not one found on two of the Islds. I have also cursorily bolstered up materials for a florula of Juan Fernandez, to compare with this, there are some striking analogies, but the J. F. flora is as peculiar as any in the world, even as St. Helena!, both are rather more so than Galapagos. J. F. hardly contains one coast plant, its Compos are wonderful & have no affinity but with a new species I turned up amongst my Fathers *dubiæ* from Elizabeth Isld. I do not see that the Galap's have the *slightest* affinity with the S. Sea Islds.—

The St. Lorenzo Lichen I can make nothing of but have sent it to Dr Taylor,[3] with no hopes however: as I could not find fructification. You notice somewhere a *blown-about-Lichen* on the Andes, at Quillota is it?—it is an *Usnea* perhaps the Antarctic *U. melaxantha* but the specimens are very imperfect.[4]

You may depend I shall give speaking lectures the moment I am able, but I should infallibly stick were I to begin so: I feel very *earth-quaquey* already. I find it hard enough to write them, I am apt to get into an inflated style, which of all things I hate in others. You will see this in a little paper on Southern Coniferæ which I have put aside for you.[5] It is mighty hard to write good English. I am now cobbling up a course, but find it far more agreeable to break off & write to you. My Edinbro address is 20 Abercrombie Place where I have taken 2 rooms, I quite forget whereabouts that is. Forbes tells me that he can trace a connection between the Botany & Geology of W. Ireland with both these features of Portugal, the Bot relation is notorious.[6] There is a similar relation mentioned in "Asie Centrale" between the Bruyeries of Holland across to Tobolsk?[7] & I thought when reading it that that stretched into Norfolk & Suffolk, Newmarket heath &c.. Brown shewed me a funny thing. Some Liverpool Parson, after reading "Vestiges", had written to all Geologists for proofs on the contrary, & rather coolly, printed all the answers.[8] Every one, but Delabeche, referred said parson to their own works!— I could not get the thing. I suppose you have read Bosanquets answer,[9] it is not half so *nice* as Vestiges. Do not growl at this long letter, I shall not trouble you again for some time—I go on Wednes⟨day⟩ & commence on Monday—

Farewell | J D Hooker

Postmark: AP 28 1845
Incomplete
DAR 100: 48

CD ANNOTATIONS
1.1 should . . . liked.] *crossed pencil*
2.7 185] '$\frac{40}{225}$' *added pencil below*[10]
2.8 42 Cryptogamic] '0' *added pencil over* '2'[11]
3.2 distributed] '((I do not understand))' *added pencil*
3.3 Jas Isld] '(other isld of group)' *added pencil above*
3.3 Jas Isld] '8' *added pencil before*
3.5 Chas Isld] '8' *added pencil before*
3.6 Albem] '4' *added pencil before*
3.7 Chatham] '4' *added pencil before*
CD created a new column in the table 'ie other parts of world (inhabited)' *to go after Hooker's* 'which are confined to the archipelago'. *The figures he calculated for each row are* '33', '39', '18', '16' 'Leguminosæ (not too much confidence)' *added to table*
4.13 its Compos . . . S. Sea Islds.— 4.15] *scored pencil*
4.14 Elizabeth Isld.] *underl pencil*
5.1 The St Lorenzo . . . that is. 6.8] *crossed pencil*
6.12 & Suffolk . . . Hooker 7.1] *crossed pencil*

1 Hooker presented the first results of his Galápagos work at three meetings of the Linnean Society, on 4 March, 6 May, and 16 December 1845, but the paper was not published until 1851 (J. D. Hooker 1845d). He sent CD a copy (Darwin Pamphlet Collection–CUL).

2 J. D. Hooker 1846.

3 Thomas Taylor, an expert on mosses and lichens, who provided much of the material on cryptogams in J. D. Hooker 1844–7.

4 *Journal of researches*, p. 444. Hooker was attempting to chart the geographical range of *Usnea melaxantha*, see J. D. Hooker 1844–7, pp. 519–21.

5 J. D. Hooker 1845a. See letter to J. D. Hooker, 22 [January 1845].

6 Edward Forbes was to announce his theory in June that year (E. Forbes 1845).

7 Humboldt 1843, 1: 54–5, in which the distribution of heaths across Europe is discussed.

[8] Hume 1845, a response to [Chambers] 1844.
[9] Bosanquet 1845.
[10] CD's '$\frac{40}{225}$' was written directly below Hooker's '185'. Thus he could have been adding the number of flowering plants to the cryptogamic plants to get the total number of species. However, it is unclear exactly what Hooker meant, as CD pointed out later (see enclosure with letter to J. D. Hooker, [11–12 July 1845]).
[11] See Hooker's annotation on the enclosure with letter to J. D. Hooker, [11–12 July 1845], n. 22, and letter from J. D. Hooker, [after 12 July 1845], where Hooker corrects this figure to '40'.

To J. D. Hooker [May 1845][1]

Down Bromley Kent
Thursday.

My dear Hooker

Every letter of mine to you ought & must begin with thanks for all the trouble you take for me. Imprimis I return all the documents on the Confervoid bodies: when you speak of my Trichodesmium 390, 391 being a "South Sea" species, you are aware that it was collected near the Abrolhos in the Atlantic; in my Journal I refer to another closely allied, rather smaller species seen off C. Leewin in Australia.[2]

If you find my Journal & can give me the names of any plants, I sh^{d} be very glad of it: if it be lost I beg you will let me have the *pleasure* of giving you a copy of the new Edit.

I shall not trouble myself about the Van Diemen's leaves: & with respect to the Beech leaves, I have already your opinion, & before Christmas, perhaps you w^{d} have time to look at them once again.

I have had several other communications with Ehrenberg, who has been uncommonly kind & obliging: he finds in the Atlantic dust 67 organic forms; but as none of them are those characteristic of Africa, he is very unwilling to believe it comes from Africa; though as I find the dust always falls with the wind between NE & SE & during the very same months, & only during these months, when the Harmattan from NE & ENE is known to blow clouds of dust from the Sahara, it appears to me absurd to doubt about it.

The loose lichen was lying on the sands of the real deserts of Iquique: Henslow said it was a Cladonia; I suppose he sent it with rest of the Crypt: to Berkeley.— By the way B. says he now thinks the Van Diemen Cyttaria a distinct species.[3]

I am very much obliged for your sketch of the Galapagos flora; it really turns out a most interesting case: I shall be very anxious to hear the final result, when the Legum: are worked out.— Albemarle differs from the others isl^{ds}. in having (at least in the parts which we visited) no damp fertile mountain-tops.—

I am glad there are materials for the Botany of Elizabeth isl^{d}; for it must be an interesting point, as one of the islands, nearest to America, of the not very low Pacific islands.—

I hope you have settled yourself pretty comfortably in Edinburgh, & that you take walks, like a good boy, & do not overwork yourself: I shall be glad to hear how the lectures go on & that you did not find them very terrific: the very thought of such a deed, as lecturing to a whole class, makes me feel awe-struck.

I beg you not to think of writing merely to answer this letter; but if at any time the spirit moves you to give me a little news about yourself, I shall be much pleased.— Depend upon it it is a very easy thing, though never believed till too late, to work the brain too much at the expence of the stomach; I dread the thought of your breaking down in your labours.

Ever my dear Hooker | Most truly yours | C. Darwin

P.S. I suppose I shall not have the Pamphlet &c till your return, & this will do *quite equally* as well as sooner for me: I sh^{d} not be surprised, that I have misunderstood you about D'Urville: I have seen the voyage before the very last & I suspect the last one has never been published: I have read the voyage in which he visited Vanikoro & found Peyrouse's wreck.[4]

DAR 114.1: 33

[1] Date based on Hooker's arrival in Edinburgh, see letter from J. D. Hooker, [28 April 1845].

[2] Cape Leeuwin, *Journal of researches*, p. 15.

[3] See letter from J. D. Hooker, [2–6 April 1845], in which Miles Joseph Berkeley's previous opinion that the *Cyttaria* of Tierra del Fuego and Tasmania are identical species is cited. See Berkeley 1845 for his final conclusions.

[4] Jean François de Galaup La Pérouse was shipwrecked at Vanikoro, New Hebrides, in 1788. Jules Sébastien César Dumont d'Urville commanded the expedition (1826–9) sent out to ascertain the fate of La Pérouse and his crew, and it is to this voyage that CD refers (see Dumont d'Urville 1832–3, vol. 5, ch. 34). Hooker had offered to lend CD the first parts of Dumont d'Urville's subsequent voyage in the *Astrolabe* and *Zélée* (Dumont d'Urville [1841–54]), see letter to J. D. Hooker, 22 [January 1845], and letter from J. D. Hooker, [late February 1845].

To G. B. Sowerby [May 1845][1]

Down Bromley | Kent
Thursday

My dear Sir

I had intended calling on you the last time that I was in London, (but was prevented) to ask you for the list of the Bahia Blanca fossil shells, which I daresay you have gone through by this time & which I am anxious to refer to.—[2] Would you be so good as to send it me as soon as you conveniently can by Post.—[3]

I am preparing a 2^{d} Edit of my Journal & therefore am compelled to put off going through the fossil shells for another month or six weeks.[4]

Believe me | dear Sir | Yours very faithfully | C. Darwin

New York Botanical Garden Library (Charles Finney Cox collection)

[1] Date based on the period when CD worked on the first number of *Journal of researches* 2d ed. (see n. 2, below).

[2] CD wanted an exact count of the recent and extinct species of shells found at Bahia Blanca in order to date the *Toxodon* and other large fossils found in the same formation. He had earlier claimed that the shells were mostly of existing species (*Journal of researches*, p. 97), but now he wanted a precise enumeration for the second edition (p. 83). The Bahia Blanca shells were discussed in the first number of the second edition, which CD worked on from 25 April until approximately 31 May. Revisions to the entire volume took him until 25 August ('Journal'; Appendix II, and letter to John Murray, [31 May 1845]).

[3] Sowerby's identifications of CD's South American shells are in DAR 43.1: 6–45, but there is no list for Bahia Blanca.

[4] The fossil shells CD refers to are presumably those to be described in *South America*. An appendix to that volume is devoted to Sowerby's description of sixty species of Tertiary fossil shells.

From W. B. Carpenter 5 May 1845

61 St Martins Lane. | London—
May 5— 1845.

My dear Sir

I should have written to you long since, respecting the results of my examination of the specimens you forwarded to me soon after Christmas;[1] had I not been prevented by unforeseen occurrences from subjecting the sections, which were then prepared, to as close an examination as I wished to give them— This, however, I have now done; and I have also made a couple of sections of the last specimen you forwarded to me,—that of the Coquimbo deposit; and the results of this more extended investigation are sufficiently interesting, to make me think it worthwhile to go over the whole subject with you again.

You are of course aware, in the first place, that any calcareous matter, in a state of crystalline aggregation, is *transparent* under the Microscope; whilst any *amorphous* deposit is quite opaque, even in a *very* thin layer. The two conditions are well seen in ordinary Shells of Mollusks, and in the Egg-shell. A tolerably thick section of the former is generally transparent, or at least translucent; the thinnest possible section of the latter is quite opaque. Not unfrequently there is in the Shells of certain Mollusks (e.g. the common Oyster) a layer of amorphous Carbonate of lime, which, altho' of immeasurable thinness, renders the portion of the section, in which it presents itself, quite opaque.

How far it is possible for any *mechanical attrition* to reduce the crystalline to the amorphous form, so completely as to change the microscopic character in the manner I have stated,—I can scarcely venture an opinion. That *Chemical precipitation* will do it,—we well know; but I am inclined to believe that a sufficiently prolonged action of attrition might also effect it.

Now supposing a deposit to have been thus reduced to the *amorphous* state, the question arises, whether, under any circumstances, it can resume, more or less a *crystalline* condition. I am strongly inclined to believe that the simple percolation of water charged with a small quantity of carbonic acid (and thereby possessing a solvent power for carbonate of lime) may have this result; and that this is the history of the formation of the *Oolites*, in which the condition of the calcareous matter,—both that forming the concentric deposits around the nuclei of the egg-like particles,—and that forming the cement which unites them together,—is decidedly *crystalline.* In this case, the crystallization has taken place, from some unknown cause, around nuclei, which commonly consist of minute Foraminifera, more or less perfectly preserved. In other instances it may occur, I should suppose, in irregular *patches* or in more distinct *strata*—

Now to apply these views to the case before us—

In regard to the *Chilian* deposit, it is, of course sufficiently evident to the unaided eye, that a great part of it consists of comminuted fragments of Shell.[2] The Microscope shows, that these are mingled with still more minute particles *of the same kind*; and that there is also a quantity, varying in proportion, of the amorphous substance, which I have alluded to as possibly resulting from the mechanical disintegration of the larger fragments.— I think I can also trace some *Spongioid* remains in the sections I have made of the specimen you sent me,—about 12 in number.

In the specimen of the *Coquimbo* deposit which you last sent me,[3] I can find *no definite appearance of organic structure*; it appears to be made up of the amorphous particles of the preceding; in some parts of which a sort of stratification presents itself, some of the layers showing indications of an incipient metamorphosis into the Crystalline condition.

In the various sections I have made of the *Pampean* deposit, amounting in all to about 25 in number, I trace a considerable variety of appearances. In a large part, I find the same appearances as in the preceding,—namely, a tolerably uniform amorphous character, with traces of incipient crystalline metamorphosis; but I also find the presence of minute rounded concretions,—resembling those of Oolites in size, but *not* in the arrangement of their particles,—a very frequent character. The *substance* of these concretions is still *amorphous*; but that of the connecting cement is often crystalline; as if water, charged with carbonate of lime, had percolated through the deposit, *after* (from whatever cause) the concretions had been formed. Of organic structures in this deposit, I find a greater variety than in either of the preceding. I have been able distinctly to recognize fragments of Shell,—though rarely. I am fully satisfied, however, that Spongioid bodies were common at the time of its formation; though their remains are by no means distinct. Mr Bowerbank, to whom I have shown the most characteristic specimens, fully agrees with me in this. The remains of the organic fibres, of the Spicula, and I think too of the Gemmules, may be distinguished more clearly than in the *Chilian* deposit. I am satisfied, also, of the existence of remains of Corals and of Polythalamia, in this deposit. But, as I said in a former communication, the great mass of it is composed of the *amorphous* matter already referred to;—sometimes uniformly diffused,—sometimes in concretions.

This is, I think, all the light that the Microscope can throw on the history of these deposits; and if it is not entirely satisfactory, nevertheless I hope it may be regarded as a contribution of some value towards the elucidation of their relations.

I shall preserve the series of Sections; and shall have great pleasure in showing them to you, or to any other Geologist who may wish to examine them, at any future time. I shall be in my present domicile for about two months longer, after which I shall return to Ripley until Michaelmas; but it is probable that I shall then remove permanently to the immediate vicinity of London.

I hope that you will not come to Town, without giving me the pleasure of seeing you. I have now a very large collection of preparations of Shell-Structure; and am anxious to make my investigations known to Scientific Men.

Your specimens (which I have brought to Town) shall be forwarded to the Geological Society. I have not been able to make sections of *all* of them, some being too friable.

Believe me to be, Dear Sir | yours very sincly | William B. Carpenter.

DAR 39.1: 36–41

CD ANNOTATIONS
1.1 I should . . . you again. 1.8] *crossed pencil*
4.1 Now supposing . . . before us— 5.1] *crossed pencil*
8.3 namely,] *opening quotation marks added pencil before*[4]
9.1 This is . . . Carpenter. 13.1] *crossed pencil*

[1] See letters from W. B. Carpenter, 21 December 1844 and 2 January [1845].
[2] CD considered it important to demonstrate that limestone formations were largely the product of disintegrated organic forms, and he quickly adopted Carpenter's conclusion (*South America*, p. 77).
[3] CD had asked Carpenter to compare rock specimens collected in the Pampas with those collected at Coquimbo, Chile (*South America*, p. 77).
[4] CD presumably marked the text in this way to indicate the beginning of the section that he subsequently paraphrased in *South America*, p. 77.

From Edward Forbes [9 May 1845][1]

Friday.

Dear Darwin

I have been too busy to look over your list with requisite consideration until yesterday.

I have marked on the lists their *probable* depths judging from the associations of the genera.[2] My marks are within wide ranges.

I could come however to more likely conclusions were two points noted

1st—the average size of the specimens & species found in each locality.

2d The comparative abundance of *specimens* of each species, & their state, as indicating life & death &c

Thus the Huafo list might indicate twice the depth if the species are all very small & the specimens of Turritella few.

All such calculations, at any rate, must be only provisional, until we know the range in Depth of mollusca in the neighbouring seas.

AL incomplete
DAR 43.1: 50

CD ANNOTATIONS
1.1 I have . . . yesterday. 1.2] *crossed pencil*
3.2 locality] ‘ $\begin{array}{r}552\\538\\\hline 14\end{array}$ ’ *added pencil*

[1] Dated from CD's reply, 13 May [1845].
[2] The lists of shells from different localities are in DAR 43.1: 51–2. The localities are Port Saint Julian, Santa Cruz, Huafo Island, Navidad, Copiapó, and the Cordillera of central Chile. Fuller identifications of the shells are given in lists throughout *South America.* CD used Forbes's estimates of the depths at which these species lived to demonstrate the amount of sea-bed subsidence that must have occurred at Copiapó and the Cordillera of central Chile (*South America*, pp. 193, 196).

To Edward Forbes 13 May [1845]

Down Bromley Kent
May 13th

Dear Forbes

I am extremely obliged for your information on depths: I fear I shall not be able to give the additional data, which you require. Should I hereafter be able to do so, I will apply to you again.[1]

I am very glad you have so soon found a naturalist for the Californian expedition I presume Mr Edmonstone[2] is actually appointed; otherwise I was going to have suggested Dr Dieffenbach, who travelled in New Zealand & who, I have little doubt, would have gladly gone.[3]

I do not at all know the route of the expedition, either outwards or homewards otherwise I would gladly make any suggestions which might occur to me on points worthy of observation. Have you, or do you know those who have, any power in settling any point, where the ship might call; for I have little doubt that Cocos isl^{d}., north of the Galapagos Archipelago, from its insulated position, & judging from the Galapagos Arch. would have a most peculiar flora & fauna.

The Revillagego (I forget how to spell it) group[4] off the coast of Mexico, I believe, has never been visited by a Naturalist. Suggest to Mr Edmonstone to take a good stock of small steel rat-traps, with which he can easily take many rodents.

I can think of only one special point of geological interest on the NW. & Californian, coast viz to ascertain in what latitude southward *great* **angular** erratic boulders have been transported by floating ice over *plains* or across wide valleys; this point w^{d} be very interesting, in comparison with the southern limit of boulders in the United States.

Of course Mr Edmonstone will attend to all facts relating to the elevation of the land,—a class of phenomena so grandly displayed in southern America.

If Mr Edmonstone liked to take the trouble of informing me, of the route of the Ship outwards & homewards, I daresay, I c^{d} suggest a few points for his observation.

Will he visit the Pacific coral-isl^{ds}.; if so I hope he will apply to the Admiralty for my Coral-Volume.

I have had a circular from the Ray Society[5] & I do not know to whom to pay my guinea: I shall be up next council & perhaps you will then inform me.

With many thanks for your information | Believe me | Ever yours sincerely | C. Darwin

Will Mr Edmonstone come home round the world, there is some most interesting pure geology to be done at the Mauritius.—

L. D. Edmondston

[1] CD was evidently unsuccessful; the figures cited in *South America* (pp. 193 and 226) are those provided by Forbes previously.
[2] Thomas Edmondston. On 21 May 1845, he joined H.M.S. *Herald* as naturalist, a position secured for him by Forbes. He was accidentally shot in Peru in 1846.
[3] Ernst Dieffenbach was apparently considered for the position by the Admiralty, but was not interested (Bell 1976, pp. 110–11).
[4] Revilla Gigedo Islands, an archipelago of four islands about five hundred miles off the west coast of Mexico.
[5] A society formed in 1844 to publish by subscription works on natural history, which would not otherwise be printed. Both Forbes and CD were members.

To J. S. Henslow 16 May [1845]

Down Bromley Kent
May 16th

My dear Henslow

The Lyells have been staying here & we have just heard from them that Mrs. Henslow has lately been much out of health. Their account referred to some little time ago, & I sincerely hope it does not now apply: do pray sometime before very long let me have a line to say how she is.

I am at work with a second edition of my Journal for Murray to bring out cheap, viz at 7s"6d in his Colonial Library: I find a good deal to alter in the scientific part: doing this work reminds me regularly of your great kindness in undergoing the wearisome labour of looking through all the proof-sheets. I hope, also, this autumn to get out my last Geological part & right glad shall I be, for I am wearied with S. America; your words', which at first astounded me, viz that it wd take me twice the number of years of the voyage to publish its results, will be more than verifyed.

I heard lately from Hooker,[1] who gives me a wonderful account of the Galapagos plants—12 new genera of the Compositæ, all confined to the group & no one species found on two islands!

I was telling my Father at Shrewsbury (where I have lately been) that you had gone into the Barberry versus corn-question & that you were a disbeliever:[2] on which he told me he had once had his attention called by a farmer to a very large field of corn, in the one of the hedges of which there were at almost regular intervals Barberry bushes, & my Father declares, that from *each* of them a wedge, pointing obliquely into the field, of discoloured corn, was most conspicuous. Next year the Barberries were all grubbed up.

When at Shrewsbury I read an article in the last Edinburgh Review on the Claims of Labour,[3] which interested me but I do not know how far you would agree with the writer.— I have often thought that I was foolish to trouble you with my remarks on this subject; but your letter set me thinking, & thoughts whether right or wrong, new or very old, like to make their escape.

Believe me my dear Henslow, ever yours | C. Darwin

American Philosophical Society

[1] Letter from J. D. Hooker, [28 April 1845].

[2] Barberry is a second host to the wheat rust fungus. CD had sent specimens of wheat rust to Henslow the previous year (see letter to J. S. Henslow, [25 July 1844]).

[3] Mill 1845. Henslow was actively concerned with the problems of agricultural labourers in the parish of Hitcham (see Jenyns 1862, pp. 84–93).

To C. G. Ehrenberg 21 May [1845]

Down near Bromley Kent
May 21

Dear and highly honoured Sir

I do not know how to thank you enough for your last most valuable letter.[1] I am particularly obliged for your examination of the Pampas mud, which I fear you undertook entirely for my sake. Your results have quite confirmed my view of its origin.[2] Considering the number & strangeness of the mammifers embedded in it (more different from our present quadrupeds, than are the Eocene fossil mammifers of Europe) the history of the Pampæan deposit is really interesting. I presume you did not examine the infusoria sufficiently to know whether the species are extinct or recent; as that would throw light on its age, which, as well as the manner of its formation, is disputed by the French Geologists.[3] In case you should ever further investigate the subject, I hope you will kindly inform me of the result; & I will endeavour to send you a few more & good samples. The Bahia Blanca is the *least* interesting part of the Pampæan Formation.

—. In the many volcanic districts, which I visited, I found obsidian only at Ascension; I will send some chips, though probably you will already have got them from this well-known island.—

As you seem interested with the great, chalk-like, Patagonian tufaceous formation, I will send soon to you through the Chevalier Bunsen,[4] a set of specimens from every station which I visited; & I will label them outside, so as to give you little trouble by sending you duplicates of those already sent. Santa Cruz is the most southern spot, where I *collected* specimens, though I *saw* the formation still further S.: Rio Negro is the most northern part where it occurs, & there only in thin beds in sandstone. The gypsum occurs in vast quantity in plates or veins, running in every direction, especially at P. St Julians: some gypsum is disseminated throughout the mass.— My specimens are small, but I will send you half of my largest, which I think will be an inch square as you desire. I shall publish a small volume on the geology of S. America this ensuing winter & how greatly will your Researches add to its value!

I have looked through all the accounts given by those on board vessels, where the Atlantic dust has fallen; & I find the wind has generally been between NE & East, & sometimes SE.—and it has always fallen during the same four months, when the harmattan blows from NE & ENE.—[5] I will detail all the facts in my little paper, which I will immediately send to the Geological Soc.— The dust to which I allude, which fell on the Princess Louise, was **not** during Meyen's voyage, but subsequently:[6] I doubt whether Meyen's dust was our infusorial dust, for it fell during a different month from any other case, & (as far as I can judge) further off the African coast.— I am sorry to find that I have misled, though in so unimportant point, as hardly to be worth your correcting. The packets marked by you IA & IB are same: I had so little expectation of the the first packet, which I sent, interesting you, that I did not describe it fully & gave only the approximate latitude & did not specify that it was not collected by myself. That collected by myself, fell on our ship, when 10 miles NW of St. Jago, (& is marked in your list 'St. Jago') as I stated in the M.S. which I sent you. But these inaccuracies are quite unimportant.

Believe me, Sir, with many thanks | Yours very faithfully & obliged | C. Darwin.

Museum für Naturkunde der Humboldt-Universität zu Berlin

[1] Letter from C. G. Ehrenberg, 8 April 1845.

[2] Ehrenberg's finding that the infusoria of the deposit probably lived in brackish water confirmed CD's view that the deposit was estuarine. See *Journal of researches* 2d ed., pp. 129–30.

[3] See especially Orbigny 1835–47, vol. 3, pt 3: *Géologie*, pp. 81–7.

[4] Christian Karl Josias Freiherr von Bunsen, Prussian ambassador in London.

[5] See letter from C. G. Ehrenberg, 8 April 1845.

[6] CD was referring to a voyage which took place in 1839–40; Franz Julius Ferdinand Meyen was botanist aboard the *Prinzess Louise* from 1830 to 1832. See 'On falls of dust on vessels traversing the Atlantic', *Edinburgh New Philosophical Journal* 32 (October–April 1842): 134–6, and 'An account of the fine dust which often falls on vessels in the Atlantic Ocean', *Collected papers* 1: 199, 202 n. 6.

From G. R. Waterhouse 21[–22] May 1845

British Museum
May 21 /45

My dear Darwin

Every spare minute I could find since I saw you has been spent in examining your Galapagos insects with a view to furnishing you with an account of them & of sending descriptions to Taylor's Mg..[1] I am not quite prepared yet but have nearly done— The Carabideous insects which I thought belonged to our European genus Calathus *do not*—they form a new genus nearly allied to Calathus but have the claws of the tarsi simple instead of being pectinated— The Heteromera form three or four distinct genera—one genus possibly is identical with a genus described by Eschscholtz from California, and is **certainly very near** to it but Eschscholtz's description is not quite detailed enough—[2] Hope[3] may have the genus in question & if so I shall in a minute decide— Another genus contains three Galapagos species and they belong to a genus containing only 3 species to my knowledge & which is found in Chile & Peru—[4] A third genus I have not yet worked out, but there is something very like it in California & also in Chile— About these Galapagos insects I will let you know more in a few days, finding I was mistaken about the Calathus I will look very closely to them—

The Butterfly which makes a clicking noise Doubleday (who it was made the remarks at the Ent. Soc.) informs me "is remarkable for having a sort of drum, at the base of the fore wing, between the costal nervure & the subcostal— these two nervures more over have a peculiar screw-like diaphragm or vessel in the interior—"[5] I send you a wing in which you will see the little drum at the base & how large the 2^d. nervure from the anterior margin is— you will clearly see, also, the spiral vessel within— mind you, I thought this latter (spiral vessel always existed in the nervures of the wings which seem to be but an extension of the the lungs or rather air vessels, in the same way as the air vessels are extended into the bones of birds, but Doubleday knows more about these matters than myself having been working hard at the Butterfly nervures lately & having published (or sent to be published) a long paper on the subject in the Linn. Soc.[6]

So far I had got yesterday when I was interrupted and detained for two hours and a half—kept 'till six o'clock—

I have found the St. Paul's Island Insects— they are all right, even the Feronia but the feronia is a parasitic fly— The name Feronia was given by D^r. Leach[7] to a genus of flies closely allied to *Hippobosca* (the species of which are parasitic upon Cattle &c—) on the one hand and to *Stenopteryx* on the other— This last is the generic name for those narrow and sharply pointed winged flies which are found in the Swallow's nest— It would be well in your new Edition to alter the name *Feronia* to *Olfersia* a name given by Wiedemann[8] to the same genus because Leache's name had been previously occupied.— The only notice of the habits I can find of *Olfersia* is in these words—(speaking of one of the species.) "M. Al. Lefebre l'a trouvée en Sicile sur le Heron"—[9] Your's appears to be a new species— there are but two or three known.

Lund's list of recent & fossil species I have copied out—[10] I should be very careful, judging from all I can learn of his specific determinations (& even generic) in putting too much faith in the list in question— I believe there are far too many species made—

I enclose a list of the fossil genera which I have discovered amongst M. Claussen's specimens[11] & give what appears to me to be the probable number of species of each, contained in our collection— I will write again soon, and in the mean time look over my notes about Lund's species &c

Believe me | Faithfully yours | Geo. R. Waterhouse

The other questions you left with me I will endeavour to answer in my next— I hope you are better— pray do not bother yourself for form sake &c &c to answer or acknowledge this

DAR 181

CD ANNOTATIONS

2.1 The Butterfly . . . this 8.3] *crossed pencil*

[1] Waterhouse 1845a.

[2] *Eurymetopon*, described in Eschscholtz 1829–33, 4: 8.

[3] Frederick William Hope, who also described parts of the *Beagle* insect collection.

[4] That is, the genus contained three other species as well as the three described by Waterhouse. The genus was *Ammophorus* (Waterhouse 1845a, pp. 19, 30–2).

[5] A butterfly, *Papilio feronia*, captured by CD in Brazil. Edward Doubleday later announced his discovery at a meeting of the Entomological Society on 3 March 1847 (*Transactions of the Entomological Society* 5 (1847–9): xii). CD reproduced Waterhouse's quotation in *Journal of researches* 2d ed., p. 33 n.

[6] Doubleday 1845.

[7] Leach 1811–16, pp. 552, 557–8.

[8] Wiedemann 1828–30, 2: 605–8. CD mentioned the 'Feronia' in *Journal of researches*, p. 10, and corrected it in the second edition, p. 10.

[9] Macquart 1834–5, 2: 640. The original reads 'sur un héron', and the observer was Alexandre Louis Lefebvre.

[10] Waterhouse enclosed a handwritten copy of the list as published in Lund 1841–2, pp. 197–200. Waterhouse's list is in DAR 205.9: 173–5.

[11] The collections of Peter Clausen were purchased by the British Museum in 1841 and 1844. See British Museum 1904–6, 1: 278. Clausen, owing to some dishonest affairs, left his native Denmark and settled in Brazil under an assumed name. Peter Wilhelm Lund and Clausen found the fossil quadrupeds on Clausen's farm in Brazil. CD briefly described Clausen's collection of fossil quadrupeds in *Journal of researches* 2d ed., p. 173. Waterhouse's list is in DAR 205.9: 172.

To Paul Edmund de Strzelecki [25 May 1845]

Down Bromley Kent
Sunday

My dear Sir

I received a few days since your kind & valuable present:[1] I am exceedingly obliged to you for it, though I feel that I have no claim on so magnificent a present.

I congratulate you on having completed a work which must have cost you so much labour & I am astonished at the number of deep subjects which you discuss. I must be permitted to express my sorrow that there are not far more copious extracts from the 'M.S. Journal': I hope some day to see it fully published.— You speak of your unidiomatic English; I heartily wish that one quarter of our English authors could think & write in language one half as spirited yet so simple.

Once again allow me to thank you very sincerely & believe me My dear Sir | Yours very faithfully | C. Darwin

You were so good, when I last saw you, as to say, that you would take the trouble of informing me (as a guide for myself) what you paid for the engraving of the *shells* alone.[2] The plates appear to me admirable.

Beinecke Library, Yale University

[1] Strzelecki 1845. A copy in the Darwin Library–CUL is inscribed 'To Charles Darwin Esqr M.A. from the Author 19th of May.' It is lightly annotated, with a note 'Abstract March 57' (the time CD began to write his 'big book', *Natural selection*).

[2] Strzelecki's work contains plates illustrating specimens of shells and corals, several of which had also been collected by CD in Van Diemen's Land (Tasmania). Strzelecki named one species of shell *Spirifer darwinii* in honour of CD's contributions 'to the advancement of physical geology and natural history generally' (p. 280). He recorded that he examined some of CD's collection in London.

To John Murray [31 May 1845]

Down near Bromley Kent
Saturday

Dear Sir

I had intended writing to you to say that I w^d^ send the M.S. for the first number to you by our weekly carrier on Thursday morning. I have much condensed, added to, & improved the scientific part: As nearly as I can possibly calculate; there will be 168 pages of your volume.

I have well considered & consulted others on title-page: I think from honesty & policy (as the work has been much quoted) we must keep to nearly the same title: I hope you will approve of the enclosed, which is strictly accurate & does not give the idea of exclusive science, as the 'Journal' is so prominent. I do not know how far you value Mr Lyells' judgment, but he approves of it.[1]

Should you [not] object to the enclosed short Dedication, it w^d^ much gratify *me* to insert it?—[2]

I enclose the map, of which I have the copper-plate. To my own taste & that of every person, whom I ever heard speak of the subject, a map in a volume of Travels is **very** agreeable: please inform me of your decision before Wednesday night, that I may send it or not with the M.S.[3]

Please return the Map and Title & Dedication: of the latter I beg you not say a word to M^r^. Lyell..

I have borrowed a few woodcuts from my Geological volumes: & I have had one made at 12^s^. as I thought it very desirable to illustrate a description of a curious Bird:[4] I will direct Mr Lee to take it & the account to you, if you approve, if not I will pay it myself.—

Will the Table of Contents ie Chapters belonging to the First Number be published in the First Number, or the whole Contents in the 3^d^ number?—

The condensing the scientific parts & additions have cost me infinitely more time than I anticipated, otherwise I sh^{d} have sent the M.S. sooner.

Finally—I take the liberty of calling your attention to the fact, that when you offered me 100£, it was for the two numbers. I did not choose to throw any obstacle in our arrangements; nor of course will I now do so; but for the copyright of my Journal in three numbers, I think 100£ is ⟨ ⟩[5] sum. Do you think you could afford to allow me an additional fifty? I assure you I have taken the utmost pains with this new Edition

Believe me dear Sir | Yours very faithfully | C. Darwin

To | J. Murray E^{qre}

Postmark: MY 31 1845
John Murray Archive (Darwin 24–5)

[1] The title eventually decided upon was *Journal of researches into the natural history and geology of the countries visited during the voyage of H.M.S. Beagle round the world.* The main change from the first edition title was the reverse order of 'geology' and 'natural history'.
[2] The dedication read:

> To Charles Lyell, Esq., F.R.S., this second edition is dedicated with grateful pleasure, as an acknowledgment that the chief part of whatever scientific merit this journal and the other works of the author may possess, has been derived from studying the well-known and admirable *Principles of Geology*.

[3] No map was included in the volume (see letter to John Murray, [4 June 1845]).
[4] The scissor-beak, *Rhynchops nigra*, is illustrated on p. 137. See also *Birds*, pp. 143–4.
[5] A word was torn out when the seal was broken.

From G. R. Waterhouse [*c.* June 1845][1]

Galapagos *Coleoptera*

Section—Carabides

genus *Feronia*, Dejean— Two species; new—

The genus Feronia is found nearly all over the world, but is chiefly confined to the temperate zones— It is a genus, as constituted by Dejean[2] containing an enormous number of species which have been grouped by many into distinct genera, but the characters are so slight, and it so difficult in most cases, to define them that Dejean, merely uses the so called genera as sections of his genus *Feronia* & in so doing he admits he has frequently been obliged to draw the lines of separation arbitrarily— Now be these sections important or not I cannot well associate the two Galapagos Feronias with any of them— Had I not known where they come from, I should has said they were from a temperate climate—they are black and will not otherwise associate with the brilliant Pæcilli (a section of

Feronia) some of which are found in tropical climates—Brazil & Peru for instance— I do not know which of the islands the specimens came from

Harpalidæ— In your collection are two specimens, of distinct species; unfortunately they are both females, and this prevents my working out their affinities in a satisfactory manner— One I think will prove to be a species of the genus *Selenophorus*—a genus the species of which are found in N. America, South America (as far South to my knowledge as Rio de la Plata), and the West Indian Islands— The other species appears to approach *most* nearly to the genus *Amblygnathus*—though I doubt if it actually belongs to that genus and it also approaches *Cratacanthus*— Both South American genera— the latter occurs on both sides of the Andes— Although, as I have said, I cannot determine the affinities with accuracy for want of the male sex, from which some of the generic characters are drawn, I may remark that I have pretty strong grounds for approximating these insects to the genera mentioned for both the Galapagos Harpali want the tooth to the mentum a character which combined with some others links your insects very closely with the American forms— No genus of Harpalidæ found exclusively in the Old World, and in which the mentum is destitute of tooth, can be confounded with them—

Family *Bembidiidæ*
genus *Notaphus*—one species; new— Notaphus is found in both Hemispheres—

Of Water beetles there are three species, one belongs to the *Dytiscidæ* and belongs to a genus which I know to be found in South America— Babington described a species from your collection found at Rio, which is very nearly allied to the Galapagos insect which I suspect is new—[3] I have the Rio species alluded to from Colombia—

Whether the genus is confined to South America I am not yet certain— You shall know hereafter— The other two water beetles belong to the *Hydrophilidæ*, one is the *Tropisternus lateralis*, an insect which is found both in the United States and in the West Indies— The other is a very small insect of the genus *Philhydrus*,— I have a species so like it from North America that I can find no distinguishing character; a second found in England which can *just* be distinguished by its being *rather* more distinctly punctured—

Family *Staphylinidæ*— The three specimens you found under a dead bird in Chatham Island, constitute a species which is **very** closely allied to our English *Creophilus maxillosus* and to the North American *Creophilus villosus*, but I think it is distinct from either—

Section Xylophagi
Genus *Apate*— The wood feeding insects you found in the branches of a dead Mimosa tree (I forget at this moment in which Island) agree perfectly with an Insect which I have from Colombia—

Then I find specimens in the Collection both of *Corynetes rufipes* and *Dermestes vulpinus*—insects which feed upon the skins of dead animals & various substances and are found every where— If a collection of subjects of Natural History (dried specimens) have not been well prepared, are sent from abroad to this country whether from the Himmalayas or Chile, they are sure contain these insects—alive!—

Family *Elateridæ*— one specimen of an Elater appears to me to approach most nearly in its characters to a Brazilian genus, from which, however, it will bear separating— I have not yet done with it—

Order *Heteromera*

It is in this order that your Galapagos collection is most rich in species— Of the genus *Ammophorus* there are three new species, one is from James' Island, another from Chatham Island—the third is not labelled,— The genus Ammophorus contains four known species all of which are found in Peru and one of which (at least) extends down to Chile— I have an insect in my collection from Mexico which is certainly very closely allied to *Ammophorus* & perhaps will not bear separation—

Of the Family *Tentyriidæ* there is a new genus containing three species— The insects of this family are found in both Hemispheres— many forms occur in the Southern parts of Europe and N. Africa; some in India, and some in the Southern parts of South America on both sides of the Andes— You have species of this family in your collection from St. Jago species are also found on Madeira[4]

Of the Family *Pedinidæ* there is also a new genus containing three species— the genus has its nearest affinities as it appears to one which is found in N. America, Mexico & both sides of Andes in South America—extending as far South as Chile & Rio de La Plata—

Of the *Lamellicornes* there is a species of *Oryctes* which *appears* to be new—

Of the section *Cyclica*—but one species a *Haltica*, but it will not associate well with any of the subdivisions of that group with which I am acquainted—

Of the section Rhynchophora—but one species—belongs to the *Anthribidæ*—not yet made out—*appears* to be a new genus—

Lastly I have to notice one of the *Coccinellidæ* a minute species perhaps a member of the genus *Scymnus*—

The great locust, contained in the Collection, Mr White[5] & I endeavoured to find out the the works upon the group but did not succeed—it belongs to the genus *Acridium* & I think is undescribed—

On the whole I do not see among the Galapagos insects, any which decidedly indicate a tropical climate— they are all of dull colouring & all small—if we except the *Oryctes* and we have a much better one in Europe as far as size is concerned—it is a small species of the tribe—I am not forgetting that *Ammophorus* *is* a tropical genus, but nobody would say so unless he knew where it came from—a small black insect—

DAR 46.2 (ser. 2): 3–5

CD ANNOTATIONS
1.4 The genus] '2' *added pencil*
2.1 *Harpalidæ*—] '2.' *added pencil*
3.2 genus *Notaphus*—] '5' *added pencil*
4.2 Babington . . . Colombia— 4.5] *scored pencil*

[1] The date is based on the period when Waterhouse was preparing an article on CD's Galápagos Coleoptera (Waterhouse 1845a), which appeared in July 1845. See also the two following letters from Waterhouse, also dated [*c.* June 1845].
[2] Dejean 1825–38, 3: 200–5.
[3] *Copelatus elegans*, described in Babington 1841–3, p. 11. Waterhouse named CD's Galápagos species *Copelatus(?) galapagoensis* (Waterhouse 1845a, pp. 23–4).
[4] The sentence 'You have . . . Madeira' was added at the bottom of the page and its position in the text indicated by *.
[5] Adam White, a naturalist in the zoology department of the British Museum. White described CD's *Beagle* arachnids.

From G. R. Waterhouse [*c.* June 1845]

British Museum
Saturday

My dear Darwin

I have just received your note, and am glad to learn from it that I shall see you soon— In the short account I sent you of the Galapagos Insects, I omitted to notice two Curcaleos which had been removed by me from the box containing the rest of the collection— I had indeed separated all of the tribe & put them together for the convenience of working them out— One of these belongs to a genus which is very characteristic of the West Indian Islands & Colombia, the other is a member of a genus which has a very wide rage—

I am just about to send my paper to Taylor for his Magazine,[1] and have written a line or two to stick on the top of it, which I will copy out, & if it strikes you there is any thing very stupid in what I have said I am sure you will let me know— I have said this (and perhaps being shorter it may answer your purpose better for your abstract than the account I sent before)—[2]

"The insects about to be described are nearly all of small size, and none of them display any brilliant colouring. Some of the species are referable to a little group, found in Chile and Peru—the genus *Ammophorus*, a genus hitherto found only in those parts; others appertain to a genus (*Anchonus*) which is almost (I have put in the word "*almost*" but I will look when I go home—I think *quite* would do—

confined to the West Indian Islands and the Northern parts of South America— Again, in the collection under consideration, are species of genera which are found all over the world, or nearly so—such as *Feronia*, *Notaphus*, and *Oryctes*,* and lastly there are species which cannot be located in any known genus, but which appertain to families having representatives in most parts of the world—such as the *Pedinidæ*, *Tentyriidæ*, and *Anthribidæ*—

"But three species amongst the Galapagos Coleoptera occur (so far as I have been able to ascertain) in any other quarter, and of these, two (*Dermestes vulpinus* and *Corynetes rufipes*) are transported to all parts of the globe visited by ships,— feeding as they do upon dried flesh &c. The third is a wood-feeding insect (*apate*) and might be transported considerable distances by floating trees—

"Some of the insects in Mr. Darwin's collection have labels attached to them from which may be ascertained the particular island of the Galapagos group in which the specimen was found, and where this is the case I have not met with any species which is common to two or more of the islands in question"

Believe me | faithfully yours | Geo. R. Waterhouse

I hope I am not boring you!

*(*here I put this foot note*)[3]

"It is from genera like these, which have a very wide geographical range, that the minor local groups appear, as it were, to radiate. Those genera which are confined to comparatively limited districts—often containing but few species, and also often presenting very remarkable abnormal modifications of structure—are in most cases referable to some family which has representatives in most parts of the world. Groups of high value, such as classes, and orders, are never confined to any particular quarter of the globe; and, even when we descend to families, restricted as they now are by Naturalists, it is comparitively rare to find them so defined as not to embrace species from widely separated localities— *Genera* may be arranged under three principal categories as regards their geographical distribution— First may be noticed those of universal range, such as *Cicindela*, secondly those which occur in both hemispheres, but affect particular zones, such as *Megacephala*, which is confined to the tropical zone; and thirdly those which are restricted to a comparatively small district such as *Manticora*, which is confined to South Africa— These genera all belong to the same family of beetles—the *Cicindelidæ*, and of this family *Manticora* presents certainly one of the most aberrant forms; and, although I am not prepared to say that those genera which have a very limited range *always* present very aberrant forms yet it is certain that such is very frequently the case

DAR 181

[1] Waterhouse 1845a, which was published in *Annals and Magazine of Natural History* edited by Richard Taylor. CD's copy is in the Darwin Pamphlet Collection–CUL. His notes about the paper are in DAR 197.3.

[2] CD used this information in *Journal of researches* 2d ed., pp. 391–2, 395.

[3] The following paragraph was printed as part of a footnote to Waterhouse 1845a, pp. 19–20. There are some minor changes in the printed text.

From G. R. Waterhouse [*c.* June 1845]

British Museum
Monday

My dear Darwin

I have been hesitating for some time whether I should bother you with another note, but as I cannot be easy without doing so—here goes— When I wrote to you last I was at the Museum where I had not my notes about the Galapagos Insects & the moment I got home I found I had made an omission—that of not noticing two species of Water beetles— The one is the *Tropisternus lateralis* of [authors] an insect which is found in the United States, in Mexico & the some of the West Indian Islands— The other, also a Water-beetle, is a minute species of Hydrophilus which agrees very closely with a North American species— these little Hydrophili however are so extremely difficult to determine that I will not pretend to give a decided opinion upon them— The *Tropisternus* is a large & very distinct species of which there can be no doubt—[1]

I have added considerably to that foot note,[2] and have been obliged to make a *foot note* of it, *and much other matter*, for my paper comes to 30 or 40 closely written pages & besides I wished as much as possible to keep such extraneous matter out of the body of the paper where it would only bother people who would consult it for the matter which it *professes* to contain— It is I hope improved & when printed of course I will forward a copy to you—and, if you do not disapprove of the additional parts I shall be much pleased—

I have in working out these insects taken the opportunity to describe several other of the more interesting insects contained in your collection—from Patagonia &c—[3] The descriptions are quite prepared for press, but I am afraid of frightening Taylor & mean to let him print what he has first—

I am particularly pleased that you do not disapprove of the notions conveyed in my last

Believe me | Sincerely yours | Geo R Waterhouse

DAR 181

CD ANNOTATIONS
1.1 I have . . . about the 1.3] *crossed pencil*
2.1 I have . . . Waterhouse 5.1] *crossed pencil*

[1] In *Journal of researches* 2d ed., p. 392, CD noted that the two species of water-beetle and one other insect (*Apate*) were the only Galápagos insect species previously known. All the others were new.
[2] See preceding letter, n. 3.
[3] Waterhouse 1845b.

To J. D. Hooker [4 June 1845][1]

Down Bromley Kent
Wednesday

My dear Hooker

I hope you are getting on well with your lectures & that you have enjoyed some pleasant walks during the late delightful weather. I write to tell you (as perhaps you might have had fears on the subject) that your books have arrived safely, viz., your paper,—M. Gerard's one on Species,[2] which I am particularly glad to see & have sent to endeavour to get one—Boston Journal—Canary Is^{d}—D'Urville[3]— I am exceedingly obliged to you for them, & will take great care of them: they will take me some time to read carefully.

I send to day the corrected M.S. of the first number of my Journal in the Colonial Library,[4] so that if you chance to know of any gross mistake in the first 214 pages, (if you have my Journal), I sh^{d} be obliged to you to tell me. I have just heard from Sulivan, that he has made a grand collection of fossil quadrupeds from the R. Gallegos in the southern part of Patagonia.[5]

Do not answer this for form sake; for you must be very busy. We have just had the Lyells here, & you ought to have a wife to stop you working too much, as Mrs Lyell peremptorily stops Lyell.

Ever yours | C. Darwin

DAR 114.1: 34

[1] Dated by the relationship to the letter to John Murray, [6 June 1845].
[2] Gérard 1844.
[3] *Boston Journal of Natural History*, containing Hayes 1844; Webb and Berthelot 1836–50; Dumont d'Urville [1841–54].
[4] Actually sent on Thursday 5 June, see letters to John Murray, [31 May 1845] and [6 June 1845].
[5] Letter from B. J. Sulivan, 13 January – 12 February 1845.

To John Murray [4 June 1845]

Down Bromley Kent
Wednesday

Dear Sir

I am exceedingly obliged for the manner in which you have acceeded to my request for the additional 50£.

I will reconsider the title when set up; at present I can think of no improvement.

I give up the map: I will direct the woodcut to be delivered to you. I do not suppose I shall want others.

I presume the proofs are to be returned by me direct to the Printer, who must send me his address.

I ought to have stated that my calculation of contents viz 168 or 169 pages in this number, & about 160 in next has been founded on the size of pages & kind of type used in Father Ripa.[1]

Your's very faithfully | C. Darwin
I will put the running Headings as you require

John Murray Archive (Darwin 26)

[1] Ripa 1844. A title that appeared in Murray's Colonial and Home Library.

To John Murray [6 June 1845]

Down Bromley Kent
Friday

Dear Sir

I sent the M.S. to you yesterday. I write now to tell you, that I find so much to add about the Fuegians,[1] & consequently so much *more* to condense & almost rewrite in the scientific part, that I cannot help fearing that possibly I may not have the second number ready having to correct press of 1st number for press at the end of this month. I **think** I shall, but not being sure, I thought it w^{d} be better not to take you by surprise. It will depend on my health, which lately has been very good; I *earnestly* wish to complete it as soon as possible. I shall be able to judge certainly by the 20th of this month:

I presume you have something else which, if required, c^{d} come out, between two of my numbers

Yours very faithfully | C. Darwin

John Murray Archive (Darwin 29–30)

[1] Because Robert FitzRoy had given an extended description of the Fuegians in *Narrative* 2: 175–224, CD elected to say little about them in the first edition of the *Journal of researches*. Much of the material introduced into the second edition was based on FitzRoy's account.

To John Murray 20 [June 1845]

Down Bromley Kent
Friday 20th

Dear Sir

I send a line to report progress.— For the two last posts I have received no proof sheets: before they came regularly & I returned them as regularly within 12 hours: there have been so few corrections, that I have not had occasion to see any revise.—

I have got through the most difficult part of the M.S. of the next number & shall have it all ready by 4th 5th or 6th of next month.— Would you kindly inform me, what your intentions are about publishing the three numbers consecutively? Anyhow I hope it will not be inconvenient to you to allow me to go on printing off, at the rate at which I am now at work.— I am anxious to get the job finished &

out of my hands I hope & think I have much improved my Journal: you may perhaps be glad to hear, that I could not possibly have condensed my Journal to go into two number's & w^{d} not have undergone the loss of time for even double the sum you have so kindly given me.—

Believe me dear Sir | Yours very faithfully | C. Darwin

P.S. | Should you object to an advertisement of the 'Zoology & Geology of the Beagles Voyage (& of course of nothing else), being printed at Mrs Smith & Elder's expense & inserted at one end of the first number?[1] They proposed it to me; they have been liberal in letting me have the full use of any of the woodcuts out of my Geological volumes.—

John Murray Archive (Darwin 33–34)

[1] The advertisement arrived too late to be included in the first number of the three-part issue (see letter to John Murray, [26 June 1845]). However, such an advertisement does appear on p. 520 of the second issue, which is in book form (Freeman 1977, p. 36).

To John Murray [23 June 1845][1]

Down Bromley Kent
Monday

Dear Sir

I am sorry to trouble you now that you are at Cambridge;[2] but I sh^{d} like you to look at the title.— M^{s} Clowes have set it up rather differently from what I intended— I had thought to have put the "Journal" alone, so that scientific part might be less prominent: please, if you are not contented, make any alterations & send it to M^{ess} Clowes,—requesting them to let me have a revise.—

I beg to call your attention to the *Table of Contents*: is it to be inserted?? if so sh^{d} it not be called Contents of Part I or Vol I?? Please inform Mrs Clowes of your determination of this point: My M.S. has *most* unfortunately extended to 180 pages.— I intend dividing the last Chapt. & putting some of it in the next number; I will write to Mrs Clowes explaining this & asking him how much must be left out to end with a *whole* sheet, for I suppose ending with a broken sheet is no œconomy.—

Yesterday (ie Sunday) I received nearly 4 sheets & a Revise.— it is impossible for me to correct so much in one day & I cannot return them till tomorrow; so that if this Part be not ready on the 30th it will *really* not be my fault. Excuse this untidy note, but I am far from well.—

Yours sincerely | C. Darwin

Mrs Clowes have printed all the sheets with wonderful correctness. I think the type looks well proportioned.—

John Murray Archive (Darwin 21–2)

[1] The date is based on Murray's deadline of 30 June for the publication of the first part of *Journal of researches* 2d ed. See also CD's subsequent letters to Murray of [26 June 1845] and [28 June 1845].
[2] The British Association met in Cambridge from 19 to 23 June 1845.

To John Murray [26 June 1845]

Down Bromley Kent
Thursday

Dear Sir

I have adopted & am much obliged for, your criticism on the Preface. It will be now too late to insert the advertisement of the Zoolo. & Geol. of the Voyage in the First Part.—

I am sorry to say that this Part runs to 176 pages: I found I could not shorten so much as not to leave a broken sheet, so that I have managed to get all in this Part, by running two chapters together & by throwing away two pages of type: I *have* written to M^rs. Clowes, begging them to send me a little account for the setting up & distributing of these two pages & for the corrections consequent on it:

I will try to make the next Part only 160 pages, but I find it almost impossible, with so many alterations to keep any accurate calculation, as is shown by the last M.S. having run out 13 pages too much.

Dear Sir | Your's very faithfully | C. Darwin

PS. To night. M^rs. Clowes will receive title & every sheet finally corrected.— The title, Preface Contents &c all been put by Mrs Clowes into half a sheet

John Murray Archive (Darwin 17–18)

To J. D. Hooker [27 June 1845]

Down near Bromley Kent
Friday

My dear Hooker

I have been an ungrateful dog for not having answered your letter sooner, but I have been so hard at work correcting proofs, together with some unwellness, that I have not had one quarter of an hour to spare. I finally corrected the first ⅓ of the old volume, which will appear on the 1st of July: I hope & think I have somewhat improved it.

Very many thanks for your remarks, some of them came too late to make me put some of my remarks more cautiously; I feel, however, still inclined to abide by my evaporation notion to account for the clouds of steam, which *rise* from the wooded Valleys after rain.[1] Again I am so obstinate that I sh^d require very good evidence to make me believe that there are two species of Polybirus in the Falkland Is^ds:[2] Do the Gauchos there admit it?. Much as I talked to them, they never alluded to such a fact. In the Zoology I have discussed the sexual & immature plumages which differ much.[3]

I return the enclosed agreeable letter with many thanks; I am extremely glad of the plants collected at St. Paul's & shall be particularly curious, whenever they arrive to hear what they are: I dined the other day at Sir J. Lubbock & met R. Brown & we had much laudatory talk about you: he spoke very nicely about your motives in now going to Edinburgh.— He did not seem to know & was much surprised at what I stated (I believe correctly) on the close relation between the Kerguelen & T. del. Fuego floras. Forbes is doing apparently very good work about the introduction & distribution of plants: he has forestalled me, in what I had hoped w^{d} have been an interesting discussion, viz on the relation between the present alpine & Arctic floras, with connection to the last change of climate from Arctic to temperate, when the then arctic lowland plants must have been driven up the mountains.[4]

I am much pleased to hear of the pleasant reception, you received at Edinburgh: I hope your impressions will continue agreeable: my associations with auld Reekie[5] are very friendly. Do you ever see Dr Coldstream?[6] if you do, w^{d} you give him my kind remembrances.— You ask about Amber, I believe *all* the species are extinct, ie without the amber has been doctored) & certainly the greater number are.—

If you have any other corrections *ready* will you send them soon; for I shall go to press with 2nd Part, in less than a week.— I have been so busy, that I have not yet begun d'Urville & have read only 1st Chapt of Canary Is^{d}![7]— I am most particularly obliged to you for having lent me the latter; for I know not where else I c^{d} have ever borrowed it.— There is the Kosmos to read[8] & Lyell's Travels in N. America:[9] it is awful to think of how much there is to read.—

What makes H. Watson a renegade? I had a talk with Capt. Beaufort the other day & he *charged* me to keep a book & enter anything which occurred to me, which deserved examination or collection in any part of the world, & he w^{d} sooner or later get it in the instructions to some ship.— If anything occurs to you, let me hear, for in the course of a month or two, I must write out something: I mean to urge collections of all kinds on any isolated isl^{ds}.— I suspect that there are several in the northern half of the Pacific, which have never been visited by a collector.— This is a dull untidy letter.

Farewell | Ever yours | C. Darwin

As you care so much for insular Floras, are you aware that I col⟨lected⟩ all in flower on the Abrolhos isl^{ds} but they are very near the coast of Brazil: nevertheless I think, they ought to be just looked at, under a geographical point of view.—

Postmark: JUN 29 1845
DAR 114.1: 35

[1] In *Journal of researches*, p. 27, CD suggested that the steam was caused by evaporation from sun-warmed leaves.

[2] CD claimed there was only one species of *Polyborus* on the Falklands, *Polyborus novae zelandiae* (*Journal of researches*, p. 66).

[3] *Birds*, p. 11.
[4] E. Forbes 1845. CD's discussion of Arctic–alpine distribution is in his essay of 1844 (*Foundations*, pp. 162–8).
[5] Edinburgh, literally 'old smokie'.
[6] John Coldstream, a friend of CD when he attended Edinburgh University (see *Correspondence* vol. 1).
[7] Dumont d'Urville [1841–54] and Webb and Berthelot 1836–50. For CD's earlier attempts to borrow the latter work see *Correspondence* vol. 2, letter to J. S. Henslow, [21 January 1838].
[8] See letter to J. D. Hooker, [11–12 July 1845], n. 8.
[9] C. Lyell 1845a.

To John Murray [28 June 1845][1]

Down Bromley Kent
Saturday

Dear Sir

I take the liberty of suggesting that if you send any copies of my Journal to the Periodicals that you would send one to the Gardeners' Chronicle; as I have sometimes contributed to it, I think it not unlikely that they would notice it favourably.—[2]

I shall have the M.S. of the Part II ready in 4 or 5 days.— I presume you will give me a copy of my Journal for myself & shall you think me very unreasonable if I ask you to let me have one dozen copies at prime cost to give to my neighbours & some naturalists who have assisted me.?— Anyhow w^d^ you be so kind as to let me have two copies (one for self & one for Mr. Lyell) as soon as out: if ready before Thursday Morning at **noon**, send them directed to

"C. Darwin Esqre.
care of G. Snow
Nag's Head
Borough"

I am sorry to trouble you with so many notes.

Yours sincerely | C. Darwin

John Murray Archive (Darwin 19–20)

[1] The date is based on CD's previous statement to J. D. Hooker that part one of *Journal of researches* 2d ed. was to be published on 1 July 1845, a Tuesday (see letter to J. D. Hooker, [27 June 1845]).
[2] The *Journal of researches* 2d ed. was favourably reviewed in the *Gardeners' Chronicle and Agricultural Gazette*, no. 32, 9 August 1845, p. 546; no. 40, 4 October 1845, p. 675.

To John Murray [3 July 1845][1]

Down Bromley Kent
Thursday

My dear Sir

I must write to thank you for your really magnificent present of the twelve copies: I assure you I think I have not the smallest claims for them, after your other liberality.

The M.S. will go on Monday or on furthest Tuesday to Mrss. Clowes: I will use your name, in urging them to set it up soon, so as not to run any risk of being too late. I presume Mrss. Clowes will keep the wood cuts safe, as one of them in the last number, belongs to Mrs Smith & Elder.

With my sincere thanks | Yours very faithfully | C. Darwin

John Murray Archive (Darwin 27–8)

[1] Dated by the relationship to the preceding letter.

From B. J. Sulivan 4 July 1845

Philomel Monte Video
July 4. 1845

My dear Darwin

After all my hopes of returning home I find we are to remain out to work in this river[1] and shall not certainly be home for a year, at first I thought of sending you all the fossils,[2] but I fear the casks being knocked about, unless one of the steamers should go direct home to Woolwich when I would perhaps send them. That you may be able to form an Idea of the formation in which the fossils are, and their probable age, I send by this packet a few in a small box which were by accident left out of the casks. the large piece you will see has three marks on it, it is part of a piece in the casks also marked with three cuts, which will show you the pieces likely to contain fragments of the same animal, some large pieces are broken in three or four pieces, but each piece was marked with the same number of cuts at the time so they can easily be fitted together again— Most of the small fragments came from the same piece, but there is a piece of white paper in the box containing a small piece of the armour like the Megatherium, and a small piece of bone petrified— they were found at the foot of the cliff among some shingle that had fallen from the surface at the summit of the cliff and are I suppose of the same age as those you found. You will also find in the box two pieces containing casts of shells, which are from the Falklands. I suppose they are similar to those you have but they come from Saunders Island near Port Egmont further to the Westward than any I had before found.[3]

I was much surprised yesterday at seeing Rugendas,[4] who you will recollect at Valparaiso (the German artist & traveller) walk into the house. he is on his way from Buenos Ayres to England having been ever since we left him in Chili Peru and Mexico. he has sent to London about two thousand drawings and is going to publish when he returns[5]

I hope I shall have an answer the next packet to the letter I wrote to you from the Falklands as I am anxious to know if the Fossils surprise you as much as they did me.

we are all daily expecting to be involved in hostilities with Rosas[6] certain demands have been made on him by the English and French Ministers which it is said he will not listen to, if so it is supposed we shall compel him, he threatens to seize British Merchants and property and take them up the country, and I fear he is trying to make us fear being firm least our countrymen suffer, for a scotch Family of nine, Father Mother & Children have all had their throats cut near Buenos Ayres. This Place still holds out. Most dreadful butcheries in the country Rosas troops lately killed above 500 prisoners in cold blood near Maldonado, principally Indians who had been serving Reveira[7]

Mrs. Sulivan & chicks go home in two months. our boy (14 months old) broke his arm six weeks since but is now well again[8] I hope Mrs. Darwin & young ones are well. Mrs. Sulivan joins me in kind rembrances to her

Believe me dear Darwin | Your sincere friend | B J Sulivan

DAR 46.1: 87–8

CD ANNOTATIONS
1.1 After all . . . Sulivan 6.1] *crossed pencil*

[1] Sulivan was engaged in hydrographic surveys along the Rio de la Plata.
[2] Fossils collected in the Falkland Islands and on the coast of South America. See letter from B. J. Sulivan, 13 January – 12 February 1845.
[3] Possibly those described in Morris and Sharpe 1846.
[4] Johann Moritz Rugendas.
[5] The sketches were published in Sartorius 1855–8.
[6] Montevideo was then under siege by forces backed by the Argentine government of General Juan Manuel de Rosas. British and French naval forces supported the city. In August 1845 a joint Anglo-French squadron forced the passage of the Parana, engaging with Rosas's forces at Obligado. Sulivan served as pilot for the expedition (Sulivan ed. 1896, pp. 66–7).
[7] Fructuoso Rivera, president of Uruguay.
[8] See letter from B. J. Sulivan, 13 January – 12 February 1845, nn. 11, 19.

From J. D. Hooker 5 July 1845

20 Abercrombie Place.
July 5th. 1845.

My dear Darwin

On the arrival of your last welcome letter I did determine to answer it that same day, but I went to a nice musical party in the evening & on my return remembered one of my next days lectures, which proved such a soporific that my answer to you was postponed sine die.

Here are two or three more notices on passages of the Journal, really from the progress I make (it is any thing but flattering) one would suppose that I found it the dullest book in life. As it is these must be too late, for which I can readily console myself in the full conviction of their futility.

May we not be at cross purposes anent the Fog on the Corcovado. I have no idea of denying the cause of its ascent, but still am obstinate enough to attribute its presence to the *coolness* & not *heat* of the leaves of the trees. My notion is, that leaves when living could not be heated to the xtent your statement leads one to suppose these ought to have been.[1] I should have expected the moisture to hang about the branches, exactly as on the grass in the morning: now after a thunder-plump we never see the vapor steaming up from a grass plot, & still less should it from the far more broken surface of a wood. However you know far more of the matter than I do, but one cannot help having his notions.

Do tell me where to expand a little that about the plants driven up the mountains from the change of temperature: I think I understand you, but not sufficiently to explain myself, will not Antarctic Botany illustrate the subject any way?

I called Watson a renegade for starting with the motto "omne ex ovo" which I took in its vulgar sense of "species are constant" & finishing almost an avowed believer in Progressive developement,[2] as enunciated & upheld in the already defunct "Vestiges".

My Students like Physiological Botany far better than any other branch, & it is no joke coaching up lectures on these subjects, though a very useful employment. Do not suppose I am overworking myself, I am not, & as the garden is almost a mile from my home I get a good walk twice a day. I do not know Dr Coldstream, if I meet him I shall not forget your message.

The more I ponder upon Insular Floras the less inclined I am to admit the mutation of species to any very great amount, it is no doubt an ever active agent, but I should look upon it & upon Hybridizing &c as the perturbing causes of our difficulties in assigning limits to species, that have in many cases rendered it hopeless to search for specific characters. The species of Insular Floras are I think peculiarly well defined & I can hardly conceive a peculiar genus to have 5 peculiar species in an isolated spot only a few miles in circumference, if these are the offspring either of progressive developement or of variation. The absence of Insects limits the operation of hybridizing in these cases. Are not Hymenoptera particularly rare in these oceanic specks—?

Would you connect the absence or rather the destruction of Kerg. Land. wood i.e. vegetation with any supposed alteration of the temperature of the Antarctic ocean.[3] Fuegia is rising, but I should think the mean temp. of those Lats. would rise with the increase of land to the N^d^ of 60, & the contrary with land to the Southwd of 60, as the present configuration of that ugly corner of the world remains.

I am sorry to hear that you have been still suffering a little. I would willingly take a little bad health (temporarily only) to let you work a bit in comfort, you do so richly deserve a little peaceful working. I am finishing this, or rather writing almost all in poor Grahams[4] sick room. he is very ill, in constant agony of pain, the mind is still clear, but all the bodily vigor gone & himself reduced to the lowest stage of emaciation, too weak to converse & sometimes to speak even, for hours together.

Lest I forget, what particular information is it you want about the plains of India,? is it about boulders,? I write there every month & my friend is a tolerably good observer.[5] What a curious mixture of the Cape & Fuegian flora Tristan d'Acunha presents. Certes Kerg. Land is allied to Fuegia in the Flora though it has some very peculiar species of its own, as the Cabbage & one or two other things, still out of 18 genera I think all but the two are S. American. Thus [1]Ranunculus 1 sp. probably identical, [2]Acæna, sp new?— [3]Montia [4]Callitriche [5]Limosella 1 species of each & all Fuegian;— [6]Juncus do.— [7]Aira? do,— [8]Agrostis do,— [9]Festuca do,— [10]Poa one sp peculiar & [11]another species peculiar?— [12]Galium sp.?.— [13]Azorella sp?— [14]Leptinella a Fuegian genus, the species & Auckld Isld.— [15]*Bulliarda* Fuegian;— [16]Cabbage & [17]another new genus peculiar to Kerg. Land.— [18]*Colobanthus* Fuegian genus.. I think that 10 at least, probably 12, are thus or will prove Fuegian species; 6 are Auckland's group species, all but one of these latter common to all antarctic countries.

As to the *Polyborus* I am no ornithologist, but do hope to shew you the two birds I mean, & if not distinct I will eat them,—bad as "Johny Rooks" are. I think I have three Hawks proper & 2 Polybori from the Falklands, both Polybori I think are Fuegian species. The piece of Amber I referred to is perfectly beautiful, enclosing innumerable ants, as beautiful as when alive, many of them had been carrying their eggs about when enclosed. Your Abrolhos plants I remarked along with the general Tropical collection, they looked uninviting, but must be done something with: the Islands are, as far as plants are concerned, too recently populated to be *very* interesting as a Botanical station, but highly interesting as illustration of what plants are easiest transported & what agents are most active in transportation.

Do intreat collectors in your instructions[6] to distinguish between native & introduced plants, there is scarcely a pure flora known to me except Kerguelens lands. Even Lord Aucklands group had Poa annua & a chickweed on the tomb of a French sailor.

Journal 192.[7] Luminous patches exactly as you described I fished up often with the net loaded by a *deep-sea-lead* & found there very large Pyrosoma.

200 *Warm-blood* sucking Insects: the Midge is another instance & some horrors at New Zealand.

235. in note— Was not the V. D. L. savage lower than the Fuegian,? he had hardly the canoe. Have we any instance of any extra Europæan & Asiatic nations improving their implements? domestic or otherwise? Good as the Chinese are, they do not improve their junks. Have we signs of progressions improvements even in the most useful implements in any savage nation? I should like a good essay on this subject, do you know any?—

257 Ant. geese, is the not the differently colored legs of the sexes an anomaly in Birds.

304. The *Fucus giganteus* or *Macrocystis pyrifera* goes N. to Rio San Francisco in California, there is but one species I think, or if 2 one is very rare indeed. I think it is also a Kamschatka plant.[8] It is curious that it, the Albatross

AL incomplete
DAR 100: 51–4

CD ANNOTATIONS
1.1 On the . . . but I should 7.3] *crossed pencil*
7.5 The species . . . defined 7.6] *underl pencil*
7.7 peculiar species . . . variation. 7.8] *scored pencil*
7.8 The absence . . . cases. 7.9] *scored pencil*; '?(? Sprengel— *[*proves*]* too much=' *added pencil*[9]
8.1 Would . . . observer. 10.3] *crossed pencil*
10.3 What . . . presents. 10.4] *scored pencil*
10.13 6 are . . . species] *underl pencil*
11.1 As to . . . enclosed. 11.6] *crossed pencil*
12.1 Do intreat . . . known to 12.2] *crossed pencil*
13.1 Journal 192 . . . Birds. 16.2] *crossed pencil*
16.1 257 . . . Albatross 17.3] *crossed ink*

[1] See letter to J. D. Hooker, [27 June 1845], n. 1. CD did not change his previous opinion that the steam was caused by heat, see *Journal of researches* 2d ed., p. 24.
[2] Watson 1845.
[3] J. D. Hooker 1844–7, pp. 219–20, 240.
[4] Robert Graham, for whom Hooker was acting as substitute.
[5] Thomas Thomson, see letter from J. D. Hooker, 30 December 1844.
[6] This may be a reference to the instructions for collectors CD referred to in his letter to J. D. Hooker, [27 June 1845], or CD's 'advice to collectors' as printed in the first edition of *Journal of researches*, pp. 598–602. The advice to collectors did not, however, appear in the second edition.
[7] Page numbers refer to *Journal of researches*.
[8] CD had identified the kelp as *Fucus giganteus* and suggested that it did not exist any further north than Chiloé, but was replaced by a different species (*Journal of researches*, pp. 303–4).
[9] Sprengel 1793. See letter to J. D. Hooker, [11–12 July 1845], for CD's comments on Christian Konrad Sprengel's work.

To Charles Lyell [5 July 1845][1]

Down Bromley Kent
Saturday

My dear Lyell

I send you the first Part of the new Edition, which I so entirely owe to you.— You will see that I have ventured to dedicate it to you[2] & I trust that this cannot be disagreeable. I have long wished, not so much for your sake as for my own feelings of honesty, to acknowledge more plainly than by mere references, how much I geologically owe you.— Those authors, however, who like you, educate people's minds as well as teach them special facts, can never, I should think, have full justice done them except by posterity, for the mind thus insensibly improved can hardly perceive its own upward ascent.— I had intended putting in the present acknowledgment in the Third Part of my geology, but its sale is so exceedingly small that I should not have had the satisfaction of thinking, that as far as lay in my power, I had owned, though imperfectly, my debt.— Pray do not think, that I am so silly, as to suppose that my dedication can anyways gratify you, except so far as I trust you will receive it, as a most sincere mark of my gratitude & friendship.—

I think I have improved this edition; especially the 2^d Part, which I have just finished; I have added a good deal about the Fuegians & cut down into half that mercilessly long discussion on climate & glaciers &c.—[3] I do not recollect anything added to this 1st Part, long enough to call your attention to: there is a page descriptive of a very curious breed of oxen in B. Oriental.—[4] I sh^d like you to read the few last pages; there is a little discussion on extinction, which will not perhaps strike you as new, though it has so struck me & has placed in my mind all difficulty with respect to the causes of extinction, in the same class with other difficulties, which are generally quite overlooked & undervalued by naturalists: I ought, however, to have made my discussion longer & shown by facts, as I easily could, how steadily every species must be checked in its numbers.[5]

I received your Travels[6] yesterday; & I like exceedingly its external & internal appearance: I read only about a dozen pages last night (for I was tired with Haymaking) but I saw quite enough to perceive how *very* much it will interest me & how many passages will be scored: I am pleased to find a good sprinkling of Nat. History: I shall be astonished if it does not sell very largley.—

Remember me most kindly to Mrs Lyell & tell her that Emma remains in her most wearisome statu quo.—[7] How sorry I am to think that we shall not see you here again for so long:[8] I wish you may knock yourself a little bit up before you start, & require a day's fresh air before the ocean-breezes blow on you.—

Will you please to remember me to Miss Lyell & give my respectful compliments to M^r. Lyell[9] | and believe me my dear Lyell | Ever yours | C. Darwin

American Philosophical Society

[1] Dated on the basis that CD received his presentation copies of the first part of *Journal of researches* 2d ed. on 3 July (see letter to John Murray, [3 July 1845]).

[2] See letter to John Murray, [31 May 1845], n. 2.

[3] *Journal of researches*, pp. 268–98; 2d ed., pp. 242–51.

[4] *Journal of researches* 2d ed., pp. 145–6.

[5] *Journal of researches* 2d ed., pp. 173–6, in which CD discusses the Malthusian idea of checks and balances in nature. This passage was one of the most notable changes that CD made to the first edition.

[6] C. Lyell 1845a. CD's annotated copy is preserved in the Darwin Library–CUL.

[7] The birth of Emma's fifth child was imminent; George Howard Darwin was born on 9 July.

[8] The Lyells were to depart on 2 September 1845 for a second visit to the United States.

[9] Marianne Lyell, whom CD had met when she visited the Lyells in 1838, and Charles Lyell Sr of Kinnordy House.

To Ernst Dieffenbach [before 9 July 1845]

[Down]

'. . . It is evident that you have not time now to pay me a visit, & indeed as Mrs. Darwin is in daily expectation of her confinement[1] I could hardly have asked you . . .

When I saw your name & that of so many other naturalists at Cambridge,[2] I wished much to have been there; but my strength so often fails me, that I expected more mortification than pleasure . . .

I should have liked to have heard the Crater-of-Elevation discussion; after having read both sides, I cannot subscribe to that view; but I think there remains something unexplained about those many vast circular volcanic ruins . . .[3]

I presume it is very unprobable that there will ever be a second German Edition of my Journal[4] . . . I have largely condensed, corrected & added to the Second English Edition, & I am sure have considerably improved & popularised it . . .'

J. A. Stargardt, Marburg (catalogue 574, 11–13 November 1965)

[1] George Howard Darwin was born on 9 July 1845.

[2] The British Association met in Cambridge in June 1845. Dieffenbach read a paper on the geology of New Zealand, see *Report of the 15th meeting of the British Association for the Advancement of Science held at Cambridge in 1845* Transactions of the sections, p. 50.

[3] For a report of the discussion see *Athenæum*, no. 923, 5 July 1845, pp. 675–6.

[4] A new translation by Julius Victor Carus was published by Schweizerbart, Stuttgart, in 1875 (Freeman 1977, no. 189).

From G. R. Waterhouse [11 July 1845]

Brit. Mus.
Friday—

My dear Darwin

I have not yet had the separate copies of my paper[1] sent me & fear more than a week has passed since you wrote— The following are the only species of Beetles in your collection of which I have been able to obtain the habitats

Creophilus—? apparently a new species (from under a dead bird in Chatham Isd.
Ammophorus galapagoensis, from Chatham Island
———— bifoveatus from James Island
There is a third species *Amm. obscurus* which was pinned, perhaps it was from a difft Island
Pedonœces galapagoensis James Island
———— pubescens Chatham Isd.
———— costatus (a third species not labelled)
Apate spe—? Chatham Isd.
Ormiscus variegatus Charles Isd
Otiorhynchus cuneiformis Charles Isd.

In none of these cases—though of some of the species there are several specimens—do I find the same species from different Islands—[2]

M. De Selys Longchamps,[3] a great man for the Neuroptera has been here— I got him to look at the dragon fly from the Galapagos—he said he thought it was a new species & that it was an American form—i.e. Brasilian—

I write in great haste, having an appointment to attend to immediately

Believe me | faithfully yours | Geo. R. Waterhouse

Of a new genus of Heteromera to which I have given the name *Stomion*, there are three species— I wonder whether they came from diff[t] Islands?

Postmark: JY 11 1845
DAR 181

CD ANNOTATIONS
2.1 In none of these cases] 'ie in only 2 genera' *added ink*
Top of first page: 'Solidified' *ink*

[1] Waterhouse 1845a.
[2] CD used this information in *Journal of researches* 2d ed., p. 395.
[3] Michel Edmond de Selys-Longchamps.

To J. D. Hooker [11–12 July 1845]

Down Bromley Kent
Friday

My dear Hooker

I sh[d] have written to you a few days ago, as I had some questions to ask & several points in your last letter, which I should much enjoy discussing with you: but on Wednesday an upsetting event happened in the fact of a Boy-Baby being born to us—may he turn out a Naturalist.[1] My wife is going on most comfortably.— First I have got a few questions about the Galapagos Plants, as I am now come (not in correcting press, but first time over) to that Chapter: I will put these questions on a separate paper & some of them you can answer by a word or two on the paper on its back & return it to me, pretty soon, if you can so manage it.[2] I cannot tell you how delighted & astonished I am at the results of your examination; how wonderfully they support my assertion on the differences in the animals of the different islands, about which I have always been fearful: I see the case excites the interest even of R. Brown.—[3]

I am sorry to plague you with so many questions but I do not think they will take you long to answer.— Of course I shall give all these results as from you.—

I could, if I had the vigour, fill half a dozen letters in discussing the many most interesting points you allude to in your letters, which always delight me & tell me much.— I am glad to have your criticisms: thank you much for your note about the so-called Fucus: I can correct my error in the range in press:[4] I wonder what sea-weed I saw that looked different & a representative.— Do you know whether the Tasmanian savage used the Boomerang & Throwing Stick?—

With respect to the non-improvements of savages, your line of argument is the same with Arch. Whatelys[5] & has I believe been developed by Dr. C. Taylor,[6] but I have not seen his book: I have always felt opposed to this, but I fancy only general arguments can be advanced against it. They maintain civilization is a miracle, requiring the interposition of God. Do you know Whatelys capital explanation of the origin of the Boomerange, from the laterally flattened first rib of the Whale.— There is much to read: Lyell's book is very good, I think;[7] I have not yet got the Kosmos, for I want to know which is the best Translation, can you tell me?[8] Strezleckis Book has greatly disappointed me.—[9] I am now reading a wonderful book, for facts on Variation—Bronn Gesichte der Natur: it is stiff German: it forestalls me, sometimes I think delightfully & sometimes cruelly.[10] You will be ten times hereafter more horrified at me, than at H. Watson: I hate arguments from results, but on my views of descent, really Nat. Hist. becomes a sublimely grand result-giving subject (now you may quiz me for so foolish an escape of mouth)— I am particularly interested by your remarks on insular Floras: your statement about the definiteness is *exactly* the reverse of that of that old logger-head B. St. Vincent.[11] A genus having *several* good species in the same small island is new to me & very remarkable & as you well observe hostile to descent: can you enlarge I shd particularly be obliged on this *sometime*, to me: *are such genera peculiar to the islands*? How differently people view the same subject, for I look at insular Floras (though not overlooking from ignorance *all* the grave difficulties) as leading to an opposite view to yours. (I must leave this letter till tomorrow, for I am tired; but I so enjoy writing to you, that I must inflict a little more on you).

Have you any good evidence for absence of insects in small islands: I found 13 species in Keeling atoll. Flies are good fertilisers; & I have seen a microscopic Thrips & a Cecidomya take flight from a flower in the direction of another with pollen adhering to them. In Arctic countries a Bee seems to go as far N. as any flower.— Not that I am a Believer in Hybridising to any extent worth mention;[12] but I believe the absence of insects wd. present the most serious difficulty to the inpregnation of a host of (not diœcious or monoœcious plant) plants:—have you ever seen C. Sprengels curious book on this subject; I have verified many of his observations: doubtless he rides his theory very hard.—[13]

Without knowing the age of the Kerguelen tree no one would, I presume, guess about any change of climate since they grew: S. America was once hotter, then much colder, than now: in N. America, within Tertiary epochs the series, has been.—hot—warm—very cold—a little warmer than now—present climate.—

I am particularly glad to observe in your former letter that you have plants in your Fathers collection from Elizabeth Isd: how I do long to see your results on the Pacific & indeed every-where else. I find just the same relation in a curious group of birds from the Galapagos to one species from near Elizabeth isl^{d}, which you do with some plants from Juan Fernandez: I find in sea-shells the Galapagos is a point of union for two grand & otherwise most distinct concho-geographical divisions of the world.—[14]

I am not aware that I want any geological information from India; but if your friend resides near those parts where the Chetah is used for hunting I am *particularly* anxious to know, whether they *ever* breed in domestication; & if never or seldom, whether they copulate, & whether it is thought to be the fault of the male or female.— Again if he reside in the silk-worm districts, any information whether the moths, caterpillars or cocoons vary at all,—whether the inhabitants take any pains in selecting good individuals for breeding—whether there is any traditional belief in the origin of any breed, ie if different breeds of the same species are found in different districts.—or any analagous information.— This w^{d} be eminently valuable to me.

I was surprised to observe that in the short report in the Athenæum of Brit: Assoc: that there was no allusion to Forbes ice-driving-plant notions:[15] it would take up a whole sheet to discuss this with you; but I sh^{d} particularly like to discuss it with you.— How I do hope we may see you here this summer; though I am fearful our plans may clash. About the 20th of August we go from home for about six weeks: is there any chance of your coming before? or w^{d} there be time in October, before your return to Edinburgh? do cogitate on this.

I am sincerely sorry to hear of the grievous state of Prof. Graham.— What a letter I have written you! God Help you!

Ever yours | C. Darwin

I know nothing, (I wish I did) on sexes of Irish Yew, but I observe all my young trees bear berries; I had thoughts of going to the Nursery & looking to the trees. I will try to get enquiries made at Florence Court.[16]

[Enclosure]

Galapagos

Was my collection your main material? **roughly** what percentage is my collection to that of others Whose other collections have you described that I may mention their names, if they form any considerable proportion. I think I collected few at Albemarle Isl. & those certainly rather near the coast.

I see[17] in all the islands, except Charles, the extra-Galapageian plants are less than half of the total number; but in Charles Is^{d} there are 39 extra-Galapageian & only 29 Galapageian: now Charles Is^{d} alone has been settled & cultivated & I collected in the upper, cultivated region: can you remember,[18] whether a good many of the plants here had a tropical weedy character:[19] when giving your results, I sh^{d} like to put in a caution about the anomalous result in this one island.[20]

Owing to your having struck out one of two brackets, I cannot be sure whether there are 185 flowering plants[21] **and** 42[22] cryptogamic plants: or does the 185 include the cryptogamic.?

Can you say how many species of Ferns there are; are they not remarkable?[23]

Is *Icaleria*[24] the name of an arborescent Composit: if so, I presume it is a tree which I noticed in the upper damp region.—

I collected leaves & unripe fruit of a tree, on which the tortoise feeds: do you chance to know the genus: it is quite unimportant.

I presume there must be a good many cases, where a genus (whether confined to the Galapagos or not) has different peculiar Galapageian species on different islands:—for if it be not so, the *genera* must be wonderfully different on the different islands.—

If the collection had been put into your hands: sh[d] you have known that it came from the American quarter: w[d] not Opuntia have told this? would other genera have told the same story?

Postmark: 12 JY 12 1845
DAR 114.1: 36–36b, DAR 100: 43–7

[1] George Howard Darwin, born 9 July 1845.

[2] See enclosure, transcribed following this letter, and Hooker's reply (letter from J. D. Hooker, [after 12 July 1845]).

[3] See *Athenæum*, no. 910, 5 April 1845, p. 337.

[4] See letter from J. D. Hooker, 5 July 1845, in which he corrects CD's identification of the giant kelp, and adds further information on its geographic range. CD made the correction in *Journal of researches* 2d ed., p. 239 n.

[5] R. Whately 1831, p. 119: 'the progress of any community in civilization, by its own internal means, must always have begun from a condition removed from that of complete barbarism; out of which it does not appear that men ever did or can raise themselves.'

[6] W. C. Taylor 1840.

[7] C. Lyell 1845a.

[8] There were two translations of Alexander von Humboldt's *Kosmos* (1845–62) under way in 1845: an unauthorised translation by Augustin Prichard (Humboldt 1845–8), and a later translation by Elizabeth Juliana Sabine (Humboldt 1846–8). The latter is in the Darwin Library–CUL.

[9] Strzelecki 1845.

[10] Bronn 1841–9. CD possessed a later reprint of the first two volumes (Stuttgart, 1842–3). His copy is in the Darwin Library–CUL.

[11] Jean Baptiste Georges Marie Bory de Saint-Vincent, who argued that plants on isolated islands were polymorphic (Bory de Saint-Vincent 1804). See letter to J. D. Hooker, 25 December [1844], and Hooker's reply, 30 December 1844.

[12] CD consistently considered hybridisation in nature to be a rare event, i.e., not an important source of new varieties or species (*Notebook C*: 151). CD clarifies his meaning in letter to J. D. Hooker, [22 July – 19 August 1845].

[13] Sprengel 1793, in which Christian Konrad Sprengel argues that all floral structures are designed to bring about pollination by insects. Sprengel's observations on the relationship between flowers and insects were important to CD as evidence to support his belief in the necessity of cross-fertilisation. CD found Sprengel's explanation of his observations unsatisfactory, and in his copy (Darwin Library–CUL) he wrote: 'How poor! Has no notion of advantage of intermarriage Seems to think fact of insects being required at all, does not deserve any explanation, & how poor a one of Dichogamy for convenience of insects—!!' (p. 18).

[14] *Journal of researches* 2d ed., p. 391.

[15] Edward Forbes's paper (E. Forbes 1845), read in 1845 and published 1846, was referred to in *Athenæum*, no. 923, 5 July 1845, p. 678.
[16] The seat of William Willoughby Cole, 3d Earl of Enniskillen. See letter from Philip de Malpas Grey-Egerton, 5 May [1844], for CD's interest in the Irish yew.
[17] CD is referring to the letter from J. D. Hooker, [28 April 1845].
[18] At this point CD added 'V. map of Pacific' in pencil.
[19] Hooker added in pencil 'Yes'.
[20] Hooker added in pencil 'I nosed it all along.'
[21] Hooker added in pencil 'Yes'.
[22] Hooker added in pencil 'should be 40 I accidentally added 2 Malden Isld' and deleted the last part of the sentence 'or does . . . cryptogamic.?'
[23] Hooker added in pencil 'No'.
[24] *Scalesia.*

From J. D. Hooker [after 12 July 1845][1]

[Edinburgh]

Your collection was the main material—amounting to 153 species of flowering plants and (& +) 40 of Cryptog. i.e. 153 + 40 not 42 = 193 total, your,

Yours Total
193 : 225 :: 1 :

Macræ a collector of the Hort. Soc. formed the best part of the Albemarle Isld Collections ie. 42 phænogamic plants, you collected 7. Phænogamic there 4 of which are not in Macræ's— Macræs total Galap are these of Albemarle Isld & one Fern. Douglas & Scouler on their way to the Columbia river collected some plants on James Isld about 15 species.[2] Of these 15, 6 you did not get at all, & 5 you got on other Islds, (not on Jas.)— the other 5 both of you got on Jas Isld. Thus though you both collected on one small Isld. your collections are very different, 11 out of his 15 you did not collect & he only 5 out of the 60 that you got, this is terribly unsatisfactory.

The only other collector is Cuming who got one Scalesia which you did not but hardly another species of plant, if you see him ask him what Isld he landed on, I did ask him once but forget. I saw at Paris a few Galap.s from Petit Thouars I think late voyage.[3]

In all the *numerical* estimates I would exclude the Cryptogs. Even the Ferns except the prop. these bear to Phænog.—there are 28 species—Chas Isld: 10, James 20, Albem. 1. I make 6 new species, only one however is very remarkable it is from Jas. Isld— The rest 22 are almost all W. Indian— There are no limits to the diffusion of Ferns. (beasts)

Scalesia is a peculiar Galapagæan arborescent Compos containing 6 species— one from Cuming. 1 Chatham Isld:—1 Albemarle, 1 Chas Isld & two James Isld.— ! — it is no doubt a damp region tree, *analogous* no affinity to the arborescent Comp. of Juan Fernandez, S.t Helena & I think Mauritius has some too.

I cannot remember the tortoises tree, but I think you noted it, if so I have also in the paper sent to Linn. Soc.[4]

I remarked also what you say of Chas' Isld & took notice of the plants at the time. Compos Ageratum conyzoides, Nicotiana glutinosa, Teucrium inflatum, Salvia tiliæfolia, Scoparia dulcis, Lantana canescens?— Verbena littoralis, Boussingaultia baselloides, Brandesia echinocephala, Sandwich Islds. only previously Amaranthus celosioides & Caracasanus, Phyllanthus obovatus, Urtica canadensis & divaricata, Hypoxis erecta, Cyperus strigosus, Panicum colonum— are all plants of S. America **not found** in the other Galapagos, those underlined are no doubt **introduced by man**, the rest I would expect in the other Islands.. & may or may not be introduced by man, I incline to think not. Take these 10 from the 68 Chas. Isld plants & the peculiar balance the common, is not this funny, upon my honor I have not *cooked the result*!—

Thus, as far as the collections go, the Florula of each Islet is ½ peculiar, but mark—; the very few species of Douglas & Macræ so disturb the results drawn from your special Herbarium, that there is no saying positively what a third collector might produce.

The instances of representative species on the several Islets may be divided into 2 groups, **1st** of peculiar genera, as *Scalesia* see p. 2[5] & Galapagoa one Alb. & one Chas & Alb. & **2nd** Extra Galapageian genera having peculiar Galapageian species. Thus, the

Euphorbiæ are very peculiar, only one species, *pilulifera* (Jas. Isld.) is mundane, but of 7 others not one is common to 2 Islets.—

Acalypha has 6 species, in the same predicament, (here the species of 2 **very** mundane genera are not widely dispersed)

Cordia has 5 species,—2 Jas' Isld. 1 Alb. & Chas (an α & β however) 1, Albem. & Jas. 1 Chat. & Alb.

Compos

Lerontea has one Chat. & 2 species Albemarle.
Erigeron?. 1, Chas & Jas. & 1, Albemarle
Milleria? 1. Chas & 1 Jas.
Spilanthes? 1, Albem. 1 Jas.
Nov Gen. Compos, 1. Alb. & Jas;—1 Chat & Chas.
3 other unknown Compos are each confined to one Islet—
These compos, so far, very *wonderfully* peculiar to seperate Islets, add Scalesia & more so still.

Borreria is a mundane genus with 7 species all peculiar, only one of which is found on 2 Islets, & yet this is a genus of *canaille* all the world over.

Chiococca 3 species 1 Chat, 1 Jas & 1 Albem.

Ipomea 2 peculiar species both on Jas. one mundane on Chatham

These are the most marked instances of peculiarity in the distribution of the peculiar species of genera presenting more than one representative

There is still an enormous deal to be done with the materials—a comparison of the Islets with all the extra Galapageian species eliminated. A comparison of each with the coast Flora. The proportion of driftable & portable xtra Galap. plants in each:—those that fly or *are flown* by birds, those that salt water does & does not kill: that birds do & do not digest &c &c &c.

The collection is *out & out* S. American, & W. coast, but from the peculiarity of some genera & most species, I should not have known where to put it, supposing Galapagos not to xist. I know so much of the Flora of the coast as not to expect so much novelty from any 100 miles of it, if forced to assign the place it wd. probably be Panama

The opuntia is nowhere in the coll.[6] The Flora is S. American throughout in character. What is the Flora of St. Felix Islds on the S. Tropic.

I hope you can read this: pray ask me about any difficulties without ceremony I will write ere long.

Ever yours Jos D Hooker

DAR 100: 43–7

CD ANNOTATIONS

1.4 193 . . . 1:] 'ie (32) not mine' *added ink*

1.4 225] '$\frac{193}{32}$' *added pencil below*

6.1 I cannot . . . time. 7.2] '6.' *added brown crayon, underl brown crayon*

[1] These answers to CD's inquiries about Galápagos plants were written on the back of CD's questions (enclosure with letter to J. D. Hooker, [11–12 July 1845]). See *Journal of researches* 2d ed., pp. 392–3, for CD's use of these notes in his much revised Galápagos chapter.

[2] David Douglas and John Scoulter touched at James Island on the Hudson's Bay Company's expedition to the Columbia river, 1824–5.

[3] Abel Aubert Du Petit-Thouars, who circumnavigated the globe, 1836–9. See Du Petit-Thouars 1840–3, 2: 313–22, for an account of his visit to the Galápagos Islands.

[4] The tortoises feed on the lichen *Usnea plicata* which hangs from the branches of trees in the upper damp regions of the islands (J. D. Hooker 1845d, p. 164, and *Journal of researches* 2d ed., p. 382).

[5] Hooker numbered the pages of his reply to CD and refers back to paragraph five.

[6] CD's Galápagos *Opuntia* specimens were described in Henslow 1837.

From J. D. Hooker [mid-July 1845][1]

[Edinburgh]

offshoot from it such as I have attempted to portray— I suppose there can be no doubt that the population of some of the Islands lately discovered was fast disappearing from sensuality &c: it is equally evident that under such circumstances it did not attain its maximum: If the accession of civilization is a miracle so must the decline of it be also, for the protracted miracle would become a 2d. nature.=

I have heard nothing about Kosmos, Bailliers I suppose to be a species of Piracy.[2] Humboldt had agreed, that Murray should have the publishing of the translation & passed me the compliment of asking who would be the best translator (for I cannot suppose he intended me the high honor of asking for *information*) I said Mrs Sabine as translator of Wrangel,[3] & he commissioned me to tell her how much he wished she would take a part in it—, consequently, with Murray's sanction, (who wished Mrs Austin to have it,[4] which H. did not like at all) I told Mrs S. & Col. S. wrote to Baron H. about it: this is all I know, I hope Murray's is Mrs Sabine's translation,[5] I will ask when I come up to London. I feared for Strzlecki's book,[6] I am very glad he did not send me the Bot. mss as he promised:, he is a nice fellow— Bronn's Gesichte I know nothing of.[7] Bother variation, developement & all such subjects,! it is reasoning in a circle I believe after all. As a Botanist I must be content to take species as they *appear to be* not as *they are*, & **still less** as **they were** or ought to be. You see I am amazed at my own incapacity to fathom or follow the subject to any good purpose (open confession is *good* for the soul).

I think I can give you plenty of instances of peculiar genera with several good species in very small Island. E.G.

St Helena. *Commidendron* (arb. Compos.) 5 reported species, certainly 3 good.
Lachanodes (arb. Compos.) 2 good species.—

Juan Fernandez.
Dendroseris (arb. Compos.) 7 species I do not know all, but suppose them all good.—
Robinsonia (arb. Compos.) 4 species, I do not know them well.

Madeira
Sinapodendron, 3 species—a peculiar genus of Cruciferæ.

Sandwich Islands.
Schiedia (Caryophylleæ) 3 species
Peteria 3 species } I think both these genera are quite peculiar } Rubiaceae
Kadua 9 species } very nearly so at any rate
Dubautia (Compos) 2 species
Microchæta (D°) 4 species
Clermontea 5 species } Lobeliaceæ I am not sure that the
Delissea— 7 species } genera are all peculiar, but nearly so.
Rollandia 2 species }
Phyllostegia 5 spec in Owhyhee,[8] 2 in Oahu } Labiatæ. I think
Stenogyne— 4 sp in Owhyhee } both are peculiar.

These Sandwich Islander's are not positive evidence, as I do not know how far some of the genera may not have solitary representative species in seperate Islets.

Of confined genera there are many examples of solitary Islands having more than one well marked species & still more of Islets having well marked species which together form a group of a mundane genus,

I have always felt opposed to Bory's (who is a great Gascon! but not altogether to be despised) views of the variableness of Insular species

I certainly have no good evidence in favor of the loose statement I made & which corresponded with a vague idea I held, of Insects being scarce on Islands:

yet 13 species is surely very **few** for Keeling if size is to be regarded, how often may you not find 13 on your own window?— Kerguelens Land has only 3.. New Zealand & V. D. L. are certainly poor—in Trinidad (of Brazils) I saw only 3, I think, a *Hemerobius* & the House-flie & cockroach, introduced from a wreck: Canaries & Madeira are poor, I think: Cape de Verds are too dependent on the W. coast of Africa to judge from— nothing struck me as so marvellous as the appearance of 4 Insecta & many Arachnida you mention as on St. Pauls rocks.— Still I agree with you on the main point that such few as there are, wd. be enough for impregnation, if they only went to work about it..

I cannot prove that there is *much* hybridizing in nature, but do not see why there should not be, as we do not doubt that species require the pollen of other individuals exactly as in the higher animals you must not breed in (I think the term is.—

I cannot hook my Kerguelens tree or climate on to the vacillating temperature of S. America, many thanks for the information though.

Do you connect the union of Conchogeographic districts at the Galapagos with the currents?—

Every young *Irish* yew bears berries, there is a sort of Irish yew in Ayrshire which I believe like the goddess Diana of the Ephesians dropped down from heaven & picked itself up in a garden, when I hear whether it bears berries I will tell you if she be equally chaste If the yew had been Italian & bows made it wd. have been dedicated to Diana.[9]

After class I go to old Chas. Lyells at Kinnordy,[10] & shall not be home till middle of August, I hope we may meet in October, I often think of our two meetings, & long for another.

And now to bother you for the last time, The reappearance of plants in certain situations is a curious phenomena of which instances are multiplying daily in this neighbourhood: there are doubtless series of seeds in some grounds lying dormant but not dead: what a curious principle life must be & what an uncomfortable abode it must often have. Cutting open rail-ways causes a change of vegetation in two ways, by turning up buried live seeds, & by affording space & protection for the growth of transported seeds: so that it is often very difficult to determine to which cause the appearance or superabundance of a plant is attributable The Dutch Clover case is constantly quoted but the Stirling castle one is more curious. The Kings Park was dug up in about 1650? during the 1st rebellion, wherever the cuts were made for encampments, the broom appeared, but in a year or two disappeared.. In rebellion of 1745, it was again encamped upon & again Broom came up & & disappeared: it was afterwards ploughed & immediately became covered with Broom, which has all for the 3d. time vanished.

I am still *talking over* the Students. Every Saturday I take them on excursion & walk them 20 miles gathering plants, about 20 or 30 generally go of the class. I have thoughts of taking them into the Grampians of Forfarshire for a week or 10

Alexander von Humboldt. Daguerreotype by Hermann Biow, 1847.
(Published in Beck, Hanno, *Alexander von Humboldt* (Franz Steiner Verlag, Wiesbaden: 1961). By permission of the Syndics of the Cambridge University Library.)

William Jackson Hooker. By S. Gambardella, *c.* 1843.
(Courtesy of the Council of the Linnean Society of London.)

days, if enough would or will come forward. I wish you could be induced to come down & join us.

To conclude, (I have been reading Scotch Sermons) how curious that water plants should be so widely diffused, water must have been a mighty agent in dissemination not only though are these diffused, but diffuseable. Aponogeton, a Cape plant, not native of cold regions bears a freezing every winter in our ponds: no one would have dreamt of it

Ever your's | Jos D Hooker.

Incomplete
DAR 100: 49–50

CD ANNOTATIONS
1.1 offshoot . . . soul). 2.16] *crossed pencil*
5.1 Islands] 'Islands' *superimposed in pencil to clarify Hooker's writing*
8.1 I cannot . . term is.— 8.4] *scored pencil*
9.1 I cannot . . . though. 9.2] *crossed pencil*
10.1 Do you . . . currents?— 10.2] *scored pencil*
12.1 After . . . time, 13.1] *crossed pencil*
14.1 I am . . . join us. 14.5] *crossed pencil*
15.2 should be . . . dreamt of it 15.5] *scored brown crayon*
15.3 not]'no' *superimposed in pencil to clarify Hooker's writing*
15.3 are] 'are' *superimposed in pencil to clarify Hooker's writing*
15.3 diffused,] *comma superimposed in pencil to clarify Hooker's writing*
15.3 but diffuseable] 'are' *interl in pencil after* 'but'; *underl brown crayon*

[1] Dated on the assumption that this letter falls between letter from J. D. Hooker, [after 12 July 1845], and letter to J. D. Hooker, [22 July – 19 August 1845].
[2] Humboldt 1845–8 was an unauthorised translation by Augustin Prichard, published by Hippolyte Baillière.
[3] Wrangel 1840, translated by Elizabeth Juliana Sabine, wife of Colonel Edward Sabine.
[4] Sarah Austin.
[5] Murray's translation (Humboldt 1846–8) is by E. J. Sabine. Sarah Austin had declined John Murray's proposal that she take on the work (correspondence in John Murray Archive).
[6] Strzelecki 1845.
[7] Bronn 1841–9.
[8] Hawaii.
[9] Artemis of Ephesus, subsequently identified with the Hellenic goddess of the same name and the Roman Diana, was associated with trees, agriculture, the hunt, and chastity. Diana was armed with a bow and arrow; yew is the wood from which the English long-bow was made.
[10] Charles Lyell Sr of Kinnordy, a friend of the Hooker family since their residence in Glasgow.

To John Murray 16 [July 1845]

Down Bromley Kent
16 Wednesday | (after post-time)

My dear Sir

I sent the M.S. to Mess. Clowes on the 7th & since then I have had only three sheets. I urged him to send me a sheet a day, stating the reason that my health is uncertain & that I would not on any account suffer, what I did last time, when, being unwell, I had to look over three or four sheets the last day.— Now I see Mr Clowes is running the same course this time & I write to inform you, that if the number sh^d^ fail being ready it will be *exclusively the* fault of M^ess^ Clowes for I cannot & will not exert myself again, if I chance to be unwell

Will you be so kind as to take the affair in your own hands with M^ess^ Clowes.

My dear Sir | Yours very faithfully | C. Darwin

John Murray Archive (Darwin 13–14)

To J. D. Hooker [22 July – 19 August 1845][1]

Down Bromley Kent
Tuesday

My dear Hooker

I am particularly obliged for your facts about solitary islands having several species of peculiar genera: it knocks on the head some analogies of mine: the point stupidly never occurred to me to ask about.— I am amused at your anathemas against variation & co; whatever you may be pleased to say, you will never be content with simple species, "as they are"—I defy you to steel your mind to technicalities, like so many of our brother naturalists.— I am much pleased that I thought of sending you Forbes article: I confess I cannot make out the evidence of his *time*-notions in distribution, & I cannot help suspecting that they are rather vague.[2] Lyell preceded Forbes in one class of speculation of this kind; for instance in his explaining the identity of the Sicily Flora with that of S. Italy, by its having been wholly upraised within the recent period;[3] & so I believe with mountain-chains separating floras.— I do not remember Humboldts fact about the Heath regions.— Very curious the case of the broom; I can tell you something analogous on a small scale: my Father when he built his house sowed many broom-seed, on a wild bank which did not come up, owing, as it was thought, to much earth having been thrown over them: about 35 years afterwards, in cutting a terrace, all this earth was thrown up, & now the bank is one mass of broom.—[4]

I see we were in some degree talking to cross purposes; when I said I did much believe in hybridising to any extent,[5] I did not mean at all to exclude crossing. It has long been a hobby of mine to see in how many flowers such crossing is probable: it was, I believe Knights view, originally that every plant must be occasionally crossed:[6] I find, however, plenty of difficulty in showing even a vague probability of this; especially in the Leguminosæ, though their structure is inimitably adapted to favour crossing, I have never yet met with but one instance of a **natural** *mongrel* (nor mule?)[7] in this family

I shall be particularly curious to hear some account of the appearance & origin of the Ayshire Irish Yew. And now for the main object of my letter; it is to ask, whether you would just run your eye over the proof of my Galapagos Chapter, where I mention the plants, to see that I have made no blunders or spelt any of the scientific names wrongly.[8] As I daresay you will so far oblige me, will you let me know a few days before, when you leave Edinburgh & how long you stay at Kinnordy, so that my letter might catch you.—

You ought to find out if possible, on what part of Albemarle Is[d] Macræ landed; it could not have been on the sterile northern part where I did: the SW. end is high & green. I am not surprised at my collection from James isl[d] differing from others, as the damp upland district (where I slept 2 nights is 6 miles from the coast, & no naturalist except myself probably ever ascended to it. Cuming had never even heard of it—. Cuming tells me that he was on Charles, James & Albemarle islands, & that he cannot remember from my description, the Scalesia, but thinks he could if he saw a specimen.—

I have no idea of origin of the distribution of the Galapagos shells, about which you ask: I presume (after Forbes' excellent remarks on the facilities by which embryo-shells are transported)[9] that the Pacific shells have been borne thither by currents; but the currents all run the other way.—

Farewell, my dear Hooker with many thanks for your long letter, always most interesting to me. Ever yours | C. Darwin

DAR 114.1: 37

[1] Dated by the relationship to the letter from J. D. Hooker, [mid-July 1845], and the letter to J. D. Hooker, [15 or 22 August 1845].

[2] Probably E. Forbes 1843. Forbes discusses the distribution of marine invertebrates through time on pp. 173–5. See also letter to J. D. Hooker, [8 September 1844]. It is unlikely that CD is referring to E. Forbes 1845, since this was available only in abstract form in the *Athenæum* (see letter to J. D. Hooker, [11–12 July 1845], n. 15).

[3] C. Lyell 1830–3, 3: 115–16. In his annotated copy of C. Lyell 1837, 3: 445, CD has marked the passage and added 'Capital!'. Both works are in the Darwin Library–CUL.

[4] See *Correspondence* vol. 2, letter from Susan Elizabeth Darwin, [early December 1837?].

[5] CD has apparently omitted the word 'not'. See letter to J. D. Hooker, [11–12 July 1845], in which CD claims he does not believe in hybridisation to any great extent.

[6] Knight 1799.

[7] CD is distinguishing between crossing varieties (mongrels) and species (mules).

[8] *Journal of researches* 2d ed., ch. 17.

[9] E. Forbes 1844, p. 326.

From Louis Fraser 23 July 1845

Zool. Soc. Gardens
July 23, 1845

My dear Sir

Judging from single specimens[1] I should say than *Larus hæmatorhynchus* was one eighth smaller than *fuliginosus* The 3 Galapagos species of *Orpheus* are much duskier colored than those from the continent— The continental species vary considerably in size, some running much larger & some much smaller than those from the Archipelago. Zenaida Galapagoensis is about $\frac{1}{5}$ smaller than *aurita*

The *Totanus fuliginosus* is much more dusky than any other species we have in our collection— it is much the same size as other species of the genus[2]

The Breeding list is in progress[3] Do you require the particulars of the Hybrids

I have the honor to be | My dear Sir | Your obedient Servant | Louis Fraser

DAR 164

[1] The bird specimens referred to had been presented to the Zoological Society by CD on 4 January 1837 (see *Correspondence* vol. 2, letter from Emily Catherine Darwin, 15 [January 1837], n. 8).

[2] For CD's use of this information see *Journal of researches* 2d ed., pp. 378–81. Sulloway 1982 discusses CD's collection of Galápagos birds.

[3] Possibly a list of animals that had bred successfully in captivity in the Zoological Society's gardens, see Appendix IV.

From Louis Fraser [24? July 1845]

Z. S. Gardens
Thursday

My dear Sir

The *Zenaida Galapagoensis* is duskier than the *Z. aurita* in this instance, I mean that the coloring is darker although the whole, is not so sombre

I have the honor to be | My dear Sir | Your most obedient Servant | Louis Fraser

DAR 164

To J. S. Henslow 25 July 1845

Down Bromley Kent.
Friday. 25 July 1845.[1]

My dear Henslow.

Very many thanks for your ten notes, and enclosures: I had seen the Paragraph otherwise I should have been much interested in the death of (as he styled himself) "Comte Thierry, King of Nukahiva and Sovereign Chief of New Zealand"[2] I wonder what has become of his wretched wife. I sincerely hope that your allotments will succeed;[3] all that I have read in favour of them sounds most encouraging, and I have never been convinced by what has been written against them. I have bought a Farm in Lincolnshire and when I go there this Autumn, I mean to see what I can do in providing any cottage on my small estate with gardens— It is a hopeless thing to look to but I believe few things would do this Country more good in future ages[4] than the destruction of primogeniture,—so as to lessen the difference in land wealth & make more small freeholders.— How atrociously unjust are the stamp laws which render it so expensive for the poor man to buy his ¼ of an acre,[5] it makes one's blood burn with indignation. Have you seen Lyells Travels?[6] He says the poorer classes in Canada complain of the timber duties! so that our Cottages are badly built under pretence to benefit a few rich merchants really no doubt for our own landowners.

Thanks for the slip about the Crag— I am astonished that stones containing 50–60 per cent of Phosphate of Lime are not most valuable.[7]

A fortnight ago we had born a little boy, our fourth child.—[8] He is to be called George; & I believe I have pleasant associations with that name from formerly playing with your Boy.—[9] I hope M^rs. Henslow is better.

Farewell. | C. D.

Copy
DAR 145

[1] The date is given as it appears on a copy of the letter in the Cambridge University Library (the original has not been found). Francis Darwin in *LL* 1: 343 n. refers to it as of 'July 4th', but the reference to George Darwin's birth (on 9 July) makes that impossible. It does suggest, however, that the original may have been dated simply 'Friday'.

[2] Charles Philip Hippolytus, Baron (not Comte) de Thierry. The report of his death was false. As 'Sovereign Chief of New Zealand' he threatened to take by force land which he claimed to own (see *New Zealand encyclopaedia*). Robert FitzRoy records that the *Beagle*'s arrival in New Zealand caused alarm because it was mistaken for Thierry's ship (*Narrative* 2: 567).

[3] Henslow had begun a scheme to provide the poor country labourers of his parish with gardens by sub-letting small plots to them. He met with strong opposition from the farm-owners. See Jenyns 1862, pp. 88–92, and Russell-Gebbett 1977, pp. 30–1.

[4] 'in future ages' is interlined in the copyist's hand and was probably also interlined in the original.

[5] Stamp duty payable on a conveyance on sale was made up of a sum charged for the deed and a sum calculated upon the value of the property transferred. For a large sale, e.g., CD's farm at Beesby, this would amount to about 1% of the selling price, but for a small sale, a ¼ acre of land for, say, £20–30, it would be nearer 10%.

[6] C. Lyell 1845a.

[7] Henslow had discovered beds of phosphate nodules in the Suffolk Crag while on holiday. Analysis showed them to be rich in phosphate of lime, which Henslow pointed out made them a useful source of fertiliser. (Henslow 1845b, Jenyns 1862, pp. 201–2, and Russell-Gebbett 1977, pp. 95–6).

[8] CD and Emma's third child, Mary Eleanor, had died in infancy; George was therefore their fourth surviving child.

[9] George Henslow.

To John Murray [27 July 1845][1]

Down Bromley Kent
Sunday

My dear Sir

I sent the last sheet back yesterday to the Printers. By putting 3¼ pages into the next Part, I have kept this Part to 10 sheets, and I hope to get index and all in another 10 & ½ sheets.

I shall send half of the M.S for the next Part in 4 or 5 days to the Printers.—

In this 3^d Part, I have a rather long description of a very curious lizard (Amblyrhynchus) of which Mr Lyell has a woodcut (No 229, p 395 Elements of

Geology) & he desires me to say, that he is quite willing to let me have the use of it, if you have no objection.[2] In this case would you be so good as to have it looked out & sent to Mess. Clowes: I have indulged myself in another woodcut at my own expence.—[3]

If the 12 copies, which you so very kindly give me are ready by Thursday morning at 10. o clock wd you be so good as to send them to the Athenæum Club, directed to me. *If not ready by 10 oclock on Thursday*, would you send them thus

"C. Darwin Ee
Leaves Green
to be forwarded to Down
by Westerham Coach
Bolt-in-tun
Fleet St"

all this direction is necessary }

Believe me | my dear Sir | Yours very faithfully | C. Darwin
I really hope my Journal is greatly improved in this 2d Part.—

John Murray Archive (Darwin 15–16)

1 The Sunday before publication of the second number of *Journal of researches* 2d ed. on 2 August 1845 (Freeman 1977, p. 35).
2 *Journal of researches* 2d ed., p. 385. The woodcut was originally published in C. Lyell 1838, p. 395, accompanying an extract from the first edition of CD's *Journal of researches*. John Murray was the publisher of Charles Lyell's works as well as those of CD.
3 This is the woodcut of the size of the beaks of different species of Galápagos finches (*Journal of researches* 2d ed., p. 379) which illustrates CD's transmutationist speculation (p. 390) that, 'Seeing this gradation and diversity of structure in one small, intimately related group of birds, one might really fancy that from the original paucity of birds in this archipelago, one species had been taken and modified for different ends.' In CD's Account Book (Down House MS) he recorded on 5 August 1845 a payment of £1 2*s* for 'J. Lee woodcuts Journal'.

From Hugh Cuming 28 July 1845

80 Gower Street | Bedford Square | London
28th July 1845

My dear Sir

I was out on Saturday when your letter arrived, otherwise I would have answered it by that day's post—

The word *Oniscia* is correct—and *Stylifer*, is also correct—[1] I did not find either of these Genera in the Pacific Ocean they are both to be found in the West Indies— Oniscia is found also in China (two or three Species)— Stylifer is found in the Philippines and the Mauritius and according to Turton[2] it is also British.

I have never seen a Monoceros any where but on the West Coast of America and the Galapagus Islands I have seen upwards of 50 parcels of Shells from the Sandwich Islands, but never one of the family amongst them. there is a constant

Trade between those Islands and California, and no doubt they might have been carried there by Sailors pay no attention to the information you have gained from Books—

I have never found any of the following Genera, which I found both at the Galapagus and the West Coast of Americ⟨a⟩ in the Pacific Ocean, nor have I ever received them, from any Collector, viz: Fissurella, Monoceros, or Cancellaria

Should you require any further information, I shall be most happy to give it, if in my power.

I remain my Dear Sir | Very truly yours | H Cuming

C Darwin Esqre

DAR 205.3 (Letters)

[1] Shells collected by Cuming on the Galápagos before the voyage of the *Beagle*. CD was struck by Cuming's discovery that seashells from the Pacific and from both coasts of the Americas that do not normally occur together are found on the Galápagos (*Journal of researches* 2d ed., p. 391). With this letter are three lists of shells in CD's hand analysing the geographical distribution of the ninety shells collected by Cuming on the Galápagos; one list has brief additions made by Cuming (DAR 205.3 (Letters)). The results are summarised in *Journal of researches* 2d ed., pp. 390–1.

[2] Probably Turton 1840.

From William Yarrell 29 July 1845

Ryder Street | St James
29 July 1845.

Dear Sir,

I have much pleasure in sending you the following report in answer to your enquiries of yesterday— Genus *Prionotus*, 4 species— Coast of New York 2— Co. of Carolina 1— Co. of Brasil 1—. these are independant of the new species made known by Mr Jenyns under the name of P. miles— from the Galapagos—

Pristipoma—about 35 species— —

The places are arranged in the order of the number of species.

India—11
Africa—8
Brasil—6
West Ind.—4
Red sea—2
Chili—1
localities unknown—3

Besides the species of Mr Jenyns—P. cantharina—Galapagos

Latilus—4 species— India 1— Isle of France 1— Brasil 1— Valparaiso 1—

In this case also Mr Jenyns L. princeps from the Galapagos—is not included—

Cossyphus. 15 species— India—10
Red sea—3
Rio—1
Sandwich Isles—1

C. Darwini—Galapagos—not included—

Gobiesox 3 species— Freshwater S. Amer—1
(Lepidogaster bimaculatus)— Britain—1
New Zealand—1

Besides Mr Jenyns's G. marmoratus and
—pœcilopthalmus—[1]

If this should not meet your object, let me know how I can mend it.
Yours very truly | W^m^. Yarrell.

Chas. Darwin Esqr.

DAR 183

CD ANNOTATIONS
1.1 I have . . . included—1.21] *crossed pencil*
1.25 G. marmoratus . . . pœcilopthalmus— 1.26] 'Chiloe & Galapagos' *added ink*
2.1 If . . . Yarrell. 3.1] *crossed pencil*

[1] The species ascribed to Leonard Jenyns are those collected at the Galápagos by CD (see *Fish*). CD was evidently interested in the further distribution of other members of the Galápagos genera. In *Journal of researches* 2d ed., p. 390, he remarked that the Galápagos fish 'belong to twelve genera, all widely distributed, with the exception of Prionotus, of which the four previously known species live on the eastern side of America.'

To Charles Lyell [30 July – 2 August 1845]

Down Bromley Kent
(Saturday)[1]

My dear Lyell

I have been wishing to write to you for a week past, but every five-minute's worth of strength has been expended in getting out my Second Part. Your note pleased me a good deal more, I daresay, than my dedication did you,[2] and I thank you much for it. Your work[3] has interested me much, & I will give you my impressions, though as I never thought you would care to hear what I thought of the non-scientific parts, I made no notes nor took pains to remember any particular impression of ⅔ of the 1st Vol. The first impression, I sh^d^ say would be with most (though I have literally seen not one soul since reading it) regret at there not being more of the non-scientific: I am not a good judge, for I have read nothing ie non-scientific about N. America, but the whole struck me as very new, fresh & interesting. Your discussions bore to my mind the evident stamp of matured thought, & of conclusions drawn from facts observed by yourself & not from the opinions of the people whom you met; & this I suspect is comparatively rare.— Your slave discussion disturbed me much;[4] but as you would care no more for my opinion on this head, than for the ashes of this letter, I will say nothing, except that it gave me some sleepless most uncomfortable hours.—

Your account of the religious state of the States particularly interested me: I was surprised throughout at your very proper boldness against the clergy. In your university chapter, the clergy & not the state of Education are most severely & justly handled; and this I think is very bold, for I conceive you might crush a leaden-headed old Don, as a Don, with more safety, than touch the finger of that corporate animal, the Clergy. What a contrast in education does England show itself! Your apology (using the term, like the old religionists who meant anything but an apology) for lectures struck me as very clever: but all the arguments in the world on your side are not equal to one course of Jamieson's Lectures on the other side, which I formerly for my sins experienced.[5] Although I had read about the coal-fields in N. America, I never in the smallest degree really comprehended their area, their thickness & favourable position: nothing hardly astounded me more in your book.—

Some few parts struck me as rather heterogenous, but I do not know whether to an extent that at all signified. I missed, however, a good deal some general heading, to the chapters, such as the two or three principal places visited. One has no right to expect an author to write down to the zero of geographical ignorance of the reader; but I, not knowing a single place, was occasionally rather plagued in tracing your course. Sometimes in the beginning of a chapter, in one paragraph your course was traced through a half-dozen places; anyone, as ignorant as myself, if he could be found, would prefer such a disturbing paragraph left out. I cut your map loose & I found that a great comfort: I could not follow your engraved track. I think in a second edition, interspaces, here & there of one line open, w[d] be an improvement. By the way, I take credit to myself in giving my Journal a less scientific air in having printed all names of species & genera in Romans: the printing looks, also, better. All the Illustrations strike me as capital; & the map is an admirable volume in itself. If your Principles had not met with such universal admiration, I sh[d] have feared there w[d] have been too much geology in this for the general Reader: certainly all that the most clear & light style could do, has been done. To myself, the geology was an excellent, well condensed well digested resumé of all that has been made out in N. America; & every geologist ought to be grateful to you. The summing up of the Niagara chapter appeared to me the grandest part: I was, also, deeply interested by your discussions on the origin of Silurian formations; I have made scores of *scores* marking passages hereafter useful to me.[6]

All the coal-theory appeared to me very good: but it is no use going on enumerating in this manner.— I wish there had been more Nat. Hist; I liked *all* the scattered fragments.—

I have now given you an exact transcript of my thoughts; but they are hardly worth your reading. I have a few remarks on particular passages, which; however, are very much in the same predicament.

Vol I.

p. 81. Are you sure of the resemblance of the corals, shells, & insects of Van Diemen's Land & the N.? Surely the insects at least are different; & are not shells too similar all over the world to offer a good standard of comparison. You speak also of the analogy of the Arctic & Antarctic *Faunas*. In T. del. Fuego, the most southern land, certainly the mammals, birds, fish, insects & I should have thought shells and corals show little signs of relations with the N.— In plants there is not only an *analogy*, but some of the species are *identical*.— Perhaps you have better authority, however, than I am aware of: I sh[d] very much like to know what your authority is about the insects, shells & coral of Van Diemen's Land.[7] (N.B. you spell it V. Dieman's L.) and this is not, I think, usual.

p. 138 Would you please to tell me, whether there are any extinct species of Fulgur & Gnathodon in the U.S. or elsewhere?[8]

p. 150—bottom paragraph strikes me as obscure, in fact, I cannot understand it. I do not see the reason (& it ought to be made very obvious) why you do not mention in your present sources of Carbonic A. the breathing of all animals.[9] Is it correct to say one gas *absorbs* another? I presume you are certain that the putrefaction of *animal* matter yields carbonic acid; I had fancied it yielded little: I think Liebig w[d] quarrel with you for calling the slow combustion or *decay* (or erecaumosis or some such word) of vegetable matter, *putrefaction*.[10]

Might you not in this discussion, bring more prominently forward the absurdity of arguing from one quarter of the globe, without knowing what was going on in other parts; for instance, whether or not, peat was forming over 1000's of miles in both the N. & S. hemispheres at that period.?

p. 181 Would it not have been better, if you had explained by what means rain-water could carry away a seam of carbon; for I, for one, do not understand how.—[11]

Vol II. p. 37 D[r] Morton is so far wrong, that the Fuegians have never used a hollowed tree, but bark sowed together: the Indian on W. coast have used immemorially planks sewed together.[12]

Vol II. p 54. Although it may be strictly true that we seldom **meet** with wood or fruits floating on the sea, yet this cannot be at all in effect true: for on the Falklands, the Galapagos, the Radack & the Keeling islands drift-wood & **fruit** & seeds are thrown up abundantly: at Keeling the fruit &c &c almost certainly must have been transported 2000 miles. Heaven knows how many thousand the *northern* firs-trees must have travelled which are cast (together with bamboos & Palms) on the Radack Is[d] not far N. of the Equator.—

p. 65. Hearne in his Travels[13] gives a grand account of the Buffaloes in the Prairies pushing each over the cliffs of the rivers when rushing to drink. Will the fact, which I give in my Journal bear on this subject, viz that in the droughts, the animals, which drink of the saline streams all perish on the spot.[14]

p. 189. Does not the present Arctic Flora, afford a parallel in extent of distribution, with your carboniferous Flora?[15]

These are my few & unimportant conjectural criticisms or rather queries,—to some of which I sh^d^ be very much obliged for answers.— I am doubtful how far you will think my very long letter worth the reading.

My wife & Baby are going on very well. Thank Mrs Lyell for her beautiful letter; Kinnordy must be a quite charming looking place. I have got a fender-stool, in imitation of Hart St^t^—[16] I hope you will find time to let me have a line: I am anxious to know when you return to London.

Farewell, I congratulate you on having brought out so capital a book as your Travels: I sincerely trust after your return that I shall have, for one, the great pleasure of reading another volume | Ever yours | C. Darwin

P.S. *Have you any of my volumes of Lamarck*??[17]

American Philosophical Society

[1] CD first wrote 'Wednesday', then deleted it and wrote '(Saturday)'. Saturday, 2 August 1845, seems to be the date that the second number of *Journal of researches* 2d ed. appeared (Freeman 1977, p. 35): the previous Wednesday was 30 July.

[2] See letter to Charles Lyell, [5 July 1845].

[3] C. Lyell 1845a.

[4] Though he criticised American racial attitudes, Lyell disapproved of the Abolitionist movement and took a pessimistic view of the possibility of emancipation (C. Lyell 1845a, 1: 181–95).

[5] CD attended Robert Jameson's lectures at Edinburgh University in 1827 (see *Autobiography*, pp. 52–3, and Ashworth 1935, pp. 99–101).

[6] CD's annotated copy is preserved in the Darwin Library–CUL.

[7] Lyell emphasised the essential similarity existing among organisms of a given era, regardless of geographical distribution, asserting particularly the resemblance between the corals, shells, and insects of Tasmania and those of the northern temperate zone. The passages are marked in CD's copy (Darwin Library–CUL) with '? ! ?' and '? !! no' (C. Lyell 1845a, 1: 81).

[8] In a somewhat ambiguous passage, Lyell seems to say he has seen fossil shells identical to living members of these American genera.

[9] Lyell argued against the theory of Adolphe Théodore Brongniart that in the earliest geological times the atmosphere had been very rich in carbon. Brongniart believed that excess carbon had been absorbed by plants and buried in the great deposits of the Carboniferous, at which time higher forms of animal life were introduced (Brongniart 1828, pp. 251–4). Lyell refuted this view by showing that modern plants do not alter the composition of the atmosphere since there is a constant replenishment of carbon from other sources.

[10] Justus von Liebig employed the term 'eremacausis' for slow oxidation of organic matter (see Liebig 1840, p. 261).

[11] Lyell was thinking of the carbon content of topsoil that had been subsequently covered by other deposits.

[12] Lyell (1845a 2: 37) had cited Samuel George Morton's mistaken assertion that canoes hollowed from logs were standard throughout the Americas. In his copy of Lyell's work CD wrote 'No' against the passage.

[13] Hearne 1795. Samuel Hearne was an explorer and Canadian administrator.

[14] *Journal of researches* 2d ed., p. 134.

[15] Lyell (1845a, 2: 188–9) had commented on the similarity between coal plants in North America and Europe, suggesting there was no parallel case at the present day.

[16] Lyell's address in London.

[17] The Darwin Library–CUL contains only volumes two and four of the seven-volume first edition of Lamarck's *Histoire des animaux sans vertèbres* (1815–22). They are annotated by CD. All eleven volumes of the second edition (1835–45) are preserved, but are not annotated.

From Charles Lyell [after 2 August 1845]

which I have devoted my life, & usually expect when I depart from the rule to have some grand flaws detected in my reasoning or facts, which may be pardoned—if the subjects do not pretend to form the staple commodity of the cargo.

I agree with you that interspaces would have been useful where the subjects are so heterogeneous.

A little more Nat: History might have easily been inserted & would not perhaps violate my rule.

I remember when some zoophytes, Sertularia, Retepora Sponges & many others arrived at the British Museum from Tasmania, J. E. Gray agreeing with me, how like they looked to those on our shore, & he said it would require minute comparison to distinguish the species. I have a collection of insects made at Hobart Town, a great many of which have so English an aspect, that any one would see the collector had crossed the line and got into a temperate climate like our's. Dr Beck[1] once pointed out to me how the Antarctic conchological fauna approached the Arctic. The number of species of Margarita for example was an instance.

There is an extinct species of Gnathodon in the Miocene of Virginia, G. Grayii, & I think I collected there two extinct species of Fulgur, but my Miocene M.S. has gone to the press for the next journal.[2]

I asked R. Brown what recent Flora had the widest range, & he said part of the Australian, but I will put your query as to that of the arctic region to Charles Bunbury from whom I have just got a long & excellent letter on the climate of the carbonif^s^ Flora, which would do well to read to the G. S. showing that the temperature cannot be inferred to be tropical, but only damp, moist equable & without coal, all the plants save the ferns being too wide from existing analogies to be reasoned from with safety.[3] I much desire your soon becoming more intimate with him, as he has so profound a knowledge of species, as well as powers of generalization & he takes to you very much, though he requires a person to go two thirds of the way towards him, which I mention as knowing you, & being aware that your besetting sin is modesty, a rare one in this world, for Sidney Smith truly said it had no ordinary or natural connexion with merit except that of the allitteration. So as we are to lend our House in Hart S^t^ to the Bunburys[4] when we are away, I shall hope to hear of your availing yourself of this propinquity.[5]

L(A) incomplete[6]
DAR 205.3 (Letters)

CD ANNOTATIONS
1.1 which . . . rule. 3.2] *crossed pencil and ink*
6.4 read to . . . propinquity. 6.14] *crossed pencil and ink*

[1] Henrick Henricksen Beck.
[2] C. Lyell 1845b.
[3] Possibly the letter referred to in F. J. Bunbury ed. 1891–3, *Middle Life* 1: 68.
[4] Frances Joanna Bunbury was Mary Elizabeth Lyell's sister.
[5] CD met Charles James Fox Bunbury at Bedford Place, the home of Leonard Horner, on 23 November 1845. Bunbury records: "He avowed himself to some extent a believer in the transmutation of species, though not, he said, exactly according to the doctrine either of Lamarck or of the 'Vestiges'. But he admitted that all the leading botanists and zoologists, of this country at least, are on the other side." (F. J. Bunbury ed. 1891–3, *Middle life* 1: 77). CD had previously met Bunbury in June 1842 at an inn in Capel Curig when CD was investigating the effects of glaciation in North Wales (F. J. Bunbury ed. 1891–3, *Early life* 1: 367). They may also have encountered each other as students at Cambridge.
[6] The letter is in Mary Lyell's hand, except for the last six lines, which were written by Lyell.

To J. D. Hooker [15 or 22 August 1845][1]

Down Bromley Kent
Friday

My dear Hooker

I have just received your note: I am sincerely sorry to hear of the state of your venerable grandfather:[2] I trust his suffering is not great.—

I enclose the proofs; would you please look over the whole of the Galapagos Ch. as the vegetation is incidentally mentioned in two or three places: you can skip about the tortoises & lizards, tameness of birds which is as before; all the rest is much altered.— I have tryed to make it as little purely scientific as possible— I hope there are no material errors in the Botany part: the proofs have been revised once, but I have not time to look them over again before sending to you.— W^{d} you please return them soon, as the Press waits for them.—[3]

I grieve to hear that the labels are displaced in my plants: I took the greatest care in this respect, whilst they were in my care.— I collected **everything** in flower in *Patagonia* & *B. Blanca*, which latter place I saw was intermediate in character between the N. & S.—

Pray **of course** send my C. Verd plants & any others whither you like.— I have no separate catalogue of plants, but if you will send me the numbers, in a row I will gladly fill up the localities.—

I know nothing of the C. Pigeon.—

You are most kind to think of giving me a copy of your work:[4] but seriously there must be many botanists, on whom it w^{d} be better bestowed, though none, who will be better pleased at the offer: now I beg you to have no scruples, & keep it & give it to some Botanist if such sh^{d} occur to you: again I thank you.—

I am so stupid, that I forget which way you prefer parcels being sent to you.— I will then return d'Urville &c.— Can you allow me to keep a fortnight or 3 weeks longer Webb & Berthelot;[5] I have been so worked with my Journal, that I have not had strength to read it.—

Will you also excuse me having put some pencil-scores to the pamphlet on l'Espèce,[6] & will you leave them unrubbed out: so that I c^d^ *hereafter* borrow it again: the passages do not strike me as worth copying & yet I sh^d^ like hereafter to refresh my memory with them.—

Remember you shall have a copy of my Journal, when complete.

Ever my dear Hooker | Most truly yours | C. Darwin

I heartily wish you success at Edinburgh;[7] though it will carry you so far away.—

What a wonderful deal of work you are about!—

DAR 114.1: 38

[1] The conjectured dates are the two Fridays between the announcements of the deaths of Robert Graham and Joseph Hooker, see nn. 2 and 7, below.

[2] Joseph Hooker died on 24 August 1845.

[3] CD's Galápagos chapter (*Journal of researches* 2d ed., pp. 372–401) was due to be published in the third and final number.

[4] J. D. Hooker 1844–7.

[5] Dumont d'Urville [1841–54], Webb and Berthelot 1836–50, vol. 3, pt 1.

[6] Gérard 1844.

[7] Robert Graham, whose students Hooker was teaching, died on 7 August 1845. Hooker became the Crown's official nominee for the vacant professorship at Edinburgh University, competing against John Hutton Balfour, the Town Council's nominee (Huxley ed. 1918, 1: 204–5).

To William Jackson Hooker [23 August 1845][1]

Down Bromley Kent
Saturday

Dear Sir William

I shall have the greatest pleasure in sending you my high opinion of your son; & if I c^d^ flatter myself that any testimonial from one so little known as myself, could be of the smallest service, I really do not know anything which would so highly gratify me.— I will write tomorrow a letter addressed to you, & if it does not turn out the right thing, I will alter its *form*, in any way you can suggest. I am truly sorry to say, that my acquaintance with Sir John Herschel is not sufficient to allow me to write to him: I never saw him before at the C. of Good Hope;[2] & have dined only once since with him.— I have, however, taken a step, which I trust you will not think meddling or presumptuous; I have written to Lyell (with whom, I am very intimate) to ask him, whether he could write to Herschel, mentioning his own & my own high opinion of your son's acquirements & talents.— I have not the *smallest* idea, how far Lyell's acquaintance would allow him to write; but I feel sure he will excuse my suggesting it to him, if you will do the same. I have been sincerely sorry to hear of the illness in your family.

Believe me | Yours truly obliged | C. Darwin

P.S. | We are much indebted for your kind invitation to Kew, which at some future time, would afford the highest pleasure to Mrs. Darwin & myself.—

Kew (English letters (1845) 23: 147)

[1] Dated by the relationship to the letters to W. J. Hooker, [25 August 1845] and 25 August 1845.
[2] See *Correspondence* vol. 1, letter to J. S. Henslow, 9 July 1836, for CD's meeting with John Frederick William Herschel.

To John Murray [23 August 1845]

Down Bromley Kent
Saturday

My dear Sir

I am really mortified to find that I have transgressed my limits again— I had in truth thought this part[1] with the index wd have been 168 pages; I now find *without* the index it has run to 174. I am astonished at this & much vexed. I have had to rewrite two Chapters, & these have, I presume, been lengthened, though I have struck out many whole pages & discussions.[2] I now send this proof sheet to ask, whether you wd like to strike out, my rather dull little account p. 486 of the C. of Good Hope (as explained for the Printers in the enclosed bit of paper) & this, with the closing up, will save four pages;[3] & I think if the index, is in very small type, it will go in the six last pages of this last sheet.

The contents must, of course be on a separate half sheet.—

Please decide freely as you like, & send the sheets **early** on Monday to the Printers, requesting them to send me a revise, if possible, by the general Post on Monday night.—

I know how late it is, but I have worked like a slave & have not lost one half hour: I sent the index on last Thursday to the Printers & hope to have a proof tomorrow morning—

I will communicate with you again in a few days.

Part I has 176 pages
—— II has 160
—— III has, if index will go in & you strike out the C. of Good Hope 176
so that I have exceeded by one whole sheet, what I promised you, & very sorry I am.

Yours very faithfully | C. Darwin

I was gratifyed by seeing two *most* favourable notice, & an extract in the Gardeners Chronicle,[4] evidently by Lindley, with whom I am not acquainted, except by correspondence.—

John Murray Archive (Darwin 31–2)

[1] The third and final number of *Journal of researches* 2d ed.
[2] The chapters in question were those on the Galápagos and Keeling Islands. The first included new

information on the flora from Joseph Dalton Hooker and on the insects from George Robert Waterhouse. The Keeling Islands chapter was expanded by a detailed summary of CD's theory of coral reefs.

[3] The passages were omitted, but this number nevertheless came to 182 pages (Freeman 1977, p. 39).

[4] *Gardeners' Chronicle and Agricultural Gazette*, no. 32, 9 August 1845, p. 546; no. 33, 16 August 1845, p. 563.

To W. J. Hooker [25 August 1845][1]

Down. Bromley | Kent
Monday Morning

Dear Sir William

I beg of you, as the most particular kindness, that if the enclosed letter[2] can anyway be improved, that you will kindly return it to me.— I c^d. have made it, with entire truth, stronger, but I thought the effect w^d. not have been so good. Would you like it shorter? or not addressed to you?

I am sure you will believe, that if my letter was to be shown to a body of scientific men, I w^d. not have dreamed of giving my opinion so authoritavely & presumptuously.[3]

The only possible good my letter can do, is adding one more in number.—

I most cordially wish your son success, though at the heavy loss to me of his residence in Edinburgh. His correspondence has been to me the greatest pleasure & use.

What a disgrace it is to our Institutions, that a Professor sh^d. be appointed by a set of men, who never heard of Humboldt & Brown.—

I shall have to write to your son in a few days & to thank him for some lynx-eyed corrections of two sheets of Journal, which he looked over.

Believe me | Yours faithfully & obliged | C. Darwin

Norwich Castle Museum (Natural History Department)

[1] Date taken from the enclosure, see next letter.

[2] See next letter.

[3] The appointment to the professorship at Edinburgh was in the power of the Town Council.

To W. J. Hooker 25 August 1845

August 25th. 1845

Dear Sir William

I have heard with much interest that your son, D^r. Hooker, is a candidate for the Botanical Chair at Edinburgh. From my former attendance at that University, I am aware how important a post it is for the advancement of Science, and I am therefore the more anxious for your son's success, from my firm belief that no one will fulfill its duties with greater zeal or ability.— Since his return from the famous

Antarctic Expedition, I have had, as you are aware, much communication with him with respect to the Collections brought home by myself, and on other scientific subjects, and I cannot express too strongly my admiration at the accuracy of his varied knowledge & at his powers of generalization.— From Dr. Hooker's disposition, no one, in my opinion, is more fitted to communicate to beginners, a strong taste for those pursuits, to which he is himself so ardently devoted.— For the sake of the advancement of Botany in all its branches, your son has my most earnest wishes for his success.[1]

Believe me, dear Sir William | Yours very faithfully | Charles Darwin

Down House | Farnborough.
To | Sir William Hooker | Royal Botanic Gardens, Kew

Kew (J. D. Hooker's testimonials)

[1] CD's letter was printed in J. D. Hooker 1845c, a volume of 153 testimonials printed in series.

To Charles Lyell 25 August [1845]

Down. Bromley Kent
Aug. 25th.

Please read this before you go.[1]

My dear Lyell

This is literally the first day, on which I have had any time to spare; & I will amuse myself by beginning a letter to you, which you can read when you have leisure; for now you must be very busy.— Firstly for the last subject, mentioned between us, viz the radiation of snow, on which you say you are interested: my Brother says I am wrong about colour, but I cannot yet give up a clear impression, I have opposed to the colour-doctrine.— I find from Leslie,[2] (in Ure)[3] the radiating (& highest) power of Lamp-black being called 100, and gold, silver, copper being 12; writing paper is 98, plumbago 75 and ice is 85. From Wells,[4] it appears, that when swan-down (white enough, & the best known radiator) exposed to open sky falls 16°; grass falls 15°; & snow falls between 12° & 13°: gravel & flag-stone (ie naked rock) are inferior to grass, but how much is not said. Dr. Wilson,[5] however, found on one occasion snow 16° colder than the air two feet above it, & this seems greater than the average coldness of grass on the clearest nights. Therefore I conclude, that a country covered with snow radiates its heat, but little less than the most favourable land.

I was delighted with your letter, in which you touch on slavery; I wish the same feelings had been apparent in your published discussion.—[6] But I will not write on this subject; I shd. perhaps annoy you & most certainly myself.— I have exhaled myself with a paragraph or two in my Journal on the sin of Brazilian slavery:[7] you perhaps will think that it is in answer to you; but such is not the case, I have remarked on nothing, which I did not hear on the coast of S. America. My few sentences, however, are merely an explosion of feeling. How could you relate so placidly that atrocious sentiment about separating children from their parents; & in the next page, speak of being distressed at the Whites not having prospered; I assure you the contrast made me exclaim out.—[8] But I have broken my intention, & so no more on this odious deadly subject.—

There is a favourable, but not strong enough review on you, in Gardeners Chron: I am sorry to see, that Lindley abides by the Carbonic-acid-gas theory.[9] By the way, I was much pleased by Lindley picking out my Extinction paragraphs & giving them uncurtailed:[10] to my mind, putting the *comparative* rarity of *existing species* in the same category with extinction has removed a great weight; though of course it does not explain anything, it shows that, until we can explain comparative rarity, we ought not to feel any surprise at not explaining extinction.—

Have you seen Kosmos,[11] I think you w^{d} probably find the subject of multiple & single Creations there discussed: at least. H. discussed subject with Hooker & Humbolt is a multiple man.— You speak about Craters of Elevation; I heartily hope your next excursion will be to Sicily: I do not feel at ease on the subject, I cannot swallow, such areas as those of St. Jago in the C. de Verds & Mauritius having been produced by the summit of a volcano having been swallowed up: my suggestion is, perhaps, hardly worth attending to, yet I cannot doubt that the circumferential mountains of St. Jago & Mauritius, have undergone angular elevation of some kind.[12]

—I am much pleased to hear of the call for a new Edit. of the Principles:[13] what glorious good that work has done.— I fear this time you will not be amongst the old rocks;[14] how I sh^{d} rejoice to live to see you publish & discover another stage below the Silurian; it w^{d} be the grandest step possible, I think.

—I am very glad to hear, what progress Bunbury is making in fossil Botany: there is a fine Hiatus for him to fill up in this country: I will certainly call on him this winter. How I shall miss my break-fast calls on Mrs Lyell & yourself.— From what little I saw of him, I can quite believe anything which you say of his talents.[15]

I will tell Murray to send the 3^{d} Part of my Journal to you on Monday, though I suppose you

AL incomplete
American Philosophical Society

[1] CD has drawn a line connecting this message to the sentence, in the first paragraph, beginning 'Firstly for the last subject,'.

[2] John Leslie.

[3] Ure 1823, pp. 262–3, in which Andrew Ure described Leslie's experiments on radiation and gave the figures as presented by CD. CD's copy is in the Darwin Library–CUL.

[4] Wells 1815.
[5] P. Wilson 1788.
[6] C. Lyell 1845a, 1: 181–95.
[7] In *Journal of researches* 2d ed., pp. 499–500, CD expanded the passages on Brazilian slavery from the first edition (pp. 27–8).
[8] C. Lyell 1845a, 1: 184–5. In his copy CD has marked the first passage with '!' and underlined the second in pencil.
[9] *Gardeners' Chronicle and Agricultural Gazette*, no. 34, 23 August 1845, p. 578. John Lindley, the editor, upheld Adolphe Théodore Brongniart's carbonic-acid-gas theory. See letter to Charles Lyell, [30 July – 2 August 1845], n. 9.
[10] *Gardeners' Chronicle and Agricultural Gazette*, no. 33, 16 August 1845, p. 563. The discussion of the causes of extinction of species is in *Journal of researches* 2d ed., pp. 173–6. See letter to Charles Lyell, [5 July 1845], n. 5.
[11] Humboldt 1845–62.
[12] *Volcanic islands*, pp. 93–6; *Journal of researches* 2d ed., pp. 484–5.
[13] C. Lyell 1847.
[14] That is, on his forthcoming visit to the United States.
[15] Charles James Fox Bunbury. See letter from Charles Lyell, [after 2 August 1845], n. 5, for CD's meetings with him.

To John Murray 27 August [1845]

Down Bromley Kent
Aug 27th.—

My dear Sir

I returned *everything* to the Printers on Tuesday night.— I am much obliged for your note: the method by which you propose to pay me the 150£ will suit me, and I will acknowledge the note whenever I receive it.—

Will you please *particularly* to see, that one of my 12 copies is sent *not later than Monday* to Mr Lyell at 16 Hart St.; as he wishes to take it to America with him on Tuesday.—

Please to send the other 11 copies to Mr. Bain's 1. Haymarket my Bookseller & *Binder* (marking outside copies for Mr. Darwin) not later than Tuesday or Wednesday night.

With respect to the Woodcuts: that of the Lizard to be put back to Mr Lyell's Elements: four of them are your property, viz. that at p. 15/ p. 137/ p. 236/ p. 246/.[1] All the remaining ones, please see **carefully** packed & returned to me, (if done with by Tuesday with the Books to Mr. Bain; if not then done with, directed to me at the Geological Soc: Somerset House; marked "not to be forwarded".): Should you ever want a reprint; I can have no doubt, that I can borrow them from Mrs Smith & Elder again.—[2]

I am much obliged for the pleasant manner, in which you have transacted the business with me.—

I beg as you as an especial favour, that should my volume sell well, that you will take the trouble to inform me; both to gratify my vanity, as Author, & what I care equally for, that I may know, that you have had no cause to repent undertaking this little work.

Believe ⟨me my d⟩ear Sir | Yours very faithfully | C. Darwin

J. Murray Esqre

P.S. Should you ever wish to publish old Books of Travel; I strongly recommend you to think of Hearne's Travels,[3] (strongly praised by Wordsworth);[4] they are to my mind admirable & little known.— Drury's Madagascar[5] is also little known for its merits—

John Murray Archive (Box 37)

[1] Three of the original woodcuts for the first edition, which John Murray had presumably obtained from Henry Colburn, with one further woodcut of the scissor-beak bird (*Rhynchops nigra*) made at Murray's expense (see letter to John Murray, [31 May 1845]).

[2] The remaining woodcuts were borrowed from Smith, Elder & Co., the publishers of *Coral reefs* and *Volcanic islands*. See letters to John Murray, [31 May 1845] and [27 July 1845].

[3] Hearne 1795.

[4] In the headnote to 'The complaint of a forsaken Indian woman', which appeared originally in William Wordsworth's *Lyrical ballads* (1798).

[5] Drury 1729.

To J. D. Hooker [29 August 1845][1]

Down Bromley Kent
Friday

My dear Hooker

I shall be extremely anxious to hear how your prospects are getting on, and busy as you will be, I trust you will, if successful let me know pretty soon.— Will you please to tell Sir William, that Lyell in a note to me[2] says, that "Professor Forbes has just written to say, that he forwarded my letter, a very strong one, to the Ld Provost of Edinburgh & he, Forbes, says it will not fail to be useful" Lyell adds, that according to Mr Horner's account the Provost is a well informed man.—[3]

It is delightful yet grievous to me to think of your success.— You must have had a multitude of letters to write, so don't write to me, till you have a little leisure, & then no one will be more anxious to hear of the progress or result of things.

Do not forget sometime, to remind me of the best way to return your books & a copy of my own Journal, which thank all the Stars in Heaven, I have at last finished.—

Ever my dear Hooker | yours | C. Darwin

All our plans have been upset, by my wife, I am sorry to say, having made or rather still making a very ⟨s⟩low recovery:[4] otherwise we sh^{d} have been in Staffordshire & I perhaps in Scotland ere this: as it is I hardly know when we can move.—

DAR 114.1: 39

[1] The Friday after the letter to W. J. Hooker, [25 August 1845].
[2] Possibly an excised portion of the letter from Charles Lyell, [after 2 August 1845].
[3] Adam Black was lord provost of Edinburgh 1843–7. Lyell refers to James David Forbes, professor of natural philosophy at Edinburgh University.
[4] From the birth of George Howard Darwin.

From J. D. Hooker 1 September [1845]

[West Park, Kew]
Monday Morning | 1st. Sept.

My dear Darwin

I think that I informed you of my venerable Grandfather's death but had so many to write to on the subject, that I may have trespassed on your good nature in favor of others who would not take the neglect so well.

About a fortnight ago I was called off hurriedly to Edinbro', to canvass the Town council for the chair of Botany: & to undergo the detestable ordeal of writing to all my friends for testimonials for them to judge me by: thanks to their kindness & blindness they have supplied me with about 50 already & I expect 40 more—[1] I had hoped that one from Brown one from Humboldt with three from Professors proving that my lectures gave complete satisfaction would have been enough, but it was not so. Balfour[2] had 60 or 80 & I must get the same. The election comes on in 1st week in October. I expect that the Crown and Town will come to a private understanding before—that, & let the nominated candidate be elected after telling the councillors who to support, & who not *at their peril.*[3] There will be an immense deal to do at Edinbro with the Garden & to reform (between ourselves) a certain sink, called the Bot. Soc. of Edinburgh, which I must of course join & either do nothing to or remodel

I returned this day week & have been busy ever since with private affairs & snatching all moments for the unlucky Flora, Antarct.

I was reading Cosmos in the railway carriages, have you seen it? the translation is never to be sufficiently execrated, I cannot understand many *pages* of it at all.[4] I can send you my copy if you have it not, for such a translation is never worth buying I should think, but I may be all wrong in my judgement.

I have been thinking more & more upon Forbes Botanico Geological remarks.[5] I can account in no other way for the similarity of the Irish & Portuguese floras, or of the Cape Horn & Kerguelens Land; Migration as an agent is all very well, but it has been ridden to death, in atempts to account for such similarities in vegetation. Now I see what Strzelecki was at, when he told me & I told you of the identtity of the Geology & Botany of Illawarra & V. D L. or something of that sort which I told you of at the time.[6] Cunningham long ago remarked that some Tasmanian plants suddenly appeared on the Blue Mts. but not (I think) at an elevation analagous to the differences of Latitude[7] If we are to account for these things geologically what an antiquity it gives to vegetation & how eminently true it must be that the Geological changes must have been slow & gradual not to have obliterated all traces of vegetation during the removal of the intervening land.

Please send any thing to Hiscock's Kew boat, Hungerford stairs: he has an office there & receives all parcels for us. I want a number of your copy of Fl. Ant. or I would have sent it ere this.

I am truly sorry that Mrs Darwin does not recover, could you not come over here for a day,? my Sisters would be delighted to make her acquaintance as w^d my mother, but she is going with my Father to Norfolk tomorrow for 10 days. You live in such an impracticable part of the country that we ie. the Ladies cannot call or even send to enquire, but you are right to be down there. I fear I shall have to go to Edinburgh again.

Many thanks for your excellent testimonial & all you have done Ever most truly yours | Jos D Hooker.

Guaya villa is in Span S. Am a species of Eugenia, so I daresay the Gallapago berry called Guaya vita is my Psidium Galapageum (allied to Guava, same genus) native of Jas. Island only, found also by Scouler there.[8]

There is an excellent paper in the Ann. Sc. Nat for April 1845 & chart,[9] on the distrib. of Conif. in Europe, it looks rather too *precise* to be accurate, (a very ungenerous remark) the replacement of species along the Italian coast is curious & what I have been working out with regard to some genera of S. American plants along the Chilian coast.

DAR 100: 14–15

CD ANNOTATIONS
1.1 I think . . . judgement. 4.4] *crossed ink*
2.10 There will be . . . Flora, Antarct. 3.2] 'Kosmos' *added pencil*
4.1 I was . . . carriages] *scored pencil*
5.5 Now I . . . time. 5.7] *scored pencil*
5.7 Cunningham . . . Latitude 5.9] *scored brown crayon*
5.10 an antiquity . . . Hooker. 8.2.] *crossed ink*
6.1 to Hiscock's . . . stairs:] *square brackets added pencil*
7.1 I am . . . day,? 7.2] *scored pencil*
9.1 Guaya villa . . . Scouler there. 9.3] *scored pencil*
10.1 There . . . coast. 10.5] *scored pencil*

[1] J. D. Hooker 1845c, in which 153 testimonial letters were eventually printed.
[2] John Hutton Balfour stood for election to the professorship as the Town Council nominee.
[3] The Town Council retained the right to appoint to the valuable University professorship (the 'chair') while the Crown appointed to the less valuable Regius professorship and the curatorship of the botanic garden. On previous occasions both parties had agreed on a single candidate for the positions; however, the Town Council had not been consulted about the Crown's nomination of Hooker and attempted to block his election. See Huxley, ed. 1918, 1: 204–5.
[4] Humboldt 1845–8. The translation was by Augustin Prichard.
[5] E. Forbes 1845.
[6] See letter from J. D. Hooker, 30 December 1844, and letter to J. D. Hooker, [7 January 1845].
[7] Cunningham 1827.
[8] See J. D. Hooker 1845d, pp. 224–5, for his description of *Psidium galapageum*.
[9] Schouw 1845.

To John Murray 2 September [1845]

Down near Bromley | Kent.
Sept 2d.—

My dear Sir

I beg to acknowledge your kind note & return you the receipt signed.

I hope on some future occasion, before long, we may have the pleasure of meeting— I am partilarly obliged for your promise information on the success on my Journal.

Yours sincerely | C. Darwin

[Enclosure]

London Sept 1st. 1845

Received of John Murray Esq. a Promissory note for One hundred and fifty pounds—due March 4, 1846—for the entire Copyright of my Journal on Natural History & Geology during the Voyage of HM.S. Beagle round the World and I hereby promise a further assignment if required—

£150"—"— Charles Darwin

John Murray Archive (Box 37)

To Susan Darwin 3[–4] September 1845

Down
Wednesday 3rd Sept., 1845.

My dear Susan

Please to thank Jos[1] for the Railway Dividend; and further ask him how it comes, that as additional shares were bought in our three Railways in July of this year, the last Dividends in all three have been the same as hitherto. It is long since I have written to you, and now I am going to write such a letter, as I verily believe no other family in Britain would care to receive, viz., all about our household and money affairs; but you have often said that you like such particulars. First, however, I am sorry to say, that poor Emma is more uncomfortable to-day than before: but her teeth are better than two days: she really has had a most suffering time and it has been so provoking that no one could come here to comfort her: Elizabeth[2] would have been such a pleasure to her. When we shall move, and what we shall do, must remain in the clouds.[3] Erasmus is here yet; he must have found it woefully dull for I also have not been up to my average: but as he was to have gone on Saturday and then on Monday and willingly stayed, we have the real pleasure to think, wonderful as it is, that Down is not *now* duller to him than Park St. I have taken my Bismuth regularly, I think it has not done me quite so much good, as before;[4] but I am recovering from too much exertion with my Journal: I am extremely pleased my Father likes the new edition.

I have just balanced my ½ years accounts and feel exactly as if some one had given me one or two hundred per annum: this last half year, our expenses with some extras has only been 456£, that is excluding the new Garden wall; so that allowing Christmas half year to be about a 100£ more, we are living on about 1000£ per annum: moreover this last year, subtracting extraordinary receipts, has been 1400£ so that we are as rich as Jews. Caroline[5] always foresaw that our expenditure would fall. We are now undertaking some great earthworks; making a new walk in the K. Garden; and removing the mound under the Yews, on which the evergreens, we found did badly, and which, as Erasmus has always insisted was a great blemish in hiding part of the Field and the old Scotch-firs; and now that we have Sale's corner, we do not want it for shelter. We are making a mound which will be excavated by all the family, viz., in front of the door out of the house, between two of the Lime Trees: we find the winds from the N. intolerable, and we retain the view from the grass mound and in walking down to the orchard. It will make the place much snugger, though a great blemish till the evergreens grow on it. Erasmus has been of the utmost service, in scheming and in actually working; making creases in the turf, striking circles, driving stakes, and such jobs. He has tired me out several times.[6]

Thursday morning. I had not time to finish my foolish letter yesterday, so I will today: Emma intends lying in bed till Luncheon, so that I shall not be able to say how she really is. Our grandest scheme, is the making our schoolroom and one (or as I think it will turn out) two *small* bedrooms. Mr Cresy is making a plan and he assures me all shall be done for 300£. The servants complained to me, what a nuisance it was to them to have the passage for everything only through the Kitchen: again Parslow's pantry is too small to be tidy, and some small room is terribly wanted to put strangers into (as you have often insisted on) and all these things will be effected by our plan; and besides there is another advantage equally great. If it is done for 350£, which with Murray 150£ I can pay out of my income I shall think it worth while. It seemed so selfish making the house so luxurious for ourselves and not comfortable for our servants, that I was determined if possible to effect their wishes; and had we not built a schoolroom and bedroom; we should have had only two spare bed-rooms; so that for instance, we could never have had anyone to meet the Hensleighs[7] and their children. So I hope the Shrewsbury conclave will not condemn me for extreme extravagance: though now that we are reading aloud Walter Scott's life,[8] I sometimes think that we are following his road to ruin at a snail-like pace. We have had some more turmoil in the village (though I have not yet been involved): old Price has been agitating building a wall across the pool, but thank Heavens he has at last aroused everybodies anger, except Sir Johns: Capt. Crosse told him the old women would hoot him through the village:[9] and Mr. Smith cut short his usual rigmarole of his "having no selfish motives" by asking him, "if it is not for yourself, who the devil is it for?" Mr. Ainslie, the new Methodist resident at old Cockle's[10] house is also litigious and has been altering the road illegally; and defies us all, casting in our teeth that we allowed Mr. Price

Copy incomplete
DAR 153

[1] Josiah Wedgwood III, Emma Darwin's brother.
[2] Sarah Elizabeth (Elizabeth) Wedgwood, Emma's elder sister.
[3] CD and Emma had been planning a visit to Shrewsbury followed by a tour to York and Lincolnshire, intending to start at the end of August (see letter to J. D. Hooker, [11–12 July 1845]). They eventually left on 15 September.
[4] See Colp 1977, p. 37, for an account of the therapeutic uses of bismuth.
[5] Caroline Wedgwood, CD's sister.
[6] Atkins 1974, pp. 24–6, gives further details of the work undertaken at this time at Down. CD's Account Book (Down House MS) records payments made to Isaac Laslett for a shed and walls, to William Sales for 'Permission to build wall' and sums for extra labour between July and September.
[7] Hensleigh and Fanny Mackintosh Wedgwood.
[8] Lockhart 1837–8.
[9] Edward Price, Sir John William Lubbock, and Captain Thomas Crosse. See *Correspondence* vol. 2, letter to [Susan? Darwin], [1843 – 8 March 1846].
[10] Edgar Cockell, surgeon and apothecary of Down.

To J. D. Hooker [3 September 1845][1]

Down Bromley | Kent
Wednesday

My dear Hooker

I had not heard of your grandfather's death: coming at the time it did, it must have been an additional blow, when you were involved in all the disagreeable annoyment about the Professor's Chair.— I shall be most anxious to hear of the result, but I shall no doubt see it in the Papers & you will have just then a multitude of letters to write.

I sh^d. like to hear, sometime whether you had other competitors besides M^r. Balfour. I do not think you have sufficiently plumed yourself on the high honour (& assuredly it is so) of being elected at your age to so great a Chair. I think there is the smallest possible chance of the good things of this life, making you fat & idle, as has too often happened in your new University.

I hope & suppose you will spend your summers in England, so that we may not be quite separated. Many thanks for your invitation to Kew; but as soon as my wife can move, we shall go into Staffordshire, & I do not now like to leave her. I heartily wish Kew & Down were not so remote.—

I will return the Astrolabe[2] (which I have only very slightly skimmed) the Boston Journal,[3]—Red-Sea Conferæ[4]—and L'Espece[5] (NB. please allow my pencil marks to stand) I will retain, for I have as yet made no progress in Webb & Berthelot.[6] I will send my Journal in some future packet. Are you really sure you can spare Cosmos:[7] I am very anxious to read it, & till knowing whether worth while not anxious to buy it. I beg you not to think of sending it, without you & any others in your family are sure they have quite done with it. I shall indeed be proud of the Antarctic Flora, as you are so kindly determined to give me a copy.—

By the way, I have never thanked you for your sharp-sighted corrections of the two sheets: I fear I sh^d. have overlooked several of them.— I abide by Pernety, as in an old French Edition it is so spelt: it is also spelt Pernety.—[8]

I am very glad that Forbes has one such thorough apperciator as you: I fancied I was bold enough in upheaving & letting down our mother earth, but Forbes beats me hollow, without any proof to speculate on Ireland & Portugal having been once connected: it staggers me. All this boldness is undoubtedly the direct consequence of Lyell's Principles.—

I shall be particularly interested to see what you make out of species taking each others places along the shores of S. America: I was particularly interested with similar facts in the Birds &c: Do the representative species actually join on a neutral territory? Are both species, or one, rare in such neutral territories? It is a curious & to my mind very interesting subject.—[9]

With my most hearty wishes for your success, ever my dear Hooker Your's | C. Darwin

DAR 114.1: 40–40b

[1] Dated by the relationship to letter from J. D. Hooker, 1 September [1845].
[2] Dumont d'Urville [1841–54].
[3] *Boston Journal of Natural History*, containing Hayes 1844.
[4] Montagne 1844.
[5] Gérard 1844.
[6] Webb and Berthelot 1836–50, vol. 3, pt 1.
[7] Humboldt 1845–8.
[8] Pernety 1769, discussed in *Journal of researches* 2d ed., pp. 196, 399. Pernety is also spelt Pernetti.
[9] The question of a 'neutral' ground where the geographical ranges of closely allied species overlap had long interested CD and was raised in what may be CD's earliest transformist statement, recorded in or about March 1837 in his *Red notebook*: 127: 'Speculate on Neutral ground of 2 ostriches: bigger one encroaches on smaller.— change not progressif ['e' *del*]: produced at one blow, if one species altered:'.

From J. D. Hooker [4–9 September 1845][1]

translation, though it is rather cool of me to ground my censure upon my own inability to understand parts of it.[2]

I am exceedingly glad that l'Espece has interested you,[3] & will try & get you a copy from Montagne,[4] through whom my father received this. I am not inclined to take much for granted from any one treats the subject in his way & who does not know what it is to be a specific Naturalist himself. Those who have had most species pass under their hands as Bentham, Brown, Linnæus, Decaisne & Miquel, all I believe argue for the validity of *species* in nature: they all direct attention to the cases where *salient* characters are unimportant, though taken advantage of by the narrow-minded studiers of overwrought local floras, & these facts, thus noticed as cautions to others, are taken up by such men as Gerard, who have no idea what thousands of good species their are in the world. Nature may have both made & muddled species, we never shall know what are species in some genera &

what not. Generally cultivation will prove the validity of a species, Gerard says that, "varieties of apples &c are more distinct than many species", but how soon all revert to crabs: again, the wheat is always adduced as a permanent variety of some unknown plant which ought on that account to rank as a species, but I do not think so, because it will never run wild: it is to me very marvellous that the wheat seed is destroyed by being left in the ground of our country & that we see so little next year on a field that has supported millions of ears, during the present.

Gerard evidently is no Botanist, he talks of having found both *Prunus spinosa* & *Rubus fructicosus* without spines: now spines are only abortive branches, & their absence or presence is never, of itself, a Botanical character; as a spine is not an organ per se; & again, no *Rubus* ever had or ever will have *spines*: the *prickles* of *Rubus* are mere *appendages of the cuticle* & have no organic connection (like spines) with the pith & wood of the plant: species vary in *pricklyness* just as they do in *hairiness*, according to the amount of spines or hairs produced; but they vary in *spinyness* according to the number of branches that are checked in growth which is much affected by want of moisture.

You are right then to Query that bit about plants *developing* spines in bad soil; for they only lose the power of nourishing the new leaf buds sufficiently & do not develope a new organ (hence hairyness is of *more* importance than spinyness in specific dist.). The *Persicaria* becoming hairy when removed from moist places is natural: hairs are believed to be provided as hygrometric appendages, to modify respiration & transpiration, water plants don't want them. It is facts such as the Irish yew presents that afford fair grounds for argument on such a topic.[5]

Quoting instances by tens or hundreds of variation in individual species is nothing new,[6] few have an idea of the labor required to establish or destroy a species of a mundane genus. You have a *Senebiera* from Tres Montes, its capsules are much larger than the common *S. pinnatifida*, but that is so universally diffused a plant & so variable in the size of its *leaves* that at first sight no one would be inclined to grant specific dignity to the Tres Montes plant from the capsules. It struck me to put this subject to a Geographical test; the result is, that the *S. pinnatifida* is probably a native of the *Plate alone*, whence it has spread by ships over all East & W. America, all West Europe near the coast, in fact both shores of the Atlantic from Britain to the Cape & from Patagonia to Canada, wherever ships touch & cultiv. ensues, & on E. from Valp to California, wherever ships go, but through many hundreds of specimens there is no variation *whatever* in the size of the pods. & I therefore conclude that the Tres Montes plant is the W. coast representative of the E. coast plant.

Now though D. C. had hinted that S. pinn. was an American plant, he did not define its limits & retained two or three identical plants as different species which came from out of the ways localities:[7] to define its limits I had not only to consult all floras where it was described, but all where it was not, for such a mundane plant creeps into every flora. My troubles did not end here, for I had no Valparaiso *Senebiera* & Bertero has an undescribed one from that port, which is alluded to as *S. diffusa*, Bert. mss. I naturally concluded your's was this, but thought I would go to Brit. Mus to confirm it for fear of accident, but Bertero's was genuine *pinnatifida*, he gave it a new name *taking for granted* it was a new species. So as *S. pinnat.* does not at Valp. vary into big pods, I am more persuaded that yours is a rep. species of W. coast of Am.—[8] That Neutral territory of rep. species you ask about is just what I want to work out but it needs great materials

Ever yours most truly | J D Hooker

DAR 104: 209, 208

CD ANNOTATIONS

1.1 translation . . . to take 2.3] *crossed pencil*
2.4 Those . . . nature 2.6] *scored brown crayon*
2.13 "varieties . . . species"] '?' *added pencil*
3.3 a Botanical . . . per se; 3.4] *underl brown crayon*
4.6 It is . . . topic. 4.7] *scored brown crayon*; 'Irish Yew' *added brown crayon*

[1] Dated on the assumption that this letter comes between CD's letters to J. D. Hooker, [3 September 1845] and [10 September 1845].
[2] Probably a reference to Humboldt 1845–8. See letter from J. D. Hooker, 1 September [1845].
[3] Gérard 1844.
[4] Jean François Camille Montagne.
[5] The Irish yew is an apparently natural and spontaneous variant that retains its distinctive form in subsequent generations regardless of external conditions. For an account of its origin see letter from Philip de Malpas Grey-Egerton, 5 May [1844].
[6] CD had a higher opinion of this exercise than Hooker. On p. 10 of his copy of Gérard 1844 (Darwin Pamphlet Collection–CUL) CD noted 'shows in great detail that vars. differ in same points as species.'
[7] A. P. de Candolle 1818–21, 2: 524.
[8] J. D. Hooker 1844–7, pp. 241–2.

To J. D. Hooker [10 September 1845]

Down Bromley Kent
Wednesday

My dear Hooker

I write to say that we are going on Monday for a month to my Father's at Shrewsbury:[1] when, therefore, you can **quite** spare Cosmos[2] w^{d} you send it directed to me. "Shrewsbury": please not send anything else with it.— I will not trouble you for the Asiatic Journal, as I can see it at the Club.— I send my Journal today for you.— N.B. If by chance you sh^{d} have sent Cosmos either to the Athenæum or Geolog. Soc. I can get it forwarded, if you will inform me.

Many thanks for your letter received yesterday, which, as always, sets me thinking: I laughed at your attack at my stinginesss in changes of level towards Forbes, being so liberal towards myself; but I must maintain, that I have never let down or upheaved our mother earth's surface, for the sake of explaining any one phenomenon, & I trust I have very seldom done so without some distinct evidence. So I must still think it a bold step, (perhaps a very true one) to sink into depths of *ocean, within the period of* **existing species**, so large a tract of surface. But there is no amount or extent of change of level, which I am not fully prepared to admit, but I must say I sh^d. like better evidence, than the identity of a few plants, which *possibly* (I do not say probably) might have been otherwise transported. Particular thanks for your attempt to get me a copy of L'Espece[3] & almost equal thanks for your criticisms on him: I rather misdoubted him & felt not much inclined to take as gospel his facts. I find this one of my greatest difficulties with foreign authors, viz. judging of their credibility.

How painfully (to me) true is your remark that no one has hardly a right to examine the question of species who has not minutely described many.[4] I was, however, pleased to hear from Owen (who is vehemently opposed to any mutability in species) that he thought it was a very fair subject & that there was a mass of facts to be brought to bear on the question, not hitherto collected. My only comfort is, (as I mean to attempt the subject) that I have dabbled in several branches of Nat. Hist: & seen good specific men work out my species & know something of geology; (an indispensable union)[5] & though I shall get more kicks than half-pennies, I will, life serving, attempt my work.— Lamarck is the only exception, that I can think of, of an accurate describer of species at least in the Invertebrate kingdom, who has disbelieved in permanent species, but he in his absurd though clever work has done the subject harm, as has M^r. Vestiges, and, as (some future loose naturalist attempting the same speculations will perhaps say) has M^r. D.—

Is not Aug. St. Hilaire a good Botanist? I presume he follows his father & Brother's views?[6]

It is a shame to plague you with this long note; but I did not sit down, with malice prepence.

Farewell; I shall long to hear the result in October.— | Ever your's | C. Darwin

DAR 114.1: 41–41b

[1] CD left for Shrewsbury on 15 September. He also visited William Herbert and Charles Waterton. See 'Journal' (Appendix II).

[2] Humboldt 1845–8.

[3] Gérard 1844.

[4] Hooker's criticism of Gérard 1844 (see previous letter) clearly struck CD as an implied criticism of his own position. Hooker denied this intention (see next letter), but CD's response here and in the letter to J. D. Hooker, [18 September 1845], shows his sensitivity on the point. CD subsequently undertook a systematic study of cirripedes and 'minutely described' many species (see *Correspondence* vol. 4), thus establishing his 'right to examine the question of species'.

[5] See letter to Emma Darwin, 5 July 1844, in which CD makes such a union of talents a criterion for choosing the literary executor to work up and publish his essay of 1844.

[6] Auguste Saint-Hilaire was not related to Etienne Geoffroy Saint-Hilaire or his son Isidore, who both favoured transformist views.

From J. D. Hooker 14 September 1845

West Park Kew
Septr. 14. | 1845.

My dear Darwin

I am indeed pleased with the last Edition of the Journal, which you have so kindly sent me; both outside & inside are engaging. How doubly interesting you have made it to the common reader! I was so taken with the account of the poor Fuegians that I sat up I cannot say how long, till you bade farewell to Woollya & I went to bed quite melancholy at the fate of poor Fuegia Basket, doubtless too truly told in the cruel footnote at p. 229.[1]

You have puzzled me beyond measure with one of your plants, it is a beautiful pea-flower from Cape Tres Montes, the only one of the kind from there; (another handsome pea-flower *Lathyrus pubescens*, is from Chonos Archip.) this is nothing more nor less than *Lathyrus maritimus*, a rare English plant found on all the N.W. coast of Europe, from the Channell to Archangel, Iceland, Shetlands Greenland & always littoral, in Asia it nowhere appears on *this* side the Okhotsk M[ts], i.e. at Kamschatka, it being replaced throughout Siberia & Russia (barring Archangel) by a similar but very distinct species. In America it commences at Oregon, runs up to Kotzebues sound inhabits the shores of all the great lakes & banks of the great rivers across the continent & on the E. coast is found from Labrador to N. York: behold its range! now what on earth brought it to Cape Tres-Montes of all places the only one South of 40° N.?. It is not a pot-herb, & the idea of its introduction by man is almost untenable. Our Chilian collections are so extensive, that I am sure it is nowhere in S. Am. but Tres Montes. Your specimens are good: read me this riddle. Its Geog. distrib. in the North cost me many hours hunting & comparing specimens as it has brethren apt to be mistaken for it in the N., but nothing like it in South.[2] The "Anne Pink" cannot have brought it?, as Cook did the coral to Endeavour River (have you heard of that?)[3]

Are you sure that it is established "that plants propagated by buds, all partake of a common duration of life.—? v. p 203 of Journal.[4]

An now for species. To begin, I do think it a most fair & most profitable subject for discussion, I have no formed opinion of my own on the subject, I argue for immutability, till I see cause to take a fixed post.[5] A knowledge of Botany *alone* will never clear up the question & alas I can bring nothing else to bear upon it, my Geology is nil: & thus you see I am ever ready to make it subservient to Botany instead of Botany to it, as must be the true relation. Do not think I meant to insinuate that you could not be a judge from not having worked out species, for your having collected with judgement is working out species: what I meant I still maintain, that to be able to handle the subject at all, one must have handled hundreds of species with a view to distinguishing them & that over a great part,—or brought from a great many parts,—of the globe. These elements your toils have fulfilled & well, M[r] Gerard is neither a specific naturalist, nor a collector, nor a traveller, what the —— is he then? nothing but a distorter of facts; or what is as bad; a compiler without judgement. His work always has its value though, as a collection of *instances* which you will make better use of I doubt not.

I am sure your Geology is most moderate, & I would safely subscribe to the Patagonian plains being once under water & to the "Strait of Darwin" up by the river St Cruz. Your evidence to my unskilled eyes at any rate is positive, Forbes' is theory, but untill I[6]

I am happy to think that this letter won't cost you much thinking.

Incomplete
DAR 100: 55–6

CD ANNOTATIONS
1.1 I am . . . the only 2.2] *crossed pencil and ink*
2.15 read me . . . think I 4.6] *crossed pencil*
4.6 meant . . . untill I 5.4] *crossed ink*
6.1 I am . . . thinking.] *crossed pencil and ink*

[1] *Journal of researches* 2d ed., p. 228 n., in which CD records that Fuegia Basket was last seen on board a sealing ship. CD continues: 'She lived (I fear the term probably bears a double interpretation) some days on board.'
[2] J. D. Hooker 1844–7, pp. 260–1.
[3] The *Anna Pink* was one of a squadron commanded by George Anson, sent by Britain in 1740 to attack Spain's South American possessions; she anchored off Cape Tres Montes for some two months (Anson 1748, pp. 138–55). James Cook took the *Endeavour* to an estuary in New South Wales for repairs after running against reefs in the Pacific Ocean. He found a giant coral boulder embedded in the hull, which had fortuitously prevented water from entering the ship (Hawkesworth 1773, 3: 559).
[4] *Journal of researches* 2d ed., pp. 202–3. In the first edition (1839) CD had linked this idea to the possibility of a fixed life span for species, thus explaining at least some extinctions (*Journal of researches*, pp. 212, 262). The suggestion did not appear in the second edition.
[5] At this time Hooker was preparing the introductory pages to the second part of J. D. Hooker 1844–7 (pp. 209–23), which appeared some time before 7 October 1845 (Wiltshear 1913, p. 357). The introduction assumes that species are immutable.
[6] This letter breaks off here; the final paragraph was written above the salutation.

To J. D. Hooker [18 September 1845]

Shrewsbury
Thursday

My dear Hooker.

I write a line to say that Cosmos arrived quite safely (NB one sheet came loose in P^t I) & to thank you for your nice note. I have just begun the introduction & groan over the style, which in such parts is full half the battle.— How true many of the remarks are (ie as far as I can understand the wretched English) on the scenery;[1] it is an exact expression of ones own thoughts.—

I wish I ever had any Books to lend you in return for the many you have lent me.—

What a curious case that of the Lathyrus; I do not think you c^d pick out on the whole American Coast, a spot so unlikely for any plant to have been introduced by man: it is inconceivably wild & most seldom visited. The maps before the Beagle's survey were all vilely inaccurate.—

All which you so kindly say about my species work does not alter one iota my long self-acknowledged presumption in accumulating facts & speculating on the subject of variation, without having worked out my due share of species. But now for nine years it has been anyhow the greatest amusement to me.—

Farewell my dear Hooker, I grieve more than you can well believe, over our prospect of so seldom meeting.

I have never perceived but one fault in you, & that you have grievously, viz modesty;—you form an exception to Sydney Smith's aphorism, that merit & modesty have no other connexion, except in their first letter[2]

Farewell | C. Darwin

DAR 114.1: 42

[1] Humboldt 1845–8, 1: 6–13.
[2] See letter from Charles Lyell, [after 2 August 1845], in which Lyell quotes the same aphorism.

From John Higgins 2 October 1845

Alford
2nd October 1845

My Dear Sir,

I enclose an Estimate of the different additional works, which I recommend to be done at Beesby, in order to put the Farm into permanent good condition; so as to command at all times a Tenant of respectability, and to require no further Outlay for many years to come, and I propose the additional Money expended, shall yield Interest at 3¼ Pr. cent the same as the purchase Money thus—

	£
Purchase Money — —	12'400 — —
Buildings &c———	920. 10. 0
Law Charges (Estimated)	150 — —
Total Outlay	13'470. 10. 0

This Outlay taken at 3¼ Pr. Cent will produce a Rental of £437.10.0 and I have no doubt but any good Tenant will be able to pay it for a term of years; and improve the Estate very considerably.[1]

You seemed much interested in the use of liquid Manure (a practise yet in its infancy); and fancying my son who is learning the practical parts of Agriculture, with a Gentn of some eminence in this County; might be able to throw some light on the subject; I asked him to do so; and I enclose a copy of his Letter.—[2] I have also fully estimated the size most suitable and the Cost of a Tank, amongst my Extra Charges.—

Edward Forbes. Calotype by Hill and Adamson, 1844.
(Courtesy of the Scottish National Portrait Gallery.)

Exterior and interior views of the conservatory at Chatsworth, Derbyshire.
(*Illustrated London News*, 31 August 1844. By permission of the Syndics of the Cambridge University Library.)

I hope to get the Purchase Deeds completed by the 10th Inst being Old Mich's day,[3] which regulates all our proceedings in this County; and I will deposit the Deeds in my Iron Chest until I go to London and can deliver them up to you.—

I shall be glad to hear you accomplished your journey without feeling any ill effects; and with my best Respects to Dr Darwin

I remain Dear Sir | Your faithful Servant | John Higgins

C. R. Darwin Esq

DAR 210.25

CD ANNOTATIONS
3.1 You seemed . . . Higgins 6.1] *crossed pencil*
On cover: 'Papers, Plan, Letters about Beesby Estate Oct.— 1845.—' *ink*

[1] Enclosed with this letter is an itemised list of the proposed improvements. CD's Investment Book (Down House MS) records that a total of £1026 0*s*. 5*d*. was advanced to Higgins during 1846 for farm buildings at Beesby. See also letter to John Higgins, 27 May [1846].

[2] This letter has not been found. The use of liquid manure (urine) had been strongly recommended by the Duke of Richmond, president of the Royal Agricultural Society, and others at the Royal Agricultural Society's annual meeting, held in Shrewsbury, 15–18 July 1845 (*Gardeners' Chronicle and Agricultural Gazette*, no. 30, 26 July 1845, p. 522).

[3] Old Michaelmas day, one of the quarter-days of the English business year. Papers enclosed with this letter indicate that the purchase was completed on 10 October 1845.

To J. D. Hooker [8 October 1845]

Shrewsbury
Wednesday

My dear Hooker

I have just received your note, which has astonished me, & has most truly grieved me.— I never for one minute doubted of your success, for I most erroneously imagined, that merit was sure to gain the day.—[1] I feel most sure, that the day will come soon, when those who have voted against you if they have any shame or conscience in them, will be ashamed at having allowed politics to blind their eyes to your qualifications & those qualifications vouched for by Humboldt & Brown! Well those testimonials must be a consolation to you.[2] Proh pudor, I am vexed & indignant by turns.— I cannot even take comfort in thinking that I shall see more of you & extract more knowledge from your well-arranged stock.— I am pleased to think, that after having read a few of your letters, I never once doubted the position you will ultimately hold amongst Europæan Botanists— I can think about nothing else, otherwise I shd like discuss Cosmos with you.— I trust you will pay me & my wife a visit this autumn at Down.— I shall be at Down on the 24th & till then moving about.

My dear Hooker, allow me to call myself | Your very true friend | C. Darwin

DAR 114.1: 43

[1] Hooker had lost the election for the chair of botany at Edinburgh University to John Hutton Balfour. See the *Scotsman*, 8 October 1845.
[2] J. D. Hooker 1845c.

To Charles Lyell 8 October [1845]

Shrewsbury
October 8th.

My dear Lyell

I have long been purpoting to write to you, but have not done so, from having seen hardly anyone & done little, & therefore having hardly anything to say.— I cannot think of any other questions about negro-crosses: but I may mention (however unlikely you may be to take up so disgusting a subject) that it has been asserted that on the negros born in N. America, the lice are larger & of a blacker colour, than the common species; & that the Europæan lice will not live on negroes. From some analogous statements made to me with respect to the men of the Sandwich islands, I am inclined to believe there may be some truth in these statements.[1] Mr. Denny (to whom I communicated specimens & this information)[2] wd. be most grateful for specimens, if you cd get them in *spirits*, through some medical man, who cd get them through some nurse to some Hospital &c &c I suggest this as a feasible means, without disgusting yourself much.—

I see Long in his Hist. of Jamaica says he has never known two mulattos have offspring!!!!!—[3] Can you obtain any comparative information on the crosses between Indian & Europæans & Negros & Europæans?—[4]

I have lately been taking a little tour to see a Farm, I have purchased in Lincolnshire: & thence to York, where I visited the Dean of Manchester,[5] the great maker of Hybrids, who gave me much curious information.— I also visited Waterton at Walton Hall & was extremely amused at my visit there.[6] He is an amusing strange fellow; at our early dinner, our party consisted of two Catholic priests & two Mulattresses! He is past 60 years old & the day before run down & caught, a Leveret in a turnip field. It is a fine old House & the Lake swarms with water-fowl. I then saw Chatsworth, & was in transports with the great Hot-house:[7] it is a perfect fragment of a Tropical forest & the sight made me thrill with delight at old recollections. My little ten-day tour made me feel wonderfully strong at the time; but the good effects did not last.— My wife, I am sorry to say does not get very strong; & the children are the hopes of the family, for they are all happy life & spirits.—

I have been much interested with Sedgwick Review;[8] though I find it is far from popular with non-scientific readers. I think some few passages savour of the dogmatism of the pulpit, rather than of the philosphy of the Professor chair; & some of the wit strikes me as only worthy of Broderip in the Quarterly. Nevertheless it a grand piece of argument against mutability of species; & I read it with fear & trembling, but was well pleased to find, that I had not overlooked any of the arguments, though I had put them to myself as feebly as milk & water.—

Have you read Cosmos yet: the English Translation is wretched,[9] & the semi-metaphsico-poetico-descriptions in the first part are barely intelligible; but I think the volcanic discussion well worth your attention; it has astonished me by its vigour & information.— I grieve to find Humboldt an adorer of Von Buch, with his classification of volcanos, craters of Elevation &c &c & carbonic-acid gas atmosphere. He is, indeed a wonderful man.—

I hope to get home in a fortnight & stick to my wearyfull S. America, till I finish it.— I shall be very anxious to hear how you get on from the Horners, but you must not think of wasting your time by writing to me. We shall miss, indeed your visits to Down & I shall feel a lost man in London, without my morning "House of Call" at Hart St.—

Emma desires to be most kindly remembered to you both & Believe me, my dear Lyell | Ever yours | C. Darwin

Postmark: OC 8 1845
American Philosophical Society

1 'The surgeon of a whaling ship in the Pacific assured me that when the Pediculi, with which some Sandwich Islanders on board swarmed, strayed on to the bodies of the English sailors, they died in the course of three or four days' (*Descent* 1: 219).
2 See letter to Henry Denny, 3 June [1844].
3 Described in [Long] 1774, 2: 336:

> Some examples may possibly have occurred, where, upon the intermarriage of two Mulattos, the woman has borne children; which children have grown to maturity: but I never heard of such an instance; and may we not suspect the lady, in those cases, to have privately intrigued with another man, a White perhaps? . . . The subject is really curious, and deserves a further and very attentive enquiry; because it tends, among other evidences, to establish an opinion, which several have entertained, that the White and the Negroe had not one common origin.

4 Lyell reported that mulattos were fully fertile, and CD agreed with this conclusion (*Descent* 1: 221).
5 William Herbert.
6 Charles Waterton. He maintained a bird preserve on his estate, Walton Hall, West Yorkshire.
7 Chatsworth was the estate of the Duke of Devonshire, located in the parish of Edensor, Derbyshire. Joseph Paxton, superintendent of the gardens and woods at Chatsworth, erected the great conservatory, 300 feet in length, which was completed in 1840.
8 Sedgwick 1845, an attack on the transformist views put forward in *Vestiges of the natural history of creation* ([Chambers] 1844). CD's response to Sedgwick's review is analysed in Egerton 1970–1.
9 Humboldt 1845–8, translated by Augustin Prichard. Volume one deals with astronomy and geology. Volcanoes are discussed on pp. 213–38.

To J. S. Henslow 28 October [1845]

Down Bromley Kent
Oct 28th.

My dear Henslow

I have to thank you for several printed notices about the Potatoes &c &c—[1] What a painfully interesting subject it is; I have just returned home, & have looked over my potatoes & find the crop small, a good many having rotted in the ground, but the rest well.— I am drying sand today in the oven to store with the greatest care in baskets my seed-potatoes.—[2] I think it a very good suggestion of yours, about gentlefolk not buying potatoes, & I will follow it for one. The poor people, wherever I have been, seem to be in great alarm: my labourer here[3] has not above a few weeks consumption & those not sound; as he complains to me, it is a dreadful addition to the evil, flour being so dear: sometime ago this same man told me, that when flour rose, his family consumed 15^{d} pence more of his 12^{s} earnings per week in this one article. This would be nearly as bad, as if for one of us, we had to pay an additional 50 or 100£ for our bread: how soon in that case, would those infamous corn-laws be swept away.[4]

At Shrewsbury we tryed the potato flour; how very curiously soon the starch separates; it really is quite a pretty experiment.[5]

I have been taking a little tour, primarily to see a farm, which I have purchased in Lincolnshire as an investment (& on which I have told my agent to arrange allotments for every labourer) & then I went & saw York & visited the Dean of Manchester, & had some hours hybrid talk.— I then visited Mr Waterton at Walton Hall, & was exceedingly amused with my visit, & with the man; he is the strangest mixture of extreme kindness, harshness & bigotry, that ever I saw.— Finally I visited Chatsworth, with which I was, like a child, transported with delight.— Have you ever seen it? Really the great Hot house, & especially the water part, is more wonderfully like tropical nature, than I could have conceived possible.— Art beats nature altogether there.[6]

I have been most sincerely grieved at Hooker's disappointment at Edinburgh: I cannot but think he will make a great Botanist; it is admirable what a stock of general & accurate knowledge, he appears to have on all such subjects, as geographical range &c &c.—

We are all flourishing here, with the exception of my wearifull stomach.— I hope Mrs Henslow is better: pray remember me very kindly to her.

Ever my dear Henslow | Yours truly | C. Darwin

Smithsonian Institution (Special collections, Dibner)

[1] In 1845 potato blight (caused by the fungus *Phytophthora infestans*) destroyed much of the European crop, causing widespread famine, particularly in Ireland. The 'notices' may have included a report by the Hadleigh Farmers' Club, in which Henslow was active ('The potato crop', *Gardeners' Chronicle and Agricultural Gazette*, no. 38, 20 September 1845, p. 648) dealing with methods of preparing the diseased potatoes for food. Jenyns 1862, pp. 204–6, discusses Henslow's interest in the potato crop failure and his schemes for alleviating local distress. Salaman 1985 documents the history of potato blight.

[2] Kiln drying and dry storage were widely suggested means for inhibiting the blight (see Lindley 1845a and Prideaux 1845). John William Lubbock, CD's neighbour in Down, recorded that he also stored potatoes under sand in the autumn of 1845 (*Gardeners' Chronicle and Agricultural Gazette*, no. 11, 14 March 1846, p. 163).

[3] Probably a man named Brooks who regularly received payments of £1 4*s*. a fortnight and appears to have been responsible for carrying out general labouring jobs at Down House and for hiring and supervising extra labour as required (CD's Account Book, Down House MS).
[4] British corn duties were reduced to a nominal figure in June 1846. See D. G. Barnes 1930, pp. 276–8.
[5] Described by Henslow in *Gardeners' Chronicle and Agricultural Gazette*, no. 41, 11 October 1845, pp. 688–9.
[6] See preceding letter for further details of CD's tour.

To J. D. Hooker 28 October [1845]

Down. Bromley Kent.
Oct. 28th.—

My dear Hooker

I have returned home now three days.— You offered to send me a copy of your testimonials; I shd like to see them (for the purpose of mentally abusing the corporation),[1] for I have read with much interest & sympathy in the Gardeners' Chron. the speech of the Provost.—[2] I cannot get over my surprise at the result, so confident did I feel about it, knowing who your competitors were.

I have finished Cosmos & you must excuse my having sent it to be half-bound, for I was really ashamed to return it, with the *out*side (not inside) in so tattered a condition. On the whole I am rather disappointed with it; though some parts strike me as admirable; there is so much repetition of the Personal Narrative, & I think no new views, in those parts on which I can at all judge.— His occasional notice of my Journal ought to turn my head.—[3]

I have been taking a little tour, partly on business, & visited the Dean of Manchester & had very much interesting talk with him on hybrids, sterility & variation &c &c.—[4] He is full of self-gained knowledge, but knows surprisingly little what others have done on same subjects.— He is very heterodox on 'species': not much better, as most naturalists would esteem it, than poor Mr. Vestiges. I also visited Chatsworth, & was absolutely delighted with the great Hot-house.

When you feel inclined (*but not before*) write to me, about yourself & tell me what your intentions are for the ensuing winter: I suppose you have lots of work in hand. I trust from my bottom of my heart, that this election will not discourage you in the noble scientific career which is open to you.— I was much amused with one of the Baillies remarks, that if Balfour had gone on the Antarctic Expedition, he would have done as well as you;[5] & no doubt if he had gone to S. America, he would have done as well as Humboldt! How they do crow over their Professors; in my day at least, it wd. have been hard to have picked out a poorer set.

I see the "Antarctic Expedition" advertised:[6] if Sir J. Ross' face & maners do not belie him, it will not be a very entertaining work.— How long & earnestly I have wished, that you wd. publish a Naturalist-Journal of the Expedition.—

Have you anything you wish to send to Ehrenberg? I am going in a few days to send a parcel to him

Will you be so good as to remember me kindly to Sir William. Yours ever | C. Darwin

DAR 114.1: 44–44b

1 Appointment to the chair of botany at Edinburgh was under the control of Edinburgh Town Council, or Corporation (see letter from J. D. Hooker, 1 September [1845], n. 3).

2 Adam Black. There is no account of Black's speech in the *Gardeners' Chronicle and Agricultural Gazette*, although John Hutton Balfour's election was announced in the issue of 18 October 1845 (p. 704). CD was perhaps referring to the account in the *Scotsman*, 8 October 1845.

3 Humboldt 1845–8, 1: 302, 319.

4 CD had corresponded with William Herbert, Dean of Manchester, on plant breeding in 1839 (see *Correspondence* vol. 2).

5 Baillies are municipal officers of Edinburgh, next in rank on the Town Council to the lord provost. The remark was made by baillie Gray, and quoted in the *Scotsman*, 8 October 1845. Hooker had accompanied James Clark Ross on the Antarctic expedition of 1839–43, and was at this time publishing the botanical results in J. D. Hooker 1844–7.

6 Ross 1847.

To C. G. Ehrenberg 29 October [1845]

Down. Bromley Kent
Oct. 29.—

Dear & highly honoured Sir

I send you a few specimens, through Chevalier Bunsen, which possibly may interest you, as they come from volcanic islands:— I beg you to observe that I do not want any information about them for *my own* publications.

There is, however, two specimens from the R. Gallegos (100 miles South of the Santa Cruz) which I should be *very much* obliged, if you could spare time to look at them. I am extremely desirous to know, whether they contain infusoria & *resemble* those many specimens of the great white infusorio-pumiceous deposit of the coast of Patagonia. If so, these specimens would extend the bed 100 miles; but they interest me especially, because the Officer who has sent me these specimens, has found in this bed numerous bones of great extinct Edentata.[1]

Should you have time to examine this specimen, & would take the trouble to inform me of the result, might I, also, beg you to inform me, what meaning you attach to the word "Fluthgebiete," which you use, when describing the Pampæan earth which I sent you. A French translation seems to express, that you attribute this deposit of the Pampas to a great flood or debacle; whereas I understood from your letter, that you considered it, as I do, as an estuary deposit.—[2]

I hope you will excuse my troubling you, & believe me dear Sir | Yours with much respect | C. Darwin

Museum für Naturkunde der Humboldt-Universität zu Berlin

1 See letters from B. J. Sulivan, 13 January – 12 February 1845 and 4 July 1845.

2 See letter from C. G. Ehrenberg, 8 April 1845, and Ehrenberg 1845b. His eventual reply to CD's query (letter from C. G. Ehrenberg, 11 March 1846) confirms CD's interpretation as given here. See *South America*, p. 248 n., where CD corrects the mistranslation of Ehrenberg's term 'Fluthgebiete'.

To John Lort Stokes [November–December 1845][1]

Down, Bromley, Kent.
Sunday.

My dear Stokes.

I do not think the most sensitive person has the smallest right to take offence at what you have said. You could hardly have corrected, as you were bound to do, what apparently has been a gross error, with more delicacy.[2]

Poor Grey has made a very amusing book, but what a catalogue of mishaps & mismanagements.[3] The whole expedition was that of a set of School Boys.

Ever Yours, | C. Darwin.

Copy
DAR 144 (FitzRoy Letters)

[1] In some unknown way this letter fell into the hands of George Grey, see letter from George Grey, 10 May 1846. Since letters took about seven months to travel between England and New Zealand (see letter from George Grey, 10 May 1846, received by CD on 3 November 1846), it seems probable that this letter was written late in 1845.

[2] Stokes had asked CD to read some of the proof-sheets of his account of the *Beagle* survey of the north-west coast of Australia (Stokes 1846). These included passages that contradicted remarks in Grey 1841 (see letter to J. L. Stokes, 3 November 1846). Stokes differed from Grey on the height of land observed from Discovery Bay and on the course of the Glen Elg River (Stokes 1846, 1: 201-3). See also Wickham 1838, pp. 464–5. Ironically, Stokes was mistaken in believing that the Glen Elg flanked Collier Bay and emptied into mangrove swamps bordering Stokes Bay on King Sound, while Grey was more correct in thinking that the river emptied 'somewhere between Camden Sound and Collier's Bay' (Grey 1841, 1: 271). In fact, the river empties into Collier Bay at the head of Doubtful Bay.

[3] Grey 1841 is the account of an attempt to explore the interior of Australia.

To J. D. Hooker [5 or 12 November 1845]

Down Bromley Kent
Wednesday

My dear Hooker

I had intended not writing to you, until I had looked through your Botany,[1] but I must at once thank you very much for it & for your two letters. I began reading the first number last night, & turned over the pages of some of the others, & saw quite enough to show me how much there will be of the very highest interest to me. How different your remarks make it to most systematic works! but I will say no more about it now, except to thank you once again very heartily for it, though I know full well how unworthy I am as a naturalist for such a present, yet I am proud of it, not to mention its real practical use to me.

It was indeed most absurdly unjust to speak of you, as a mere systematist. You speak of your printed letter, as being "bilious", I do assure you, as far as my judgment goes, I see no signs of such feelings.—[2] You must forgive me for alluding to such a subject, but I must say I admire from the bottom of my heart, the manner in which you have borne your disappointment & illiberal treatment. Your noble (& really interesting) set of Testimonials must be a consolation when you think of the Baillie's speeches.—

I am glad to hear that you are hard at work again & continue to find interesting geographical results: assuredly, as you say in your Preface, geographical distrib: will be the key which will unlock the mystery of species.[3] By the way I have written to Capt. Beaufort some queries, & amongst others urged him to direct attention to the Floras of *all isolated* islands.— I presume of course, you have specimens of the junction of the Beech-parasite, in spirits;[4] I gave some to Brown, who, I daresay, w[d] give them up, if you want more specimens.—

Many thanks for l'Espece;[5] could you lend me sometime, your former copy that I may transpose my marks (or rather exchange copies) as I do not want the trouble of looking it over again. I shall be glad to see the other pamphets; though I do not expect much, if they are by Gerard.— I am sorry to say I have sent my very small packet to Ehrenberg: I did not give you a fair chance & ought to have retained it longer; but I am in a hurry for Ehrenberg's answer.— I will return the Testimonials to you;—I shall not, however, send to the Geolog. Soc. for another week..— You ask about my health: I have been unusually well for a week past, owing, I believe, to what sounds a great piece of quackery, viz twice a day passing a galvanic stream through my insides from a small-plate battery for half an hour.—[6] I think it certainly has relieved some of my distressing symptoms.— My wife is not as strong as she ought to be.

—If you want to read a zoological book, I think Waterhouse's Mammalia (now publishing by Bailliere) w[d] interest you;[7] I can lend it you at some future time, when several of the numbers are out.— I hope this next summer to finish my S. American geology;[8] then to get out a little zoology[9] & hurrah for my species-work, in which, according to every law of probability, I shall stick & be confounded in the mud.—

I wish I could get you sometime hence to look over a rough sketch (well copied) on this subject,[10] but it is too impudent a request.

Farewell my dear Hooker | Yours most sincerely | C. Darwin

DAR 114.1: 45–45b

[1] J. D. Hooker 1844–7. Apparently this refers to part one. Although all ten numbers of part one were available by May 1845 (Wiltshear 1913) Hooker had not sent it earlier because he did not have a complete set for CD (see letter from J. D. Hooker, 1 September [1845]). CD's annotated copy is in the Darwin Library–CUL.

[2] Hooker had written a strong letter to the *Caledonian Mercury*, 27 October 1845, on the subject of the Edinburgh elections.

[3] Hooker was far more circumspect than CD implies: 'Hence it will appear, that islands so situated furnish the best materials for a rigid comparison of the effects of geographical position and the various meteorological phænomena on vegetation, and for acquiring a knowledge of the great laws according to which plants are distributed over the face of the globe' (J. D. Hooker 1844–7, p. xii).

[4] J. D. Hooker 1844–7, pp. 452–3, discusses the apparent differences in structure of *Cyttaria* when dried or preserved in spirits of wine. CD is referring to the 'disc' formed when this fungus emerges from the beech tree's bark.
[5] Gérard 1844.
[6] The application of a current from a voltaic battery was commonly known as 'galvanisation', and was a popular therapeutic treatment during the first half of the nineteenth century (Colwell 1922; Rowbottom and Susskind 1984). CD does not mention galvanic treatment after 1846.
[7] Waterhouse 1846–8, published in parts. For CD's review of volume one, see *Collected papers* 1: 214–7. CD's copy is in the Darwin Library–CUL.
[8] *South America*. CD had recommenced work on this book on 29 October 1845 and finished correcting the last proof on 1 October 1846 ('Journal'; Appendix II).
[9] CD's *Beagle* invertebrate collection had not been described in *Zoology*.
[10] CD's essay of 1844 (*Foundations*, pp. 57–255).

To Edward Hitchcock 6 November [1845]

Down Bromley Kent
Nov. 6th.

Dear Sir

Absence from home has prevented me sooner acknowledging your truly generous present of the Final Report on the Geology of Massachusetts.—[1] I assure you I feel sensibly the honour & kindness you have done me. I have as yet read only a little, but I see that there will be much that will interest me greatly; I allude more especially to your detailed accounts of the alluvial deposits, ice & water action, &c. &c. Your's is indeed a magnificent work with its numerous & striking illustrations. I am delighted to possess the excellent plates on the footsteps,[2] & I daresay I shall find some further information, though I have carefully read your several papers.[3] In my opinion these footsteps (with which subject your name is certain to go down to long future posterity) make one of the most curious discoveries of the present century & highly important in its several bearings. How sincerely I wish that you may live to discover some of the bones belonging to these gigantic birds: how eminently interesting it would be know, whether their structure branches off towards the Amphibia, as I am led to imagine that you have sometimes suspected. The finding the bones of the Rhynchosaurus in the pure hard sandstone of Grindshill in Shropshire (where there are some Reptile footsteps) may give one hopes.[4]

I am preparing a little volume on the geology of S. America, which, when published next summer, I will beg you to do me the kindness to accept; though it is a miserable acknowledgment for your grand work.

With my sincere thanks and much respect. Pray believe me, dear Sir | Yours faithfully & obliged. C. Darwin

Postmark: 6 NO 6 1845
Amherst College Archives (President Edward Hitchcock papers)

[1] Hitchcock 1841. CD's copy of the first edition in two volumes is in the Darwin Library–Down.

[2] Plates 31–49 depict fossil footprints that Hitchcock thought were made by birds. They were later shown to be dinosaur tracks (see Dean 1969).

[3] Beginning in 1836, Hitchcock had published a series of papers in the *American Journal of Science and Arts* about the footprints.

[4] Richard Owen discussed the fossil bones and footprints from the sandstone of Grinshill quarry, near Shrewsbury, in R. Owen 1841, pp. 145–6.

To G. B. Sowerby 12 [November 1845][1]

Down Bromley Kent
Wednesday 12th.

My dear Sir

I intend being in town this day week,[2] & with your permission should be very glad on Thursday the 20th to come to your house as early as convenient to yourself & begin our work of going through the shells, it will probably take two or three or more mornings.[3] If you can possibly so arrange it, would you oblige me by agreeing & send me note soon & please inform *me at which hour*; (the earlier the better for my health, after 9 oclock) it will suit you to receive me.— I will bring the specimens & M.S. with me.— You will remember, that you were so good as to acquiesce in my proposal (for in no other way could I permit myself to trespass on your valuable time) of keeping a memorandum of the number of hours I consume & allowing me to in some degree remunerate you for the loss.—

Pray believe me with apologies for troubling you. Yours very faithfully | C. Darwin

Supposing it impossible for you to receive me on Thursday morning the 20th, would you kindly inform me, what three or four *first* consecutive mornings you could spare; as I am pressed to get on & cannot proceed, till my species are ultimately worked out.

Houghton Library, Harvard University

[1] For the basis of the date see nn. 2 and 3, below.

[2] CD attended a council meeting of the Geological Society on 19 November 1845 ('Journal'; Appendix II).

[3] CD's Account Book (Down House MS), for 24–5 November 1845, records expenses for a five-day trip to London, with a note of 1*s*. 'present to Sowerby's man'. Presumably the meeting concerned Sowerby's appendix for *South America*, pp. 249–64. Sowerby's identifications of CD's South American shells are in DAR 46.1: 6–45.

To Smith, Elder & Co. 13 November [1845]

Down near Bromley. | Kent
Nov 13

Dear Sir.

I leave entirely to your better judgment the size of the advertisement, & the extent which you may think it worth while to advertise; I am doubtful of its doing much more good than making early subscribers complete their copies.—[1] I have made & suggested a few alterations in the advertisement The line about "a general sketch of the Zoology &c &c" *must* be omitted.[2]

Would it not be worth while to introduce the names Owen, Waterhouse Gould Jenyns & Bell of the contributors after each part, they are good & known names,?

If you require space to do this, you might omit the sentence in italics, about the Lord Commissioner & the following sentence, & make a general Heading to the whole advertisement in large type, of "With the Approval &c &c &c."

As the advertisement now stands, it is not evident, that the Geological Observations were written by me; my name had better be introduced, where I have marked it.

The title of the Coral-Volume had better, I think, be a little fuller, as I have introduced it.[3]

I am very much obliged to you for having sent me the advertisement to look at & | Believe me | Yours very faithfully | C. Darwin

Harry Ransom Humanities Research Center, University of Texas at Austin

[1] The letter refers to an advertisement by the publishers of the *Zoology* and the geology of the voyage of the *Beagle* that was printed on p. 520 of the second issue of *Journal of researches* 2d ed., which appeared in December 1845, see Freeman 1977, pp. 35–6.

[2] An 1844 advertisement, which appeared in *Volcanic islands*, described the *Zoology* as including 'a general Sketch of the Zoology of the Southern Part of South America'. This part of the original plan was abandoned because the Treasury grant expired.

[3] All CD's suggested revisions were carried out: the contributors' names were added to each part of *Zoology*, CD's name was added after 'GEOLOGICAL OBSERVATIONS', and the title of *Coral reefs* was filled out.

To J. D. Hooker [17 November 1845]

Down Bromley Kent
Monday

My dear Hooker

I cannot find out anything more about the other species of cactus: it grew on Albemarle Is^d^ & I *think* on the other islands. The specimen of O. Galapageia described by Henslow came from James Isl^d^. Have you two or three of the briefest miserablest notes about the Galapageian trees, such as, "this is commonest kind" "red flowers" &c &c If not I w^d^ copy them for you.—[1]

I hope you have received Cosmos. if not, please inform me.— What am I do with your parcels for Ehrenberg & Humboldt: I am going to London for a few days to my Brothers "7. Park St Grosvenor Sq^r^." & will take them with me & I will forward them per steam-boat or leave them till called for, or bring them back with me & return them with the pamphlets which you have so kindly lent me, but wh. will take me some time to read. I have sent to see if I can buy Geograph. Part of Canary Is^d^.[2] for I am ashamed to say I have not yet got on with it.—

I have just got as far as Lycopodium in your Flora[3] & in truth cannot say enough how much I have been interested in all your scattered remarks. I am delighted to have in print many of the statements which you made in your letters to me, when we were discussing some of the geographical points.— I can never cease marvelling at the similarity of the Antarctic Floras: it is wonderful.— I hope you will *tabulate* all your results & put prominently what you allude to (& what is preeminently wanted by non-botanists like myself) which of the genera are & which not found in the lowland or in the highland Tropics, as far as known.— Out of the very many new observations to me, nothing has surprised me more, than the absence of Alpine floras in the S. islands: it strikes me as most inexplicable. Do you feel sure about the similar absence in the Sandwich group: is it not opposed quite to the case of Teneriffe & Madeira? & Mediterranean islands?? I had fancied that T. del Fuego had possessed a large alpine flora!—[4] I sh^d. much like to know whether the *climate* of N. New Zealand is much more insular than Tasmania; I sh^d. doubt it from general appearance of places & yet I presume the Flora of the former is far more scanty than of Tasmania: do tell me what you think on this point.— I have also been particularly interested by all your remarks on variation, affinities &c: in short your book has been to me a most valuable one & I must have purchased it, had you not most kindly given it & so rendered it even far more valuable to me.—

When you compare a species to another, you sometimes do not mention the station of the latter (it being I presume well-known), but to non-botanists such *words* of explanation w^d add greatly to the interest, not that non-botanists have any claim at all for such explanations in professedly botanical works.— There is one expression which you Botanists often use (though I think not you individually often) which puts me in a passion, viz calling polleniferous flowers[5] "sterile", as non-seed-bearing. Are the plates from your own drawings, they strike me as excellent. So now you have had my presumptuous commendations on your great work.

Ever yours | C. Darwin

P.S. I must sometime beg your copy of l'Espece[6] to copy my marks, as I by no means want to wade through so poor a performance again.

Postmark: NO 1⟨7⟩ 1845
DAR 114.1: 46–46b

[1] CD's plant notes were originally given to John Stevens Henslow, before the Galápagos plant collection was transferred to Hooker (see *Correspondence* vol. 2, letter to J. D. Hooker, [13 or 20 November 1843]). The notes are described in D. M. Porter 1981. Henslow described the *Opuntia* in Henslow 1837.

[2] Webb and Berthelot 1836–50.

[3] J. D. Hooker 1844–7, pt 1.

[4] See *Correspondence* vol. 2, letter to J. D. Hooker, [13 or 20 November 1843].

[5] Male flowers, in plants with separate sexes.

[6] Gérard 1844.

From J. D. Hooker [19 November 1845][1]

Kew
Wednesday

My dear Darwin

I doubt not you are very busy & I shall therefore be brief in answers to your last.

There are not more than two numbers to all the Gal. collection that I can find. I have often tried to make your notes hinge on to the species. I wish you would come & take a look at them before I return them to Henslow. "Cosmos" came all right, I am quite annoyed at your binding it, but obliged *truly* all the same. Please send back my things for Humboldt by bearer, or else to Hiscock's Kew boat Hungerford Stairs or by P. Deliv. Coy, *not* by Steamer.

What on earth do you want to buy Webbs Geograph part of Canary Islds for.?[2] I had rather you kept mine altogether I can no more use it now than I can 2 pair of spetacles; you may really & truly keep it altogether if you will. I have my Father's copy here. Would you like his historical & *Anthropological* portion Zoology &c..?—

I cannot make out that Sandwich Isld have any thing of an Alpine Flora, there are I know a few alpine things but nothing to what 12000 ft should give. I expect the higher regions are desperately barren, bad soil earthquaquey perhaps &c.— Terra del Fuego at 56° has no alpine Flora that I could make out, or next to none, I diligently explored the hill tops, 1000–1700 ft.[3]

I think the climate of Tasmania infinitely more xcessive than New Zeald. but am no Meteorologist, if any one would calculate them I have lots of Tasmanian temperatures

I hope never again to call ♂ flowers sterile. Botanical nomenclature is quite a science per se & you must be forbearing with us, we having nothing familiar like stomachs nerves eyes &c to talk of & by; all is a new language I will pay particular attention to your most val. hints especially as coupled with such pretty comps. The Plates are all done by Mr Fitch except the Dissections, which are almost invariably my own, copied by him on the stone.[4]

Could we arrange to turn over the Gal. plants together,?— I fear you are come to Town on your health & dread bothering you. Henslowe is to be in Town Friday & I intended returning them then to him, he will be at 44 Queen Square Bloomsbury on Saturday.

Ever yours most truly | J D Hooker

PS. I append a few comparison of what I took as standards of *excessivity* in climates. V. D. L & London are tolerably close, I wish I had New Zeald. I used to calculate that mean of Max & Min differences of the months, of the days & of each day (I think) gave tolerable results. I am rather confused now

P.S. I have 9AM, 3PM, 9PM temps. throughout our cruize including 8 Months at Tasmania & 4 at New Zeald., or rather my dear delightful Capt R. has them all[5] but a very few scattered results I amused myself working out of V. D. L. & London.

Sept V D L.		March London		
		1820	1821	1822
Max. Monthly var.	31	36.	26	33
Max Daily var	—24	—21.	—22	—23
Min Daily var	—2	—3	—4	—5

For 15 days Aug. V. D. L. Mean diff. of daily Max & Min

		& London. 1822	—1821	—1820
Max Monthly var	—28	—20	—18	—26
Max. daily var	—20	—15	—11	—12
Min daily var	—6.	—6.	—1.	—2.

V D L. October 1840		London 1820	1821	1822
Max. var in Month	36	—38	—40	—33
Max. in a day	—26	—27	—20	—20
Min in a day	—8	—3	—5	—8

V D.L. Nov[r] 12 days begin. 1840 London May (whole Month)

		1820	—1821	—1822
Max Month var	—30	—37	—30	—27
Max daily var	—25	—27	—25	—19
Min daily var	—6	—10	—10	—5.

Kerguelens Land July 19 first days—& same days in London January

	K.G. Land.	Lond. 1820	1821	1822
Max. Month var	17	—34,	—28,	20
Max. daily var	—10	—18	—11	—15
Min. daily var	—2	—2	—0	—3.

DAR 100: 57–8

CD ANNOTATIONS

1.1 I doubt . . . Zoology &c..?— 3.4] *crossed pencil and ink*
6.1 I hope . . . Hooker 8.1] *crossed ink*

[1] Dated on the basis that Hooker wrote on the Friday before CD's trip to London, 19–24 November 1845, recorded in CD's Account Book (Down House MS) on 24 November. See also letter to G. B. Sowerby, 12 [November 1845].

[2] Webb and Berthelot 1836–50.

[3] CD kept the following note with this letter:

> Hooker. Antarctic Work.. Nov. 45. The absence of Alpine plants in the antarctic isl[ds]: & especially in S. America shows that the manufacture of such species (& of all species) is a hard effort.— H. attributes the uniformity & paucity of species in S. America & Antarctic [*interl*] islands to uniformity of climate. Opposed to this is the very fact of same species inhabiting Alpine & lowland situations. Surely there is no great difference in climate between N. New Zealand & Tasmania & yet great diff: in species. Surely S. S. America including Patagonia

presents greater range of climate than S. Africa, & yet wonderful difference in number of plants.— Surely some other cause of difference in number: in isl^{d} we can understand this on my theory: & theoretically understand it, by oscillations ['ha' *del*] not having been many.— (DAR 100: 59).

[4] Walter Fitch, the official draughtsman for Kew publications. He drew the lithographs for J. D. Hooker 1844–7.

[5] James Clark Ross, who commanded the Antarctic expedition and used some of Hooker's notes in composing J. C. Ross 1847.

To J. D. Hooker [21 November 1845][1]

7 Park St
Friday

My dear Hooker

Thanks for your two notes & kind invitations to Kew: I am sure you will believe me, when I say what true pleasure it w^{d} have given me & at present I am unusually well, but I have such an accumulation of work, (partly common business & partly scientific shell-work for Geology every morning) that I doubt whether I shall get it done much under a week.— On my return home, I was (& am) going to ask you, whether you will not spare a Sunday & pay us a visit I w^{d} send my Phaeton to meet you at the Sydenham Station in time for our dinner on Saturday & send you back on Monday or Tuesday.— I intend asking Forbes & Falconer if they will spare the time, & I do heartily hope you will be able to come, we c^{d} have some fine Nat. Hist talk.

Pray never answer my Nat. Hist. questions & discussions, except when at leisure & so inclined, otherwise my conscience must prevent my writing.— Thanks for your most generous offer about Webb;[2] I wished for a copy to save myself the labour of writing out notes: I have sent to Paris to know price, & if dear, I will *so far* accept your offer, as to make faint pencil marks & then in a year or two beg the loan of it again just to skim through my marked passages.— I am particularly obliged for your information about Henslow, & I will beat up his quarters, if I can, today.— I feel sure I sh^{d} not be able to tell you anything about my Galapagos plants, as my numbers are gone, otherwise, of course, I w^{d} make a point of coming to Kew on purpose.— I think of returning either on Tuesday or Thursday home.—

Very many thanks for your information & again, thanks for your kind invitation Yours ever in Haste | C. Darwin

DAR 114.1: 47

[1] Dated from CD's trip to London 19–24 November, see letter from J. D. Hooker, [19 November 1845], n. 1.

[2] Webb and Berthelot 1836–50.

To J. D. Hooker [25 November 1845]

Down Bromley Kent
Tuesday

My dear Hooker

You held out some hopes that you would pay us a Sunday visit: will you come on Saturday 6th of December. If so my phaeton shall be at the *Sydenham Station* to meet the Train which leaves London Bridge at 2°. 20′, & this will just bring you in time for our dinner. If the 6th does not suit you will the ensuing Saturday the 13th be more convenient. As soon as I get your reply, I will ask Forbes Falconer & perhaps Waterhouse[1] to come by the same Train, & I hope we shall have some good Nat. Hist talk & I flatter myself I shall have the four most rising naturalists in England round my table & much I shall enjoy it.—

I returned yesterday from Town, after having had four very active successfull days in London.—[2] I failed to catch Henslow at home.—

Farewell, I hope you will come. Ever yours | C. Darwin

DAR 114.1: 48

[1] Edward Forbes, Hugh Falconer, and George Robert Waterhouse.
[2] See letters to G. B. Sowerby, 12 [November 1845], and to J. D. Hooker, [21 November 1845].

To J. D. Hooker [29 November 1845][1]

Down Bromley Kent
Saturday

My dear Hooker

We are delighted that you will come.— I write now to tell you that the Croydon Railway per Sydenham is much the best way here. I believe coaches go only in the morning & late in evening to Bromley & then you have 6 & ½ miles here in a Fly.— If Waterhouse does not come, my phaeton will hold three perfectly well with one on the Box.— Coming this way you will have the morning till 2°. 20′ in London for business.— If Waterhouse comes, I intend sending a Fly as well as my phaeton; now do you oblige me & come in & say not a word about payment, because Waterhouse (*between ourselves*) is very poor & it w^{d} vex me *exceedingly* if he (or indeed anyone) had to pay hard cash for coming here, when I derive so infinitely more pleasure than any one else can have from coming here.

Now be you goodnatured & help me.—

In some respects, I sh^{d} have enjoyed a visit from you yourself better than with others, but for several reasons (besides being heartily glad to see them) I was obliged to ask Falconer, & I had long wished to ask Forbes, but I doubted whether he w^{d} like to come.

Ever yours | C. Darwin

DAR 114.1: 49

[1] The Saturday before Hooker was expected at Down, see previous letter.

To G. B. Sowerby [1 December 1845]

Down Bromley Kent
Monday

Dear Sir

I have found the two pectens,—but I cannot find the two Turritellas.—[1] According to my memory I put out all my specimens of this genus from Navedad to aid you in your determinations. Are you sure that the Turritellas, which you want are from Navedad?[2]

Will you look through once again **carefully** all the specimens which I left with you & send me a line *tomorrow* by return of Post & then I will look through my **whole** collection & see if they have got in the wrong packet: which will cost me some trouble— I think I left them with you.—

Yours faithfully | in Haste | C. Darwin

Postmark: DE 1 1845
American Philosophical Society

[1] Species of shells described by Sowerby in the appendix to *South America*, pp. 249–64.

[2] Four *Turritellae* are described, one of which is from Navidad, Chile. CD placed great emphasis on the identification of the *Turritellae* as extinct species closely resembling those of today. In *South America*, pp. 132–7, he used this information to support his argument that the fossiliferous deposits of coastal Chile and Peru were laid down at the beginning of the Tertiary period, at a time roughly comparable to the Eocene formations of Europe (p. 133) and when the climate of South America was considerably warmer.

To G. B. Sowerby [3 December 1845]

Down Bromley Kent
Wednesday

My dear Sir

I am sorry you have had so many notes to write.—

I send the Pecten *centralis* & *germinatus* & Terebratula variabilis(?) from P. St. Julians. (also Terebratula variabilis(?) from S. Josef) Also a few specimens from S. Cruz, which I *suspect* I somehow overlooked the other day when I was with you. You will be able to see by comparing them with the M.S.

I also send some large Pectens, which were collected on board the Beagle, but I do not know where, but strongly suspect from P. St. Julians; they may aid you.—

I have all my Tertiary shells **here**, & I have looked through them all & the two Turritellæ are not amongst them.— I *know* I intended to leave them with you. I now remember which Box they were in, when I brought them to your house, & they are not in it now.— I fear they must have got mingled with the other Turritellæ. I believe I remember that one was in a **very small** paper parcel. It will be most vexatious if they are lost.— I think all the Turritellæ were put together. Can you tell by examining by the matrix, whether two specimens from two places have not been put in one lot.— I shall be very much grieved if they do not turn up.— I have looked so carefully they must be somewhere at your house.

My dear Sir | Yours very Faithfully | C. Darwin

New York Botanical Garden Library (Charles Finney Cox collection)

To G. B. Sowerby [9? December 1845]

Down Bromley Kent
Tuesday

My dear Sir

I am very much obliged to you for having so promptly & kindly finished your undertaking.—

The large Pectines may certainly be left undescribed.—

I am not likely to be in town for a fortnight or more & therefore, without you think you can explain any points vivâ voce better than on paper, it would perhaps be best to have all the specimens & M.S. packed up together & I will send our Carrier for them on Thursday week which is the first day that I can.— But *please* be so kind as to send me a line, if you think it **at all** *adviseable* that I sh^{d} see you on the subject & I will make a point of calling the very first time I come up.—

Will you when at leisure oblige me by informing me how many hours of your time you have been so good as to give to me.[1]

It is very strange & provoking about the Turritellæ: I shall have to go through all, when I begin to calculate how far I can go in engraving the new spec.[2] & then possibly they may somehow turn up.

Your's very faithfully | C. Darwin

New York Botanical Garden Library (Charles Finney Cox collection)

[1] CD's Account Book (Down House MS) has the following entry, dated 15 December 1845: 'Sowerby cheque work at shells £4··10s'.

[2] The engravings of the Tertiary shells were made by George Brettingham Sowerby Jr (see letter to G. B. Sowerby Jr, 31 [March 1846]).

To J. D. Hooker [10 December 1845][1]

Down Bromley Kent
Wednesday

My dear Hooker

I have nothing particular to say, but the spirit moves me to write to you to say how much I enjoyed your visit & all our raging discussions.— I somehow feel that I had no talk with you individually & this I much regret, as I wanted to hear about your future plans & to talk over many things: you really must pay us a Sunday visit early in the Spring & you shall come as late as you like on Saturday.— I have finished all the numbers, of the A. F. with undiminished pleasure.[2] I *suspect* p. 241 (4 lines from bottom) there is false print of S. "Australis" for pinnatifida???—[3] I have been pleased to observe how many species my Fuegian collection contains: I think you ought somewhere to explain where C. Negro & Elizabeth islands are, as few would be able to find out without the best charts.

I presume you can trust Capt. Kings localities; I ask, because in his zoological collection, the localities have been infamously mingled; African species marked St of Magellan!

After you were gone, Wedgwood[4] was suggesting, that you might not unlikely get a collection made on summit of Fernando Po, through Jameson,[5] the great African merchants. They are so liberal, that it really does not seem improbable, if the importance of the object were explained to them that they w^d^ request some surgeon or other person to make the attempt. I want to know, whether any insects were collected on Kerguelen's Land: if so how interesting it w^d^ be to compare them with those from Fuegia: Waterhouse w^d^. be the man for the comparison. Are there any land-birds there?— By the way can you inform me, how many numbers of the Antarctic Zoology have come out;[6] as Gray, in his own & Richardson's name was so magnificent as to present me with the few first numbers.

How I should like such another reunion, as we had the other Sunday; for my own part, I learn more in those discussions than in ten times over the number of hours readings.

Ever my dear Hooker | Yours truly | C. Darwin

DAR 114.1: 50–50b

[1] The first Wednesday after J. D. Hooker's visit to Down on 6 December 1845.

[2] J. D. Hooker 1844–7.

[3] CD was correct. The error is rectified in the *corrigenda* to J. D. Hooker 1844–7 (p. 548). The page referred to was in the thirteenth number of Hooker's work, which appeared sometime before 4 December 1845 (Wiltshear 1913).

[4] Hensleigh Wedgwood.

[5] Robert Jamieson, a London merchant instrumental in opening up parts of Africa through his shipping concerns.

[6] Richardson and Gray 1844–75, in which the zoological results of James Clark Ross's Antarctic expedition were described.

From Roderick Impey Murchison [1846?]

Dear Charles,

As I am in the neighbourhood of the "Statistical Account" I thought I might as well find Chapter & verse for my parallels. They are p *446* of the vol for Inverness. See also p *507*.[1]

Yours Ever | R. I. M.

Wednesday Night.

DAR 171

CD ANNOTATIONS
1.3 See also p *507*.] *crossed pencil*; 'Gravel-mounds on the Moray Frith in valley of Nairn' *added pencil*

[1] CD had apparently asked for more detailed information on the Scottish terraces mentioned in Murchison, Verneuil, and Keyserling 1845, 1: 550–1 (see Thackray 1978 for the history of the publication of this work). *The new statistical account of Scotland* vol. 14 (1845) (Inverness— Ross and Cromarty) refers (p. 446) to parallel roads in the valley of the River Nairn; p. 507 gives an explanation of their formation and refers to CD's theory of the formation of the roads of Glen Roy.

From G. B. Sowerby [1846]

Note for Mr. Darwin relative to his addition to Struthiolaria ornata App. p. 12.

I see no objection to this addition being inserted, provided that the fact of Mr. Cuming having found a living species of Struthiolaria at Arica be given on the authority of Mr. Cuming himself and not on mine.[1]

Mem: In my opinion the latin of Dar*w*in is Dar*v*inius—that of So*w*erby—So*v*erbius—as w is not a latin letter—Mr Darwin must use his own discretion.[2]

I percieve Mr Darwin has altered the spelling of Nav*e*dad to Nav*i*dad— it is a subject with which I am entirely unacquainted & only mention it for Mr Darwin's consideration.

Trigonocelia p. 4. The observation may properly stand as it is.[3]

A memorandum 1p
DAR 43.1: 5

[1] *South America*, p. 260, where Sowerby stated that this was 'the only fossil species of this rare genus that he has seen'. Hugh Cuming is not mentioned.
[2] The forms *Darwinii* and *Darwinianus* are used in *South America.* Sowerby named two species of shells after CD.
[3] *South America*, p. 252.

To John William Lubbock [16 January 1846][1]

Down
Friday

Dear Sir John

I cannot forbear sending you our cordial thanks for the kind manner in which you have acceded to all our wishes about the little piece of land. If you were to feel how exposed we are to every wind under Heaven, you would understand our strong wish to have one sheltered walk, and I look forward to considerable amusement in tending & pruning the trees.

Pray believe me | dear Sir John | Your's very faithfully | C. Darwin

The Royal Society, London (LUB.D.16)

[1] In an agreement with J. W. Lubbock dated 12 January 1846 (DAR 210.15), CD agreed to pay a rent of £1 12*s*. annually for a term of 21 years for 1½ acres of land adjoining the Down House property. CD further agreed to plant the land with 'Underwood, Shrubs and Trees' and to construct a fence. The famous sandwalk, CD's 'thinking path', is in this strip (see Atkins 1974, pp. 25–6).

From G. B. Sowerby 17 January 1846

My Dear Sir

You will probably have expected this sooner, the points to be settled were however difficult— After the most careful examination of all the specimens the result will appear in the separation of two sorts from T. ambulacrum this I have done, not because I am myself fully satisfied of its propriety, but because I found characters which would be considered satisfactory (as distinguishing characters) by some (D'Orbigny inter alia) characters which some would regard as specific, others as only variations—however my opinion is that wherever a combination of characters distinguishing one subject from another exist, they should always be noticed, whether they be considered as the characters of a species or only of a variety.[1]

I will now explain my operations

1st those of Huafo & Mocha I have described as T. Chilensis

2d that of Port Desire, together with the Navedad (T. carinifera) and several specimens without locality, but numbd *632* I have described as T. Patagonica.

3d the St Julian must be regarded as neither more nor less than T. ambulacrum

4th I wd strike out from the Navedad series the one resembling T. carinifera

5th Also the T. Ambulacrum with somewhat variable external characters from the Port desire list.

6th Of the additional specimens one (numbered 632) is already disposed of; the two others are such mere casts or fragments that they cannot be brought into the account.

7th I must now add with respect to the fragments from Navedad formerly said to resemble carinifera, but which I now recommend you to strike out, that, one fragment also belongs to T. Patagonica, the rest are probably Ambulacrum, but are very indistinct.

The specimens shall be packed up on Monday or Tuesday and sent to the Geol: Society for you—

I am My Dear Sir | Your very obliged | G B Sowerby

17th Jan^{y} /46—
N.B. *Too late for post.*

DAR 43.1: 1–2

CD ANNOTATIONS
2.4 *632*] 'Port Desire' *added pencil*

[1] Refers to shells brought to Sowerby by CD in November 1845. See CD's letter to G. B. Sowerby, 12 [November 1845], and intervening letters. CD may have queried Sowerby's species determinations, or the additional work on *Turritella* may indicate that the two specimens from Navidad, reported missing in CD's letter of [1 December 1845], had been found. Sowerby had already been paid for his descriptions, see letter to G. B. Sowerby, [9? December 1845], n. 1. Two fragments of Navidad *Turritellae* are included among the descriptions Sowerby gives in the appendix to South America (pp. 256–7), but it is not known whether these are the missing specimens.

To Royal Geographical Society [28 January or 4 February 1846][1]

Down Bromley Kent
Wednesday

My dear Sir

I am very much obliged to your kindness in having sent me the enclosed paper.[2] As it is chiefly directed against my views, it does not become me to make any remarks on what I think of its value. I must, however, protest against several quotations, which anyone w^{d} suppose were from my writings, but which appear to be extracted from some Indian Review.— I may be permitted also to protest against the statement at p. 3, which is *absolutely* the *contrary* to what I have imagined & stated in *many places* in my writings to have been the case.—

I do not believe Mr C. has ever taken the trouble to read my Book[3] or even my own abstract in my Journal:[4] if he has, he has overlooked *all* my chief data & gives as new in his paper, what he might have found in my work.

As I do not treat of the zoology of coralls, I make free to urge you, before his Paper is published, to get some zoologist to skim it over, as his errors are very great & would be discreditable to any scientific publication to include.— I have marked a few places with pencil for *your* inspection: I have made this note more critical than I intended, but I hope you will excuse me.[5]

With my renewed thanks | Believe me | Yours sincerely | C. Darwin

Lieut Nelson R. J. has published a theory in Geolog. Trans vol 5. p. 103 on Bermuda,[6] something like this, of Lieut Christophers.

Royal Geographical Society

[1] The date is based on the reading and review of William Christopher's paper (see nn. 2 and 5, below).
[2] William Christopher's paper was read 26 January 1846 at the Royal Geographical Society and submitted to George Bellas Greenough for review (Council Minute Book, Royal Geographical Society). No mention is made in the minutes of CD as a reviewer.
[3] *Coral reefs.*
[4] *Journal of researches* 2d ed., pp. 465–82.
[5] The council minutes of 9 February record the decision not to publish the paper.
[6] Richard John Nelson. See Nelson 1840.

To J. D. Hooker [31 January 1846][1]

Down Bromley Kent
Saturday

My dear Hooker

I sh^d^ have written to you some time ago (though I do not exactly know I had anything particular to say) but I have been very unwell for a fortnight with a cold which affected my detested stomach.

I have at last finished Webb & Berthelot,[2] & carefully packed it up; shall I return it you or keep it? If you will be so good as to leave my few pencil marks (that I may hereafter skim through it) it is absolutely the same to me whether returned now or hereafter; I have been a good deal disappointed with it, & think it much spun out with empty remarks & generalities: I see he says he will give a list of all species at end of the Descriptive Part; so that I am at present no wiser than when I begun, how far in proportional numbers the Flora is peculiar & how closely the different islands of the Western division resemble each other, or the character of the Alpine plants, or indeed in any of the grand features, which I think w^d^ naturally interest every philosophic naturalist. How different are your results as given in your Flora & letters to me!

I see Berthelot quotes & concurs with Gaudichaud (who is a good man, is he not?) that the plants in volcanic islands are polymorphous ie variable:[3] this is directly the reverse, I know, of your opinion.—[4]

I am delighted to hear of more species from the Galapagos; what a wonderful spot it is! I am surprised to hear of the W. Indian character of the Flora, though as this Flora is common to Panama, as you say, it makes it less surprising.[5] It is an odd coincidence that the one shell common to the two coasts of Panama, is found, also, at the Galapagos.

What an odd chance it was the discussion in the Gardeners Chronicle about the longevity of fruit-trees;[6] I cannot say I am convinced; surely Oak-trees attain a greater age than lime-trees & it is a mere assumption to say it is due to mechanical causes. What a stride from Annuals to Eternity of life! I have seen none of our Party here, since we met here: what a grievous thing poor Forbes'[7] illness has been. I have had one note from him, saying he was much better. What a loss he w^d^ be to science: it is grievous to think of it.—

Farewell my dear Hooker excuse this dull note & don't write till you feel inclined. | Ever yours | C. Darwin

DAR 114.2: 53–53b

[1] Dated from the relationship to the letter from J. D. Hooker, 1 February 1846, and on CD's noting that he finished reading Webb and Berthelot 1836–50 on 30 January (DAR 119; Vorzimmer 1977, p. 134).
[2] Webb and Berthelot 1836–50, vol. 3, pt 1.
[3] Sabin Berthelot did not in this instance directly quote Charles Gaudichaud but cited Gaudichaud 1826 immediately following the reference to polymorphic plants (Webb and Berthelot 1836–50, vol. 3, pt 1: 74–5).
[4] See letters from J. D. Hooker, 5 July 1845 and [mid-July 1845].
[5] J. D. Hooker 1846, pp. 255–6.
[6] Lindley 1845b.
[7] Edward Forbes.

From A. C. V. D. d'Orbigny 31 January 1846[1]

Monsieur,

J'ai tardé beaucoup plus que je ne l'aurais voulu à vous envoyer la liste des déterminations de vos coquilles de la Bahia Blanca,[2] mais des circonstances indépendantes de ma Volonté sont venues m'en empècher, malgré mon grand desir de vous envoyer promptement ce petit travail, veuillez je vous prie m'excuser et ne pas me taxer de négligence.

Toutes les espèces marquès d'une + en marge, sont décrites par moi dans la *Paléontologie de mon Voyage*.[3] Celles qui n'ont pas ces marques sont également décrites Dans les *Mollusques de mon Voyage*,[4] ainsi vous pouviez les mentionner et vous dispenser de les décrire.

Toutes ces espèces se trouvent vivantes sur les côtes voisines et appartiennent bien à la faune Actuelle. Elles sont comme vous l'avez pensé, du même age que les coquilles fossiles de la Bahia de San-Blas, et ont toutes leurs identiques vivans sur le même lieu.

Pour les espèces de l'Uruguay, elles sont marines, et ont aussi leurs analogues vivans sur la côte. elles sont toutes les deux décrites dans les *Mollusques de mon Voyage*, dont le fin s'imprime en ce moment.[5]

Je vous remercie beaucoup, Monsieur, des renseignemens que vous voulez bien me donner sur les personnes avec lesquelles je pourrais correspondre pour avoir des fossiles d'Angleterre.[6] Je vais m'occuper a prèparer mes doubles, et je suivrais vos bons conseils en leur adressant des collections de fossiles. Si en attendant vous en voyez quelques uns veuillez, je vous prie me dire, s'ils sont disposés a accepter mon proposition.

Si vous avez quelques doubles des fossiles des Iles Falkland, ils me seraient bien utiles pour mes publications futures.

Veuillez croire, Monsieur, aux sentimens les plus devoués avec lesquels j'ai l'honneur d'etre | Votre très humble serviteur | Alcide d'Orbigny

Paris a 31 janvier 1846.

DAR 43.1: 66–7

[1] For a translation of this letter see Appendix I.

[2] The list has not been preserved with the numerous other lists of shells, identified by Orbigny, in DAR 43.1. In *South America*, p. 5, CD cited Orbigny on the shells of Bahia Blanca and San Blas.

[3] Orbigny 1835–47, vol. 3, pt 4.

[4] Orbigny 1835–47, vol. 5, pt 3.

[5] See *South America*, p. 2.

[6] Orbigny was writing Orbigny 1845–7 and planning work on Orbigny 1850–2. See letter from A. C. V. D. d'Orbigny, [June–July 1846].

From J. D. Hooker 1 February 1846

West Park Kew
Feby 1. 1846.

My dear Darwin

Forbes is better I am exceedingly glad to hear, not able yet to leave his bed-room. His attack I suppose you know was ulceration of a (or the) kidneys— he has discharged a great deal of pus during micturation, but no mucus, so that there are yet considerable hopes. Really I do fear that his must be a very dangerous complaint & am deeply greived to hear of his suffering; poor fellow. Falconer told me he was very patient. I should wish much to see him, but do not like being obtrusive.

I asked about your numbers at the Brit. Mus. (of the Antarct. Zool.)[1] which is always there for you, & you are requested to send for your numbers when convenient, taking care to let them know what numbers you already possess, as they loose count. Also Reeves has one or more numbers of my flora for you,[2] I told him not to send them to me, as I have only to send them into Town again, if you like I will get those at the Britt. Mus. & leave them at Reeves or send them both any where you like in the course of the week.

Pray keep Webb & Berthelot for the present if you won't for alltogether. The book wants more tabulated results greivously, but they are so difficult before the species are worked out & Webb is very cautious. Gaudichaud is a very good Physiologist, & a clever fellow, also a very nice man, but not a systematist, & I should not take his *word* for Volc. Insular floras being polymorphous. It is very easy saying so, & my contradicting it is only making matters worse if it is only because I do not see it; but I must confess I think all are following Bory.[3] I hardly know which is easiest, for one who has not properly studied the subject to see evidences of the statements of another like Gaudichaud or to remark their absence as myself. I would not at all have you take my word for it; but I must say that I cannot see it to be the case, but the contrary. nor can it well be, if true that genera bear a large proportion to the species of Islands. It is possible that we may not mean the same thing by variable. Volc Islands are often lofty & steep & small, as Mauritius &c &c the consequence of which is that every species is found in several states at different heights; & from the confined area this at once strikes the observer, but I do not admit a Mt. state of a plant solely dependent on climacteric causes to be properly speaking a Variety, it is a state produced by assignable causes, & does not tend much to confusion of species. Nothing would be more impracticable than to make more than 16 species of the 16 Kerguelens' Land plants. I should not call one of them polymorphous. Lord Auckland & Campbell's Isld. possess but one variable genus Coprosma,. The St Helena plants are remarkably invariable, even the Ferns. The addition of other Galapagos plants to yours did not cause me either to do or undo any of the species founded on your materials. I will write to Webb about it;[4] it looks very conceited to say so, but I really believe the fact to be that the Frenchmen do not know what Continental Floras are, they only study Insular ones (Bory, Gaudichaud Webb). Extended tracks of country with a large flora are full of polymorphous genera, even take Great Britain & there is scarce a Nat. Ord. without its bone of contention. However I will write to Webb tomorrow at any rate & take the opportunity of asking him to explain himself a little. Pray what is the case with animals?.

Is the shell common to both coasts of Am. confined to Panama on the W.? I alluded to the fact in my notes on Galapago Botany, it struck me as most curious & confirmatory of migration. Do you know any thing of a gulf stream from the Gulf of Panama running S.W. & meeting the great S. Polar current at the Galapagos,? there *ought* to be one.[5] Have you seen Pet. Thouar's Voy. of Venus & acct: of the different. temps. of the straits between the several Isld^{s}?[6] Is it true what Douglass[7] says that the *Cactus* of Jas Isld. grows 40 feet high?. he was apt to pull the long bow.

Anent the longevity of Trees, I was amazed at Lindley's Doctrine, that they may live for ever, it is contrary to all analogy in the Veg. Kingdom; why do not herbs do so? no doubt it is very rare with Trees as it is with us, to die of *sheer old* age; there is generally an immediate cause. I am now however endeavoring to solve the mystery of an Ichaboe Plant that kills itself by turning all its bark into wax!— when the axis perishes & leaves a wax cast behind. Still I have no notion of apples &c dying out simultaneously, surely the evidence against that was reasonably conclusive. See how it is, that every one thinks all *Evidence* that goes with his preconceived opinion & all *trivial* that opposes it.

At present I endeavor to hold aloof from all speculations on the origin of species, & wish to till at any rate this part of my flora is finished. When that is the case I should like to have much talk about it with you, at present I go on the old assumption that each species has one origin is immutable & migrates. I am sore puzzled with two closely allied mundane genera (that trouble Volc. Islands very little) *Gnaphalium* & *Senecio* each are singularly widely diffused, but are differently affected with regard to varieties or rather to the relation the Sp. of one country bear to those of another. *Sencio* is made up of closely allied polymorphous species, but the species of no two countries are the same (generally speaking), the further you go from any centre, the more different the species become, the more they swerve from the typical forms of the country started from, there is no tendency in them to return either by variation to the old types, or by presenting analogous species, they display a centrifugal force. *Gnaphalium* on the other hand presents us in every country with many identical species & the typical species of each country always show a centripetal tendency to vary into one or two common types. Thus, the Chilian species of *Senecio* are unlike English, but the Fuegian & Patagonia are more so, but both the Chilian & Fuegia species of *Gnaphalium* are ever trying hard to become English, & succeeding too. This is an old story of Mundane genera &c, in one case a Mundane genus has mundane species the other has local species.

Please tell me again what you objections are to my use of *sterile* or *abortive* in Botany. I would wish to be accurate in the use of terms, but am at times hard up.

My Father says that the plants of Islands are not polymorphous as far as his recollection of Insular Floras goes, but that it is in extended areas that variable plants chiefly occur. The Andes or rather Cordillera flora is very bad, from its great amount of species, It probably also appears worse, from the actual area being so much greater than it appears when projected on a map. The more I consider the difference in trouble, between the Floras of Galapagos (or Auckland Isld) & of Fuegia (which is continental) the more convinced I am that the latter are the more polymorphous by far: trouble I mean in determining the limits of species, I still think there is no comparison in that respect:—the question is one of comparison.

I daresay you will think me very impertinent for suggesting to one of your experience on a subject of Geology; but as you are going to publish on that of S. Am. I would be so bold as to say, that I wish Geologists would give more maps & cuts, & especially maps of what the country is supposed to be during the epoch under consideration. It takes nearly a page to describe what country is under water & what above it when such & such a formation was depositing, all of which to such a tyro as I am, would read much simpler on a duodecimo map. In my tenderest years I used to pore over a map of the world before the flood! that gave me a wonderfull insight into geology as I then thought. How I should like a little map of the country between Plata & Horn before the 400 ft. were raised, another of the time when Icebergs roamed over the plains—another of 2 or 4 of the 8 periods of rest expressed by intensity of colors:—but I am rather out of my reckoning; only if you did think of it lithography would do it very cheap indeed.

My acct: of Kerg. Land Cabbage is extracted with notes in Chambers' Journal,[8] they conclude with rather pitying my ignorance in supposing there can be any reasonable mystery about its origin or that of any other plant, it being doubtless only a state of one of the seaweeds on the shore!— That's Progressive Developement with a vengeance, or rather Developement per Saltum

Ever my dear Darwin Most truly Your's | Jos D Hooker

Have you any tendency to subscribe £1"1, to a man who has been xloring Africa (Duncan)[9] & is going again; I promised Capt Beaufort to ask some of my friends & want friendly help.

DAR 100: 60–2

CD ANNOTATIONS
1.1 Forbes . . . subject 3.8] *crossed pencil*
3.16 but I do not . . . species. 3.18] *underl pencil, ink and pencil crosses in margin*
3.18 Nothing . . . founded 3.23] *scored brown crayon*
3.24 I will write . . . migrates. 6.4] *crossed pencil*
4.5 Have you . . . Isld^s? 4.6] *scored ink*; 'No' *added pencil*
6.9 the further . . . started from 6.11] 'Ch. 4.' *added and circled pencil*
6.19 local species.] *cross added pencil*
7.1 Please . . . up. 7.2] *crossed pencil*
8.6 in trouble] 'in' *added pencil to clarify Hooker's writing*
8.9 & of Fuegia . . . comparison. 8.10] *scored pencil*
9.1 I daresay . . . help. 12.3] *crossed pencil*
Top of first page: 'This is merely that Insular plants not more variable' *pencil*

[1] Richardson and Gray 1844–75.
[2] Reeve Brothers, King William Street, Strand, publishers of J. D. Hooker 1844–7.
[3] Bory de Saint-Vincent 1804.
[4] Hooker had met Philip Barker Webb, co-author of Webb and Berthelot 1836–50, in Paris in February 1845.
[5] The description in J. D. Hooker 1846, p. 255, is based on observations by Robert FitzRoy in *Narrative* 2: 505.
[6] Abel Aubert Du Petit-Thouars, commander of the *Vénus* 1836–9. See Du Petit-Thouars 1840–3, 2: 313–22.
[7] David Douglas.
[8] An extract from Hooker's *Flora Antarctica* (J. D. Hooker 1844–7, pp. 238–41) was printed as 'The Kerguelen's Land Cabbage' in *Chambers's Edinburgh Journal* n.s. 5 (1846): 76–7.
[9] John Duncan. CD was one of many solicited to finance Duncan's trip from the Cape of Good Hope to Timbuktu in 1846.

To J. D. Hooker [5 February 1846][1]

Down Bromley Kent
Thursday

My dear Hooker

I got your note, just as I was setting out for the Geolog. Council.—[2] Your invitation has tempted me much; owing to the Horners' coming here Tuesday or Wednesday is the first day I c^d. come. But will you oblige me by sending me *one* line to say whether Gay[3] speaks English, for to my shame & utter disgrace I cannot speak French. When in Chile I talked with him in Spanish & I now have forgotten that! So that if he cannot speak English, it w^{d} be quite disagreeable to meet him & I w^{d} much prefer visiting you in Spring, when the 4 to 5 hour's drive would be pleasanter.

If he does not speak English & therefore I do not come, pray remember me very kindly to him (though perhaps he has quite forgotten me) & say I remember with much pleasure our short interviews.— I sh^{d} also much like to see Mr Miers.—[4] If I come, I will, (if convenient to you) come about 4 oclock on Wednesday & return the middle of the next day.— You cannot well imagine what a bold step I feel this, I *literally* have not slept out of *near* relations house or inn for five years!!

I have not thanked you for a long & as usual interesting letter, received some time since

Ever most truly yrs | C. Darwin

DAR 114.2: 51

[1] For the basis of the date see n. 2, below.
[2] CD's Account Book (Down House MS) records a payment on 6 February for a trip to London.
[3] Claude Gay, who had travelled extensively in Chile during 1830 and 1831.
[4] John Miers, who worked as a botanist and engineer in South America, 1819–38. CD's copy of Miers' account of his travels (Miers 1826) is in the Darwin Library–CUL.

From G. B. Sowerby 7 February 1846

My Dear Sir

On the other leaf I have copied your list and stated what I believe will be found correct relative to each genus. My conclusion would be that the fossil shells of Navedad have *not* a particularly tropical character. The Cassis is small—a larger species is found in the Med^{n}. though the large sorts are tropical. It may be doubted if Sigaretus be really found in the Med^{n}. it is probable that it may belong to S. Australia. Perna abounds in tropical latitudes—and if any are found on the S. Australian coast—they are scarcely known—fossil species are found in Piedmont & Normandy &c. Harpa is one of the most decidedly tropical genera. Conus, Cypræa, Ovulum, Mitra, Terebra—Meleagrina, Perna, Voluta—Fusus—Triton &c may be regarded for the most part as tropical—though there are perhaps some exceptions in each genus. There is scarcely a genus of any extent that has not some species in temperate and extra-tropical zones. Thus there are Cones, Cowries, Mitres, Fusi, Tritons &c in the Med^{n}:

Concerning Trigonocelia I believe there is not known any recent species: fossil species exist in our Europæan tertiary beds. How can D'Orbigny admit Trigonocelia if he turn out Cucullæa? Crassatella is not peculiarly characteristic of the Australian seas, some species are found in the Atlantic and Indian seas.

You know the recent Struthiolariæ belong to N. Zealand—and the Trigonia abounds at or near to Sydney, N. S. Wales: the genus is found fossil almost everywhere—

Thus have I, to the best of my abilities answered your queries—[1] I hope satisfactorily and remain | My Dear Sir Your very obliged | G B Sowerby

7th Feby. 1846—
Charles Darwin Esqr.

Gastridium. I know no recent species; other fossil species are found "*aux environs de Paris*".
Monoceros, the greater number belong to the Southern parts of S. America—
Voluta, ranges as you say to "Str. of Magellan" there are also some fine South Australian species (magnifica, fluctuata, Turneri, Zebra, pulchra &c) though I believe the greater number and those with most brilliant colours are tropical.
Oliva—abounds in tropical latitudes; though some fine species are found to the northward: it can scarcely be regarded as in any degree characteristic of other than the tropics.
Pleurotoma—few species belong to temperate or Mediter^n. zones.
Fusus—ranges to high northern latitudes, though I believe the finer species are intertropical.
Turritella } Your Navedad species might belong to temperate latitudes as far as
Trochus } they are concerned.
Cassis—in general tropical, some few *small* species are found north of the tropics—there are also some S. Australian species.
Pyrula—tropical
Triton—some large species are found in M: Medn: & S. Australian.
Sigaretus—not entirely confined to tropical climates
Natica—has a most extensive range.
Bulla—also.
Terebra—though not entirely confined to tropical latitudes, is nevertheless very characteristic of them.
Dentalium—very nearly as Natica, though the large & fine species belong to the tropics.
Corbis—I believe tropical—as there is some uncertainty about your Navedad species, it would not be right to draw any conclusion from it.
Cardium—has a very extensive range.
Venus—do.
Pectunculus—do. its greatest developement in the Medn.
Cytherea—has a very extensive range.
Mactra—do.
Pecten—do.

DAR 43.1: 3–4

[1] For CD's use of this information see *South America*, ch. 5.

To J. D. Hooker [8? February 1846]

[Down]
Sunday

My dear Hooker

Thanks for your franck note: we shall have the greatest pleasure in visiting you in the Spring.—

Even if M. Gay had talked English, I believe I c^{d} not have come, for going to London on Wednesday has brought on in spite of Galvanism,[1] three days bad sickness.— Emma yesterday to thank Lady Hooker[2] & yourself for the Goldsmith Hall good-things, which will delight the children: I am going to send to Bromley tomorrow & will enquire at the Coach offices.

Will you ask Gay for me, what Birds, Reptiles or Mammifers (not introduced) inhabit Juan Fernandez? I have always been curious about it—

In Haste. I will write again before long.

My dear Hooker | C. Darwin

P.S | The good things have just arrived & they look so very good that the poor children will be deluded out of some of them

Do you take in the New Horticultural Journal.— Have you seen the Dean of M^{s} Paper?[3]

DAR 114.2: 52

[1] See letter to J. D. Hooker, [5 or 12 November 1845], n. 6.
[2] Maria Hooker, J. D. Hooker's mother.
[3] W. Herbert 1846. William Herbert was the Dean of Manchester.

To J. D. Hooker [10 February 1846][1]

Down Bromley Kent.
Tuesday

My dear Hooker

I ought to have written sooner to say that I am very willing to subscribe £1·^{s}1 to the African man (though it be murder on a small scale) & will send you a Post-office-order payable to Kew, if you will be so good as to take charge of it.[2]

Thanks for your information about the Antarctic Zoolog. I got my numbers when in town on Thursday: w^{d} it be asking your publisher to take too much trouble to send your Botany to the Athenæum Club: he might send two or three numbers together: I am really ashamed to think of your having given me such a valuable work; all I can say is that I appreciate your present in two ways, as your gift & for its great use to my species-work.— I am very glad to hear that you mean to attack this subject some day: I wonder whether we shall ever be public combatants: anyhow, I congratulate myself in a most unfair advantage of you, viz in having extracted more facts & views from you than from any one other person.

I daresay your explanation of polymorphism on volcanic isl^{ds} may be the right one: the reason I am curious about it, is, the fact of the birds on the Galapagos being in several instances very fine-run species;—that is comparing them, not so much one with another, as with their analogues from the continent.— I have somehow felt like you, that an alpine form of a plant is not a true variety; & yet I cannot admit that the simple fact of the cause being assignable ought to prevent its being called a variety: every variation must have some cause so that the difference would rest on our knowledge in being able or not to assign the cause. Do you consider that a true variety should be produced by causes acting through the parent? but even taking this definition are you sure that alpine forms are not inherited for one, two or three generations? Now would not this be a curious & valuable experiment, viz to get seeds of some alpine plant, a little more hairy &c &c than its lowland fellow, & raise seedlings at Kew: if this has not been done, could you not get it done? Have you anybody in Scotland from whom you c^{d} get the seeds.?

I cannot answer your question about the Purpura or about the currents: western currents flow strongly across the southern half of the group: I have not heard of the observations on the temperature of the sea in the different channels; but I know it varies wonderfully.—

I don't believe a Cactus ever was seen at the Galapagos 40 ft. high.— 20 or w^{d} be nearer the mark.—

I have been interested by your remarks on Senecio & Gnaphalium: would it not be worth while (I sh^{d} be **very curious** to hear the result) to make a short list of the generally considered variable or polymorphous genera, as Rosa, Salix Rubus &c &c, & reflect whether such genera are generally mundane & more especially whether they have distinct or identical (or closely allied) species in their different & distant habitats.—

Don't forget me, if you ever stumble on cases of the same species being **more** or **less** variable in different countries.—

With respect to the word "sterile" as used for male or polleniferous flowers, it has always offended my ears dreadfully, on the same principle that it would to hear a potent stallion ram or Bull called sterile, because they did not bear, as well as beget, young.

With respect to your geological-map suggestion, I wish with all my heart I c^{d} follow it, but just reflect on the number of measurements requisite; why at present it could not be done even in England, even with the assumption of the land having simply risen any exact number of feet..— But subsidence in most cases has hopelessly complexed the problem: see what Jordan-hill-Smith says of the dance up & down, many times, which Gibraltar has had *all within the recent period.*[3] Such maps as Lyell has published of sea & land at the beginning of the Tertiary periods[4] must be excessively inaccurate: it assumes that every part on which Tertiary beds have not been deposited, must have then been dry land;—a most doubtful assumption.—

Kerguelen Land cabbage (*Pringlea antiscorbutica*).
(Plates XC–XCI of J. D. Hooker 1844–7. By permission of the Syndics of the Cambridge University Library.)

Robert Chambers. Calotype by Hill and Adamson, 1844.
(Courtesy of the Scottish National Portrait Gallery.)

I have been amused by Chambers V. Hooker on the K. Cabbage:[5] I see in the Explanations[6] (the spirit of which, though not the facts, ought to shame Sedgwick)[7] that Vestiges considers all land animals & plants to have passed from marine forms; so Chambers is quite in accordance. Did you hear Forbes when here,[8] giving the rather curious evidence (from a similarity in error) that Chambers must be the author of the Vestiges: your case strikes me as some confirmation.—[9]

I have written an unreasonably long & dull letter, so farewell. C. Darwin

Did you extract anything about J. Fernandez from Gay?

Do you take in the Hort. Journal;[10] I want much to see first Number.—

DAR 114.2: 54–54b

[1] Dated on the basis of n. 2, below.

[2] See letter from J. D. Hooker, 1 February 1846, n. 9. In CD's Account Book (Down House MS) there is an entry dated 12 February: 'Subscription for Duncan of Africa per Hooker'.

[3] James Smith of Jordanhill. See J. Smith 1846.

[4] 'Map shewing the extent of surface in Europe which has been covered by water since the deposition of the older Tertiary strata', C. Lyell 1830–3, vol. 2, facing p. 304.

[5] See letter from J. D. Hooker, 1 February 1846, n. 8.

[6] [Chambers] 1845. CD recorded that he had read this on 6 February 1846 (DAR 119; Vorzimmer 1977, p. 134).

[7] Adam Sedgwick published a scathing attack (Sedgwick 1845) on *Vestiges of the natural history of creation* ([Chambers] 1844), to which [Chambers] 1845 was a partial answer.

[8] Edward Forbes had joined Hooker, Hugh Falconer, and George Robert Waterhouse at Down House on 6 December 1845, see letters to J. D. Hooker, [25 November 1845] and [10 December 1845].

[9] The identity of the author of *Vestiges of the natural history of creation* was not officially revealed as Robert Chambers until 1885, although unofficially known from 1854 (A. Desmond 1982, p. 210).

[10] *Journal of the Horticultural Society of London.*

To J. D. Hooker [15 February 1846][1]

My dear H.

I have been obliged to make the order payable to Dr. Joseph D Hooker at X cross,[2] as they told me there is no order office at Kew—

I am sorry thus to trouble you.— | C. D.—

Sunday

Does Sir William know the Dean of Manchester's address in *London*.?—

DAR 114.2: 54c

[1] Dated from the relationship to the previous letter.

[2] Presumably CD meant Charing Cross post office. A money order made possible the safe transmission of small sums through the post by specifying the name of the recipient and the office at which payment was to be made.

To William Thompson 18 February [1846?][1]

Down Bromley Kent
Feb. 18th

Sir

I am much obliged for your kindness in sending me the note on the Atlantic Dust, as in case I ever get many more facts together, I may perhaps publish an additional note.—

With respect to the migration of Birds, it would be a sincere pleasure to me to aid, even in the smallest degree, one whose writings I have for several years been accustomed to read with much pleasure & instruction; but I really have nothing to say: I apprehend Forbes alluded to a mere speculation of mine (not grounded on facts & therefore quite useless to anyone) that birds probably followed lines of now lost & sunken land. I merely alluded to this notion of mine, when talking with Forbes on his views on the distribution of plants on land since subsided.— In some future year I intend publishing on the variation of plants & animals in the domestic & natural state, & I shall then (I fear) not be able to refrain from some speculations on this & allied subjects, but, as I have said, I really have no facts, or speculations of sufficient importance to be at all worth communicating in detail.— I am sorry that you shd. have had the trouble of writing for nothing; but may I be permitted to hope that our communication on paper may some day lead to our personal acquaintance.

I shall look forward with interest to your work containing your observations: I beg to remain, dear Sir | Your faithfull & obed: sevt. | C. Darwin

Ulster Museum, Botanic Gardens, Belfast

[1] Dated on the assumption that Thompson wrote soon after CD's 'An account of the fine dust which often falls on vessels in the Atlantic Ocean' was published in February 1846 (*Collected papers* 1: 199–203). H. C. G. Ross 1979, p. 364, gives the information that the paper is watermarked 'J. Whatman 1844'.

From Edward Forbes [25 February 1846][1]

3 Southwick Street | Hyde Park.
Wednesday

Dear Darwin

To answer your very welcome letter so far from being a waste of time is a gain for it obliges me to make myself clear & understood on matters which I have evidently put forward imperfectly & with obscurity. I have devoted the whole of this week to working & writing out the flora question, for I now feel strong enough to give my promised evening lecture on it at the R. Institn. on Friday[2] & moreover wish to get it in printable form for the Reports of our Survey. Therefore at no time can I receive or answer objections with more benefit than now. From the hurry & pressure which unfortunately attends all my movements & doings I rarely have time to spare in preparing them for publication to do more than give brief & unsatisfactory abstracts, which I fear are often extremely obscure.

Now for your objections[3]—which have sprung out of my own obscurities.

I do not argue in a circle about the Irish case, but treat the botanical evidence of connection & the geological as distinct. The former only I urged at Cambridge,[4] the latter I have not yet publickly maintained.

My Cambridge argument was this—that no known currents whether of water or air, or ordinary means of transport would account for the little group of Asturian plants—few as to species but playing a conspicuous part in the vegetation—giving a peculiar botanical character to the South of Ireland. That as I had produced evidence of the other floras of our Islands (i.e. the Germanic—the Cretaceous & the Devonian ((these terms used topographically not geologically) having been acquired by migration over continuous land (the glacial or alpine flora I except for the present—as ice-carriage might have played a great part in its introduction)) I considered it most probable & maintained that the introduction of that Irish flora was also effected by the same means. I held also that the character of this flora was more southern & more ancient than that of any of the others,—& that its fragmentary & limited state was probably due to the plants composing it having (from their comparative hardiness—heaths, Saxifrages &c) survived the destroying influence of the glacial epoch.

My geological argument now is as follows: Half the Mediterranean Islands or more are partly—in some cases (as Malta) wholly—composed of the upheaved bed of the Miocene sea: so is a great part of the south of France from Bourdeaux to Montpellier: so is the west of Portugal, & we find the *same* corresponding beds with the same fossils (*Pecten latissimus* &c) in the Azores. So general an upheaval seems to me to indicate the former existence of a great post-Miocene land, the region of what is usually called the Mediterranean flora. (Every where these Miocene islands &c bear a flora of that type) If this land existed it did not extend to America (for the fossils of the Miocene of America are representative & not identical): where then was the edge or coastline of it, Atlantic-wards?

Look at the form & constancy of the great fucus-bank & consider that it is a *Sargassum* bank & that the *Sargassum* there is in an abnormal condition & that the species of this genus of fuci are essentially ground-growers.—& then see the probability of this bank having originated on a line of ancient coast. (see diagrams) I cannot admit the *Sargassum* case to be parallel with that of *Confervae*[5] or *Oscillatoria*

Now, having thus argued independantly 1st on my flora & 2^{d} on the geological evidences of land in the quarter required: I put the two together to bear up my Irish case.

I think I have evidence from the fossils of the boulder formations in Ireland that if such Miocene land existed, it must have been broken up or partially broken up at the epoch of the glacial or boulder period.

This diagram may explain my meaning better than a hundred words

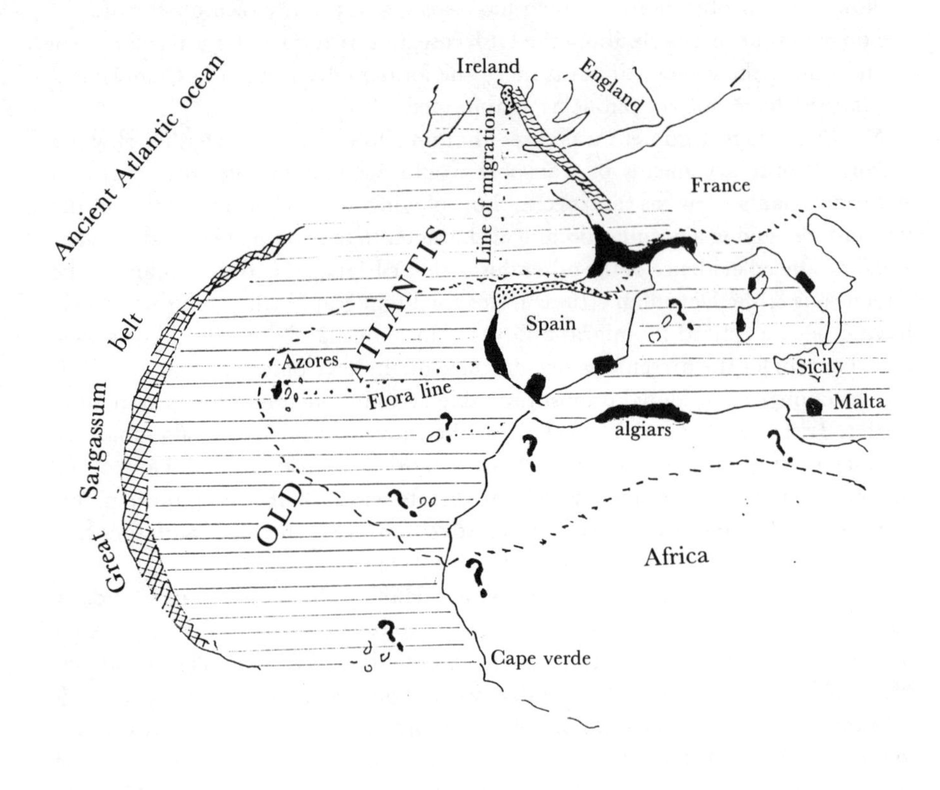

present boundary line of Schouw's 3[d] Botanical region.[6]
I believe there is no relation between alpine floras south of this line & those north of it—this however is another subject.

Area of my Alantis or ancient post-miocene land.

Fragments of the upheaved miocene now remaining

Sargassum belt constant between 15° & 45° N

Line of ancient barrier, possibly existing during glacial period. This is the line of my Devon. Flora—which is common to the Channel Isles, Devon & Cornwall & S.E. of Ireland.

regions of the "Asturian flora" in Ireland & the peninsula.

—all objections thankfully received:—[7] ever most sincerely | Edward Forbes.

You'll be sorry to hear that poor Cumming[8] has been suddenly struck with paralysis & is in a very precarious state.

DAR 164

CD ANNOTATIONS
1.1 To answer . . . obscure. 1.10] *crossed pencil*
5.6 to indicate . . . post-Miocene land] *scored pencil*
5.7 (Everywhere . . . type) 5.8] *scored pencil*
Beside diagram: 'Why not drifted away?'
'Azores not actually united'
'is there Post-eocene' *pencil*

[1] Dated from the reference in the letter to Forbes's Royal Institution lecture on 27 February 1846.
[2] E. Forbes 1846.
[3] CD's letter to Forbes is missing, but see letter to J. D. Hooker, [13 March 1846], for some of his objections to Forbes's theories.
[4] E. Forbes 1845. The paper was read at the British Association meeting of 1845 in Cambridge.
[5] This was an answer to one of CD's objections, see letter to J. D. Hooker, [13 March 1846], on the question of *Fucus* and *Confervae*. The sentence was written in the margin.
[6] Joakim Frederik Schouw, who divided the vegetation of Europe into twenty-two botanical regions in Schouw 1823, pp. 501–24.
[7] With Forbes's letter in DAR 164 is a note by CD:

> Forbes arguments from several Spanish plants in Ireland, not being transported not sound, because sea-currents and air-d[itt]o and migration of birds in *same lines*.— I have thought not-transportation the greatest difficulty—now we see how many seeds every plant & tree requires to be regularly propagated in its own country, for we cannot think that the great numbers of seeds superfluous, & therefore how small is the chance of here & there a solitary seedling being preserved in a well-stocked country.—

[8] Hugh Cuming.

To J. D. Hooker [25 February 1846][1]

Shrewsbury | (next Monday back at Down)
Wednesday

My dear Hooker

I came here on account of my Fathers health, which has been sadly failing of late, but to my great joy he has got surprisingly better.— I write now on account of the enclosed note: do you wish for the scraps, if so of course they are at your service; I presume I asked formerly for you.—[2] Let me have a line in answer some time, & I will write to Henslow.— I had not heard of your Botanical appointment & am very glad of it, more especially as it will make you travel & give you change of work & relaxation.[3] Will you not some time have to examine the Chalk & its junction with London Clay & greensand &c? if so our house w^{d} be a good central place, & my horse w^{d} be at your disposal: could you not spin a long week out of this examination? it would in truth delight us, & you c^{d} bring your Papers (like Lyell) & work at odd times.—

Forbes has been writing to me, about his subsidence doctrines; I wish I had heard his full details, but I have expressed to him in my ignorance my objections, which rest merely on its too great hypothetical basis; I shall be curious, when I meet him, to hear what he says— He is also speculating on the gulf-weed. I confess I cannot appreciate his reasoning about his miocene continent, but I daresay it is from want of knowledge.—

You allude to the Scicily-flora, not being peculiar, & this being caused by its recent elevation (well established) in main part; you will find Lyell has put forward this very clearly & well.—[4] The Appenines, (which I was somewhere lately reading about) seems a very curious case.—

I think Forbes ought to allude a little to Lyell's work on nearly the very same subject as his speculations; not that I mean that Forbes wishes to take the smallest credit from him or any man alive: no man, as far as I see, likes so much to give credit to others, or more soars above the petty craving for self-celebrity.—

If you come to any more conclusions about polymorphism, I sh$^{d.}$ be very glad to hear the result; it is delightful to have many points fermenting in one's brains, & your letters & conclusion always give one plenty of this same fermentation. I wish I c^{d} ever make any return for all your facts, views & suggestions.

Ever yours most truly. C. Darwin

Pray give my best remembrances to M$^{r.}$ Bentham[5]

DAR 114.2: 55

[1] CD was in Shrewsbury between 21 February and 3 March ('Journal'; Appendix II). The only Wednesday was 25 February.

[2] The scraps referred to were a small, miscellaneous collection of plants, some of them duplicates from CD's *Beagle* collection, offered by John Stevens Henslow to Hooker. Henslow later wrote to Hooker himself (28 February and 9 March 1846, collection of R. A. Hooker).

[3] Hooker was appointed botanist to the Geological Survey of Great Britain in February 1846 (Huxley ed. 1918, 1: 207).

[4] C. Lyell 1837, 3: 445. In CD's copy (Darwin Library–CUL) he has written 'Capital!' beside this passage.

[5] George Bentham, with whom Hooker was staying (see letter from J. D. Hooker, 2 [March] 1846, n. 1).

To J. D. Hooker [25 February – 2 March 1846]

[Shrewsbury]

Please return this;[1] I send it thinking, that you w^{d} like to read it.— I cannot see my way about his post-miocene land.

In haste. | Ever yours | C. Darwin

DAR 114.2: 56c

[1] Letter from Edward Forbes, [25 February 1846].

From J. D. Hooker 2 [March] 1846

Pontrilas,[1]
Feby 2 1846.[2]

Dear Darwin

You are very good to send me Forbes' letter, which I return with many thanks I am exceedingly glad to have seen it, especially as I could not be at the lecture. Between ourselves I cannot say that the things being put in black & white has tended to fix my ideas: though as you make no comment I may be differently acted upon to what I ought.

It is a notable fact that of all the Irish Flora, only some 10 or 15 are peculiar to that portion of the U.K., & it is the more singular that 10 of those inhabiting the W. coast should be Asturian plants, the majority of them peculiar to the W. of Ireland & Asturias or that part of the Peninsula.[3] I cannot account for this by any known probable laws of migration, nor on the other hand was I aware of the xtensive system of changes his theory invoked. It looks exceedingly pretty on paper, but would look more so, if any more of the said 10 plants inhabited the Azores & Madeira, except *one*. Now though Botanical evidence is in favor of some theory that will bring Asturias &, W. Ireland into closer connection, the same evidence appears to me against bringing Azores & Madeira into the same category, for those Islands ought to possess more than one of these 10 plants. It is true they possess hundreds of other things instead & that their Floras are Mediterranean, but these other things are not Irish & all that argument for their previous connection by land is overwhelmed by their wanting what their mountains ought to have retained & were as likely to have received as those of W. Ireland. I do not know how far my disconnecting the Islands from the same theory that includes Ireland will affect the whole question. I am still inclined to admit any theory that will appear so *Botanically* reasonable as that proposing the existence of land between Asturias & Ireland the apparent proof of which is drawn from the fact of the very 10 plants which would be likely to have availed themselves of this bridge being found at its opposite ends. But it is Botanically unreasonable to suppose that Azores & Madeira if Island in the same circumstances did not receive the same plants & retain them.

I have little doubt but that the V. D. L. plants found at Illawarra migrated thither from V. D. L. but I should not think so many species would have done so had ocean alone intervened. I may tell you candidly that my mind becomes more made up to be a migrationist than ever, that I have changed in so far that my previous indecision was the other way; but I cannot get over what perhaps you cannot so fully feel, that the *only* plants *peculiar* to the W of Ireland are *all* Asturian, & some of them confined to Asturia. I wish I could talk with you on this subject for I am sure you cannot understand what I have written above.

As for the rest.— Schouw's 3^{d} Bot. region is very good & proper.[4] The Canaries are the transition between that & the Nubian Abyssinian & Sengambia Flora, including the C. D. V.[5] To that succeeds the Asiatic Flora xtending from W. Tropical Africa to Java at least; & again to that the Cape Flora.

It is not a very good remark that there is no relation between the Alpine Flora's N. & S. of the dotted line, (I speak from memory), the fact is, there is little relation between the Alpine Floras of any of the N. Atlantic Islds & those of Europe & that is the worst part I fear of the whole theory. Till we know the Atlas Flora it is dangerous to meddle with the Azores & Madeira. The Mt. Floras of widely separated spots are more similar than the low land ones, & so the Mt. Plants of these Islds. sh$^{d.}$ be more Spanish than the others, but I fear it is not so, & that whilst the Alpigeni have come from Africa (Atlas) the lowlanders have come from Europe.

As to the great Sargassian bed, that is "all my eye" & I fear will tend to throw discredit on the rest of Forbes work. The idea that it indicates a previously existing line of coast is surely preposterous, & untenable, nor do I think it has the form he gives it in longitude, certainly not in Latitude. The fact is that floating weeds are found all the way between the Sargasso sea & England, & as soon as you enter (on leaving Sargasso) the cold current (from the Newfoundland Banks), abundance of the F. *nodones* & *serratus* & *canaliulatus* are found, & I think I found a bit of one of these (eminently cold country) algæ in company with the *Sargassum*, certainly very little to the Northward of it. I never doubted but that the Sargassum was floated off from a continent & retained in the "meeting of the waters" at the Sargasso Sea, There is however much to be done regarding that Sargasso sea, there is little doubt but that it is rapidly decreasing, if not changing its place very remarkably, or both. I would like to collect statistics regarding it: please retain any thing you may see.— I cannot object to the Azores being peopled with Europæan plants by means of the Miocene (or Post-Miocene I hardly understand which) land, that may once have united them, but am inclined to look upon their Mt: Floras as of a much more recent nature than those of the Mts. of Spain. Do not however take this as my decided opinion, the question as a Botanical one requires detailed study, which I cannot now give it.

My ideas of my future duties as Bot. Surveyor are so undefined that I cannot say whether I may or may not make an opportunity of going down to Kent, few things would give me so much pleasure as accepting your hearty invitation. As I told Henslowe I took all duplicates from the Gal. Coll. that I dared. If you have nothing else to do with them I would ask them to go to the Paris Herb., who liberally gave me every scrap they could spare of their Galap. scraps. I now remember well that my views of the Sicily flora are Lyells & no doubt unconsciously borrowed from him. I am so apt to take what I read & forget, for my own, when I do remember it, that I cannot doubt Forbes doing so too, but he that publishes *must* remember. The Appenine fact I long have known & have lately seen it stated by a man of the name of Alexander (I think) a monied dabbler in Europæan Botany, who travels much.[6]

Bentham who desires his very kind remembrances, has been talking about Polymorphism with me & we certainly do conclude wholly against Bory. I must define the term Polymorphism as applied to a Flora or genus, as indicating one whose species are difficult of separation & definition. It is in such genera & groups that Volcanic Islds are lamentably deficient, in comparison to continents. Take any Isld. or group of Islds. as St. Helena, J. Fernandez, Tasmania *even*, Sandwich, Galapagos, & their species are admirably defineable & distinct, take the Flora of the same area as any of those occupys' out of a continent, & in most cases there will be some dozen genera that will drive you crazy. I do think that I may safely dismiss the question but pray do not let me do so if you have any doubt. I have too much respect for nature to force a Botanical hypothesis if at variance with evidence derived from other branches of Nat. Hist.

I am sorry for Cuming, though he has behaved abominably to Botanists & particularly to my Father who paid him once upwards of £700 for himself & friends & with my Mother's & my aid distributed 13 sets of his S. American plants at no expense to himself, of between 2000 & 3000 species each. I hope that it is not a little that would induce me to think so very ill of anyone as I in common with all Botanists do of Cuming's *meanness* to give it the most lenient term. I am truly glad he behaved better about shells & since I heard it have not failed to couple that with my bad opinion of him, when asked.

I have a sister to be married about the 24th. in Norfolk,[7] which will take me away from my work here earlier that I expected: she is to splice a Scotch Presbyterian clergyman[8] to whom she has been long attached & our friends being all in Norfolk, it will take place there, whence she will go to Glasgow for her future abode.

I hope I may congratulate you on your fathers restoration to his usual health & the continuance of your own & Mrs Darwin's.

Ever most truly Yours | Jos D Hooker

DAR 100: 63–8

CD ANNOTATIONS

1.1 You are . . . ought. 1.5] *crossed pencil*
2.11 those islands . . . 10 plants.] *scored brown crayon*
3.1 V. D. L. plants . . . think 3.2] *scored pencil*
3.6 cannot so . . . understand 3.8] *scored brown crayon*
4.1 Canaries] 'Canaries' *added in pencil to clarify Hooker's writing*
4.2 Abyssinian . . . at least 4.4] *scored pencil*
4.4 Java] *underl pencil*
5.1 not] *underl pencil*
5.1 there is . . . Europe 5.3] *scored pencil*
5.2 relation. . . Europe 5.3] *scored pencil and brown crayon*
5.5 Mt.] 'Mountain' *added pencil*
5.6 separated] 'Alpine?' *added pencil, del pencil*
5.6 Mt.] 'Mountain' *added pencil*

5.8 Alpigeni] 'Alpigeni?' *added pencil*
6.1 As to . . . much. 7.12] *crossed pencil*
8.3 the term . . . question 8.10] *scored pencil*
8.10 doubt. . . Hooker 12.1] *crossed pencil*

[1] Pontrilas House, Hereford, was the residence of George Bentham.
[2] Although Hooker dated this letter 2 February 1846, it is clearly an answer to CD's letters to Hooker, [25 February 1846] and [25 February – 2 March 1846].
[3] Edward Forbes argued that the Asturian region of northern Spain had once been connected to the west of Ireland by a gigantic land mass that also included the Azores and Madeira (E. Forbes 1845 and 1846). See letter from Edward Forbes, [25 February 1846].
[4] Schouw 1823, pp. 512–14 (see letter from Edward Forbes, [25 February 1846], n. 6).
[5] Presumably the Cape Verde Islands.
[6] Possibly Richard Chandler Alexander, who travelled through the Alps and elsewhere in Europe in 1842 (Alexander 1846a and 1846b).
[7] Maria Hooker.
[8] Rev. Walter McGilvray. See Allan 1967, pp. 108–9, 158, for an account of his engagement to Maria Hooker.

From C. G. Ehrenberg 11 March 1846[1]

Berlin
d. 11 Maerz 1846.

Hochzuverehrender Herr

Ihre neue freundliche Zusendung von Gebirgsproben durch die Preuß. Gesandtschaft in London habe ich empfangen und ich würde mich schon längst beeilt haben Ihnen Nachricht darüber zu geben, wenn nicht eine schwere Krankheit meiner Frau mich fast ein halbes Jahr lang schon in meiner ganzen Correspondenz gestört hätte.

Folgende Mittheilungen mögen Ihnen anzeigen daß ich dessenungeachtet Ihre Wünsche nicht berücksichtigt gelassen habe.

1. Die Proben vom Gallegas Fluße in Patagonien zeigen zwar auch mikroskopische Organismen, aber ganz andere als die große weiße Tuff Masse von St. Cruz. Es sind Süßwasser—oder brakische Formen und nicht geglüht. Diese Ablagerungen schließen sich demnach an die von Monte Hermoso an. Übrigens habe ich das Organische nur in den 2 bimsteinartigen lockeren Proben besonders in der gelblichen gefunden.[2]

2 Sie wünschen ferner zu wissen ob das Wort *Fluthgebiet* einerley sey mit dem englischen Ausdruck *Estuary deposit.*[3] Beides ist wohl ziemlich gleich bedeutend, doch könnte eine *Ablagerung* (*deposit*) im *Fluthgebiete des Meeres* eine *Süßwasser Ablagerung* seyn, während *estuary deposit* wohl entschieden stets des *Meeres Ablagerung* ist *bei der Fluth.* Das *obere Fluthgebiet des Meeres im Festlande* ist *the upper district of the tide in a River or Continent*, wo durch Aufstauung des Flußwassers und dessen Mischung stets brakische Deposits erfolgen, wie z.B. bei Hamburg 18 deutsche Meilen vom Meere ein oberes Fluthgebiet ist, bei Cuxhaven das untere. Das obere wird oft nur ausnahmsweise erreicht von der Fluth.

3. Die mir gesandte Probe von atmosphaerischen Staube aus Malta habe ich sogleich analysirt und das merkwürdige Resultat in unserer Akademie der Wissensch. vorgetragen.[4] Ich habe darin 43 Organismen beobachtet, von denen 31 auch im Capverdischen Staube völlig gleichartig sind. Ihre gelehrte und so reichhaltige Zusammenstellung der Staub-Beobachtungen möchten Sie doch ja publiciren. Der Gegenstand scheint mir wichtig zu seyn und wenn auch unsere Ansichten über den Ursprung und die Luftzüge etwas differiren, so wird die dadurch angeregte fortgesezte Beobachtung doch bald die Wahrscheinlichkeits-Grenze finden. Es ist natürlich daß die bisher noch verhältnißmäßig geringe Zahl von geographischen Beobachtungen der mikroskopischen Organismen Sicherheit der Resultate nicht geben kann. Dennoch ist der Weg der Beobachtung der allein gangbare und das Resultat desselben das allein wissenschaftlich befestigende. Sie sind mitten in der herrlichsten Gelegenheit den ausfallenden Gegenstand weiter zu entwickeln.

Mit dieser oder irgend einer anderen nahen Gelegenheit erhalten Sie meine lezten Mittheilungen über diese Objecte. Meine freundlichsten Grüße schließen den jetzigen Brief.

Ich verbleibe in aufrichtiger Hochachtung | Ihr | freundlich ergebenster | Dr C G Ehrenberg

Sollten Sie gelegentlich mir einige Fragmente von jezt auf Ascension wachsenden Grasern senden können so wäre es mir sehr lieb zur Vergleichung. Nur müßten sichere Namen dabey seyn, sonst können es Blattfragmente oder Stengel Fragmente ohne botanischen Werth seyn, zoll groß.

DAR 39.1: 62–3

CD ANNOTATIONS

2.1 Folgende . . . gefunden 3.6] '(*fresh water & brackish, not like white tuff—*)'[5] *added pencil*

3.1 Die Proben . . . gefunden 3.6] *scored pencil*; 'only part useful' *added pencil*

On cover: 'Ch. IV & V' *pencil*; 'Thanks' *pencil*

'Hooker grasses; named species' *pencil, partially circled pencil*; 'V. my catalogue'[6] *pencil*

'Beaufort Dusts'[7] *pencil*

'Cordillera rock' *pencil*

[1] For a translation of this letter see Appendix I.

[2] The information in this paragraph was used in *South America*, p. 117. It led CD to conclude that 'the 200 to 300 feet plain at Port Gallegos is of unknown age, but probably of subsequent origin to the great Patagonian tertiary formation.'

[3] CD quoted this letter as his authority in correcting mistranslations of Ehrenberg's term 'Fluthgebiete'. Previously, it had been translated by Alcide d'Orbigny as a 'flood', lending support to Orbigny's view that the Pampas deposits had been laid down by a debacle. See *South America*, p. 248 n., and letter to C. G. Ehrenberg, 29 October [1845], n. 2.

[4] Ehrenberg 1845c, pp. 377–81.

[5] The information was used in *South America*, p. 117, see n. 2, above.

[6] A reference to CD's catalogue of *Beagle* plant specimens which had been sent to Joseph Dalton Hooker by John Stevens Henslow.

[7] This entry and CD's subsequent letter to C. G. Ehrenberg, 25 March [1846], indicate that CD followed up the Atlantic dust problem with the Hydrographer's Office, but no letter to Francis Beaufort on the subject has been found; nor is there any further mention of this question in the extant correspondence until 23 November 1876, when CD wrote to Julius Victor Carus stating that he did not know why he had doubted his Atlantic dust paper and he now thought it worth translating.

To J. D. Hooker [13 March 1846][1]

Down Bromley Kent
Friday

My dear Hooker

It is quite curious how our opinions agree about Forbes views; I was very glad to have your last letter, which was even more valuable to me than most of yours are & that is saying, I assure you, a great deal.— I had written to Forbes to object about the Azores on the same grounds, as you had, & he made me some answer, which partially satisfied me, but really I am so stupid I cannot remember it. He insisted strongly on the fewness of the species absolutely peculiar to the Azores—most of the *non*-Europæan species being common to Madeira: I had thought that a good sprinkling were absolutely peculiar.— Till I saw him last Wednesday[2] I thought he had not a leg to stand on in his geology about his post-miocene land; & his reasons upon reflection seem rather weak; the main one is that there are no deposits (more recent than the miocene age), on the miocene strata of Malta &c, but I feel pretty sure that this cannot be trusted as evidence that Malta must have been above water during all the post-miocene period. He had one other reason, to my mind still less trust worthy.— I had also written to Forbes,[3] before your letter, objecting to the Sargassum, but apparently on wrong grounds; for I could see no reason, on the common view of absolute creations, why one Fucus sh^{d} not have been created for the ocean, as well as several confervæ for the same end. It is really a pity that Forbes is quite so speculative: he will injure his reputation, anyhow on the Continent; & thus will do less good.— I find this is the opinion of Falconer, who was with us on Sunday & was extremely agreeable: it is wonderful how much heterogeneous information he has about all sorts of things— I the more regret Forbes cannot more satisfactorily prove his views, as I heartily wish they were established, & to a limited extent I fully believe they are true: but his boldness is astounding.—

Do I understand your letter right, that W. Africa & Java belong to the same Botan: region, ie that they have *many non-littoral species* in common; if so, it is a sickening fact; think of the distance with the Indian Ocean interposed! Do some time answer me this.—

With respect to Polymorphism, which you have been so *very kind* as to give me so much information on, I am quite convinced must be given up in the sense you have discussed in: but from such cases as the Galapagos birds & from hypothetical notions on variation, I sh^{d} be very glad to know whether it must be given up in a slightly different point of view, that is whether the peculiar insular species are generally well & strongly distinguishable from the species on the nearest continent (where there is a continent near): the Galapagos, Canary isl^{ds} & Madeira ought to answer this: I sh^{d} have hypothetically expected that a good many species would have been *fine* ones, like some of the Galapagos birds, & still more so on the different islands of such groups.—

I am going to ask you some questions, but I sh^{d} really sometimes almost be glad if you did not answer me for a long time or not at all, for in honest truth I am often ashamed at, & marvel at, your kindness in writing such long letters to me. So I **beg** you to mind, never to write to me, when it bores you.— Do you know "Éléments de Teratologie (on "monsters I believe?) Végétable par A. Moquin-Tandon"—[4] is it good book & will it treat on hereditary malconformations or varieties?— I have almost finished the tremendous task of 850 pages of A. St. Hilaire's Lectures,[5] which you set me, & very glad I am you told me to read it, for I have been much interested with parts.— Certain expressions which run through the whole work put me in a passion: thus I take, at hazard,—"la plante n'etait pas tout à fait **assez affaiblie** pour produire de veritable carpelles":[6] Every organ or part concerned in reproduction,—that highest end of all lower organisms,—is according to this man, produced by a lesser or greater degree of "affaiblissment"; & if that is not an affaiblissement of language, I don't know what is.—

I have used an expression here, which leads me to ask another question: on what sort of grounds do Botanists make one family of plants higher than another: I can see that the simplest cryptogamia are lowest & I suppose from their relations, the monocotyledenous come next; but how in the different families of the Dicotyled: The point seems to me equally obscure in many races of animals, & I know not how to tell whether a Bee or Cicindela is highest: I see Au. Hilaires uses a multiplicity of parts—several circles of stamens &c, as evidence of the highness of the Ranunculaceæ:[7] now Owen has truly, as I believe, used the same argument to show the lowness of some animals & has established the proposition, that the fewer the number of any organ, as legs or wings or teeth, by which the same end is gained, the higher the animal.[8]

One other question. Hilaire says (p. 572) that "chez une *foule* de plantes c'est dans le bouton", that impregnation takes place: he instances only Goodenia & Falconer cannot recollect any cases.— Do you know any of this "foule" of plants? From reasons, little better than hypothetical, I greatly misdoubt the accuracy of this, presumptuous as it is: that plants shed their pollen in the bud is, of course, quite a different story. Can you illuminate me?

Henslow will send the Galapagos scraps to you.[9] I direct this to Kew, as I suppose after your sister's marriage (on which I beg to send you my congratulations) you will return home.

Ever yours truly | C. Darwin

There are great fears that Falconer will have to go out to India—[10] this will be a grievous loss to palæntology.—

DAR 114.2: 56–56b

[1] The first Friday after CD's trip to London on 11 March, see n. 2, below.

[2] Charles James Fox Bunbury recorded that he met both CD and Edward Forbes at the Geological Society council meeting of 11 March 1846, where Bunbury and Forbes discussed E. Forbes 1846 (F. J. Bunbury ed. 1891–3, *Middle life* 1: 124–5). The expenses of CD's trip to London are recorded in his Account Book (Down House MS).

[3] The letter has not been found, but see letter from Edward Forbes, [25 February 1846], which is the reply.

[4] Moquin-Tandon 1841.

[5] Saint-Hilaire 1841. There is an annotated copy in the Darwin Library–CUL.

[6] Saint-Hilaire 1841, p. 480. This sentence is marked with pencil quotation marks in CD's copy (Darwin Library–CUL).

[7] Saint-Hilaire 1841, p. 617. Beside this passage CD has written: 'If most complicated & altered form is to be highest— no for worm from fish-origin might then be highest.—' (Darwin Library–CUL).

[8] R. Owen 1843a, p. 365.

[9] See letter to J. D. Hooker, [25 February 1846], n. 2.

[10] Hugh Falconer was appointed superintendent of the botanic gardens in Calcutta in 1848.

To J. D. Hooker [24 March 1846][1]

Down Bromley Kent
Tuesday

My dear Hooker

I had a letter yesterday from Ehrenberg, in which he expresses a strong wish for some specimens of the grasses from Ascension. (no doubt for comparison with the microscopical objects in the tuff)[2] but they must be *named* else they will be useless to him. He says specimen an inch in length w^{d} be sufficient, & I presume he does not want the inflorescene.— I will tell him that I ask you, & if you cannot supply him no one can.— I am going to write almost immediately to him & shall endeavour to send through the Geograph. Soc.—

Since last writing to you, I have finished Hilaire & found one of my queries about plants being higher & lower well discussed, though yet I do not feel quite satisfied.—[3]

I see he praises Moquin-Tandon's work on Teratologie Vegetable.—[4]

Ever yours | C. Darwin

DAR 114.2: 57

[1] Dated from CD's reference to having received the letter from C. G. Ehrenberg, 11 March 1846, which arrived at Down on 23 March (see letter to C. G. Ehrenberg, 25 March [1846]).

[2] Probably for Ehrenberg 1846.

[3] Saint-Hilaire 1841, pp. 791–4. The pages are heavily marked and annotated. CD commented on p. 791: 'There is no highest, there is most modified, & by mans standard high & low. The impossibility of saying what is *highest* is conformable to my theory—which is highest var of cabbage or dog?' CD recorded that he finished reading this work on 20 March 1846 (DAR 119; Vorzimmer 1977, p. 135).
[4] Moquin-Tandon 1841.

To C. G. Ehrenberg 25 March [1846]

Down Bromley Kent
March 25th

Dear & highly Honoured Sir

I received your kind letter two days ago, & beg to thank you sincerely for the information contained in it.—[1] Herewith I send a copy of my little paper on the Atlantic Dust,[2] (published in the Geolog. Journal) & which I would have sent ere this, had I supposed you would have cared to see it.—

I have asked the Hydrographer to the Admiralty (Capt. Beaufort) to call the attention of Officers to the dust & to collect specimens of it.

I have no specimens myself of grasses from Ascension, but I have written to Dr. Hooker & I well know he will proud to send you specimens if he has them: I doubt, however, whether he has *yet* named his grasses.

Sometime ago I sent you some specimens (through the Chev. Bunsen[3]) of rocks of the Secondary period from the Cordillera; shd you **have** examined them I shd esteem it a great favour to know the result.—

I regret much to hear of the long illness in your family: being a married man myself, I can appreciate your distress.

Pray believe me, dear Sir, with much respect. | Yours faithfully & obliged | C. Darwin

P.S. | I have received the Ascension plants from Dr. J. D. Hooker for you.—[4] I enclose his note, as you might like to see the scanty list of really indigenous Phaneragam: plants.— You will observe there is only one certainly indigenous grass, or at most two.— Many plants have been of late introduced there.—

Museum für Naturkunde der Humboldt-Universität zu Berlin

[1] Letter from C. G. Ehrenberg, 11 March 1846.
[2] *Collected papers* 1: 199–203.
[3] Christian Karl Josias von Bunsen.
[4] Presumably CD kept this letter open until he received the specimens from Joseph Dalton Hooker, sometime after 25 March (see next letter). The specimens were sent to Ehrenberg by 10 April, see letter to J. D. Hooker, 10 April [1846].

From J. D. Hooker [25 March 1846][1]

West Park Kew
Wednesday Night.

My dear Darwin

Thanks to the Railways a few hours has sufficed to transport me to the heart of

Norfolk whither I went for my sisters marriage, & back again. As soon as I possibly can, I hope tomorrow, I will look & see if I have any-thing worth Ehrenberg's having, from Ascension, in the shape of a Grass, including sedge I suppose, though as I did not collect with any idea of having the specimens made such a philosophical use of, my hopes are not high of proving useful.

I owe you for a famous letter & long to be at you with an answer, but shall refrain in full till I look over St Hilaire, whose notions on *affaiblissement* are not very familiar to me, & my own very crude.

When at Mr Bentham's I went through the French Galapago things & find more proof of the imported i.e. non peculiar plants being of a northern origin, in one of them being a decidedly Californian *Baccharis* (Compositæ) which was not in your Herb.. another is a new species of a California genus of Compos. (*Hemizonia*). These being done I am again ready to do a little to my notion of the distrib. of Gal. plants, though alas with hands fuller than ever. I think we are agreed on Polymorphism in the sense we did argue it, & also in that we are now about to treat it under. One of the great objects I had in view in my *notice* above alluded to, was to group the plants according to their derivation & I have a class in reserve for "*apparently peculiar species possibly the altered forms of introduced plants*" It is quite true that in most Islands there is a lot of very dubious species, by no means to be confounded with their country-men, & not polymorphous in the said Island, but woefully near certain continental congeners.

Thus I would divide the Galapago plants into 4 groups 1. Ubiquitous E.G. *Avicennia*— 2. of nearest continent as *Baccharis*. 3^{d} Possibly altered state of continental species, as [2] 4. original creations as *Pleuropetalum* or *Scalesia*. The 3^{d} group may not be a large one in the Galapagos, (according to my notions) but its acknowledged existence is a matter of some importance. In the cases of Madeira, the Canaries & Azores, said group 3^{d} must be very considerable. Such however is the difference of opinion amongst Botanists as to what should or not be a species, that the question in any shape will be a troublesome one, though not on that account to be dismissed unconsidered.

I stumbled on a splendid fact the other day, that the *Lycopodium cernuum* is only found in the immediate neighbourhood of the hot springs in the Azores. When alluding to its distribution at p. 114 of my Flora I dared not mention that it was not known to be an inhabitant of Madeira or the Canaries, as I thought it *must* turn up there, now however I do not expect it & feel sure that the presence of this *torrid* plant in the Azores is due to the hot-springs. What I am most pleased at is the apparent proof of the universal suspension of the sporules of this genus in the air & the consequent strengthening of my hypothesis, that the genus should be decimated *sparing only every tenth*!. Of course it is a strong fact for migration, & for the existence of the impalpable spawn of Fungi &c in all air.

I have been more coolly analyzing the bearings of Forbes Botanical question lately, & with the distressing result, that I fear I must haul out of all participation with him. You will think me unstable as water, & I must blame myself for speaking too much without thinking. It is not from a reconsideration of *his facts* & arguments that my faith is weakened, but from an independent examination of the Flora of the N. Atlantic Isles & W. U. Kingdom which shews, that there are plants in these regions which have been more *put to* in getting there than the Asturias ones' need have been. Such are the American plants *Eriocaulon 7 angulare* in the Hebrides & W. Ireland, An American *Neottia* in S. Ireland & *Trichomanes brevisetum*, in W. Ireland & Madira, all of them American plants not found further E. on continent of Europe or Africa. Also the Gymnogramma Totta a fern of the Cape only in Madeira & Azores & *Myrsine Africana*, which positively skips from the Cape across all intermediate Africa on one side to Abyssinia, & on the other to the Azores!. I hope to be allowed a conversation with Forbes on the subject, for really, with his Sargassum weed &c he is going too far.

Cannot Smith (Jordan Hill) give any information about the Miocene strata of Malta?—[3]

Certainly there is no objection to the hypothesis of a *Sargassum* being an absolute creation, though I see no reason to call for such an aid in this case, the species being in my opinion decidedly the littoral Atlantic one

I was too rash in expressing as I did to you the W. African & Java Floras, as belonging to one & the same region. It is true enough they are disgustingly alike, without being absolutely the same. What I should have said was, that there was no marked Botanical features to seperate them, such as there are to seperate the Cape or Australia from either or one another. From Java to Benin through India the Flora is of the same type. The latter differs only in being rather more American than any country East of it, in wanting many fine Java things, & in poorly representing the most of the rest, whilst it hardly makes up for all these deficiencies by any peculiarity in genera or Nat. Ords. of its own. I think I am right in saying (though I would not print it) that the Benin Flora is more Javanese than the Peruvian is Brazilian. I must enter one caveat, that all we know of the Benin Flora is so much *littoral* that one's judgement can hardly help being warped.

Falconer is as you say a very nice fellow indeed, I do hope he will not be sent again to India with this work of his in hand, his knowledge of Palæontology is very great I believe,

I wonder I never thought of sending you Moq. Tandon, it is a very good *systematic* work on the subject, but will not answer all your ends.[4]

I should be glad to hear your objections to S^t. Hilairs theory, I do not at all subscribe to it, though the fact always appears to me astonishing, that you may cause a young leaf bud to become a flower bud by checking the progress of the sap in it. It is very easy to explain on what sort of grounds Botanists make one class of plants higher, & as easy to prove them futile by their results. I do not however think your objection valid urged on the grounds of Owens observations on organs which are developed in the animal kingdom, but which organs are valueless for systematic purposes if present even in the Vegetable. It is upon the modifications of the sexual organs & their accessories that all the Nat. Orders are defined. The organs of loco-motion afford the Botanist no characters, those of digestion next to none: & the mode after which the various component parts of a compound body (a plant) are arranged is valuable only for the 3 highest groups, Monocot. Dicot. & Acot. & not absolute even amongst these. Generally speaking in Botany highness & lowness are synonymous with complexity & simplicity of structure. I can hardly conceive either simplicity or complexity of one particular organ indicating the rank of a being in the scale of creation.

I believe that many plants shed their pollen in the bud & should have thought that impregnation might take place there too, I don't see why not. Many erect flowers have the stigma exserted long beyond the anthers. & I do not see why impregnation should not in some precede expansion as I think it does in some water plants. I shall keep it in mind. Moq. Tandon shall go to Athenæum with Hort. Soc. Journ. next week[5]

Ever most truly Yours | Jos D Hooker

DAR 104: 188–91

CD ANNOTATIONS
1.1 Thanks . . . through 3.1] *crossed pencil*
6.7 *put to* . . . have been. 6.8] *crossed pencil*; 'transported for greater distances than from the Asturias' *added pencil*
8.1 Certainly . . . expressing 9.1] *crossed pencil*
9.1 W. African & Java] 'Birds closely similar' *added pencil*
9.4 such as . . . another. 9.5] *scored pencil*
9.5 the Cape or Australia] *underl pencil*
10.1 Falconer . . . ends. 11.2] *crossed pencil*
12.10 of loco-motion . . . none: 12.11] *scored pencil*
12.14 I can hardly . . . Hooker 14.1] *crossed pencil*
Top of first page: 'not read carefully' *pencil*

[1] Dated from the relationship with letter to J. D. Hooker, [24 March 1846]. Maria Hooker married Rev. Walter McGilvray on 24 March 1846.
[2] Hooker left a space in the manuscript, presumably intending to insert a name later.
[3] James Smith, who had visited various Mediterranean locations during the winters of 1839–46.
[4] Moquin-Tandon 1841.
[5] *Journal of the Horticultural Society of London*, which included W. Herbert 1846.

To J. D. Hooker [29 March or 5 April 1846][1]

[Down]

My dear Hooker

If you can conveniently send the grasses (if not too heavy for Post) by Wednesday morning I will enclose them with some things of mine & send them through the Prussian Embassy[2] to Ehrenberg.

In Haste, | Yours | C. Darwin

Sunday

DAR 114.2: 58

[1] The dates are those of the two Sundays between the letter from J. D. Hooker, [25 March 1846], and the letter to J. D. Hooker, 10 April [1846].
[2] That is, via Christian Karl Josias von Bunsen, Prussian ambassador in London.

To Smith, Elder & Co. 30 March [1846]

Down Bromley Kent
Monday | March 30th.

Dear Sir

I am much obliged for your note & very clear Agreements, one of which I return signed. I am sorry, (though I hardly expected it) that you can not aid me in my Third Part, which I must publish, as well as I can, by commission with you.—[1]

It is provoking that the rules of the trade will not permit you to sell the Geolog. Parts at reduced prices,[2] for of course putting a new title page, as for a new Edition, would do only for such gentlemen as Mr. Colburn & Co.—[3] Whatever it costs my third Part shall be sold at not more s10. d6.— I must economise in every possible way, for many engravings are wanted for it—[4]

Allow me once again to thank you for the uniformly kind attention which I have on every occasion received from you & believe me dear Sir | Yours very faithfully | C. Darwin

New York Botanical Garden Library (Charles Finney Cox collection)

[1] This refers to *South America*, which is the third part of the geology of the *Beagle* voyage. The publication of *Zoology* had exhausted the government grant of £1000 (see *Correspondence* vol. 2, letter to A. Y. Spearman, 9 October 1843, n. 1), causing CD and the publisher to subsidise the later work (see *South America*, p. iii).
[2] Under an agreement signed in 1829, publishers and booksellers were obliged to maintain a uniform retail price for each title. See J. J. Barnes 1964, p. 1.
[3] For CD's experience with Henry Colburn see letter to John Murray, 12 April [1845]. Despite CD's reference to this 'ungentlemanly' practice, Smith, Elder & Co. went even further than Colburn in 1851, when unsold copies of *Coral reefs*, *Volcanic islands*, and *South America* were then bound as a single volume and reissued with a new title-page at the greatly reduced price of 10*s*. 6*d*. (see Freeman 1977, p. 58).
[4] CD's Account Book (Down House MS) records a number of payments for illustrations for *South America*, including those of 15 December 1845 and 5 July, 19 August, and 4 and 18 September 1846, totalling £47 13*s*. In addition, on 10 August 1847 CD paid a lump sum to Smith, Elder & Co. of £61 8*s*.; on 27 February 1848 he received £5 8*s*. 4*d*. from the same firm.

From G. R. Waterhouse [30 March 1846]

List of Mammalian remains found in the Buenos Ayres district,[1] and purchased by the British Museum[2]

Ord. **Edentata**

Gen. *Megatherium*—remains of at least three if not four individuals. The bones which we possess vary much in size, and yet the smallest do not show any indications of immaturity—possibly the sexes differed in size.

Mylodon robustus—remains of one individual

Genus *Glyptodon*, remains of at least three distinct species— cannot say whether there be more than one individual of each.

Ord. **Pachydermata**

Mastodon audium remains of three individuals

Macrauchenia—a ramus of a lower jaw, with the molar teeth, supposed to belong to this animal—the form of the jaw and teeth greatly resembles that of a Rhinoceros

Toxodon—a complete lower jaw, with many of the teeth,—supposed to differ from the *T. platensis*—, and anterior part of the upper jaw, probably of a different individual— this fragment is in a different condition to the lower jaw, and was imbedded in a blackish earth containing minute particles of iron pyrites(?) like gold dust, a leg bone of a glyptodon & some of the Mastodon remains present the same conditions

Ord. **Carnivora**

Machairodus Kaup— *Smilodon*, & formerly *Hyæna* of Lund	} Great portion of the skull & a nearly perfect lower jaw of the same individual—

I feel no doubt that these parts belonged to an animal specifically identical with one found in the Brazilian caverns— As far as we know, however, the extinct species are generally distinct in the two districts, Brazil & La Plata, or at least those found in the latter quarter are rare in Brazil & vice versa. We *have* a tooth of a Megatherium from the Brazil caverns but in those caves I have found no *Mylodon* remains, a nearly allied animal the *Scelidotherium* was there abundant, judging from the remains—which belong to 2 or 3 species. Our Glyptodons from Brazil are distinct from those of Buenos-Ayres— There must be at *least* 6 or 7 species of *Glyptodon*,

My dear Darwin

I need scarcely say if there are any other points I can help you in, I shall with pleasure do it.

Believe me faithfully yours | Geo. R. Waterhouse

British Museum
Monday morning

Postmark: MR 30 1846
DAR 39.1: 64–5

CD ANNOTATIONS
Top of first page: 'Ch. IV.' *pencil*

[1] The fossils are included in CD's short account of places in the Pampas region where mammiferous remains had been found (see *South America*, pp. 106–7). The account was intended to help geologists investigating the area in future, and included every fossil station then known. With Waterhouse's list is a note by CD, of a tooth of *Toxodon platensis* found near Buenos Aires. This also is cited in the list, and in *South America*, p. 88.

[2] The fossils were purchased in 1845 (*South America*, p. 106). The vendor was Pedro de Angelis, an antiquary and traveller in South America, see British Museum 1904–6, 1: 207.

To George Brettingham Sowerby Jr 31 [March 1846][1]

Down Bromley Kent
Tuesday 31

Dear Sir

I write merely to thank you for your note with the desired information.— As soon as I can find out how many Cordillera shells require illustration, I will immediately decide upon the number of the Tertiary & make arrangements for you to commence.[2]

Yours faithfully | C. Darwin

Postmark: MR 31 1846
American Philosophical Society

[1] The letter has a mourning border, presumably for Elizabeth (Bessy) Wedgwood, Emma's mother, who died on 31 March 1846.

[2] In the appendix of *South America*, Edward Forbes described eleven species of Secondary shells of the Cordillera and George Brettingham Sowerby described sixty species of the Tertiary formations of South America. Both sets are illustrated by G. B. Sowerby Jr.

To Robert Hutton [April 1846][1]

Down Bromley Kent
Wednesday

My dear Sir

I am very much obliged to you for the loan of the Horticultural Journal: I have read Dr. Herbert's journal with interest.[2] I will return the Journal today to the Athenæum Club & I hope it will not inconvenience you my sending it there, instead of direct to your house.— My wife joins me in kind remembrances to Mrs. Hutton & your family.

Believe me, Yours sincerely, with thanks | C. Darwin

American Philosophical Society

[1] Dated from the reference to W. Herbert 1846, which CD was anxious to read, see letter to J. D. Hooker, [8? February 1846]. Robert Hutton was at that time a vice-president of the Geological Society and may have lent CD his copy of the journal on one of CD's trips to London. The letter has a mourning border of the kind used by CD after the death of Elizabeth (Bessy) Wedgwood on 31 March 1846.

[2] CD's notes on W. Herbert 1846 are in DAR 74: 149–150. CD recorded and commented on the observations that bore upon his notions on competition and adaptation. William Herbert's observations on the effects of struggle between plants were later used in *Natural Selection*, pp. 195–6.

To Ernst Dieffenbach 6 April [1846][1]

[Down]

[On the geological work of Tschudi and Buch.] '. . . My health keeps indifferent & I do not suppose I shall ever be a strong man again: everything fatigues me, & I can work but little at my writing: this summer, however, I shall get out my geology of S America . . .

I found Bronns Geschichte,[2] which you recommended me, very useful, for references to facts on variation . . .'

J. A. Stargardt, Marburg (catalogue 574, 11–13 November 1965)

[1] Date taken from the Stargardt catalogue.
[2] Bronn 1841–9.

To J. D. Hooker 10 April [1846]

Down Bromley Kent
April 10th

My dear Hooker

I was much pleased to see & sign your certificate for the Geolog.[1] we shall thus occasionally, I hope, meet.— I have been an ungrateful dog not to have thanked you before this for the cake & books. The children & their betters pronounced the former excellent, & Annie wanted to know, whether it was the gentleman "what played with us so".— I wish we were at a more reasonable distance that Emma & myself c^d have called on Lady Hooker with our congratulations on this occasion.—[2] It was very good of you to put in both numbers of the Hort. Journ: I think D^r. Herbert's article well worth reading.[3] I have been so extravagant as to order M. Tandon, for though I have not found as yet, anything particularly novel or striking, yet I found that I wished to score a good many passages so as to reread them at some future time, & hence have ordered the book.[4] Consequently I hope soon to send back your books.— —I have sent off the Ascension plants through Bunsen to Ehrenberg.—

There was much in your last long letter which interested me much; & I am particularly glad that you are going to attend to *polymorphism* in our last & incorrect sense in your works; I see that it must be most difficult, to take any sort of constant limit for the amount of possible variation. How heartily I do wish that all your works were out & complete; so that I could quietly think over them; I fear the Pacific islands must be far distant in futurity.— I fear indeed that Forbes is going rather too quickly ahead; but we shall soon see all his grounds, as I hear he is now correcting the press on this subject;[5] he has plenty of people who attack him; I see Falconer never loses a chance & it is wonderful how well Forbes stands it.

What a very striking fact is the Bot. relation between Africa & Java; as you now state it, I am pleased rather than disgusted, for it accords capitally with the distribution of the mammifers: only that I judge from your letter that the Cape differs even more markedly, than I had thought, from the rest of Africa & much more than the mammifers do: I am surprised to find how well mammifers & plants seem to accord in their general distribution.—

With respect to my strong objection to Aug. St. Hilaire's language on affaiblissement, it is perhaps hardly rational, & yet he confesses that some of the most vigorous plants in nature have some of their organs struck with this weakness— he does not pretend, of course, that they were ever otherwise in former generations—or that a more vigorously growing plant produces organs less weakened & thus fails in producing its typical structure.— In a plant in a state of *nature*, does cutting off the sap, tend to produce flower buds? I know it does in trees in orchards.—

Owen has been doing some grand work in morphology of the vertebrata: your arm & hand are parts of your head or rather the processes (ie modified ribs) of the occcipital vertebra![6] He gave me a grand lecture on a cod's Head.—[7] By the way would it not strike you as monstrous, if in speaking of the *minute* & lessening jaws, palpi &c of an insect or crustacean, anyone were to say they were produced by the affaiblissement of the less important but larger organs of locomotion.— I see from your letter (though I do not suppose it is worth referring to the subject) that I could not have expressed what I meant when I allowed you to infer that Owens rule of *single* organs being of a higher order than *multiple* organs, applied only to locomotive, &c; it applies to even the most important organ: I do not doubt that he would say the placentata having single wombs, whilst the marsupiata have double ones, is an instance of this law. I believe, however, in most instances where one organ, as a nervous centre or heart, takes places of several, it rises in complexity; but it strikes me as really odd, seeing in this instance eminent Bot: & Zoolog.: starting from reverse grounds.—

Pray kindly bear in mind about impregnation in bud: I have never (for some years having been on the look out) heard of an instance: I have long wished to know how it was in Subularia or some such name which grows on bottoms of Scotch lakes, & likewise in a grassy plant, which lives in brackish water I quite forget name near Thames, which elder Botanists doubted whether it was a Phanerogam.— When we meet I will tell you why I doubt this bud-impregnation—

We are at present in a state of utmost confusion, as we have pulled all our Offices down & are going to rebuild & alter them— I am personally in a state of utmost confusion also, for my cruel wife has persuaded me to leave off snuff for a month & I am most lethargic, stupid & melancholy in consequence. We have just lately had a death in our family, namely my wifes mother: she has, however, long been in such a state of health, that her death was a great relief to herself, & her age was great.—[8]

Farewell | My dear Hooker. | Ever yours | C. Darwin

I know nothing about Henslow's Storm Man.—[9]

NB. You generally spell Henslow, Henslow**e**

Sh^d you ever chance to hear that Dr Herbert has come to town will you kindly inform me.—

DAR 114.2: 59

[1] Hooker was elected a fellow of the Geological Society on 6 May 1846. CD probably received the certificate by post, as he refers to it again in letter to J. D. Hooker, [16 April 1846].

[2] Maria Hooker, whose daughter Maria was married on 24 March, see letter from J. D. Hooker, 2 [March] 1846.

[3] W. Herbert 1846. After asking Hooker to lend him the first number of the *Journal of the Horticultural Society of London* in February (see letter to J. D. Hooker, [8? February 1846]) CD apparently managed to borrow a copy from Robert Hutton, see letter to Robert Hutton, [April 1846].

[4] Moquin-Tandon 1841. CD's annotated copy is in the Darwin Library–CUL.

[5] E. Forbes 1846.

[6] R. Owen 1846b.

[7] R. Owen 1846d, lecture V, pp. 84–129.

[8] Elizabeth (Bessy) Wedgwood died on 31 March 1846 at the age of 82.

[9] In a letter to Hooker, dated 9 March 1846, John Stevens Henslow wrote: 'I have been recommending Darwin to read a most interesting book by Thom— on the cause of Storms' (collection of R. A. Hooker). The reference is to Thom 1845.

From J. D. Hooker [11–15 April 1846][1]

myself entirely on your resources.

I have no time at present to answer your last long & xcellent letter.

Falconer gives me no *specific objections* to Forbes views.

Nothing can be Botanically so strong as the contrast between Cape & Rest of Africa, it is as strong as between Australia & India I should think.

Minute palpi legs & jaws &c cannot be an affaiblissement of legs, for in your old friend Chiasognathus the **legs**, Maxillæ & Palpi are all enormous for the tribe: there are surely plenty of other instances in 〈 〉

〈 〉 made you leave off Snuff if even for a week. It is always astonishing to me that you *can* go on with it; when it is to you so decided a stimulant. Do pray knock it off altogether. I am sure you will be much better if you do; it *must* hurt you, & is growing a 2^d nature.

I have only a steel pen & can hardly form my letters with it.

Ever yours most truly | Jos D Hooker.

Incomplete
DAR 104: 205

CD ANNOTATIONS
1.1 myself . . . views. 3.1] *crossed pencil*
4.1 Nothing . . . think. 4.2] '! ? !' *added pencil*
5.1 Minute . . . Hooker. 8.1] *crossed pencil*

[1] This fragment is the only remaining part of Hooker's reply to the previous letter.

To J. D. Hooker [16 April 1846]

Down Bromley Kent
Thursday

My dear Hooker

It would give me great pleasure to help you even in the construction of a sentence, though if you knew what a bad hand I am in building my own, you would apply to some better workman.— I find I am more disabled than usual in this instance from not knowing the precise facts.— In many respects I like your expression centrifugal, & it is a striking one which is a great advantage; I would use "centrifugally" & avoid the word force.— I doubt more about "centripetal", as it appears that the Gnaphaliums tend to revert to more than one centre or type. I presume in the case of Senecio you actually mean that the species differ in rough proportion to the distance from some one country inhabited by your typical form; if you mean that the groups of species differ in different countries in proportion to their distances apart, I w^{d} certainly altogether avoid "centrifugal", as it irresistibly leads the mind to one type & *tends* to the notion of *one central spot whence the species* have spread; in this case one naturally wishes to know what is your typical form, & what is its country.— The whole case strikes me as eminently curious.— Shall you elsewhere enlarge on Gnaphalium? I do not quite understand why you state that the species *return* in each country to a few typical forms, instead of supposing that the same typical forms have been originally widely spread, & have in each country varied a little.—[1]

I wish with all my heart I could aid you; I am often myself driven half-desperate over a paragraph.— I have made one or two most trifling pencil suggestions: I do not understand what you mean by "its recognized states".—

I shall be proud to append my name to your certificate on Wednesday.—[2] I shall not be able to return your Books quite so quickly as I anticipated, as Bailliere has no copy of M. Tandon.[3]

Ever yours | C. Darwin

Would it not be adviseable when you remark on the confined ranges of species of Senecio, though belonging to a genus, of univers⟨al⟩ diffusion & numerous in species,—to point out why this is remarkable, *viz* in as much as the species of most genera which are large in number & have very wide ranges have themselves wide ranges.—[4]

Postmark: AP 16 1846
DAR 114.2: 60

[1] Hooker was preparing an account for J. D. Hooker 1844–7, p. 309 (*Gnaphalium*) and p. 315 (*Senecio*). See also letter from J. D. Hooker, 1 February 1846.
[2] A reference to Hooker's forthcoming election to the Geological Society (see letter to J. D. Hooker, 10 April [1846], n.1). CD attended a council meeting of the Geological Society on 22 April (Appendix II).
[3] Moquin-Tandon 1841.
[4] CD's suggestion is taken up in J. D. Hooker 1844–7, p. 315 n.

To Richard Owen [21 April 1846][1]

Down Bromley Kent
Tuesday

My dear Owen

I am very anxious to have to have ten minutes talk with you, chiefly about the mammifers of the Plata,[2] & will call on you on Thursday morning at about ten oclock, if you will excuse so early an hour. If you *cannot* see me, perhaps you w^d^ be so kind as to send me a line to save me trouble to, "7 Park St Grosvenor Sq^r^"

I have commenced your British Fossils[3] with very great interest

Most truly yours | C. Darwin

American Philosophical Society

[1] The conjectured date is based on CD's record of having finished reading Owen's *British fossil mammals* (R. Owen 1846c) on 9 May 1846 (DAR 119; Vorzimmer 1977, p. 135) and on a visit to London recorded in CD's Account Book (Down House MS) on Thursday, 23 April.
[2] CD may have sought the interview to go over his summary of 'Localities within the region of the Pampas where great bones have been found' (*South America*, pp. 106–7). See also letter from G. R. Waterhouse, [30 March 1846].
[3] R. Owen 1846c. CD's annotated copy is in the Darwin Library–CUL.

From William Hopkins 27 April 1846

Cambridge
April 27th. 1846—

My dear Sir,

Your letter reached me just as I was leaving Cambridge, I had consequently no time to attend to your geometrico-geological Problem.[1] Since I returned however, a day or two ago, I have re-examined it. It is not possible I find by any such modification as you suggest, to make the results accord with your observations on the Northernmost side of granitic axis. The least inclination that could possibly be given to the *laminated beds* by the anticlinal elevation would be 26° on the above-mentioned side of the ridge. To produce this result, the dip which must be given by the anticlinal elevation itself (of course in a direction perpendicular to the anticlinal ridge) must equal nearly 70°, and the strike of your laminated beds would then be perpendicular and therefore their dip parallel to the anticlinal ridge.

By varying the dip of the anticlinal ridge we can vary the dip of the laminated beds from 26° to any other value up to 90, but then there would be a determinate corresponding position to the line of strike. Thus if we take the least value of the dip (26°) the direction of the dip would be a long way from the North, and if we take our condition such, that the direction of that dip shall be to the North, then the dip will be much greater than your observed dip, on the North side of the anticlinal line.

As far as this goes, it would prove, or at least would render extremely probable that the lamination was produced after the protrusion of the granitic ridge. By means of another hypothesis however, you might obtain a nearer approximation to your observed results. It is this; the geometrical crest of the anticlinal ridge, instead of being taken horizontal must be supposed to ascend in going towards the North-west, in which case the direction of the dip, produced by it instead of being perpendicular to the anticlinal line will incline more towards the East on the Northernmost side of the anticlinal line, with the corresponding change in the direction of the dip in the Southernmost side of the line. By a proper combination of this new element with the anticlinal dip on the North side of the ridge, we can satisfy the two conditions that the dip of the laminated beds shall have an assigned value, and its direction should coincide with the given direction.

This new hypothesis of a deviation from horizontality in the geometrical ridge of the anticlinal line, is frequently true, towards the extremities of anticlinal lines, as well as in the other parts where there may have been certain irregularities in the elevation. This deviation however, can never exceed a few degrees, if it be continued thro' any considerable distance. Now if we take 12° as a mean of your observed dips of the laminated beds on the Northernmost side of the anticlinal line it would require, that the deviation from horizontality just mentioned should be *more than* 14° probably not less than 20 in order to satisfy your observations. This amount I conceive to be utterly inadmissible so that I do not see how you can by any admissible hypothesis, account for your observed phenomena in the way you have suggested.

I have used the term *geometrical ridge* or *crest*; it may be considered as the line in which the two parts of the same bed respectively on opposite sides of the anticlinal line would meet if produced, assuming as an approximative case, that these two portions of the bed tho' inclined to the horizon are still plane or *flat.* The additional hypothesis above mentioned complicates the problem considerably especially in obtaining numerical results. I have not therefore worked out an example with any assumed numerical data. If however you think the additional hypothesis admissable within certain assigned limits, I will work you out an example which will give the best approximation I can make to your observations.

The problem thus generalised is one of great geometrical complexity; but still with three or four pieces of pasteboard, I could in ten minutes interview give you a distinct conception of the problem. The device I before suggested to you is only calculated to give a somewhat rough conception of the problem, and I am not sure on looking at it again whether the numerical values of the angles there given, would make the case very approximate to that which you first proposed to me I should not chose it as the means of explaining the problem generally to any one uninitiated into the mysteries of geometry. The problem is an important one in geological elevations, and occured to myself a considerable time ago, tho' I have had no occasion to make any exposition of it till your application to me.

AL
DAR 39.1: 54–6

CD ANNOTATIONS
On cover: 'Hopkins on axis. interesting Laminate District' *ink*

[1] The letter continues the discussion of the problem dealt with in letter from William Hopkins, 3 March 1845. CD sought to explain the anomalous strike and dip of mica-schist formations in the Chonos Archipelago. See *South America*, pp. 158–9.

To J. D. Hooker [May 1846]

Down *Farnborough* Kent
Thursday

My dear Hooker

I write merely to say that I this day have sent off by the Kew Boat, M. Tandon & the two nos of Hort. Journal.—[1] I shd. think more of Tandon, if his arrangement & a good many of his ideas & perhaps conclusions had not been copied from Is. St. Hilaires Animal Teratologie.—[2] I observe that he says at Thoulouse every year alpine plants are brought into the Bot. Garden.— Do you ever correspond with or know him? I shd. like to hear something about these alpine plants; Linnæus, I remember, says they are generally sterile in lowland gardens; & I shd be curious to know how this is;[3] whether the pollen is bad as it is in some cases or whether the fruit after setting fails. What fine opportunities M. Tandon would have in trying whether any alpine *varieties* of lowland plants have acquired any hereditary qualities.

Thanks for Hopkirk;[4] by an odd chance I have had this name in my note-book, to look at in Brit. Mus. for the last six months; so I shall be very glad to see what it is about.— What fellows these Germans are; I heard of Hopkirk, in a reference to Bronn's Gesicckte![5] By the way I hope some future year to get some information (which I am rather curious about) at Kew, about what plants being healthy yet are sterile in cultivation: some of the head gardeners, I daresay by walking about could call my attention to what plants will seed & what won't: I find the Pollen often affected in cultivated plants.—

I am quite delighted to hear how systematically you are going through the individual powers of transport of the Galapagos plants; I have often wished to see this done, & I have never met with such a discussion.— (Mem: there is one N. American or Mexican bird at the Galapagos.)[6]

What a pity that the shells are different on opposite sides of Panama (as Cuming declares: I wish I had cross-questioned him closely on this point) for if they had not been so, how easy it would have been to have broken down ie not elevated the isthmus & so procured new & perhaps southerly currents. Remember the only bad weather sets in, with great rollers from the north, but I fear it is not accompanied by heavy gales of wind. How interesting does the problem become, when your exact knowledge drives one to speculate on a particular course of migration & not from America in the whole, as I have always looked at it.—

Where are Petit Thouars observations; I sh^{d} like to see them?

I had never heard a word except from Forbes, about Edmondstone,[7] & am grieved at (but will never repeat) what you say: he sent the other day a long letter to Forbes with nothing original in it. I urged him by letter to collect everything at the Galapagos, & attend particularly to the productions of the *different* islands.

I sh^{d} like sometime to hear what you think of Dieffenbach. (whom I saw the other day) I never know what to think of his abilities, & rather fear they are less than his zeal: he seems very poor, & I cannot think on what he can live: he is in poor lodgings at 62 S. Molton St.—

Farewell my dear Hooker—with many thanks for all the books you have lent me— How curious I shall be for the Galapagos Paper: I have one or two numbers of your A. Flora unread & intend to let one or two more accumulate before I do read them as that gives me more satisfaction.

Yours Ever | C. D.

DAR 114.2: 61–61b

[1] Moquin-Tandon 1841 and the first two numbers of the *Journal of the Horticultural Society of London*, which included W. Herbert 1846. See letters to J. D. Hooker, [10 February 1846] and 10 April [1846].

[2] I. Geoffroy Saint-Hilaire 1832–7. CD possessed a copy of this work (Darwin Library–CUL).

[3] Linnaeus 1741. According to his reading notebook, CD read this and other articles from the *Kongliga Swenska Wetenskaps Academiens handlingar* in manuscript translation at Maer, the Wedgwood family home, during a visit there from 10 June to 14 November 1840: 'Sweedish Philosoph. Acts. vol 1 to 7. M.S. Translat.— from 1740.' (DAR 119; Vorzimmer 1977, p. 124).

[4] Hopkirk 1817.

[5] Bronn 1841–9.

[6] *Dolichonyx oryzivorus*, a lark-like finch (*Birds*, p. 106).

[7] Thomas Edmondston, naturalist on board H.M.S. *Herald*.

From William Hopkins 5 May 1846

Cambridge
May 5— 46

My dear Sir

The additional datum you sent to me yesterday is of no material service in your problem.[1] If your observations were sufficient to establish a general regularity in the lamination over a sufficiently wide area, (and such I think appears to be your conviction) we can scarcely resist the conclusion that the local irregularities in the laminated beds in the immediate neighbourhood of the granitic ridge (where *alone* I presume you found them) are due to the upheaval of that ridge. Allowing the validity of the conclusion, I should be disposed to attribute at least a considerable portion of the discordance between observed & calculated results to irregularities in the elevation of the ridge such as might arise for instance from violent *lateral* thrusts, the effects of which might be resisted by the flanks of the mass in an uncertain and irregular manner, and which might thus produce effects which it would be impossible to bring under the dominion of calculation. A good deal may safely be allowed to irregular action of this kind.

Let me again assure you that your problem has been *no trouble* to me. In a subject like Geology where it is impossible for any one to be equally acquainted with every branch of it, I hold it to be a *duty* to assist each other, and on that principle you will always find me ready to act.

Your's very truly | W Hopkins

DAR 39.1: 57–8

[1] CD evidently responded promptly to the letter from William Hopkins, 27 April 1846. As *South America*, p. 159, makes clear, CD still hoped his alternative hypothesis had some basis.

To the Admiralty 9 [May 1846][1]

Down near Bromley | Kent
Saturday 9th.

My dear Sir

I should be very much obliged if you would kindly forward the accompanying letter to Sulivan in the Philomel,[2] & should you be sending *any package*, also, the accompanying, paper,[3] which relates to a subject, on which he has made some observations at the Falklands.—

I hope you will excuse me troubling you & believe me | Yours truly obliged | Charles Darwin.

Houghton Library, Harvard University

[1] 9 May was the only Saturday the 9th in 1846.
[2] H.M.S. *Philomel*, then returning to England, arrived 19 June 1846 (Log of *Philomel*, Public Record Office, ADM 53/1025).
[3] 'On the geology of the Falkland Islands' (read 25 March 1846), *Collected papers* 1: 203–12.

From George Grey[1] 10 May 1846

Govt. House. Auckland. New Zealand.
May 10th. 1846.

My dear Sir.

The enclosed note which I believe bears your signature, having been mysteriously sent to me (by whom & for what purpose remains unexplained) I have thought it proper to mention the circumstance to you and at the same time to return the note.[2]

I ought perhaps to apologize for having read it, but it was so folded that my own name first caught my eye—and I concluded therefore that it had been sent to me with the intention that I should peruse it—[3]

Believe me my dear Sir. | faithfully yours. | G. Grey.

Charles Darwin Esq. | &c. &c. &c.

Copy
DAR 144 (FitzRoy letters)

[1] Grey was Robert FitzRoy's successor as governor of New Zealand.
[2] For the text of the enclosure, see CD's letter to J. L. Stokes, [November–December 1845]. See also letter to J. L. Stokes, 3 November 1846, and letter from J. L. Stokes, 6 November 1846.
[3] John Lort Stokes and CD decided that the note had become mixed in with the proof-sheets of Stokes 1846 and sent to the printers by mistake, from whence it was forwarded to Grey by someone unknown, see CD's letter to J. L. Stokes, [*c.* 26 November 1846]. See also letters to George Grey, 10 November 1846, and to Robert FitzRoy, 23 November [1846]. CD refers to the incident again in *Correspondence* vol. 4, letter to George Grey, 13 November 1847.

To J. D. Hooker [19 May 1846][1]

Down Farnborough Kent
Tuesday Evening

My dear Hooker

It has just flashed across me suddenly, that I brought home a very few plants in Spirits of Wine (with the colours noted) namely some sea-weeds, & 2 Orchideous plants from shady parts of Forests of T. del Fuego—a Calceolaria from Elizabeth Is^{d} St of Magellan (which at the time I thought a wonderful production of nature!) & a salt-plant from near a Salina at Port St Julians in same jar with Opuntia Darwinii from do.— Has Henslow ever given you these? He is now in Cambridge & c^{d} probably find them (if you have not seen them & would like them) & this is the reason I write today, though not well— I go to London for a few days tomorrow.

I received the other day another number of your Antarctic Work, & I have now 4 or 5 to read: Hopkirk did not come: I mention this **not at all** as wanting immediately but in case of any accident. Have you seen Bunbury's Paper in Last number of Geolog. Journal on American Coal-plants: your observations on uniformity of Floras on W. coast of S. America under equable climate bears on his remarks,[2] & I will point them out to him; that is, if I am not confounding one of your letters, so valuable to me, with your published remarks.—

Many thanks for your most critical & scientific, & I don't doubt true specific character of Dieffenbach; but I am sorry you took so much trouble about it: I asked chiefly out of simple curiosity, & partly from having been urged to recommend him as Naturalist on any occasion which might turn up, & I was quite unable to make up my mind about him.— There has always struck me as a want of originality in him.—

I am delighted that you are in the Field, geologising or palæontologising:[3] I beg you to read the two Rogers' account of the Coal-fields of N. America; in my opinion they are **eminently** instructive & suggestive: I can lend you their resumè of their own labours & indeed I do not know that their work is yet published in full.[4] L. Horner gives a capital balance of difficulties on the Coal-Theory in his last Anniversary Address, which, if you have not read, will, I think interest you.—[5] In a paper just read an Author throws out the idea that the Sigillaria was an aquatic plant, I suppose a Cycad-Conifer with the habits of the Mangrove.—[6] From *simple* Geological reasoning, I have for some time been led to suspect, that the great (& great & difficult it is) problem of the Coal would be solved on the theory of the **upright** plants having been aquatic— But even on such, I presume improbable notion, there are, as it strikes me, immense difficulties; & none greater than the width of the coal-fields. On what kind of coast or land could the plants have lived? It is a grand problem, & I trust you will grapple with it: I shall like much to have some discussion with you. When will you come here again?

I am very sorry to infer from your letter that your Sister[7] has been ill.

Ever yours | My dear Hooker | C. Darwin

I have heard today that Lyell has found Cheirotherium footsteps in true coal-measures Palæozoic rocks of America!!! Hurrah![8]

DAR 114.2: 62–62b

[1] Dated on the basis of an entry of 22 May 1846 in CD's Account Book (Down House MS) recording payment for a trip to London. The visit coincided with a meeting of the Geological Society on Wednesday 20 May (see n. 8, below).

[2] C. J. F. Bunbury 1846, see especially p. 87.

[3] Hooker, like the other naturalists on the Geological Survey, was carrying out fieldwork during the summer. He was in South Wales in May and June, examining the coal-beds for fossil plants *in situ* (Huxley ed. 1918, 1: 210).

[4] CD refers to a set of extracts from papers by William Barton Rogers and Henry Darwin Rogers in the *Transactions of the Association of American Geologists and Naturalists* for 1843, which are now in the Darwin Pamphlet Collection–CUL. They include Rogers and Rogers 1843.

[5] Horner 1846, pp. 170–81.

[6] Binney 1846, p. 393. Edward William Binney's paper was read at a Geological Society meeting on 22 April 1846. CD attended this meeting (Appendix II).

[7] Elizabeth Hooker.

[8] Charles Lyell wrote to Leonard Horner from Philadelphia on 27 April 1846 about the discovery. It was announced on 20 May at the Geological Society meeting of that date (see C. Lyell 1846a and K. M. Lyell ed. 1881, 2: 102–3). CD attended the meeting (Appendix II).

To John Higgins 27 May [1846][1]

Down Farnborough Kent
May 27th

Dear Sir

I write to acknowledge the draft for 190£· 13s· 1d & to thank you for it.—[2]

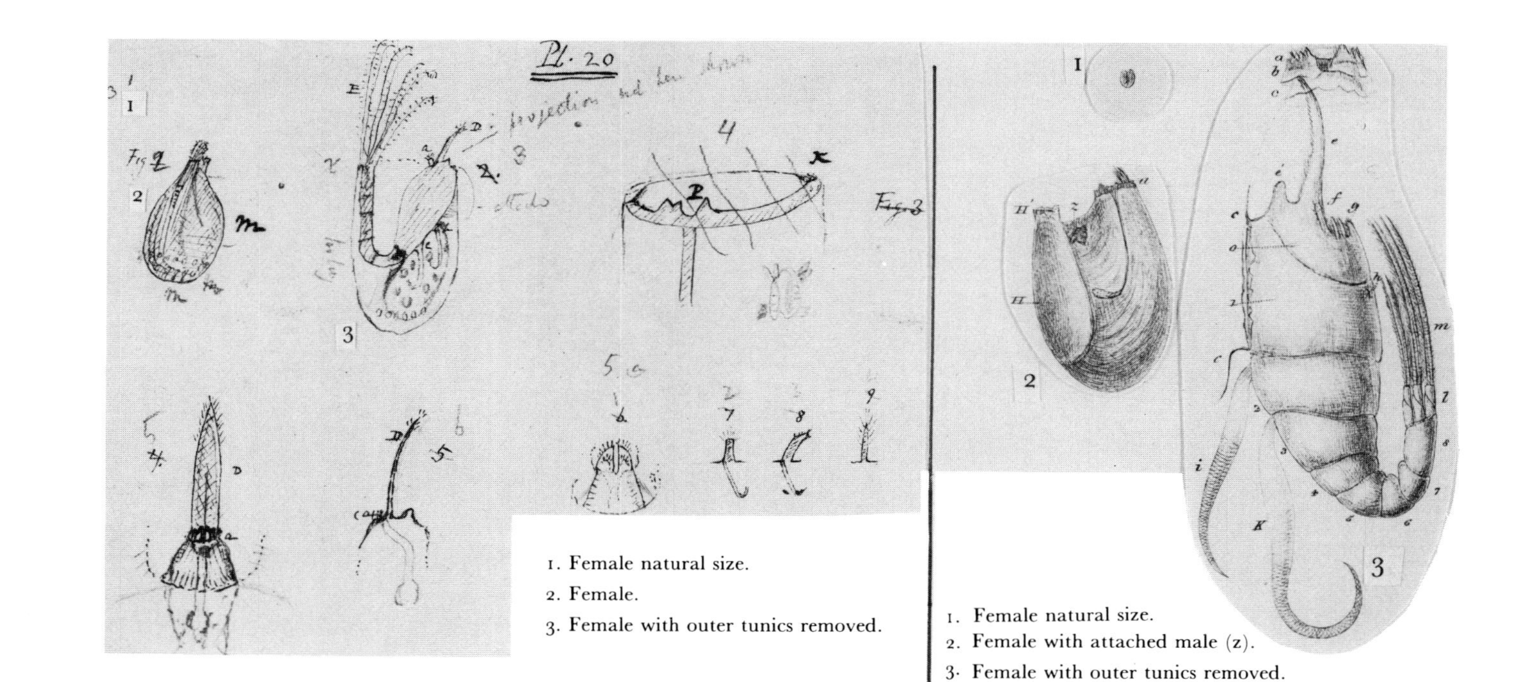

Drawing of *Arthrobalanus* (*Cryptophialus minutus*) made by Darwin during the *Beagle* voyage (DAR 29.3: 72).

Figures of *Cryptophialus minutus* from *Living Cirripedia* (1854) plate XXIII.

Alcide d'Orbigny. Daguerreotype, 1843.

(Published in Heron-Allen, Edward, 'Alcide d'Orbigny: his life and his work', *Journal of the Royal Microscopical Society* (1917): 1–109. By permission of the Syndics of the Cambridge University Library.)

I am very glad Mr Hardy[3] is going on well.

Will you be so good as to give me as long a notice as you conveniently can, when you want the money for the new Farm House, as my Father owing to some arrangements has been rather short of money this Spring, but I apprehend, very soon now all times will be equally convenient to him.—[4]

My Bankers are Mrs Robarts, Curtis & Co & I sh^{d} be glad, if, as you propose, you would pay the rent direct into their hands for the future, being so good as to notify me of the same.—

I wish I could have the pleasure of seeing you this week, but I shall not be up in London.

Dear Sir | Yours very faithfully | C. Darwin

P.S. | Do not forget the steeper pitch of roof in the new farm House & oak Lintels.

Lincolnshire Record Office

[1] The date is based on an entry of that date in CD's Account Book (Down House MS).

[2] The half-yearly rent for Beesby Farm, less Higgins' percentage as agent (CD's Investment Book (Down House MS), p. 20). For the purchase of the farm, see letters from John Higgins, 15 March 1845 and 2 October 1845.

[3] Ralph Hardy, contractor, who pulled down the old farm buildings at Beesby (R. W. Darwin's Investment Book (Down House MS), p. 74).

[4] CD had already borrowed £12,577 4*s.* from his father to purchase the farm (CD's Investment Book, p. 19). Robert Waring Darwin advanced an additional £1,026 0*s.* 2*d.* to cover the cost of farm buildings. See CD's Investment Book, p. 19, and R. W. Darwin's Investment Book, p. 74 (Down House MSS).

From A. C. V. D. d'Orbigny [June–July 1846]

"In order to arrive at satisfactory results in an undertaking so vast,[1] I shall need the assistance of all persons who are interested in the advance of geology, and I trust to your procuring for me in England many correspondents willing to exchange the fossil shells of various formations for such portions of my works on Palæontology as they may most require. I wish, for instance, to obtain, 1st, fossils of the Crag and London Clay, and 2nd, those of the Cretaceous formations of the Isle of Wight and Blackdown; but, above all, the fossils of your Carboniferous, Devonian and Silurian beds."

Incomplete
Quarterly Journal of the Geological Society of London 2 (1846): 59

[1] The excerpt appeared in a note headed 'D'Orbigny's "Paléontologie Universelle" ' announcing that Orbigny 'has undertaken to prepare a work under the above title, in which he proposes to give figures, with accompanying descriptions, of *every species* of fossil shells hitherto determined, from all geological formations in all parts of the world.' It is also stated that the excerpt is from a letter from Orbigny 'Addressed to Mr. Darwin.'

From Searles Valentine Wood 5 June 1846

28 Fortess Terrace | Kentish Town

My dear Sir

Accept my thanks for your kind communication which is so far satisfactory that the experiment has been tried altho with a result contrary to what I sho^d. have expected. a casual deviation from the normal form is easily understood which may arise from an accidental interruption of Natures Laws but for that cause to be continually in operation is not so explicable

In regard to your question respecting variations among the Mollusca[1] I can only say that as far as my observations have gone I think their range is greater among fossils than in recent species & the most extraordinary varieties are from the Mam: Crag[2] among estuary shells these are in some instances of extraordinary proportions removing the extremes of variation far beyond what is generally considered the limit of a species these variations (or perhaps more properly distortions) consist principally in the elongation or shortening of the spiral cone beyond its general proportions without affecting in any way the form of the aperture consequently there need be little or no alteration in the animal inhabitant & this I have always considered as depending upon an external cause & more particularly with regard to the Crag shells whose location was in an estuary where a sudden alteration or reduction of temperature may have partially paralized or injured the natural powers of the animal. the shells of the present day from similar localities do not appear to be similarly affected which I wo^d. attribute to a more uniform or rather a less variable temperature

I have been hunting largely among the Freshwater & Estuary deposits of the Eocene Period in Hampshire but I have not met with the like variations among their inhabitants which I wo^d attribute to the same cause & that which conduces to a healthy condition of the animal whatever may be its locality whether on land or in the water will be favourable to its uniformity[3] the Helix aspersa for example where it is in great profusion & flourishes well thousands may be examined without perceiving the slightest deviation in form

I do not therefore at all agree with Mr Forbes if he says the cause is an internal one which I cannot well understand the alteration in the shell is of course caused by the animal but the alteration in the animal is I believe produced by a cause exterior to itself either from food or temperature

I must apologize for this apparent delay in answering your note but I only obtained it a short time since for my visits to London are few & far between & your note I was told had been long at Somerset House.[4]

Believe me Dear Sir | Yours truly | Searles Wood

June 5. 1846.

DAR 181

CD ANNOTATIONS
1.1 Accept . . . question 2.1] *crossed pencil*
3.6 profusion . . . form 3.7] *scored pencil*
4.2 understand the] *altered in pencil to* 'understand. The'
Top of first page: 'Mr. W. told me [*over illeg*] that common Periwinkle is more variable in Crag than in living state.' *ink*

1 CD's query to Wood has not been located; however, CD appears to have asked about information received from Edward Forbes regarding variation in molluscs (*Natural selection*, p. 106). Wood was an expert on the fossils of the East Anglian Crag.
2 Mammaliferous crag of Norfolk. This observation by Wood was later cited by CD in *Natural selection*, p. 106.
3 Noted in *Natural selection*, p. 107.
4 The address of the Geological Society.

To Richard Owen 21 [June 1846]

Down Farnborough Kent
Sunday 21

My dear Owen

I have just heard from Capt. Sulivan R.N. that he is arrived in London with six casks of fossil bones from the *southern* part of Patagonia, as I before mentioned to you.—[1]

He is anxious to have them inspected by you & I shd. be extremely glad to be present. I send this note open through him to make any alterations. I have suggested to him to send the casks direct to the College of Surgeons, & perhaps you could direct some one to get them unpacked, & ready for inspection. He proposes to call with me on you on Monday next The 29th— at Two oclock if, as I hope, that may suit you.—[2]

He has not made up his mind to what Public Body he will present these fossils, but I apprehend you will not object to receiving them at the College temporarily, & no doubt there will be some duplicates: in case he shd. think the British Museum better than the College[3]

Yours very truly | C. Darwin

British Museum (Natural History), General Library (Owen correspondence).

1 Bartholomew James Sulivan had arrived from Montevideo on 19 June. The fossils are those collected at Rio Gallegos, Patagonia during the survey voyage of H.M.S. *Philomel* and described in Sulivan's letters of 13 January – 12 February and 4 July 1845.
2 CD left spaces in this sentence and, in accordance with CD's instructions at the beginning of the paragraph, presumably B. J. Sulivan filled in 'Monday next The 29th—' and 'Two'.
3 The fossils were presented to the Royal College of Surgeons and described by Owen at the 1846 British Association meeting in Southampton, see letter from B. J. Sulivan, 13 January – 12 February 1845, n. 4.

To William Crawford Williamson 23 June [1846][1]

Down Farnborough Kent
June 23d

Dear Sir

Absence from home has prevented me answering your letter of the 14th more promptly.—[2] Unfortunately I did not keep any list of the specimens, which I sent you, & cannot charge my memory whence they all came, & therefore am not able very fully to answer your queries.[3]

I could perceive no evidence of any particular chemical action in the Tertiary strata of S. America, certainly there has been no metamorphic action. The fossils are almost invariably calcareous & well preserved; but I do not think the strata, of which specimens were sent you, were those which contained fossil shells: I did not know your object & picked out specimens, which appeared to my eye most likely to contain infusoria.—[4] The white pumiceous mudstone from Patagonia contains no shells.— Most of such strata, as I sent you, have, I believe, resulted primarily in volcanic action, that is either from erupted ashes, or triturated volcanic rocks.—[5]

I am much surprised at what you say about the abundance of the siliceous matter; my impression was a different one.— You refer to a 'Tufaceous Layer' & speak of each fragment being of siliceous matter: is this Tufaceous layer, marked R. Negro? if so, I think, you will find that the whole is easily fusible & therefore cannot be silica; but I do not know what microscopical test you have to distinguish glassy feldspar & silica.—[6] I think I sent specimens from St. Fe,[7] I have no reason to suppose that they have originated in volcanic action; & the beds are associated with others abounding with calcareous fossil shells.— In many of the Tertiary formations of S. America, the fossils occur in sandstone concretions, which have been formed by the aggregation of calcareous matter.—

I wish it was in my power to give you more satisfactory information.

Believe me dear Sir | your's faithfully | C. Darwin

Missouri Botanical Garden Library

[1] The date is based on an endorsement '1846 (W. C. W.)' by Williamson.

[2] CD recorded payment for a trip to London in his Account Book (Down House MS) in an entry dated 17 June 1846.

[3] Geological specimens from the Tertiary strata of Patagonia. They were sent to Williamson by CD in connection with the rewriting and enlargement of a paper read by Williamson on 4 November 1845, but not published at that time. The final revised version of the paper included comments made by CD and references to his South American specimens (Williamson 1848, pp. 66–7). A separate, advance printing of Williamson's paper dated 1847 is in the Darwin Libary–Down.

[4] Williamson found no infusoria in any of the specimens, a finding confirmed by William Benjamin Carpenter (Williamson 1848, pp. 66, 128 'addenda').

[5] See letter to C. G. Ehrenberg, 23 March [1845].

[6] Williamson's original observation was corrected in the published version: 'This specimen appears to contain neither Polythalamia nor siliceous organisms' (Williamson 1848, p. 67). He noted that the specimen from Rio Negro was glassy feldspar (Williamson 1848, p. 67 n.).

[7] Santa Fé Bajada. CD discussed the fossil deposits in *Journal of researches* 2d ed., pp. 129–30. Williamson described both the limestone and the calcareous marl of the Pampas, noting that they had been altered by some agency after their deposition (Williamson 1848, pp. 94, 109).

To Emma Darwin [24 June 1846][1]

[Down]
Wednesday

My dearest old Soul

I was exceedingly glad to get your letter, with so wonderfully good an account of your voyage & of the dear little souls happiness;[2] I am glad you took them. Do you not think you had better come back by land? & had you better not stay more than a fortnight, I propose it to you in *bonâ fide* & wish you to do so, though I do long to have mine own wife back again. Yesterday was gloomy & stormy; I was sick in middle of day, but two pills of opium righted me surprisingly afterwards: however I was extremely glad that Sulivan did not make his appearance.[3] The house is getting on well,[4] though Lewis[5] had a quarrel & turned off all his carpenters: Lucy[6] was very goodnatured & took keen interest about one man, whose wife has come from a distance with a Baby & is taken very ill— The poor man was crying with misery, but we have persuaded Lewis to take him back again.

At last the flower garden is looking gay.—

I have been getting on very badly with my work as it has been extremely difficult & I have had so many letters to write.—

Etty[7] was very charming, though I did not see much of her yesterday; she is very affectionate to her dolls, but at last got tired of them, & declared with great emphasis, that "she *would* have a real live Baby" & "Mamma *shall* buy one for me"— I asked to send a message to you, "say A. B. S, say, big woman in little letter"

Give my very best love to all at Penailly I enclose A. Sarah's[8] letter; I have strongly recommended to bring Henry, but can not repeat all reasons: I have spoken doubtfully about Horse & Phaeton.[9]

Goodbye, my own old dearest. Kiss the children for me. Etty often talks about them. | Your affect. | C. D.

DAR 210.19

[1] The date is based on Emma's notation 'I at Tenby' and the notation on letter to Emma Darwin, [25 June 1846], 'June 1846 to me at Tenby'. The date also accords with the end of Bartholomew James Sulivan's voyage on 19 June 1846.

[2] Emma took William and Anne Darwin to Tenby, a seaside resort and fishing village in Wales. Several of Emma's aunts lived there: Emma and Fanny Allen, Jessie Sismondi, and Harriet Surtees. Emma and the children stayed for eighteen days, returning home by 7 July. See CD's Account Book (Down House MS) 7 July 1846 and letter to Emma Darwin, [25 June 1846].

[3] Sulivan had just returned from South America. Possibly CD had invited Sulivan to come to Down before meeting him in London on 29 June. See letter to Richard Owen, 21 [June 1846].

[4] According to CD's Account Book (Down House MS) payments of £150 and £199 for alterations were made to 'Mr Laslett' on 13 June and 27 August 1846. Isaac Laslett was the bricklayer in Down.

[5] John Lewis. See letter to Emma Darwin, [3–4 February 1845], n. 5.

[6] Presumably a servant at Down House.

[7] Henrietta Emma, who was almost three years old.

[8] Sarah Elizabeth (Sarah) Wedgwood, CD and Emma's aunt.

[9] Probably a reference to Sarah Wedgwood's plans to move to Down in 1847. Here she lived the life of a recluse and, according to Henrietta Litchfield, 'her horse and phaeton seemed to be kept entirely for our service' (*Emma Darwin* 2: 105). Henry Hemmings was one of Sarah Wedgwood's servants (*Emma Darwin* 2: 106).

To Emma Darwin [25 June 1846][1]

[Down]
Thursday afternoon

My dear wife

Today has been stormy & gloomy, but rather pleasant in the intervals, only I have been stomachy & sick again, but **not** very uncomfortable; I will take blue-pill again. A proof has come from the Printers & saying the Compositor is in want of M.S. which he cannot have & I am tired & overdone[2] I am ungracious old dog to howl, for I have been sitting in summer-house, whilst watching the thunder-storms, & thinking what a fortunate man I am, so well off in worldly circumstances, with such dear little children, & such a Trotty,[3] & far more than all with such a wife. Often have I thought over Elizabeths words, when I married you, that she had never heard a word pass your lips, which she had rather not have been uttered, and sure am I that I can now say so & shall say so on my death-bed,—bless you my dear wife.—[4]

Your very long letter of Monday has delighted me, with all the particulars about the children—how happy they seem: I will forward it to Caroline, though twice it has "my dearest N.".—[5] Trotty is quite charming, though I am vexed how little I am able to stand her: somehow I have been extra bothered & busy; this morning I sent off five letters.—

Lady L.[6] has asked me to meet on Saturday the old Griffin & the Browns, & I have accepted it doubtfully, though I do not think I shall have the heart to go.—

Remember I go to London on Monday— You do not say how Jane is.—

Trotty has just said "that rascal has not gone into Garden"—so I asked whom do you mean? "Georgy, cause he p^{s} so"

Your affect. | C. D.

DAR 210.19

[1] A notation on the letter in Emma's hand reads 'June 1846 when I was at Tenby'. CD's reference to a London trip can only mean that of 29 June recorded in his Account Book (Down House MS) on 1 July.

[2] The early parts of *South America* were being set up at this time.

[3] Henrietta Emma Darwin. The nickname 'Trotty' was taken from the nickname for Toby Veck in Charles Dickens' *The chimes* (London, 1845 [1844]).
[4] A sentiment repeated in *Autobiography*, p. 96.
[5] CD referred to himself as Emma's 'Nigger'. The letter was presumably sent on to Caroline Wedgwood, CD's sister, who had married Emma's brother Josiah.
[6] Harriet Lubbock.

To J. D. Hooker [8 or 15 July 1846][1]

Down Farnborough Kent
Wednesday

My dear Hooker

I have been a shamefully bad correspondent; but I have not been quite so well as usual of late & have been overworked in trying to finish my S. American geology, of which I am inexpressibly wearied. We are extremely much obliged to you & to Sir W. & Lady Hooker for your invitation, which I had hoped much to accept in the Spring, but fear must now give up. My father's health is rather failing & I must go there the last day of this month[2] & I have much work to do in the interval, & have just recommenced a course of Galvanism, wh. I sh^{d} be sorry to break through, as I have had a good deal more sickness than usual. My wife, moreover, has just returned from 18 days absence at Tenby.[3] I am very sorry to give it up, as we both looked forward to the visit with **much** pleasure. Shall you be at Southhampton?[4] I have some feeble thoughts of it; I sh^{d} like much to spin some scientific yarns (as you express it) with you.—

Now for your letter; by all means trust FitzRoys measurements in the Appendix,[5] the difference I have no doubt arises from recalculation with more accurate data for refraction &c.— With respect to Antuco[6] I know nothing; the little I know on snow-line I have given in my 1st Edit: of Journal.—[7]

I was very sorry to hear about poor Edmonston. The Galapagos seems a perennial source of new things; I hope you know which islands he visited.[8]

I was having some talk with Lyell about coal, when in London:[9] his fossils from Alabama being most of them identical in species according to Bunbury with the coal-plants 20° degrees N. in Europe, is an interesting fact: I told him of your remarks on the equability of climate & wide extension & he has quoted them in a Paper just despatched for Silliman's Journal.[10] The more I think on coal, the more utterly perplexed the subject appears to me. He finds the Oolitic coal resting on granite. with no underclay &c.—

I received some time since & finished Hopkirk: there is very little in it, & I will return it soon.—

I have just finished your late numbers of the A. Flora & have been in truth delighted with them: I read a good many books, but I know none, which are so suggestive as your's;— I refer, of course, to your generalizations, such as your discussion under Myrtaceæ,[11] which interested me particularly, & even more your discussion on numbers of individuals & species: your conclusions on this latter point have surprised me much.[12] Long life to you & may your Book extend to a

100 numbers— By the way, you cannot think how proud I am at seeing how many species I collected: it has often been a vexation to me, how much trouble I threw away on some collections, amongst which I formerly ranked my plants, but now they are a real source of pleasure to me.

There is one point of detail in works, like yours or the Zoology of the Beagle, which I have often regretted; namely that there is not some conventional means of showing the *general* Habitat, from the *particular* habitat of the specimens under description: thus you sometimes put, "Habitat, Chonos Isl^{d}." & in turning over the page & reading your *remarks* I find it also inhabits Chiloe & Chile &c.—

Farewell, I shall be very glad to have some talk with you again, till we meet goodbye. | Ever yours | C. Darwin

DAR 114.2: 63

[1] Dated by the reference to Emma having just returned from Tenby, see n. 3, below.
[2] CD went to Shrewsbury from 31 July to 9 August ('Journal'; Appendix II).
[3] Emma, William, and Anne Darwin were in Tenby from 19 June to 7 July 1846, see letter to Emma Darwin, [24 June 1846], n. 2.
[4] The British Association was to meet in Southampton, 10–16 September 1846.
[5] *Narrative*, appendix to volume two.
[6] A volcano near Concepción, Chile. See *Journal of researches*, p. 374.
[7] *Journal of researches*, p. 275.
[8] Thomas Edmondston was accidently shot dead on 24 January 1846 in Peru, shortly after visiting the Galápagos on board H.M.S. *Herald*.
[9] CD recorded the expenses of a trip to London in his Account Book (Down House MS) on 1 July 1846.
[10] C. Lyell 1846b, quoting Hooker on p. 230.
[11] J. D. Hooker 1844–7, pp. 275–6, which discusses the relationships between floras of the southern hemisphere. This passage is extensively scored in CD's copy (Darwin Library–CUL).
[12] J. D. Hooker 1844–7, pp. 277–8, in which Hooker claimed that the area that contained most species of a particular family or genus did not necessarily carry a large number of individuals. In his copy CD has noted: 'very odd ?once it was different | it may be if all individuals of all the species be counted' (p. 277).

To Reeve Brothers [August 1846][1]

Down: Farnborough Kent.
Saturday

Sir

The section[2] will do very well: there are a few trifling corrections. The writing is very bad; I do not allude to the mistakes which my bad-writing may excuse, though I never before had half so many made; but to the size of the lettering varying in every third & fourth name. Please just to look in No^{r} 2. at "Valley of Horcones" & at lower right hand corner at "Feldspathic Clay-slate"— Would you be so good as to look at No^{r}. 2. & consider what plan is best to separate the letters of reference from the names below.— In No 3^{d}. I sh^{d}. think brackets would keep the names together best.

The paper must be cut close (as close as I have done on left side) to your border, otherwise it will not go into my volume.—

With respect to quality of Paper it must be *thin* for folding & yet take colour: I enclose a scrap, wh. answered well for copper in a former volume, but I must leave it to you to settle for stone.—

500 to be printed only 250 to be coloured.—

Please to observe to send me in **second proof**, before you print off, as I dare not trust the writer; & I must see one copy coloured on the paper which you propose to use.—

Gentlemen | Your obliged & obed^t svt | C. Darwin

To | Mrs Reeve, Brothers

New York Public Library (Berg collection)

[1] The date is based on an entry in CD's Account Book (Down House MS) dated 4 September 1846: 'Cheque L. Reeve. Lithograph for G. of S. America £10. 2s.'

[2] A coloured plate at the back of *South America* depicts the formations of the Portillo and Uspallata passes and the valley of Copiapó, Chile. See letter to Smith, Elder & Co., [19 October 1846], for CD's response to the binding of the plate.

From Edward Forbes [7 August 1846][1]

Friday.

Dear Darwin

The shells which needed describing, are now described.[2] Shall I send you the MS?

The Conception *Nautilus* I cannot satisfactorily identify with any other. I propose to call it *N. D'Orbignyanus* if you have no objection. It comes very near some lower & middle cretaceous species.[3]

I send the M^ss. wanted.

I leave the Fuego Ancyloceras unnamed. Give D'Orbigny's supposed identif^n. of it. It is however not in a state sufficiently secure for my conscience to hang upon its horn.[4]

The Terra del fuego fragments impress me with the notion that they indicate cretaceous epoch—probably the early part.

Such also is the impression I take up (& more firmly) from those of Corderilla of Central Chili.[5] I send that M^ss. with the other. I am now ready to see Sowerby.[6] If he is not coming tomorrow, I will be in my office all Monday.

Our survey reports are now out. I wish I could send you a copy of my paper[7] but as yet have not been able to lay hands on one for myself

Ever, most sincerely | Edward Forbes

DAR 43.1: 49

CD ANNOTATIONS

1.1 The shells . . . part. 5.2] *crossed pencil*

2.3 lower & middle cretaceous] *underl pencil*

6.2 If he . . . myself 7.2] *crossed pencil*

[1] The conjectured date is the first Friday after the announcement of the publication of E. Forbes 1846 in *Memoirs of the Geological Survey* (*The Publishers' Circular*, 1 August 1846, listing the volume as 'just published' in the period 13–29 July).
[2] Forbes described CD's Secondary fossil shells in the appendix of *South America*.
[3] Forbes eventually decided it resembled two species from the Upper Greensand (*South America*, pp. 127, 265).
[4] According to *South America*, p. 152, Forbes agreed with Orbigny that the fossil was early Cretaceous.
[5] Noted by CD in *South America*, p. 193.
[6] George Brettingham Sowerby Jr, who illustrated the fossil shells for the plates in *South America*. Forbes and George Brettingham Sowerby prepared descriptions for the appendix.
[7] E. Forbes 1846.

To Charles Lyell [8 August 1846]

Shrewsbury
Saturday

My dear Lyell,

I was delighted to receive your letter, which was forwarded here to me. I am very glad to hear about the new Edit of the Principles,[1] & I most heartily hope you may live to bring out half-a-dozen more editions. There would not have been such books as d'Orbignys S. American Geology[2] published, if there had been seven Editions of the Principles distributed in France. I am rather sorry about the small type; but the first Edit, my old true-love,[3] which I never deserted for the later editions, was also in small type.— I much fear I shall not be able to give any assistance to Book III;[4] I think I formerly gave my few criticisms, but I will read it over again very soon (though I am slaving to finish my S. American Geolog) & see whether I can give you any references.—

I have been thinking over the subject, & can remember no one book of consequence, as all my materials (which are in an absolute chaos on separate bits of paper) have been picked out of books not directly treating of the subjects you have discussed, & which I hope some day to attempt: thus Hooker's Antarctic Flora,[5] I have found eminently useful & yet I declare I do not know what precise facts I could refer you to. Bronn's Gesichte[6] (which you once borrowed) is the only systematic book I have met with on such subjects; & there are no general views in such parts, as I have read, but an immense accumulation of references, *very useful to follow up*, but not credible in themselves;—thus he gives hybrids from ducks & fowls just as readily as between fowls & pheasants! you can have it again, if you like.— I have no doubt Forbes essay, which is I suppose now fairly out, will be very good under geographical head.[7] Koelreuter's *German* Book[8] is excellent on Hybrids, but it will cost you a good deal of time to work out any conclusions from his numerous details. With respect to variation, I have found nothing, but minute details scattered over scores of volumes.— But I will look over Book III again: What a quantity of work you have in hand! I almost wish you c^{d} have finished America,[9] & thus have allowed yourself rather more time for the old Principles, & I am quite surprised that you c^{d} possibly have worked your own new matter in within six weeks. Your intention of being in Southampton will much strengthen mine & I shall be very glad to hear some of your American Geolog. news.—[10]

You have pleased me much by saying that you intend looking through my Volcanic volume: it cost me *18 months*!!! work & I have heard of very few who has read it; now I shall feel whatever little (& little it is) there is confirmatory of old work or new will work its effect & not be lost. I wish my S. American volume was out for same end, & I daresay you will be heartily glad it is not, for you must with all your work in hand, grudge time for your own new materials. I sh^d have liked to have had your opinion on my facts & short discussion regarding the foliation of the metamorphic schists,[11] which I am now correcting;—and another on the absence of recent conchiferous deposits & on the tertiary formations having been deposited during subsidence:[12] but I will have mercy on you & say no more on my volume, of which I am inexpressibly weary & thank Heavens have now finally corrected $\frac{2}{3}$ of, & hope to see published this month.

I return home on Tuesday, having been here for a week to see my Father: Emma & the children have been having colds but are otherwise well.— I hope you found Mr & Mrs Lyell tolerably well.— How I shall enjoy having you for a visit to Down & I believe you cd with quiet & fresh air do more work with us than in that horrid place London.—

I must go to work to proof-sheets: my vol will be about 240 pages, dreadfully dull yet much condensed: I think, whenever you have time to look through it, you will think the collection of facts on the elevation of the land & on the formation of terraces pretty good.

Goodbye with many thanks for your letter & my kindest remembrances to Mrs Lyell. | Ever yours | C. Darwin

Postmark: AU 8 1846
American Philosophical Society

[1] The seventh edition of the *Principles of geology* (C. Lyell 1847).

[2] Orbigny 1835–47, vol. 3, pt 3: *Géologie*.

[3] CD had the first edition with him on the *Beagle* voyage; see *Correspondence* vol. 1, letter to J. S. Henslow, 18 May – 16 June 1832, and *Autobiography*, pp. 77, 101, for CD's adoption of Lyell's views. CD's annotated copy is in the Darwin Library–CUL.

[4] The third volume of Lyell's previous edition (C. Lyell 1840a) contained an extended discussion of the transmutation and first appearance of species (chapters 1–11). CD's copy is lightly annotated (Darwin Library–CUL).

[5] J. D. Hooker 1844–7.

[6] Bronn 1841–9. CD's annotated copy is in the Darwin Library–CUL.

[7] E. Forbes 1846.

[8] Kölreuter 1761–6. CD's copy is in the Darwin Library–CUL. CD's frequent citations in the *Origin* and later works make clear that Joseph Gottlieb Kölreuter's work was one of CD's major sources on hybridism and an important influence in the development of his theory. DAR 116 contains CD's abstracts and notes on thirteen papers by Kölreuter.

[9] An account of Lyell's second visit to the United States (September 1845 – June 1846) was eventually published as C. Lyell 1849. CD's annotated copy is in the Darwin Library–CUL.

[10] Lyell presented a short account to the British Association (C. Lyell 1846c).
[11] *South America*, pp. 140–68.
[12] *South America*, pp. 135–9.

To Leonard Jenyns [14 or 21 August 1846][1]

Down Farnborough Kent
Friday

Dear Jenyns

I am much obliged for your note & kind intended present of your volume, which I value much.—[2] As the Athenæum Club is near Yarrell's,[3] perhaps you could send it there, (but do not, if it happens to be inconvenient) for I invariably call there when in town, & occasionally send for parcels when they have accumulated. I feel sure I shall like it, for all discussions & observations on what the world would call trifling points in Natural History, always, appear to me very interesting.[4] In such foreign periodicals, as I have seen, there are no such papers, as White, or Waterton;[5] or some few other naturalists in Loudon's & Charlesworth's Journal,[6] would have written, & a great loss it has always appeared to me.—

I sh^d. have much liked to have met you in London, but I cannot leave home, as my wife is recovering from a rather sharp fever attack & I am myself slaving to finish my S. American Geology, of which, thanks to all Plutonic powers, two-thirds are through the press, & then I shall feel a comparatively free man.

Have you any thoughts of Southampton? I have some vague idea of going there, & sh^d. much enjoy meeting you.—[7]

My health continues pretty well; never right & seldom very wrong, as long as I live quite quietly.

The little d'^s, of whom you enquire, now number four, two of each gender.[8]

Believe me, dear Jenyns with very many thanks for your kind present | Ever yours truly | C. Darwin

Bath Reference Library (Jenyns papers, 'Letters of naturalists 1826–78', Octavo vol. 1: 43(9))

[1] The dates are based on the publication date of Jenyns 1846 (between 29 July and 14 August according to *The Publisher's Circular* of 15 August 1846) and CD's return from Shrewsbury on 9 August ('Journal'; Appendix II).
[2] Jenyns 1846.
[3] William Yarrell.
[4] Jenyns 1846 was written using material that he had collected while working on his edition of Gilbert White's *The natural history and antiquities of Selborne* (Jenyns 1843). See *Correspondence* vol. 2, letter to Leonard Jenyns, [May–September 1842]. CD's annotated copy of Jenyns 1846 is in the Darwin Library–CUL.
[5] Gilbert White and Charles Waterton.
[6] The *Magazine of Natural History, and Journal of Zoology, Botany, Mineralogy, Geology and Meteorology*, first series, volumes 1–9 (1829–36), was edited by John Claudius Loudon. The second series, volumes

1–4 (1837–40), was edited by Edward Charlesworth. Both series are complete and extensively annotated in the Darwin Library–CUL.

[7] Both CD and Jenyns attended the British Association meeting in Southampton from 10 to 16 September 1846 (see letters to J. S. Henslow, [5 October 1846], and to Leonard Jenyns, 17 October [1846]).

[8] An allusion to CD's children.

To Leonard Horner [17 August – 7 September 1846][1]

Down Farnborough Kent
Monday

My dear Mr. Horner

In following your suggestion in drawing out something about Glen Roy for the Geolog. Committee, I have been completely puzzled how to do it.—[2] I have written down what I shd *say*, if I had to meet the head of the Survey[3] & wished to persuade him to undertake the task, but as I have written it, it is too long, ill expressed, seems as if it came from nobody & was going to nobody, & therefore I send it to you in despair, & beg you to turn the subject in your mind.—

I feel a conviction if it goes through the Geolog. Part of Ordnance Survey, it will be swamped, & as it is a case for more accurate measurements, it might, I think without offence, go to the head of the real Surveyors.[4]

If Agassiz or Buckland are on the Committee, they will sneer at whole thing & declare the beaches are those of a glacier lake,—than which I am sure I cd convince you, that there never was a more futile theory.[5]

I look forward to Southampton with much interest & hope to hear tomorrow, that the lodgings are secured to us.—

You cannot think how thoroughily I enjoyed our geological talks & the pleasure of seeing Mrs. Horner & yourself here.

Ever your obliged | C. Darwin

[Enclosure][6]

The Parallel Roads of Glen Roy in Scotland have been the object of repeated examination, but they have never hitherto been levelled with sufficient accuracy. Sir T. Lauder Dick[7] procured the assistance of an engineer for this purpose, but owing to the want of a true ground-plan, it was impossible to ascertain their exact curvature, which, as far as could be estimated, appeared equal to that of the surface of the sea. Considering how very rarely the sea has left narrow & well-defined marks of its action at any considerable height on the land, & more especially considering the remarkable observations by M. Bravais[8] on the ancient sea-beaches of Scandinavia, shewing that they are not strictly parallel to each other & that the movement has been greater nearer the mountains than on the coast,—it appears highly desirable, that the roads of Glen Roy should be examined with the utmost care during the execution of the Ordnance Survey of Scotland. The best instruments & the most accurate measurements being necessary for this end, almost precludes the hope of its being ever undertaken by private individuals; but by the means at the disposal of the Ordnance, measurements would be easily made even more accurate than those of M. Bravais.

It would be desirable to take two lines of the greatest possible length in the district, and at nearly right angles to each other, and to level from the beach at one extremity to that at the other, so that it might be ascertained, whether the curvature does exactly correspond with that of the globe, or if not, what is the direction of the line of greatest elevation. Much attention would be requisite in fixing on either the upper or lower edge of the ancient beaches, as the standard of measurement, & in rendering this line conspicuous.— The heights of the three roads, one above the other & above the level of the sea, ought to be accurately ascertained. Mr Darwin observed one short beach-line North of Glen Roy & he has indicated on the authority of Sir David Brewster, others in the valley of the Spey; if these could be accurately connected, by careful measurements of their absolute heights or by levelling, with those of Glen Roy, it would make a most valuable addition to our knowledge on this subject. Although the observations here specified would probably be laborious, yet considering how rarely such evidence is afforded in any quarter of the world, it cannot be doubted that one of the most important problems in geology,—namely the exact manner in which the crust of the earth rises in mass,—would be much elucidated, & a great service done to Geological Science.

American Philosophical Society and DAR 145 (enclosure)

[1] The earlier date is based on CD's reference to the Horners' visit to Down following his return from Shrewsbury on 9 August (see letter to J. D. Hooker, [3 September 1846]). The closing date is the last Monday before the British Association meeting in Southampton.

[2] A reference to the sectional committee for geology (section C) of the British Association. Leonard Horner was president of the section at the Southampton meeting.

[3] Thomas Frederick Colby.

[4] CD's memorandum was acted upon at the British Association meeting in Southampton in September 1846. The Association recommended that 'the so-called parallel roads of Glen Roy and the adjoining country be accurately surveyed, with the view of determining whether they are truly parallel and horizontal, the intervening distances, and their elevations above the present Sea-level' (*Report of the 16th meeting of the British Association for the Advancement of Science held at Southampton in 1846*, p. xix). The survey was undertaken and the results were published in 1874 (see *ML* 2: 174 n.). By that time, however, the problem of the origin of the roads had been settled by other evidence in favour of the glacial lake theory (see Rudwick 1974, pp. 147–53).

[5] Rudwick 1974 describes the debate over the origin of the roads. In the *Autobiography*, p. 84, CD recalls his explanation as a 'great failure' .

[6] The original enclosure has not been found; the text printed here is taken from a copy in DAR 145.

[7] Dick 1823.

[8] Bravais 1845.

To Robert Mallet 26 August [1846][1]

Down Farnborough Kent
Aug. 26th.

Sir

I take the liberty of writing to thank you for your most obliging present of the Dynamics of Earthquakes, & for the much too honourable mention you make of my name.—[2] I have read your memoir with the *greatest* interest & it has much cleared my ideas, though undulations of all kinds will ever be of difficult comprehension to non-mathematical heads.—

During writing the first Edition of my Journal, I consulted Mr. Whewell,[3] & soon perceived how difficult a subject that of waves was. In the Col. Library Edit, I condensed what I had said, & now heartily wish I had said nothing for I felt at the time that I was out of my depth.[4]

I beg to send you a paper of mine, which perhaps you may never seen & may be scarcely worth your reading; my chief object in writing it, was to show the intimate connection & indeed identity of the forces, which pour forth lava & elevate continents: this to my mind is an important conclusion.—[5] At p. 621. you will see an indirect argument against Earthquakes being undulations in an underlying fluid mass & in favour, (of what your memoir will never again allow to be doubtful) of their being a vibration or oscillation in the solid crust.—[6]

Have you seen the Mrs. Rogers of N. America papers on earthquakes,[7] they push the doctrine of fluid undulations to a monstrous extent, & if I might take the liberty to suggest, it would be adviseable to send a copy of your memoir to them; for they are *excellent* geologists.

How wonderfully interesting it would be, if you could have your instrument[8] worked in Chile: in Lima, there was or is a merchant Mr Maclean[9] with a strong taste for Natural History, & earthquakes are frequent there.

I have some intentions of being at Southampton,[10] & I trust, you will allow me to introduce myself to you.

With my sincere thanks | I beg to remain dear Sir | Yours faithfully | C. Darwin

It has often occurred to me that a faithful record of earthquakes in Chile, would perhaps afford curious results from coincidences with the moon or state of tides.—[11]

Smithsonian Institution (Special collections, Dibner)

[1] The date is based on the reference to the September 1846 meeting of the British Association in Southampton.

[2] Mallet 1848a. CD received pre-publication copies of this work and Mallet 1848b, which were read at the Royal Irish Academy on 9 February and 22 June 1846, respectively. CD's copies are dated 1846, bear an abbreviated title, and are separately paginated. They are in the Darwin Pamphlet Collection–CUL. The following references are to the later published version. In the paper Robert Mallet cites CD on the twisting displacement of building stones and ornaments (Mallet 1848a,

pp. 54–5), notes CD's use of the term 'undulation' to describe earthquake motion (pp. 57–8), and quotes at length from CD's description of tidal waves associated with earthquakes (pp. 65–6).

[3] See *Correspondence* vol. 2, letter to William Whewell, 18 June [1837].

[4] CD described the Concepción earthquake in *Journal of researches*, pp. 368–81, and 2d ed., pp. 301–12. However, Mallet explained the twisting displacement of building stones as due to different centres of gravity and adhesion in the affected stones (Mallet 1848a, pp. 54–7). Mallet also explained numerous aspects of tidal waves as the result of the different velocities of wave propagation in the rock of the sea-bed, the water, and the air above (Mallet 1848a, pp. 65–74).

[5] 'On the connexion of certain volcanic phenomena in South America', *Collected papers* 1: 53–86. The passage cited by CD appears on pp. 72–3.

[6] Mallet argued that the primary motion of earthquakes had to occur in the crust since any motion imparted to the crust by the underlying fluid would be transmitted more rapidly in the crust than in the fluid itself.

[7] Rogers and Rogers 1843.

[8] Mallet 1848b describes an experimental seismograph. Though his invention was never used, elements of his design were incorporated into later instruments (*DSB* 9: 60–1).

[9] John Maclean.

[10] See n. 1, above.

[11] CD referred to this idea in *Notebook A*: 137, 153.

From Edward Forbes [September 1846]

Dear Darwin

I return you a proof of the Fossil-Descriptions corrected.[1] As D'Orbigny gave the names, had not "E. Forbes" better be left out after each species—the general heading sufficing—whilst Sowerby's name would remain attached to his descriptions[2]—they to be placed within inverted commas.

As I have been absent from town for the last 4 days, working my

AL incomplete
DAR 43.1: 46

[1] 'Descriptions of Secondary fossil shells from South America', appendix, *South America*, pp. 265–8. Eleven species are described.

[2] George Brettingham Sowerby described sixty-one Tertiary fossil shells in his part of the appendix, and two of the Secondary fossil shells in Forbes's.

From J. D. Hooker [before 3 September 1846][1]

[Clifton]

This probable fracas between the 2 Geographers distresses me, for they are almost the only 2 men who have looked on British Flora with the eyes of philosophers.[2] Watson in particular ranks in my opinion at the very head of

English Botanists, whether for knowledge of species or of their distribution; he first wrote philosophically upon them & his works are of the highest order.

Unfortunately he is touchy & very *severe* when first offended, though he never holds a grudge long.

I need hardly ask what you are about, as my Proof sheets come from Reeves enveloped with cabalistic diagrams, all your own—which I doubt not belong to the Geology of S. America.[3] When that work is over you will I suppose attack *Species* as you have long promised I wish you joy of the task: & shall be very glad to know your views— I have done all of Edmonstones Galapago plants that have been received,[4] but understand that these are only *duplicates* of a much fuller collection not yet received. As it is, they modify the results drawn from the xamination of previous collections materially; there being more Guayaquil species amongst them.

I have *at last* finished down to the Ferns of Flor. Ant. & begun the Cryptog. I am ready on my return to send you a return of the species identical & representative inhabiting N. temperate & Antarctic regions. I hope you get your numbers regularly from Reeves; but he is not the most regular of publishers. This winter I shall (entre nous) bilk the Survey & work at home. My address is as above, where I shall be for a week or 10 days,[5] longing to hear how you all are & what about

Ever my dear Darwin | Most truly yrs. | Jos D Hooker.

Incomplete
DAR 100: 79

CD ANNOTATIONS
1.1 This . . . received. 3.7] *crossed pencil*
3.5 Galapago . . . them. 3.9] '20' *added brown crayon*
3.8 materially . . . them. 3.9] *scored pencil*
4.1 I have . . . Hooker. 5.1] *crossed pencil*
4.2 ready . . . regions. 4.3] *scored pencil*

1 Dated from CD's reply, see letter to J. D. Hooker, [3 September 1846].

2 Hooker refers to Hewett Cottrell Watson's belief that Edward Forbes had appropriated Watson's ideas in E. Forbes 1845 and 1846, without acknowledging their source. Watson, like Forbes, argued that the British flora was made up from elements of other European floras which had, at one time or another, extended over Britain (Watson 1843). See Watson's editorial comments in the *Phytologist* 2 (1846): 483–4 and the appendix to volume one of Watson's *Cybele Britannica* (Watson 1847–59, 1: 465–72) for his accusations.

3 The illustrations for *South America* were printed by Reeve Brothers, publishers of J. D. Hooker 1844–7. See letter to Reeve Brothers, [August 1846].

4 Thomas Edmondston, who collected in the Galápagos. According to Hooker his collection was second only to CD's and contained several plants not found in other collections (J. D. Hooker 1846, p. 238).

5 Hooker was examining the fossil plants of the Bristol coalfields in his capacity as botanist to the Geological Survey (Huxley ed. 1918, 1: 210).

To John Maurice Herbert [3 September? 1846][1]

Down. | Farnborough Kent.
Thursday.

My dear Herbert.

I was very glad to see your handwriting & hear a bit of news about you.— Though you cannot come here this autumn, I do hope you & Mrs Herbert will come in the winter, and we will have lots of talk of old times & lots of Beethoven.

I have little or rather nothing to say about myself; we live like clock work, & in what most people would consider the dullest possible manner I have of late been slaving extra hard, to the great discomfiture of wretched digestive organs at S. America, & thank all the fates I have done ¾ths of it,— Writing plain English grows with me more & more difficult & never attainable.— As for your pretending that you will read anything so dull as my pure geological descriptions lay not such a flattering unction on my soul,[2] for it is incredible— I have long discovered that geologists never read each others works, & that the only object in writing a Book is a proof of earnestness & that you do not form your opinions without undergoing labour of some kind. Geology is at present very oral, & what I here say is to a great extent quite true. But I am giving you a discussion as long as a Chapr. in the odious Book itself.

I have lately been to Shrewsbury[3] & found my Father surprisingly well & cheerful.

With our kindest united regards to Mrs Herbert, believe me, my dear old friend. | Ever yours. | C. Darwin.

Copy
DAR 145

[1] Dated on the basis of CD's reference to his progress with the manuscript of *South America*. In the following letter to J. D. Hooker, [3 September 1846], CD states that he will have completed the book by the end of the month.

[2] Shakespeare, *Hamlet*, 3. 4. 145 (Arden edition).

[3] CD had visited Shrewsbury between 31 July and 9 August 1846 ('Journal'; Appendix II).

To J. D. Hooker [3 September 1846]

Down Farnborough Kent
Thursday

My dear Hooker

I hope this letter will catch you at Clifton, but I have been prevented writing by being unwell & having had the Horner's here as visitor, which with my abominable press-work has fully occupied my time. It is, indeed, a long time since we wrote to each other; though, I beg to tell you, that I wrote last, but what about I cannot remember, except, I know, it was after reading your last numbers, & I sent you a uniquely laudatory epistle, considering that it was from a man who hardly knows a daisy from a Dandelion to a professed Botanist.— By the way I announced Hopkirk's book being sent back, did it reach you?—

I was very glad to hear what you were about; but I fear you must feel your time rather thrown away.— I cannot remember, what papers have given me the impression, but I have that, which you state to be the case, firmly fixed on my mind, namely the little chemical importance of the soil to its vegetation.— What a strong fact it is, as R Brown once remarked to me, of certain plants being calcareous ones here which are not so under a more favourable climate on the continent, or the reverse, for I forget which; but you no doubt will know to what I refer.— By the way there are some such cases in Herbert's paper in Hort. Journal:[1] have you read it, it struck me as extremely original & bears *directly* on your present researches.— To a *non-botanist* the Chalk has the most peculiar aspect of any flora in England; why will you not come here & make your observations? *We* go to Southampton, if my courage & stomach do not fail, for the Brit. Assoc: (do you not consider it your duty to be there?), & why cannot you come here afterwards & *work*. I expect Sulivan here the first week in October & I hope to get a few more here, & how glad we should be if you c^{d} come then or at any time whatever.—

Before the end of the month, I shall have quite finished my S. American Geology, & extremely glad I shall be, for I have been pushing on & feeling jaded for the last several months by it.—

I am astonished (having felt a curiosity on the point) at the number of species on 2 square yards (or two yards square?); though I cannot read whether it is *26* or *16* to 48 species; does this include cryptogams: if you do *not* publish this, I sh^{d} like much sometime to hear more particulars about this; if you publish, where will it be?[2]

I am much pleased to hear you have worked out the identical & representative species of N. temperate & Antarctic regions & shall be exceedingly glad to see it; but as it **of course** will be published, I will not think of troubling you to send it me: I hope you will add, whenever you know, whether species of the same genera are found in the intermediate *tropical* districts, saying, whether in America or elsewhere, whether on high-lands or lowlands; this no doubt w^{d} add to your trouble, even if you gave *only* such information as you possessed without search, & surely it w^{d} add great interest to your results: Mr Gardener's list of Mountain Brazil plants[3] w^{d} thus come incidentally in, as indeed w^{d} lists from all parts of the world.

I have not yet seen Forbe's memoir,[4] but have ordered it, & will enjoy writing to you my opinion. I am very sorry to hear what you say about Watson's previous work; I feel sure that Forbe's own noble indifference to fame is the main cause of his not in some instances making proper acknowledgment.— Horner (*private*) tells me that he has just remonstrated with him, for not having mentioned Lyell's views on climatic changes, & his answer was,—"I sh^{d} as soon have thought it necessary to refer to Linnæus, as originator of specific characters".—& I have no doubt this is the simple truth.—[5] I cannot remember whether I have ever read (except a few papers) any of Watson's works: could you sometime lend me the chief? I sh^d. much like to see them.

I am almost sorry for your eternal additional labours on the Galapagos Flora; though as yet your work assuredly has not been thrown away, as many have referred to your curious geographical results on this archipelago.— I suppose you feel sure that Edmonston's Collections from the mainland have not been mingled with those of the Galapagos—[6]

Have you ever thought of G. St. Hilaire "loi du balancement", as applied to plants: I am well aware that some zoologists quite reject it, but it certainly appears to me, that it often holds good with animals.— You are no doubt aware of the kind facts I refer to, such as great development of canines in the carnivora apparently causing a diminution—a compensation or balancement—in the small size of premolars &c &c.—[7] I have incidentally noticed some analogous remarks on plants, but have never seen it discussed by Botanists.— Can you think of cases of any one species in genus, or genus in family, with certain parts extra developed, & some adjoining parts reduced? In varieties of same species, double flowers & large fruits seem something of this,—want of pollen & of seeds balancing with the increased number of petals & development of fruit.—

—I hope we shall see you here this autumn, & I will let you know when Sulivan comes—or anytime will suit us.

Ever my dear Hooker | Most truly yours | C. Darwin

Do you know whether Dieffenbach is in London, & where?

P.S. | (I do not quite understand, do you intend giving up the Museum of Economic Geology altogether? I know it is private:— I sh⟨^d be⟩ almost glad of it.—)

Postmark: SP 5 1846
DAR 114.2: 64–64b

[1] W. Herbert 1846.

[2] Although it is not clear to what Hooker's figures apply, CD's interest was probably aroused by the evidence they provided for the diversity of vegetation in small areas and under uniform conditions of life. Such evidence became important to CD in support of his principle that the maximum diversity of living beings permits the maximum amount of organic life in any area, and the corollary that natural selection, by tending to maximise the amount of organic life, will also tend to maximise the diversity of organic beings. These ideas formed the basis of CD's principle of divergence as elaborated in *Natural selection* (pp. 227–51) and the *Origin*.

[3] Gardner 1846.

[4] E. Forbes 1846.

[5] Charles Lyell evidently thought Edward Forbes's reply was 'sufficient' and had no wish to enter into controversy with Forbes, see K. M. Lyell ed. 1881, 2: 105–13.

[6] CD refers to Hooker's comment that Thomas Edmondston's Galápagos collection contained some mainland species from Guayaquil, see letter from J. D. Hooker, [before 3 September 1846].

[7] See letter from G. R. Waterhouse, 26 April 1844, n. 3.

From J. D. Hooker 28 September 1846

West Park Kew
Septr. 28. 1846.

Dear Darwin

I have delayed answering your letter till my return here, in the hopes that it might be possible to accept your kind invitation to meet Capt. Sulivan at Down. Two visitors however tie me by the leg, one of them our friend Harvey of Dublin, who I only see once a year: had it been otherwise I would have run down for the Sunday of next week, when I suppose, from your letter, Capt. S. will be with you.

Hopkirk reached me quite safe, now long ago, the work is curious as being one of the very few English ones devoted to the subject: unfortunately the author seems to have had no direct object in writing or very clear arrangement.[1]

"Plants & soils" is a long subject, I am tolerably convinced now, from a little observation of my own, that the mechanical & not the chemical qualities of the soil are the important ones. There is I expect hardly any soil that does not contain enough of the inorganic elements that any plant wants, to support a growth of such plants. Still there are assemblages of limestone-plants, & the predilection of such for that rock in England remains to be explained. If it be granted that the Limestone (& chalk) are warmer soils (& all the information I can collect goes to prove it) than the Sandstones, then the solution will be easy, for I think I can show that the Southern extension of the Scotch plants in England is upon the Sandstones & the N. extension of continental species in Britain upon the Calc. rocks. There are hundreds of exceptions, but these comprehensive questions must not be worked by exceptions, which may prove to be only apparent when the leading facts are established. The *black* traps will help me; for their dry hot sunny slopes on the Clyde present us with plants confined on the Severn to the Mt: Limestone. Sir H.[2] wants the difficulty solved by examinations of the areas squared out on his Ordnance maps; I am not capable of generalizing from such a limited field of observation, though others doubtless are who are better skilled in making the most of small means; were I to follow the subject as I would like, I should examine the sites of the local British species on the continent..

Many thanks for mentioning Herberts paper, I have it in my eye all along, it is as you say a good deal to the purpose. Another year I should like *much* to work a little at your chalk & clay floras but have no idea of what a summer may bring forth: if Sir H. will not let me work the thing as any Botanist would advise, I should throw it up; though this I have never hinted to any one: but Sir. H. though a kind, good man, has no system of his own & cannot bear Forbes & my working with books.

How glad I shall be when your S. A. is done, you will then feel freer than you have for long. I know no greater pleasure than getting a tedious job out of hand when one has really been doing the best to it.

It was very stupid of me to say 2 square yds instead of 2 yd^s. square but it is a mistake I *always* make when not thinking. I have just been told of 27 *mosses* (not

Hepatica) on one rock i.e. block of rock in Scotland & there are only some 300 sp. in Britain. Watson supposes that every county contains upwards of one half the whole British Flora.[3]

When such results of any comparisons between the representative species of the N. & S. hemispheres that my Flora will shew, shall be worked out, I shall probably publish them in the L. Soc. Trans. which will save me the labor of getting them through the press myself, even if any publisher w^{d} undertake them, which is not likely.[4] I have not seen Forbes since studying his paper & really do not know what to say when I do, for he will be sure to ask me about it, & most unfortunately he does not seem to know the Geographic Distrib. of the English Plants. I must confess to have taken his modification of Watson's types of vegetation as correct, & this for granted, but I had occasion to look closely at them the other day & find his S.E Flora, numbered III., to be altogether a fallacy: all or almost all the 20 species on whose supposed presence he founds it, being as common or more in the W. or N as in the E. or S. & some of them not existing in the S.E. at all.! or if so as introduced species. I now see the cause for Watson's being so peculiarly savage & offering me proof that all that is correct is mere plagiarism. I still however quite acquit Forbes of any intentional piracy, he has long & early understood & appreciated Watson's views & has fancied that he has grounds for modifying them. I do all I can to appease W., but in vain, he threatens to denounce F. publicly & if he does I fear that it will read awkwardly for our friend. I need not ask you to say nothing of this, except you can offer some way or means of keeping these, almost the only 2 Philosophical Brit. Botanists, out of a broil, at which all the dirty species-mongers will chuckle.[5]

I will send you one of Watson's works at once: you must judge it by what has been previously done, or even done up to this time, by any other Brit. Bot.; more than by its own intrinsic merits, which however are very high.

I have excluded all of Edmonstones plants that can have been from the Coast. A great developement of an organ in plants is often accompanied by a suppression of others E.G. Rafflesia, a *staring* xample: I should like to have the opportunity of applying St Hilaires laws *directly* to the subject. I have heard of them.

Of Deiffenbach I know nothing whatever.

I do not intend to work much more for the Geol. Mus. this year, except at home, I have a great deal on hand.

I could not get to the Brit. Assoc. I wanted much to have got to Switzerland or Germany but could not. Two more Mediterranean Algæ have been found within these few months on the W. of Ireland.

Have you seen the new Cosmos?[6] it is excellently well done & quite a different book to read— I will send or bring it you if you care. Should Capt. Sulivan stay over Sunday week with you, & it be entirely convenient for me to come, I w^{d} try & get down, if only for a day, but do not see why I should not take my usual allowance at Down..

Ever most truly Yours | Jos D Hooker.

DAR 100: 69–72

CD ANNOTATIONS
1.1 I have . . . best to it. 5.3] *crossed pencil*
3.1 "Plant . . . Calc. rocks. 3.11] 'Soils & Plants' *added pencil*
7.2 that my Flora . . . Trans. 7.3] *scored pencil*
7.3 the labor . . . Coast. 9.1] *crossed pencil*
7.6 he will be . . . at all.! 7.12] 'Forbes Geograph. Bot' *added pencil*
10.1 Of . . . could not. 12.2] *crossed pencil*
13.1 excellently . . . Down.. 13.5] *crossed pencil*
Top of first page: '20' *brown crayon*

[1] Hopkirk 1817, on botanical monsters.
[2] Sir Henry Thomas De la Beche, director-general of the Geological Survey. The 'Plants & soils' project proposed for Hooker was very much in keeping with the theoretical aims of De la Beche and his Survey staff (see Secord 1986).
[3] Watson 1835, pp. 41–2.
[4] *Transactions of the Linnean Society*. Hooker never published a separate work on representative species.
[5] See letter from J. D. Hooker, [before 3 September 1846], n. 2. Edward Forbes refused to enter into a public debate with Hewett Cottrell Watson, despite Watson's attacks.
[6] Humboldt 1846–8, a new translation into English by Elizabeth Sabine that superseded the previous translation by Augustin Prichard (Humboldt 1845–8).

To John Gould [*c*. October 1846][1]

Down Farnborough Kent
Tuesday

Dear Gould

I have seen in the Papers that you are going to Guatimala; is this actually true!![2] I am very curious to know, but my immediate object in writing is to say that I have had a letter from Dr. Dieffenbach expressing a great wish to accompany you.[3] His means of living, however, are limited & he could not travel under 200£ per annum. All his collections would be at his employer's disposal.— He has a very general & I believe sound knowledge on Chemistry, Geology, Natural Philosophy & some Botany & Zoology, so that he would be in many respects a very good man for an Expedition.— Will you kindly send me a line & tell me some news of your plans? Pray do.—

Do you know anything & would you exert yourself a little to get Dieffenbach taken on Ld. Ranelagh's expedition,[4] supposing, as I fear is too probable, that you will not want him yourself.[5]

With all good wishes. Ever yours sincerely. | C. Darwin

Cambridge University Library (Add. 4251: 329)

[1] Dated on the basis of nn. 2 and 5, below.
[2] John Gould never went to Guatemala (Sauer 1982) although it appears that he did begin to plan such an expedition, see n. 5, below. No reference to his plans in British newspapers has been traced; however, in a letter to John Gould dated 29 October 1846 (British Museum (Natural History), Zoology Library) Ernst Dieffenbach states that Gould's proposed expedition was mentioned in the continental newspaper *Galignani*.

[3] This letter has not been found, but see letters to Edward Forbes, 13 May [1845], and to J. D. Hooker, [19 May 1846], in which CD refers to Dieffenbach's wish to join a natural history expedition. In 1845 Dieffenbach was offered an opportunity to join an expedition to the west coast of North America, which he declined (see G. Bell, pp. 110–11).

[4] Thomas Heron Jones, 3d Viscount Ranelagh, announced plans for an expedition to South America in *Athenæum*, 8 August 1846, p. 819, and further details were published in *Athenæum*, 21 November 1846, p. 1191. The expedition was never carried out.

[5] Dieffenbach wrote to Gould on his own behalf, reiterating the comments made by CD, in a letter dated 29 October 1846 from Giessen (British Museum (Natural History), Zoology Library). Gould's reply, dated 25 November 1846, makes it clear that his plans were 'still in Embryo', that the newspaper reports were unauthorized, and that he was in no position to offer Dieffenbach any assistance (British Museum (Natural History), Zoology Library).

To W. B. Carpenter [October–December 1846]

Down Farnborough Kent
Thursday

My dear Sir

When at Southampton you said you would give me the address of the artist whom you employ to draw from under the microscope, & whom you pay at 5^{s} per hour.[1] An early answer would very much oblige me, if you would kindly take the trouble.— The objects which I want chiefly drawn are minute corallines, & minute articulata & mollusca & their various organs;[2] please say whether you think your artist would do such things well; it is a very different style, hard & precise, from those wonderfully beautiful sections which you exhibited at Southampton.— I hope you will excuse this trouble.— Sh^{d} you have any communication with your artist, w^{d} you mention my name as a probable applicant for his assistance

Pray believe me, my dear Sir | Yours very faithfully | C. Darwin

American Philosophical Society

[1] Probably Samuel William Leonard who prepared the illustrations for Carpenter's article in the *Report of the 14th meeting of the British Association for the Advancement of Science held at York in 1844*, see plates 1–20 in Carpenter 1844.

[2] The references indicate that at this time CD intended to describe the invertebrate specimens collected during the *Beagle* voyage. This plan had been announced earlier for the *Zoology* (see Freeman 1977, p. 26) but was not achieved, probably because the Government grant was exhausted. In the event, the Cirripedia occupied CD until 1854 and he undertook no further work on the *Beagle* invertebrates.

To Robert FitzRoy 1 October 1846

Down. Farnborough. Kent.
Oct. 1st 1846.

Dear FitzRoy.

I did not hear for more than 4 weeks after your return that you were in England, and now, though I have nothing particular to say—I cannot resist writing to congratulate you on your safe arrival after your bad passage home,[1] and to express my most sincere hope that Mrs. FitzRoy and your family are all well— I was in London yesterday—for the first time—since I heard of your return and found your address at the Admiralty & in Dover St. the servant told me to direct as I have done. I got your pamphlet[2] the other day and was very much interested by it, for I had heard comparatively little about New Zealand; I fear you must have undergone much trouble & vexation and been ill repaid except by the consciousness of your own motives, for the sacrifices which I am aware you made in accepting the Governorship.—[3]

I hope that your health has not suffered and that you are as strong & vigorous as formerly— I have hardly the assurance to ask you to spare time to write to me, but I should be very glad to hear about yourself Mrs. Fitzroy & the children. My life goes on like Clockwork, and I am fixed on the spot where I shall end it; we have four children, who & my wife are all well. My health, also, has rather improved, but I am a different man in strength and energy to what I was in old days, when I was your "Fly-catcher", on board the Beagle; I have just finished the 3rd. & last part of the Geology of the Voyage of the Beagle—viz: on S. America; I will direct a copy (in a weeks time when published) to be sent to the Carlton Club, without you prefer it being sent elsewhere.— It is purely geological & dull enough, but I hope it contains some few new facts & views— I have now with the exception of some Zoological papers on the lower marine animals completed all which I shall ever attempt on the materials collected during the voyage—[4]

I am aware how little chance there is of your having time to spare, but if ever when in Town Mrs. FitzRoy & yourself should feel inclined to spend a few days in the country—it would give my wife & myself real pleasure;—we have a tolerably comfortable house in a very quiet, retired, airy part of the country—

I think I ought to apologise for the length of this letter.— Pray give my kind & respectful regards to Mrs. FitzRoy & Believe me, dear FitzRoy | Yours truly & obliged | Charles Darwin

Copy
DAR 144

[1] FitzRoy, with his wife and children, had endured a stormy sea journey following his recall from his post as governor of New Zealand.

[2] FitzRoy 1846, an account of his governorship.

[3] FitzRoy had been obliged to deal with conflicts between British settlers and the native Maoris. He was unable to satisfy either side, and his independent approach to problems displeased the Colonial Office. He was replaced by George Grey in October 1845 (Mellersh 1968, pp. 234–5). FitzRoy evidently incurred considerable private expense as governor (see letter from B. J. Sulivan, 13 January – 12 February 1845).

[4] The *Beagle* barnacles are included in the monographs on *Living Cirripedia* (1851, 1854). CD's intention to describe other marine invertebrates was not realised. Chancellor *et al.* 1987 provides descriptions of the *Beagle* invertebrates now in the Oxford University Museum.

To J. D. Hooker [2 October 1846][1]

Down Farnborough Kent
Friday

My dear Hooker

I have not heard from Sulivan lately; when he last wrote he named from 8th to 10th as the most likely time. Immediately, that I hear, I will fly you a line, for the chance of your being able to come: I forget, whether you know him, but I suppose so; he is a real good rattling fellow. Anyhow, if you do not come then, I am very glad that you propose coming soon after. My wife is going to Shrewsbury for one week (I staying at home to look after the babbies) after Sulivan's visit, & that is the only engagement we have. I shd like to get a reunion of naturalists, but fear most will fail me, but I do trust to see you here.—

Thanks for all your curious information on distribution, & offer of books. Kosmos I will buy,[2] but I shd be very glad to borrow Watson, though I am in no sort of hurry for it, as I have several Books in hand, not having even yet had time & inclination to read Forbes;[3] as I feel that one must buckle to for such a task. I am rather low at hearing that your discussion on relations of Southern & northern forms will appear in Linn. Transt as I fear it will be almost indefinitely delayed. You must indeed have a great deal of work on hand & I heartily wish you had as much leisure as I now have.

I am going to begin some papers on the lower marine animals, which will last me some months, perhaps a year, & then I shall begin looking over my ten-year-long accumulation of notes on species & varieties which, with writing, I daresay will take me five years, & then when published, I daresay I shall stand infinitely low in the opinion of all sound naturalists—so this is my prospect for the future.

Are you a good hand at inventing names: I have a quite new & curious genus of Barnacle, which I want to name, & how to invent a name completely puzzles me.—[4]

By the way, I have told you nothing about Southampton;[5] we enjoyed (wife & self) our week beyond measure: the papers were all dull, but I met so many friends, & made so many new acquaintances (especially some of the Irish Nat.list) & took so many pleasant excursions.— I wish you had been there: on Sunday we had so pleasant an excursion to Winchester with Falconer, Col. Sabine, & Dr Robinson, Dean of Armagh, & others. I never enjoyed a day more in my life. I missed having a look at H. Watson. I suppose you heard that he met Forbes & told him he had a severe article in the Press.[6] I understood that Forbes explained to him that he had no cause to complain, but as the arcticle was printed, he wd not withdraw it, but offered to Forbes for him to append notes to it! which Forbes naturally declined.

I am extremely glad you have not given up thoughts of examining our Flora; I shd like beyond all things going part of the way on your expeditions.

My dear Hooker | Your's most truly | C. Darwin

DAR 114.2: 65

[1] The Friday before the proposed visit to Down by Bartholomew James Sulivan mentioned in the letter. See also letter to J. D. Hooker, [6 October 1846].
[2] Humboldt 1846–8. There is a copy in the Darwin Library–CUL.
[3] E. Forbes 1846.
[4] CD refers to the barnacle collected in the Chonos Archipelago, which he at first named *Arthrobalanus* and then *Cryptophialus minutus*. The descriptive work that CD anticipated taking 'some months' developed into an eight-year study of the whole sub-class, see *Living Cirripedia* and *Fossil Cirripedia*. Ghiselin 1969, Winsor 1976, Crisp 1983, and Southward 1983 describe CD's barnacle work in detail.
[5] CD attended the meeting of the British Association in Southampton, 10–16 September 1846. He was a vice-president of the committee for section D (zoology and botany).
[6] Possibly the attack published in the appendix to Hewett Cottrell Watson's *Cybele Britannica* (Watson 1847–59, 1: 55, 465–72). See letter from J. D. Hooker, [before 3 September 1846], n. 2, for Watson's complaints against Edward Forbes.

To W. D. Fox [before 3 October 1846][1]

Down Farnborough Kent
Friday

My dear Fox

The potato seeds were collected from ripe tubers in the Cordillera of central Chile, in a most unfrequented district, many miles from any inhabited spot, & where the plant was certainly in a state of nature— they were collected in the spring (ie autumn of the S. Hemisphere) of 1835 & shipped for England in May, & so probably were planted in the spring of 1836 by Prof. Henslow, from whom came the tuber which you had.—

I see *since* I wrote to you someone has urged the necessity of sending to S. America for new seed![2]

I have sometime thought of calling the attention of the readers of the Gardeners Chronicle to the remarkable difference of climate of the Chonos islands & central Chile, in both of which places the Potato grows wild— if you think it worth while to allude to this, refer to the 1st. Edit. of my Journal, if you have it.—[3]

Many thanks for your answer about the Potash.[4] I shall certainly set up a bottle & quill—

Very many thanks for your most kind invitation to us all, & I assure you it would give us much real pleasure to accept it; but the journey is fearfully long, & my wretched stomach hates visiting out, as much as the rest of the inward man enjoys seeing his old friends—

I enclose list of songs for Mrs Fox, & I wish with all my heart, I cd hear her singing them—[5] pray give our kind remembrances to her & believe me | My dear Fox | Ever Yours | C. Darwin

Christ's College Library, Cambridge (Fox 107)

[1] The information in this letter was used by Fox in a letter to the *Gardeners' Chronicle and Agricultural Gazette*, no. 40, 3 October 1846, p. 661.

[2] Continuing anxiety about the potato blight led to many different suggestions about the best way to maintain current stocks and to prevent further devastation. In his letter to the *Gardeners' Chronicle and Agricultural Gazette* (see n. 1, above) Fox asserted that there was no point in growing from seed, since CD's potatoes were also diseased.

[3] The first edition gives more details than the second, although both refer to the difference in climate, see *Journal of researches*, pp. 347–8, and *Journal of researches* 2d ed., pp. 285–6. John Lindley, editor of the *Gardeners' Chronicle*, was an outspoken proponent of the view that potato rot was caused by climatic factors.

[4] CD's letter containing the query has not been found, nor has Fox's reply.

[5] Fox's first wife, Harriet, had died in 1842 (see *Correspondence* vol. 2, letter to W. D. Fox, 23 March [1842]). He remarried on 20 May 1846; his second wife was Ellen Sophia Woodd.

To Charles Lyell [3 October 1846]

[Down]

My dear Lyell

I have been much interested with Ramsay;[1] but have no particular suggestions to offer: I agree with all your remarks made the other day.[2] My final impression is, that the only argument against him, is to tell him to read & reread the Principles, & if not then convinced to send him to Pluto.— Not but what he has well read the Principles! & largely profited thereby.— I know not how carefully you have read this paper, but I think you did not mention to me, that he *does* p. 327 believe that the main part of his great denudation was effected during a vast (almost gratuitously assumed) slow *Tertiary* subsidence & subsequent *Tertiary* oscillating slow elevation.— So our high-cliff argument is inapplicable. He seems to think his great subsidence only *favourable* for *great* denudation, I believe, from the general nature of the off-shores sea's bottoms, that it is almost necessary: do look at two pages p. 25 of my S. American Vol. on this subject, when out next week.—[3]

The foundation of his views viz of one great sudden upheaval,[4] strikes me as threefold.— 1st. to account for the great dislocations; this strikes me as the odder, as he admits that a little northwards, there were many & some violent dislocations at many periods during the accumulation of the palæozoic series.—[5] If you argue against him allude to the cool assumption that petty forces are conflicting:[6] look at Volcanos; look at recurrent similar earthquakes at same spots; look at *repeatedly* injected intrusive masses. In my paper on Volc. Phæn. in Geolog. Transact, I have argued (& Lonsdale thought well of the argument, in favour, as he remarked, of your original doctrine) that if Hopkins views are correct, viz that mountain-chains are subordinate consequences to changes of level in mass, then, as we have evidence of such horizontal movements in mass having been slow, the formation of mountain chains (differing from volcanos only in matter being **in**jected instead of **e**jected) must have been slow.—[7]

Secondly, Ramsay has been influenced, I think, by his alpine insects: but he is wrong in thinking that there is any necessary connection of Tropics & large insects; videlicet Galapagos. Arch. under the Equator. *Small* insects swarm in all parts of Tropics, though accompanied generally with large ones.—[8]

Thirdly, he appears influenced by the absence of newer deposits on the old area, blinded by the supposed necessity of sediment accumulating somewhere near (as no doubt is true) & being *preserved*;—an example, as I think, of the common error, on which I wrote to you about. The *preservation* of sedimentary deposits being, as I do not doubt, the exception, *when they are accumulated during periods of elevation or of stationary level*; & therefore the preservation of newer deposits would not be probable, according to your view that Ramsay's great palæozoic masses were denuded, whilst *slowly rising*— Do pray look at end of II. Chapt at what *little* I have said on this subject in my S. American volume.[9]

I do not think you can safely argue that *whole* surface was probably denuded at the same time to the level of the lateral patches of Magnesian conglomerate.—

The latter part of paper strikes me as good, but obvious. I shall send him my S. American Vol. for it is curious on how many similar points we enter, & I *modestly* hope it may be a half oz: weight towards his conversion to better views. If he w^{d} but reject his great sudden elevations, how sound & good he w^{d} be.— I doubt whether this letter will be worth the reading.

Ever your's | C. D.

Postmark: OC 3 1846
American Philosophical Society

[1] Andrew Crombie Ramsay, who was employed by the Geological Survey. CD's comments apply to Ramsay 1846, a copy of which is in the Darwin Library–CUL.

[2] CD had made a trip to London with Emma which he recorded in his Account Book (Down House MS) on 30 September.

[3] Ramsay 1846 dealt with evidence for marine denudation revealed by the work of the Geological Survey in Wales. In *South America*, p. 25–6, CD had argued that the cliffs on St Helena were formed by subsidence of sloping volcanic strata, allowing the ocean to cut away at successively higher levels of the slope while depositing detritus underwater. Ramsay considered slow subsidence an important aid to the process though not a necessary one (Ramsay 1846, pp. 326–7).

[4] Ramsay (1846, pp. 314–17) believed that the convoluted strata appearing over much of South Wales had been produced by a single great catastrophe occurring near the end of the Carboniferous.

[5] Ramsay (1846, p. 314) had noted pre-Carboniferous disturbances in North Wales.

[6] Ramsay (1846, p. 317) argued that in view of the enormous lateral pressure required to fold strata thousands of feet thick, no such result 'could have obtained from the conflicting action of petty forces working at various times'. Lyell was apparently considering Ramsay's argument in connection with revisions for C. Lyell 1847. Lyell did not follow CD's suggestions however, adding only a footnote to his volume commenting on the short time-span allowed by Ramsay (see C. Lyell 1847, p. 168 n.).

[7] As assistant-secretary of the Geological Society, William Lonsdale oversaw the refereeing of CD's paper. See *Correspondence* vol. 2, letters to William Lonsdale, [15 May 1838], [*c.* June 1838], and [8 March 1839]. CD had cited William Hopkins (1838) in his discussion of the slow elevation of mountain chains ('On the connexion of certain volcanic phenomena in South America', *Collected papers* 1: 76–80).

[8] Ramsay had cited the conclusion of Peter Bellinger Brodie (1845), that the small size of most fossil insects in British Secondary strata indicated a cool or temperate climate. This conclusion seemed to contradict the tropical character of many Secondary fossils, but Ramsay argued that Brodie's insects

constituted evidence for the continued existence of a temperate highland region existing side by side with tropical lowlands throughout the Secondary period. Thus, the extensive denudation which eroded the once massive formations of South Wales had not occurred until the Tertiary (Ramsay 1846, pp. 324–5). The presumed thickness of these formations had, of course, been part of Ramsay's argument that only a great catastrophe could have folded them.

9 *South America*, pp. 135–9. Here CD argued that substantial sedimentary deposits were not likely to accumulate in periods of elevation because of the denuding action of the breakers. He used this argument to account for the absence of fossiliferous Tertiary strata on the west coast of South America. He used it again in the *Origin* (pp. 172–3, 290–2) as a partial explanation of the absence of transitional forms in the fossil record.

To J. S. Henslow [5 October 1846]

Down Farnborough Kent
Monday Morning

My dear Henslow

In a few days' time my third & last Part of the Geology of the Voyage of the Beagle, viz on S. America will be published, & I want to know how I can send you a copy. I take shame to myself that the others were not sent, for I consider that you have a *right* to them.[1] Have you any house of call, where your parcels accumulate?

I am very sorry & so is my wife that your scheme of paying us, a little visit on your way to Mr Jenyns broke down.—[2] I shd. have much enjoyed having you here. I was at Southampton & saw there L. Jenyns: I did not think he was looking at all well; sadly too thin: I was very glad to hear from him, that Mrs. Henslow was a little better. L. Jenyn's new book appears a very nice one, as far as I have read.—[3] I wish you had been at Southampton; there was a capital congregation of naturalists & I saw many old friends & was introduced to many new acquaintances. Altogether we enjoyed (for my wife was with me) our week exceedingly, & took some little excursion, especially one to Winchester Cathedral. I think I shall certainly attend the Oxford meeting[4] & no doubt you will be there.

You cannot think how delighted I feel at having finished all my Beagle materials, except some invertebrata: it is now 10 years since my return, & your words, which I thought preposterous, are come true, that it wd take twice the number of years to describe, that it took to collect & observe.

Farewell my dear Henslow, how I wish that I lived nearer to you | Yours most truly | C. Darwin

Endorsement: '6 Oct 1846.'
DAR 93: 15–16

1 *Coral reefs* and *Volcanic islands*.

2 George Leonard Jenyns, Henslow's father-in-law.

3 Jenyns 1846. See letter to Leonard Jenyns, 17 October [1846], for CD's comments on the book.

4 The seventeenth meeting of the British Association was scheduled to take place in Oxford in June 1847. CD did attend (see 'Journal'; *Correspondence* vol. 4, Appendix II, and de Beer ed. 1959, p. 11).

To J. D. Hooker [6 October 1846]

Down Farnborough Kent
Tuesday

My dear Hooker

The Sulivans come here next Friday or Saturday; will you come here on Saturday if you can? I intend asking some others & my phaeton shall be at Sydenham to meet the Croydon Train, which leaves London Bridge at 3°·15′— I hope you will come; & please kindly send me a line in answer

Ever yours | C. Darwin

DAR 114.2: 66

To J. D. Hooker [8 October 1846]

[Down]
Thursday

My dear Hooker

I do not know whether you will get this; but if you do, it is to beg you, if you have & if you can **perfectly spare** it, to bring any good work on the Corallinas or Nulliporas of Lam:—[1] As they are undoubted plants,[2] perhaps you may have something. I am going to write a paper on their propagation.—[3] You are a real good man for coming

C. D.

Forbes is obliged to be off for Wales—

I have not heard from Falconer.—

A cousin of ours, Mr Allen,[4] a very tall young man of the Colonial Office will probably meet the Phaeton at Sydenham—

Postmark: OC 8 1846
DAR 114.2: 67

[1] Lamarck 1815–22. There is a copy in the Darwin Library–CUL. The calcareous secretions of corallines and nullipores form part of many coral reef structures, and CD discusses them in *Coral reefs*, pp. 9–10, 24–5, 85–6.
[2] *Coral reefs*, p. 9. Lamarck had classed them as animals (Lamarck 1815–22, 2: 199–200, 325–6).
[3] No such paper has been found. CD had been interested in the propagation of corallines since the *Beagle* voyage (*Correspondence* vol. 1, letters to J. S. Henslow, 11 April 1833 and 24 July – 7 November 1834). See Sloan 1985.
[4] Probably John Hensleigh Allen, see *Emma Darwin* 2: 100.

To John Lindley [*c.* 10 October 1846][1]

Down Farnborough Kent

Dear Sir

I have directed my Publishers to send a copy of my S. American Geology to the Gardener's Chronicle, though *perfectly* aware it is quite out of the line of books reviewed in it. Yet there is one single passage, on the ground of benefit to others, which I venture to call your attention to, more especially as the inference has been chiefly drawn from an article in the Gardeners Chronicle,[2] this is at the end of the III. Chapt & refers to the exceedingly *pure* salt of Patagonia not answering well for preserving meat.[3] I have suggested to the merchants of La Plata to add deliquescent chlorides to this natural salt; but my suggestion will never be heard, without it be backed by such authority as your's. The consumption of salt is very great in those countries & in peacable times would be immense. Will you add the information what percentage of muriate of lime ought to be added & its market cost?[4]

I hope on the above grounds you will excuse the *very unusual* step I have taken in writing to you about my own work.—

I have not forgotten your extremely kind offer of allowing me to consult Books in the Hort. Library;[5] I have lately been busy with my geology & shall be for some time employed on Invertebrate zoology, but hereafter your kind offer will be of the greatest service to me.

Pray believe me | My dear Sir, with much respect. Yours very faithfully | C. Darwin

Of course this note requires no sort of acknowledgment.

Kew (Lindley letters, A–K: 191)

[1] Dated from the references to the publication of *South America* in the letters to J. S. Henslow, [5 October 1846], and to A. C. Ramsay, 10 October [1846]. *The Publishers' Circular* of 2 November 1846 lists the date of publication as being between 14 and 30 October.

[2] In a report about the Agricultural Chemical Association, *Gardeners' Chronicle and Agricultural Gazette*, no. 6, 8 February 1845, p. 93.

[3] *South America*, p. 75, where CD erroneously cited the reference as *Horticultural and Agricultural Gazette*. See also letter to Trenham Reeks, [before 8 February 1845], and Reeks's reply, 8 February 1845.

[4] A short review of *South America* appeared in *Gardeners' Chronicle and Agricultural Gazette*, no. 5, 30 January 1847, p. 71. The additional information was not supplied in the review. Instead, the editor included CD's statements about the purity of Patagonian salt and suggested that someone might wish to experiment on reducing the purity to the standard of European salt.

[5] Lindley was vice-secretary of the Horticultural Society and effectively ran it between 1841 and 1858.

To A. C. Ramsay 10 October [1846]

Down Farnborough Kent
Oct. 10th.

Dear Sir

Having just read your excellent memoir on Denudation,[1] I have taken the liberty to send you a copy of my volume on S. America, finding that we have discussed some related questions.— I wish I had profited by your memoir before publishing my volume.— I see that we entirely agree on the sea's great power compared with ordinary alluvial action, & likewise on the frequency of grand oscillations of level & on several other points. If you had time to read parts of my volume, I should much like to discuss with you many cases, such as my notion of subsidence being necessary for the formation of high sea-cliffs, *as inferred* from the *nature of the sea's bottom off* them,[2] likewise the horizontal elevation of the Cordillera, as inferred from the sloping gravel fringes in the valleys[3]—on the non-horizontality of lines of escarpments round old bays, &c &c—[4]

I grieve to see how diametrically opposite our views are (I being a follower of Lyell) on the probability of great & sudden elevations of mountain-chains: I cannot but think, that you would have estimated existing forces, as more than "petty" & entertained some doubt about their being "conflicting"[5] had you inspected with your own eyes the wide area of recently elevated & similarly affected districts in S. America. There is much which I could say on this head, but I will not intrude on you. May I ask, whether you do not admit Mr Hopkin's views of mountain-chains being the subordinate effects of fractures consequent on changes of level in the surrounding areas;[6] & does not all the evidence, which we possess, tend to show that widely-extended elevations are slow, & may we not infer from this that the formation of mountain-chains is likewise probably slow.— I cannot see any difficulty, after a line of fracture has been once formed, in fluidifyed rock being pumped in by as many strokes, as it is pumped out in a common volcano, & yet producing a symmetrical effect.—

But I much fear that I have cause to apologise for having written at such unreasonable length: the interest excited in me by your Memoir, must plead my excuse, & trusting that you will forgive the liberty I have taken | I remain, dear Sir | Yours faithfully | C. Darwin

Imperial College Archives

[1] Ramsay 1846.
[2] See *South America*, pp. 25–6, and letter to Charles Lyell, [3 October 1846], n. 3.
[3] CD believed that gravel terraces along the valleys of the Cordillera were marine deposits laid down as the land was gradually elevated (*South America*, pp. 62–7).
[4] CD believed that the non-horizontal inclination of step-like terraces along the valleys of Coquimbo and Guasco was due to the elevation of land around bays (*South America*, pp. 41–6).
[5] See letter to Charles Lyell, [3 October 1846], especially n. 6.
[6] Hopkins 1838. See letter to Charles Lyell, [3 October 1846], n. 7.

To Leonard Jenyns 17 October [1846]

Down Farnborough Kent
Oct 17th.

Dear Jenyns

I have taken a most ungrateful length of time in thanking you for your very kind present of your Observations.[1] But I happened to have had in hand several other books & have finished yours only a few days ago. I found it very pleasant reading & many of your facts interested me much: I think I was more interested, which is odd, with your notes on some of the lower animals than on the higher ones. The introduction struck me as very good; but this is what I expected, for I well remember being quite delighted with a preliminary essay to the first number of the Annals of N. History.[2]

I missed one discussion, & think myself illused, for I remember your saying you would make some remarks on the weather & Barometer,[3] as a guide for the ignorant in prediction. I had, also, hoped to have perhaps met with some remarks on the amount of variation in our common species: Andrew Smith once declared he would get some hundreds of specimens of larks & sparrows from all parts of Great Britain & see whether with finest measurements he c[d] detect any proportional variations in beaks or limbs &c. This point interests me from having lately been skimming over the absurdly opposite conclusions of Glöger[4] & Brehm;[5] the one making half-a dozen species out of every common bird & the other turning so many reputed species into one. Have you ever done anything of this kind; or have you ever studied Gloger's or Brehm's works?[6]

I was interested by your account of the Martins, for I had just before been utterly perplexed by noticing just such a proceeding as you describe; I counted seven one day lately visiting a single nest & sticking dirt on the adjoining wall.[7] I may mention that I once saw some squirrels eagerly splitting those little semi-transparent spherical galls on the back of oak-leaves, for the maggot within; so that they are insectivorous.— A Cychrus rostratus once squirted into my eye & gave me extreme pain; & I must tell you what happened to me on the banks of the Cam in my early entomological days; under a piece of bark I found two carabi (I forget which) & caught one in each hand, when lo & behold I saw a sacred Panagæus crux major; I could not bear to give up either of my Carabi, & to lose Panagæus was out of the question, so that in despair I gently seized one of the carabi between my teeth, when to my unspeakable disgust & pain the little inconsiderate beast squirted his acid down my throat & I lost both Carabi & Panagæus![8]

I was quite astonished to hear of a terrestrial Planaria; for a year or two ago I described in the Annals of N. H. several beautifully coloured terrestrial species of the S. hemisphere, & thought it quite a new fact.[9] By the way you speak of a sheep with a broken leg not having flukes: I have heard my Father aver that a fever or any *serious accident*, as broken limb will cause in a man all the intestinal worms to be evacuated; might not this possibly have been the case with the flukes in their early state.[10]

I hope you were none the worse for Southampton; I wish I had seen you looking rather fatter: I enjoyed my week extremely & it did me good. I missed you the few last days & we never managed to see much of each other; but there were so many people there, that I for one hardly saw anything of any one.—

Once again thank you very cordially for your kind present & the pleasure it has given me, & believe me | Ever most truly yours | C. Darwin

I have quite forgotten to say how greatly interested I was with your discussion on the statistics of animals: when will Nat: Hist: be so perfect that such points, as you discuss, will be perfectly known about any one animal![11]

Postmark: 18 OC 18 1846
Bath Reference Library (Jenyns papers, 'Letters of naturalists 1817–76', Quarto vol. 2: 51(9))

[1] Jenyns 1846. See letter to Leonard Jenyns, [14 or 21 August 1846]. There is an annotated copy in the Darwin Library–CUL.
[2] Jenyns 1837. On the copy of this letter he made for Francis Darwin (DAR 145), Jenyns noted: 'Darwin has here made a mistake: the paper to which he alludes was published in the first Number of the "Magazine of Zoology and Botany." 1837.' The confusion undoubtedly arose because the *Magazine of Zoology and Botany* changed its name to *Annals of Natural History* in 1838. Jenyns' paper, dated 21 April 1836, appeared in the volume for 1838. It was entitled 'Some remarks on the study of zoology, and on the present state of the science'.
[3] Jenyns noted on the copy of this letter (DAR 145) that, 'These subjects were subsequently treated of in a separate volume "Observations in Meteorology." 1858.' See Jenyns 1858.
[4] Constantin Wilhelm Lambert Gloger. CD's annotated copy of Gloger 1833 is in the Darwin Library–CUL.
[5] Brehm 1831.
[6] Jenyns 1837, pp. 23–6, discusses the problem of determining highly variable species, citing both Brehm 1831 and Gloger 1833.
[7] Jenyns (1846, pp. 161–2) described young martins daubing mud on house walls with no apparent intention of building a nest; he interpreted the behaviour as an instinct showing itself 'before it is wanted'.
[8] Jenyns mentioned a similar incident (Jenyns 1846, pp. 234–5). For CD's recollections of *Panagaeus crux major* see *Autobiography*, p. 62.
[9] 'Brief descriptions of several terrestrial *Planariae*', *Collected papers* 1: 182–93.
[10] A phenomenon previously mentioned in letter to Henny Denny, 3 June [1844].
[11] Jenyns 1846, pp. 331–65, followed by a 'Calendar of periodic phenomena in natural history, as observed in the neighbourhood of Swaffham Bulbeck', pp. 366–412.

To Joseph Beete Jukes[1] [18 October 1846]

Down Farnborough Kent
Sunday

My dear Sir

I am sorry to say I know nothing of the fossils in question; I certainly never saw any fossils collected by Stokes,[2] & I looked through the collection sent to me.. I think some specimens were sent back to him, but I have one box now, & when you are in town, I will send it you, if you choose.

I am much obliged for your very kind enquiries about my health.— I enjoyed my week at Southampton extremely. What dreadful weather you must have had during the last fortnight amongst the mountains.—[3]

Yours very truly | C. Darwin

Postmark: OC 18 1846
University of Oklahoma Libraries (History of Science collection)

[1] CD and Jukes had corresponded in 1838, when Jukes sought an appointment as geological surveyor of Newfoundland (see *Correspondence* vol. 2, letters to J. B. Jukes, 6 December 1838 and 25 December [1838]).
[2] Jukes had recently returned from a survey of the north-east coast of Australia as naturalist aboard H.M.S. *Fly*, 1842–6. He had apparently sought information from John Lort Stokes, who had surveyed the north-west coast.
[3] Jukes was engaged in the geological survey of North Wales.

To J. D. Hooker [18 October 1846][1]

[Down]
Sunday

My dear Hooker

I sh^d very much enjoy a walk with you, as proposed, at Kew, but the infamous weather & more especially my utter ignorance how long it will take the artist to do M^r. Arthrobalanus, render it impossible to fix with certainty. I go to London tomorrow & I w^d propose to come about 1 oclock & *lunch* with you if you will give me some at West Park,[2] & afterwards have a walk in the Garden.— Will you send me a line to "7 Park St Grosvenor Square",[3] telling me honestly whether this will be **perfectly** convenient to you & whether you will allow me to take my **chance** of coming, so that if I do not come by 2. you will know that I have been prevented.

Ever yours | C. Darwin

Many thanks for your note about the Omnibuses.

DAR 114.2: 69

[1] Dated on the basis of CD's trip to London from 19 to 23 October 1846, recorded in his Account Book (Down House MS) on 24 October.
[2] The Hooker family home, at Kew.
[3] The address of Erasmus Alvey Darwin, with whom CD stayed.

To Smith, Elder & Co. [19 October 1846]

7. Park St | Grosvenor Sq^r
Monday Night

Dear Sir

I have just received my copy of my work & I write in great haste to beg you *instantly* to give instructions to the Binder not to bind the coloured Plate with its back to all the letter press; it is almost impossible to refer to it, as it is bound in my copy; I never saw such a *stupid* trick.—

You must, if you please, get the man to cut out **all** that are bound & paste them in to front the letter-press[1]

In Haste | Dear Sir | Yours very faithful | C. Darwin

I like much the looks of the volume. I have not yet seen any advertisements

Endorsement: 'Oct 20. 1846'
American Philosophical Society

[1] In CD's personal copy of *South America* the coloured plate showing three sections of the Cordillera is, as he says, bound so that it faces towards the back cover; other copies of *South America* seem to have it bound in the correct way. CD's copy is in the Cambridge University Library.

To J. D. Hooker [26 October 1846][1]

Down Farnborough Kent
Monday Morning

My dear Hooker

Your drawing is quite *beautiful*;[2] I cannot thank you enough, & I feel, as I before said guilty—your goodnature is as wonderful as mesmerism.— I have been reading heaps of papers on Cirripedia, & your drawing is clearer than almost any of them. The more I read, the more singular does our little fellow appear,[3] & as you say, looking at its natural size, a microscope is a most wonderful instrument. How different would the drawing have been, if I had employed an artist! not to mention the invaluable assistance of having my loose observations confirmed, & the several points observed only by you.— I shall of course state this in the beginning of my paper,[4] & when I have not sëen the thing, give it on your authority.

I have a few questions to ask & I sh^d. be much obliged for an answer before very long.—

(1) Is the *fold* at the posterior & lower end of stony, dentated valve, fimbriated & of the same texture, as the two inner fleshy & fimbriated valves? it is shaded in the same manner.—

(2) Regarding fig. 13 & 14, you speak as if you thought the kind of bars at the base of the two jaws were muscles; do you feel sure of this, for on the voyage & since they felt *decidedly* hard & shelly, & Burmeister has described similar hard supports to the jaws in Lepas.[5]

(3) Did you happen to notice, whether the cherotherium footsteps pointed upwards or to the base of animal? I can look, if you do not remember.—[6]

(4) I see you have not put in, any trace of the oblique articulation between the head & sternum; do you deliberately give it up? I certainly *thought* that there was one.—

(5) The head in fig. 10 seems to me hardly distinct enough dorsally from body; whereas in fig 11. it appears to me, just as I saw it; I can get artist to vary the line a trifle.

(6) Would you please return my own wretched drawing; & I sh^d be very much obliged for a copy, in same style as your others, of my fig. with legs retracted; the merest outline would do; I want it because the appearance of these larvæ are now so utterly different from what they were when alive & it shows position of legs when fully retracted.— I describe it, as pointed coffin-shaped, & twice as

long as the egg in the last previous state, (your fig. 21) that is *not* including the two projections: I send a tracing of your two figures, for the scale sake: I see I state that the two anterior club-shaped organs in state on fig 21. are *very much longer* than in fig 20.

I have not knowledge enough to discuss the nature of the limbs in this larva; & indeed I doubt whether any one has, for its relations are to various very distinct families.

I return the lens, with very many thanks, & with ditto for having written to Adie—[7] My lens have been altered (for 3^{s} 6^{d} only!) & a great comfort it is. You really are the most goodnatured man I ever knew,—too goodnatured for so true a zealot to your own science,—and I thank you cordially—

Ever yours | My dear Hooker, C. Darwin

P.S. | I find that I have one other query.— In fig. 5. the inner & third tunic, is not represented as enveloping whole animal & ovisac, but folding in & terminating at the posterior & lower edge of ovisac; hence the ovisac appears like a reduplication of this 3^{d} & inner membrane. My impression is different, viz that this 3^{d} inner tunic is continuous with muscular tunic.— Pray mind, I *beg* you not for this or any other query, to dissect any more, indeed I doubt whether you have the specimens; for I will look to this point, only I have so very much more confidence in your observations than in mine own, that I am glad to know your impression.— I am drawing up my description, which runs out much longer than I expected or like.—

DAR 114.2: 68

[1] Dated on the basis of CD's statement that he has had his lens altered. An entry in his Account Book (Down House MS) for 24 October reads: 'Alteration of Microscope 3s. 6d.'

[2] Hooker had provided drawings of barnacles, including *Cryptophialus minutus* (see n. 3, below), which CD used in *Living Cirripedia* (1854), see p. 566 n. Hooker's drawings have not been found but some of CD's original figures, dating from the *Beagle* voyage, are in DAR 29.3: 72.

[3] *Cryptophialus minutus*, which CD originally called *Arthrobalanus*.

[4] No such paper has been found although CD evidently intended to publish one (see 'Journal'; Appendix II). The results of his researches were eventually published in *Living Cirripedia* (1854): 563–86.

[5] Burmeister 1834.

[6] Possibly an allusion to the small triangular points that cover the external membrane of *Cryptophialus* (see *Living Cirripedia* (1854), plate 23, figure 3), which resemble *Cheirotherium* footprints.

[7] Alexander James Adie, an optician and instrument-maker in Edinburgh.

To Librarian 27 October [1846 or 1848?][1]

Down.
Oct. 27th

Dear Sir

I will send by Carrier, (or bring myself) on Thursday morning, the Books in hand, & sh^{d} be obliged for vol. of Phil. Trans. for 1799, the Part with paper by

A. Knight on Fecundation of Plants.[2] Also a great work descriptive of animals in Ld. Derby's menagerie.—[3]

Yours faithfully | C. Darwin

University of Michigan Library (Department of rare books and special collections)

[1] The conjectured date is based on the publication of J. E. Gray 1846 (see n. 3, below) and on the assumption that CD would not need to borrow early volumes of the *Philosophical Transactions of the Royal Society* after 1848 since he inherited his father's set. There is, however, no record of exactly when the volumes were moved from Shrewsbury to Down. This letter could not have been written in 1847 because CD was in Shrewsbury on this date in that year.

[2] Knight 1799. This and other works by Thomas Andrew Knight were frequently cited by CD in *Natural selection*, *Origin*, *Variation*, and his botanical works.

[3] J. E. Gray 1846. Cited in *Natural selection* and *Variation.* If the conjectured date is wrong then J. E. Gray 1850 may be the work referred to. Edward Smith Stanley, Earl of Derby, had formed a menagerie at Knowsley Hall, Lancashire.

To Robert FitzRoy 28 October [1846]

Down. Farnborough. Kent.
Oct. 28th.

Dear FitzRoy.

I am extremely much obliged to you for your very kind letter received a fortnight since. It was very goodnatured of you to take the trouble to write at such length and I assure you that it was not thrown away, on me, for I have been deeply interested with your letter; and have read it several times— Some time ago, I got your pamphlet[1] and your clear statement gave me the first idea, I had of the connected history of the events at poor New Zealand.—[2] What a change then, since, I passed so tranquil an evening and night at Waimata;[3] and what a far more disastrous change at Tahiti![4] I most sincerely hope that your present quiet abode will do Mrs. FitzRoy much good,[5] pray give our very kind remembrances to her, & that your spirits will recover their wonted "elasticity"—

I am astonished to hear that you have any thoughts of taking a ship,[6] considering the sacrifice of leaving your family, but you are an indomitable man— Sulivan, his wife, and two youngest children, were staying here, when your letter came; and he expressed deep interest in hearing news of you he was going on to Hammonds.—[7] I saw Stokes, several times in London, whilst writing his book,[8]—but I could not 〈 〉

〈 〉 How long you have remembered my speech about the ditch; but you would almost believe it if you had seen me for the last half month daily hard at work in dissecting a little animal about the size of a pin's head from the Chonos Arch.[9] & I could spend another month on it, & daily see some more beautiful structure! 〈 〉[10]

Farewell, dear Fitz-Roy, I often think of your many acts of kindness to me, and not seldomest on the time, no doubt quite forgotten by you, when, before making Madeira, you came and arranged my hammock with your own hands, and which, as I afterwards heard, brought tears into my father's eyes.[11]

Copy incomplete
DAR 144 and *LL* 1: 332

[1] FitzRoy 1846.
[2] See letter to Robert FitzRoy, 1 October 1846, n. 3.
[3] CD describes his trip to Waimate in *Journal of researches*, pp. 507–11; it was his only pleasant memory of New Zealand (p. 514). For FitzRoy's account of the *Beagle*'s stay in New Zealand from 21 December to 30 December 1835, see *Narrative* 2: 564–618.
[4] Both CD and FitzRoy had liked Tahiti and thought it quite civilised in contrast to New Zealand. See *Collected papers* 1: 19–38. Since 1843, when Tahiti unwillingly became a French protectorate, there had been a succession of bloody battles between the French and local inhabitants (*EB*).
[5] Mary Henrietta FitzRoy had been dangerously ill following the birth of her fourth child in New Zealand (Mellersh 1968, pp. 222, 270).
[6] FitzRoy was never given command of another vessel; in 1848 he was appointed superintendent of Woolwich Dockyard and in 1849 he was appointed to conduct the trials of the *Arrogant*, the navy's first screw-driven steamship.
[7] Robert Nicholas Hamond served with Bartholomew James Sulivan under FitzRoy in the *Thetis* and the *Beagle*. See *Correspondence* vol. 1, letters to Caroline Darwin, 24 October – 24 November [1832] and 30 March – 12 April 1833.
[8] Stokes 1846.
[9] *Arthrobalanus*, the cirripede from the Chonos Archipelago later named *Cryptophialus minutus* (*Living Cirripedia* (1854): 566–86).
[10] This paragraph is taken from a fragment separate from the page from which the preceding text was transcribed. Francis Darwin identifies the recipient as FitzRoy (*LL* 1: 349), and the reference to *Arthrobalanus* fixes the date as October 1846; therefore it is probable that both sections of text are parts of the same letter.
[11] This valedictory paragraph, which Francis Darwin dated '1846' and included in *LL* 1: 332, probably belongs to this incomplete letter to FitzRoy. The original has not been found.

To Daniel Sharpe [1 November 1846]

Down Farnborough Kent
Sunday

My dear Sir

I have been much interested with your letter & am delighted that you have thought my few remarks, worth attention. My observations on foliation are more deserving confidence than those on cleavage;[1] for during my first year in clay-slate countries I was quite unaware of there being any marked difference between cleavage & stratification; I well remember my astonishment at coming to the conclusion that they were totally different actions, & my delight at subsequently reading Sedgwicks views;[2] hence at that time I was only just getting out of a mist. With respect to cleavage-laminæ dipping inwards, on mountain-flanks I have certainly often observed it, so often that I thought myself justified in propounding it as usual; I might perhaps have been some degree prejudiced by Von Buch's remarks,[3] for which in those days I had a somewhat greater deference than I now have.— The mount at M. Video (p. 146 of my Book) is certainly an instance of the cleavage laminæ of an hornblendic schist dipping inwards on both sides; for I examined this hill *carefully* with compass in hand & note book. I *entirely* admit, however, that a conclusion drawn from striking a rough balance in one's mind, is worth *nothing*, compared with the evidence drawn from one *continuous* line of section. I read Studer's Papers carefully & drew the conclusion stated, from it;[4] but I may very likely be in an error. I only state that I have "*frequently*" seen cleavage-laminæ dipping inwards on mountain-sides; that I cannot give up, but I daresay a general extension of the rule, (as might justly be inferred from the manner of my statement) would be quite erroneous.[5]

Von Buch's statement is in his Travels in Norway; I have unfortunately lost the reference & it is a high crime, I confess, ever to refer to an opinion, without a precise reference. If you never read these Travels, they might be worth skimming, chiefly as an *amusement*; & if you like, & will send me a line by the Gen. Post on Monday or Tuesday, I will either send it up with Hopkins on Wednesday, or bring it myself to Geolog. Soc. I am very glad you are going to read Hopkins;[6] his views appear to me *eminently* worth well comprehending: false views & language appear to me to be almost universally held by geologists on the formation of fissures, dikes & mountain-chains. If you would have the patience, I sh^d^ be glad if you w^d^ read in my "Volcanic Islands" from page 65, or *even 54* to 72, viz on the lamination of volcanic rocks: I may add that I sent the series of specimens there described to Prof. Forbes of Edinburgh, & he thought they bore out my views.[7]

There is a short extract from Prof. Rogers' in the last Eding. New Phil. Journ. well worth your attention on the cleavage of the Appalachian Chain,[8] & which seems far more uniform in direction of dip, than in any case, which I have met with: the Rogers' doctrine of the ridges being thrown up by great waves I believe is monstrous; but the manner in which the ridges have been thrown over (as if by a lateral force acting on one side on a higher level than on the other) very curious, & he now states that the cleavage is ⟨parallel⟩[9] to the axis-plane of these thrown over ridges.— Your case of the limestone beds to my mind is the greatest difficulty on any mechanical doctrine; though I did not expect ever to find actual displacement, as seems to be proved by your shell-evidence.—[10] I am extremely glad you have taken up this *most* interesting subject in such a philosophical spirit; I have no doubt you will do much in it;—Sedgwick let a fine opportunity slip away.— I hope you will get out another section like that in your letter; these are the real things wanted.

Believe me Yours very sincerely | C. Darwin

Postmark: 2 NO 2 1846
British Library (Add. MS. 37725: 4–5)

[1] See *South America*, pp. 162–8. In a paper read to the Geological Society on 2 December 1846, Sharpe referred to CD's views (see Sharpe 1847, p. 105). During the *Beagle* voyage, and afterwards while he was working on the geology of South America, CD summarised his observations on cleavage and stratification, see DAR 41.

[2] Sedgwick 1835, especially pp. 469–75.

[3] Buch 1813. A copy is in the Darwin Library–CUL.

[4] Studer 1842. In *South America* (p. 164 n.) CD refers to only this paper by Bernhard Studer. The source is erroneously given as volume twenty-three instead of volume thirty-three of the *Edinburgh New Philosophical Journal* (April–October 1842).

[5] CD's discussion of inwardly dipping cleavage planes, in which he cited Buch 1813, p. 169, and Studer 1842, is in *South America*, p. 164. CD's statement there is slightly different: 'On the flanks of the mountains both in Tierra del Fuego and in other countries, I have observed that the cleavage-planes frequently dip at a high angle inwards'.

[6] Sharpe 1847, p. 100, cites three publications by William Hopkins: Hopkins 1836, 1838, and 1845–56. It is probably Hopkins 1838 that is referred to here.

[7] See letters to J. D. Forbes, 11 October [1844] and 13 [November 1844].

[8] H. D. Rogers 1846.

[9] The word 'parallel' has been added in pencil over a repair to the manuscript. It is not in CD's hand.

[10] Described in Sharpe 1847, pp. 75–87. CD cited Sharpe on this point in his article on geology (*Collected papers* 1: 237) for the Admiralty *Manual of scientific enquiry* (Herschel ed. 1849):

> Fossil shells have been found by Mr. Sharpe in slaty rocks, which have had their shapes greatly altered, and all in the same direction; here then we have a guide to judge of the amount and direction of the mechanical displacement which the surrounding slate-rocks have undergone.

To J. L. Stokes 3 November 1846

Down Farnborough Kent
Nov 3^d^ 1846

My dear Stokes

I have just received to my great surprise the letters of which the enclosed are verbatim copies.[1] That with my signature was in my hand-writing. I remember enclosing it to you with one of your proof-sheets in answer to some query, whether Capt. Grey could be offended at your manner of referring to some Bay or River.—[2] I beg you to inform me immediately how it could possibly have been sent to Sir G. Grey. It places me in the position of wishing to make myself presumptuously impertinent to him,—a position the very opposite to my feelings regarding him.— I shall of course inform Sir G. Grey that I have written to you, & I should think it would be most agreeable to yourself to allow me to enclose your entire answer, or at least a paragraph from it; & I shall enclose a copy of this note. He will then see, the whole part which I have been made by some unknown means to play in this disagreeable affair.—

I hope Mrs Stokes continues to be in a pretty good state of health | Believe me, my dear Stokes | Yours very truly | C. Darwin

Perhaps you will return me this note & I will enclose it.—

Auckland Public Library (Sir George Grey collection)

[1] The enclosures were copies of the letter from George Grey, 10 May 1846, and the letter to J. L. Stokes, [November–December 1845].

[2] Collier Bay and Glen Elg River (see letter to J. L. Stokes, [November–December 1845], n. 2).

From J. L. Stokes 6 November 1846

Haverfordwest | South Wales
Saturday | Nov. 6th. 46

My dear Darwin

Your letter of the 3d. with its enclosure has *greatly* surprised and annoyed me. I remember receiving the note of yours you have alluded to and thought I had destroyed it at the time, but how, or by what unfair means it has been most wickedly sent to Governor Grey I am quite at a loss to know[1] It gives me great concern to think, that I should in any way be the means of placing you in such a disagreeable position, and rest assured it will ever be a matter of deep regret to your very faithful friend | J Lort Stokes

P.S. | I shall endeavour to find out the mischief maker.

Auckland Public Library (Sir George Grey collection)

[1] Stokes later concluded that CD's original letter had been sent to his printers mixed in with proof-sheets. See letters to Robert FitzRoy, 23 November [1846], and to J. L. Stokes, [*c.* 26 November 1846].

To J. D. Hooker [6 November 1846][1]

Down Farnborough | Kent
Friday

My dear Hooker

As usual I have to send you many thanks. Your letter & the rough drawings came in quite good time, for I lost four days by having been unwell & am nothing to boast of now.—

I have had two mornings more of dissection & made out some points pretty well—the articulation under mouth is one of the most distinct in whole body; the cheirotherium steps *mostly* point upwards, but some downwards & some obliquely.— My greatest doubt is about the relations of the inner tunic & ovisac; I have given it up in despair, for after shelling the animal perfectly clear of its two outer tunics, it is, I find, impossible to make myself quite sure that the oviscac is not an enlarged & constricted portion of the inner tunic. I hope to send my paper to Owen next week for judgment.[2] As you say there is an extraordinary pleasure in pure observation; not but what I suspect the pleasure in this case is rather derived from comparisons forming in ones mind with allied structures. After having been so many years employed in writing my old geological observations it is delightful to use one's eyes & fingers again.—

I shall be in London on Wednesday the 18th for Geolog. Soc 19th, 20, 21: do you think you shall be up either of these days, I would meet you at Craig's Court[3] or anywhere which would suit you, for you to put a few strokes into your drawings; there are two or three *very trifling touches*, which I think would improve matters. If you can come up, wd *you give me a few days notice*, as I have to appoint meetings with several people.

Many thanks for the Sertulariæ could you bring them with you? I **particularly** beg you to remember, that should you meet with anyone who would describe them, at once to ask me for them, for it is doubtful whether I shall attempt to describe any objects, except such as I began working at in their living state, & of which I made out something of their structure.[4] In my ordinary condition it is mortifying to find myself quite exhausted after at most 2½ hours work in whole day.—

It will indeed give me great pleasure if you will come here & spend some days & bring some work & if I have anything under the microscope I will take advantage of your wicked offer of assistance: whenever it would suit you it would suit us, so do sometime early this winter propose yourself, for as I have said all times are alike to us & I should enjoy ⟨i⟩t very much— I hope to have Lyell here before long on same terms.—

I do not like the thoughts of your Bombay expedition[5]

Ever yours | C. Darwin

DAR 114.2: 70–70b

[1] Dated by the relationship to the letter to J. D. Hooker, [12 November 1846], and on CD's implication that his paper was nearly completed ('Journal'; Appendix II).

[2] See letter to Richard Owen, 25 November [1846].

[3] The address of the Museum of Economic Geology and the Geological Survey of Great Britain.

[4] *Sertulariae* are small branching polyps, and although CD investigated coral and coralline species he never undertook work on this genus.

[5] Hooker had been invited to accompany Alexander Gibson, conservator of forests in Bombay, on a four to five month tour of the Cannar (now Kanara) province of India (Huxley ed. 1918, 1: 216).

To George Grey 10 November 1846

Down Farnborough Kent
Nov. 10. 1846

My dear Sir

I beg to thank you for the courteous tone of your communication of the 10th of May 1846, considering the circumstances under which it was written. I enclose a letter which I immediately wrote to Capt. Stokes & his answer;[1] these will, I trust, exonerate us of intentional impertinence. Some most malicious person must have sent my note to you. I have been much mortified by perusing it, & though I am not presumptuous enough to suppose that you can care much for my opinion of your work on Australia,[2] it is a satisfaction to me to be enabled to name to myself many individuals, to whom I have expressed my strong opinion of the many high qualities shown in your work, of which, the amusement it afforded, was but a small part. Your account of the aborigines I have always thought one of the most able ever written.— As we are not likely to have any further communication, permit me to add that I have a most pleasant recollection of our former acquaintance.—[3]

With much respect, I beg to remain | Your's faithfully | Ch. Darwin

His Excellency | Sir G. Grey | &c &c &c

Auckland Public Library (Sir George Grey collection)

[1] Letter to J. L. Stokes, 3 November 1846, and letter from J. L. Stokes, 6 November 1846.
[2] G. Grey 1841.
[3] CD had probably met Grey in the spring of 1837 when Robert FitzRoy was awarded the Royal Premium by the Royal Geographical Society (5 May 1837). At that time plans were being made for Grey's expedition, sponsored by the Society, to leave in the *Beagle* in early June (*Journal of the Royal Geographical Society of London* 7 (1837): x–xi). CD had also visited the *Beagle* on 28 May 1837 (*Correspondence* vol. 2, letter to J. S. Henslow, [28 May 1837]).

To J. D. Hooker [12 November 1846][1]

Down Farnborough Kent
Thursday

My dear Hooker

Would you be so kind as to send me a line to say whether you shall be up in London, on next Thursday, Friday or Saturday,, & if you are naturally coming up, will you appoint some hour & day at some place & I will be there: I am anxious to know *at once* whether you will be up, as I have appointments to make for all these days & cannot do so, till I hear from you.—

I believe Arthrobalanus has no ovisac at all!, & that the appearance of one is entirely owing to the splitting, & tucking up to the posterior penis, of the inner membrane of sack.— I have just found a Cirripede with an indisputably abortive anterior penis; so that this chief anomalous feature (viz two penes) in Arthrobalanus is in some degree brought within bounds.—

Ever yours | C. Darwin

The alteration in my microscope, in accordance with your advice, has really been beyond value: the porcupine quills better than the glass tubes; the Chutney Sauce capital, so that I have many daily memorials of you.

N.B. I have cleverly invented two blocks of wood to support my wrists when dissecting under microscope a splendid invention.

Adios.—

DAR 114.2: 71

[1] Dated on the basis of CD's intention to visit London from 18 to 21 November (see letter to J. D. Hooker, [6 November 1846]).

To J. D. Hooker [14 November 1846]

[Down]
Sat. Morn.

My dear Hooker

No one of my children was ever so delighted with a new plaything as I am with the looks of my charming lens'. Many thanks for them.

I write now to propose, as you say you have nothing else to do in London, that I shall come down to Kew on Thursday or Friday.— If both days suited you *equally* well, you should leave the choice open to me, according as my stomach feels; otherwise I would come on either day fixed by *Luncheon time*.

If I do *not* hear from you, in note to Park St., I shall expect you on Friday at Park St. at 11½.—

I find the microscope work really delightful, & I am this morning going to begin with a new animal & with my new lens'.—

Farewell— what a good thing is community of tastes, I feel as if I had known you for fifty years— adios | C. Darwin

Postmark: NO 14 1846
DAR 114.2: 72

To J. D. Hooker [15 November 1846]

[Down]
Sunday

My dear H.

I stupidly forgot yesterday, that I have an engagement on Thursday, & therefore I propose to come to Kew on Friday, that is without you prefer coming up to London.

Let me have a line at 7 Park St.

Ever yours | C. Darwin

Postmark: 16 NO 16 1846
DAR 114.2: 73

To J. D. Hooker [17 November 1846]

[Down]
Tuesday

My dear H.

I am delighted, I thought of coming to Kew— I only proposed the other plan, as I thought you would probably be coming into London on your own affairs.— I will be with you on Friday a little before Luncheon time, & much I shall enjoy it.— I have not time to accept your kind offer of the *three* Beds

Yours, | C. D.

Postmark: NO 17 1846
DAR 114.2: 74

To Catherine Darwin [22 November 1846][1]

[Down]
Sunday

My dear Catherine

I am most sorry to hear how very poorly my dear Father continues: I do hope his cough will get less tight & troublesome at night. Pray write often if it be but a line; you have been very good in writing hitherto & thank you much.

Give our best congratulations to Jos,[2] (to whom however I have to write myself) on all his anxiety being over: we are strongly on the girl-side-faction.— Give my best love to Caroline.

I enclose a second letter of FitzRoys',[3] I do not know whether my Father will care to hear it, but it costs only 2^d^ sending & you can read it or not as you like: I must have it returned *soon.*

I was at tea at Lyell's on Friday night; he dined at the Millmans[4] the day before & met Ld Lansdown,[5] Macaulay[6] & other great guns: there was much talk about Dr Darwin,[7]—Dr Kaye Shutleworth[8] was telling about the prophecy of nitric acid making explosive cotton[9] & some one added about the potato disease;[10] so Macaulay remarked every one has heard of his prophecy on steam vessels & railways & he then repeated the whole passage,[11] & added "though we have not yet the navigating balloons, we are of course, *as he says so*, sure to have them someday." So his prophetic spirit is the talk of London.—

Yours affect. | C. D.

DAR 92: 1–2

[1] The date is based on the references to the birth of Lucy Caroline Wedgwood and to Henry Hart Milman's dinner party. See nn. 2 and 4, below.

[2] Josiah Wedgwood III is being congratulated on the birth of his fourth daughter, Lucy Caroline, on 17 November 1846.

[3] For CD's reply see letter to Robert FitzRoy, 23 November [1846].

[4] Henry Hart Milman, rector of St Margaret's, Westminster. Charles Lyell discusses this dinner party in a letter to his father dated 26 November 1846 (K. M. Lyell ed. 1881, 2: 114–15).

[5] Henry Petty-Fitzmourice, 3d Marquis of Lansdowne.

[6] Thomas Babington Macaulay.

[7] Erasmus Darwin, CD's grandfather.

[8] James Phillips Kay-Shuttleworth, a Poor Law commissioner.

[9] There was great interest in Christian Friedrich Schönbein's discovery, in 1846, of gun cotton, a highly explosive compound prepared by steeping cotton in nitric and sulphuric acids. Erasmus Darwin in the *Botanic garden* (Darwin 1791, p. 25 n.) had noted:

> From the cheapness with which a very powerful gunpowder is likely soon to be manufactured from aerated marine acid, or from a new method of forming nitrous acid by means of mangonese or other calciform ores, it may probably in time be applied to move machinery, and supersede the use of steam.

[10] The potato crop in Ireland had failed for two years, and there was much speculation as to the cause. See also letter to J. S. Henslow, 28 October [1845]. The *Gardeners' Chronicle and Agricultural Gazette*, no. 12, 21 March 1846, p.179, had quoted Erasmus Darwin's account of Thomas Andrew Knight's theory that extended propagation from grafts and bulbs weakened plants and that they might be restored through propagation from seed. Erasmus Darwin had applied the theory to the tuberous propagation of potatoes (Darwin 1800, pp. 95–8).

[11] Darwin 1791 (pp. 29–30), Pt 1, Canto I, lines 289–96:

> Soon shall thy arm, UNCONQUER'D STEAM! afar
> Drag the slow barge, or drive the rapid car;
> Or on wide-waving wings expanded bear
> The flying-chariot through the fields of air.
> —Fair crews triumphant, leaning from above,
> Shall wave their fluttering kerchiefs as they move;
> Or warrior-bands alarm the gaping crowd,
> And armies shrink beneath the shadowy cloud.

The associated note to line 254 (p. 26), on steam power, reads:

> There is reason to believe it may in time be applied to the rowing of barges, and the moving of carriages along the road. As the specific levity of air is too great for the support of great burthens by balloons, there seems no probable method of flying conveniently but by the power of steam, or some other explosive material; which another half century may probably discover.

To Robert FitzRoy 23 November [1846][1]

Down. Farnboroug Kent.
Novr. 23rd.

My dear FitzRoy.

Many thanks for your letter which I have forwarded to Shrewsbury, as your kind enquiries will I am sure gratify my Father and my two sisters who live with him his health lately has been a good deal shaken— The Antipodes to which you refer, has not altered the style of your letters, as I know by comparison, for a few days ago I found in a bundle of old letters two addressed to me at B. Ayres from you & I have seldom read nicer & kinder letters.—

You will be glad to hear that Cap: King's T. del Fuego plants[2] have at last been published by D^r. J. Hooker (the Antarctic Botanist) in first rate style;[3]—he has also made out a capital memoir on my Galapagos plants (the collection of which you so much favoured, by leaving me on James Is^d.) and they turn out of extraordinary interest & novelty.—[4] I have had a disagreeable incident with Governor Grey of New Zealand;—when Stokes was writing his book[5] he sent me a proof, asking me my opinion whether he had contradicted some statement of Grey's civilly to which I answered in the affirmative, & ended my note in these words— "Poor Grey, what an amusing book he has written—but what a catalogue of mishaps & mismanagement; it was an expedition of a set of School Boys."—[6] Well, some malicious scoundrel, without Stokes' knowledge, sent this note to Grey—, who returned it to me with a short, though civil note![7] Is not this disagreeable and the more so, as if I had expressed my whole opinion it would not have been so contemptuous; thank Heaven, I hope I shall never see him again.[8]

Farewell— I hope that you may soon get a ship, & be on your own element again, though it seems a cruel wish— | Ever yours very sincerely— | C. Darwin.

P.S. I sent the Yarmouth Scrap to Lyell as he has worked far more than anyone, on the changes on the English Coast.

Copy
DAR 144

[1] The date is inferred from the date of the letter to George Grey, 10 November 1846.

[2] Plants collected on the first *Beagle* expedition to South America, 1826–30, under the command of Phillip Parker King.

[3] In addition to King's collection, Joseph Dalton Hooker described CD's and his own Tierra del Fuego plants in *Flora Antarctica* (J. D. Hooker 1844–7).

[4] Hooker had given a taxonomical description of the plants to the Linnean Society in 1845 (J. D. Hooker 1845d), and read a paper on their geographical distribution in December 1846 (J. D. Hooker 1846). CD had just read a preliminary version of J. D. Hooker 1846, see next letter.

[5] Stokes 1846.

[6] Letter to J. L. Stokes, [November–December 1845].

[7] Letter from George Grey, 10 May 1846.

[8] See letter to George Grey, 10 November 1846, n. 3. Grey accepted CD's apology and explanation. See *Correspondence* vol. 4, letter to George Grey, 13 November 1847.

To J. D. Hooker [23 November 1846][1]

[Down]
Monday

My dear Hooker.

I have read your paper pretty carefully,[2] but to fully appreciate it, it ought to be read two or three times, & that I shall do when in print. In my opinion, it is without comparison, the best essay on geograph. distrib. in any class, which I have ever met with; & poor judge though I may be, I have looked far & wide for such discussions in vain. I will praise it no more, though in truth I could say with earnestness much more.—

Now for my small criticisms.—

p. 3. I say mountains from 1000 to (above) 4000 ft: I know of three of about 3700 & one of 4700.

p. 4. swampy land on *summit* of isl^{ds} not rare

p. 5. "almost absolute sterility" a good deal too strong: much (& I think you ought to *allude* to this) of the lower lands, (which by the way is almost always *rocky*, very little *soil* in any part) is thickly covered with a starved poor yet often thick brushwood.

p. 9. C. Verds quite as barren, or rather more barren than Galapagos.—

p. 7. sentence about St. Jago unintelligible to me. (how interesting your remarks on the Compositæ)

p. 18— it is not quite obvious at first, whether the 28 species of Compositæ include *all* the Galapageian species.

p. 19 Does your remark mean that the same species of Compositæ are not widely spread in the same country, or are the same species rarely found in widely separated countries, that is more rarely than the species in other families.

p 21. some obscurity in a sentence marked in a discussion to me very interesting: are there not two species to each genus at the Galapagos?! how very surprising! your ratio is 1 to 1.7.—

p. 20 just consider whether a too strong impression is not given of the affinity of the Juan F: & St. Helena; is not the case of the *really*, not *analogically* related plants in these islands, quite the exception to the rule. Is Aristida a confined genus.— Do you keep distinct your terms of analogy & affinity—just reconsider the mere wording of this sentence, & the first impression it would give. My impression from this sentence jars against what I take from rest of paper; but I sh[d] have had to have kept this paper for a week to look back to all such & other points.

p. 23. in appearance, summit-land abundantly moist enough for tree-ferns.—

p. 27. The arrangement of leading paragraph, strikes me as awkward; as the *possibly altered* plants, come chiefly under W. Indian type, *as stated* **at end** of paragraph, ought they not to be mentioned earlier in the paragraph.— Ought you not, at least in a bracket, to state that your W. Indian type includes both sides of Is[th]. of Panama; I was in a perfect fidget to know whether you meant to *exclude* Panama.— So again, when first referred to (as *first* impressions *always tell*) would it not be better to add to "Mexican type", including temperate or dry or highland parts of both Americas; no **after** explanation quite does away with first impression that you mean *exclusively* Mexican.—

p. 28. (4 lines from top) ? add after the "24" out of the "45".— as now stands not perfectly clear at *first* sight.

p. 41. No doubt owing to my noddle being occupied by some stupid puzzle, I cannot, though I tryed for $\frac{1}{4}$ of hour, understand your Table; does in 2[d]. "American" mean extra-Galapageian, or exclusively American so as to exclude those *also* found in Pacific islands.— Third column, I cannot conjecture meaning of; does "islets" refer to the Galapageian islets, if so, under Charles isl[d] would it not be $\frac{96 \; 15}{81}$ instead of 74—or do islets refer to islands in other parts of world; even then I cannot make your numbers intelligible: I again repeat I daresay this is my stupidity, but *some* others will be as stupid. Though not properly introduced, would it not be well, as a *refresher*, to give a *separate* line for total of whole group, leaving blank the columns, under which no total can be given?

p. 41. James Is[d] by no means loftiest;—Albemarles has highest mountains.

Mem. Buccaniers formerly *much* frequented James Is[d] & I believe Charles.— ? allude to absence of gales (as I have done) not dispersing seeds from island to island? I suspect whenever damp parts (for such occur & a running stream!!) on windward coast of Chatham Is[d] are explored, your numerical results will I suspect be altered. *Is not the upland Flora much more peculiar than the lowland Flora?*

Your paper is in my eyes splendid, & *I long to have a copy.*
Ever yours | C. Darwin
I return lens & many thanks
Pray give my remembrances & thanks for his kind reception to Sir William.— Please inform me about your Sisters health.[3]

DAR 114.2: 75

[1] Dated on the basis of CD's visit to Hooker at Kew on 20 November 1846 (see letter to J. D. Hooker, [17 November 1846]), and the letter from J. D. Hooker, [24 November 1846], which preceded his reading of the Galápagos paper at the Linnean Society, see n. 2, below.
[2] The final version of Hooker's paper on the geographical distribution of Galápagos plants, J. D. Hooker 1846, was read to the Linnean Society on 1 and 15 December 1846.
[3] Elizabeth Hooker.

From J. D. Hooker [24 November 1846][1]

Tuesday.

Ten thousand thanks my dear Darwin for the most ungracious of all offices executed & in the most gracious manner. I do not think we have much to quarrell about.

The St Jago sentence is a case of bad Grammar, or something very like it I have made it now to mean that a scantier examination of St Jago has produced more plants than a more careful one of the Gals.

Confound the Compositæ; I do declare & maintain that, for the largest order in the world & the most ubiquitous, its individual *species* are less ubiquitous than they should be. I have been a dozen times at my wits' end (thank goodness it's not far to go) for expressive language— It is not that the Comp. have not many widely spread species but no great proportion of them in comparison to the bulk of the order.

I have remodelled the obnoxious sentence on props. of gen to sp. please return the enclosed if approved Ratios are most deceiving in Botany: plants are so much more widely dispersed than land animals, that small numerical differences must be received with thankful hearts. Also the prop. of gen. to sp. amongst all plants is only as 1:4: which latter figure diminishes with the area.

I have *mesmerised* the sentence about affinity between Juan F. & St Helena: but I maintain that it bears upon one of my conclusions anent Insular Peculiar Floras very directly:—that they have wholly inexplicable points of resemblance. I endeavour to draw a line between affinity & analogy, surely the former is the word when congeners are concerned.

Eheu the ugly sentence on the division of the plants. W. Indies Botanically speaking includes Panama: & I have put it in so; I used only the shortest sentences to avoid periphrasis, my besetting sin. How to get in Mexicos including *[*lowld*]*

&c &c &c[2] under three *breaths* passes me, however here it goes & the L. Soc. may gasp till the windows are opened. I have also capsized or distorted the sentence to give prominence to *your* dear altered species:

I am thoroughly horrified at your not comprehending the table, but not believe me in the smallest surprized: of all the crabbed papers it is the crabbeddest: & how you got so far with any powers of comprehension left is the wonder. The 2^d. column I have made to read "*Total* xclud. those common to America".. As every Pacific Galap. plant is also American, there is no use bothering further about the Cannibal Isld^s.— Perhaps it w^d. be better to say simply "confined to Galapagos"

The third column cannot be 96–15; because 15 indicates those "confined to group but found likewise on other Islets": whereas 3^d. col. indicates the whole number of species including some *America*, but xcluding such American as are (like the Galaps.) common to other Islets. It is however I *clearly* (as clearly as such a woolly brained

AL incomplete
DAR 100: 77–8

CD ANNOTATIONS
1.1 Ten . . . Gals. 2.3] *crossed pencil*
3.1 Confound . . . should be. 3.3] 'Because easily connected into species a few wide ranges is all that is wanted.' *added pencil*
3.4 It is not . . . order. 3.6] *scored pencil*
5.3 I endeavour . . . concerned. 5.5] *scored brown crayon*
6.1 Eheu . . . brained 8.5] *crossed pencil*
Top of first page: '20' *brown crayon*

[1] The Tuesday before Hooker read the first part of his paper (J. D. Hooker 1846) to the Linnean Society on 1 December 1846.

[2] In his previous letter CD suggested that Hooker explain more fully his Mexican type by 'including temperate or dry or highland *parts of both [*interl*] Americas'. Accordingly, in J. D. Hooker 1846, Hooker remarks that the Mexican type includes 'those whose nearest allies belong to Mexico or the higher levels in Columbia, or to the lower latitudes of the Southern United States, California or Chili' (p. 250).

To Richard Owen 25 November [1846][1]

Down Farnborough Kent
Nov. 25th

My dear Owen

Having worked out pretty carefully my new articulated Balanus (to be call Arthrobalanus) I have become so much interested in the structure of the sessile Cirripedes, that I am dissecting 5 or 6 of the other genera. I work out mouths & cirri carefully, muscular structure & tunics of the sack, & a some of the structure of the viscera; whether this is worth doing I am not sure, but I have not forgotten your great kindness in acceding to my request of reading over this time my descriptions.— I have, however, strayed from my object, which was to ask, whether you could by any chance supply me with a few specimens in Spirits of any of the **Sessile** genera, for my specimens are all small & some immature. If the College had any large set of specimens, of any of the genera, perhaps it would be possible to let me have a few; I could return the parts, after having them drawn, preserved & dissected in Spirits

Pray excuse this trouble | Ever yours truly | C. Darwin

P.S. | Most, perhaps all of the shells, about which I more than once tormented you, I have found at Sowerby's.—[2] proh pudor on me.

British Museum (Natural History), General Library (Owen correspondence)

[1] CD worked on *Arthrobalanus* during October 1846 and began dissecting other cirripedes in November (see 'Journal' (Appendix II) and letters to J. D. Hooker, [2 October 1846] and [26 October 1846]).

[2] Possibly the shells referred to in letters to G. B. Sowerby, [1 December 1845], [3 December 1845], and [9? December 1845].

To J. L. Stokes [*c.* 26 November 1846][1]

Down Farnborough Kent
Thursday

My dear Stokes

Many thanks for your letters, I enclosed the proper one to Sir G. Grey & this will show him, that we did not intentionally mean to insult him.— It has been a vexatious affair; for what I remember of him, I like much.— I have very little doubt that your explanation is the true one, viz that my note went in your proof-sheet to the Printers.—

I return you the S. Australian letter with thanks; I was glad to see it.—

I congratulate you heartily on the great success of your Book;[2] if I were in your place, I should be prouder of having been introduced to the old Duke[3] (as I heard was the case) than for a hundred pistols from Joinville,[4]—extraordinary as that compliment is.

Farewell my dear Stokes

AL
Sir Tom Ramsay

[1] This letter must have been written after CD's letter to George Grey, 10 November 1846. In his letter to Grey, CD did not mention the possibility that his original note went to the printer with Stokes's proof-sheets. Neither did CD mention this explanation in his letter to Robert FitzRoy, 23 November [1846]. Presumably CD received a letter from Stokes after 23 November to which this is his reply. The 26th is the earliest possible Thursday following 23 November.

[2] Stokes 1846 was reviewed in the July 1846 issues of the *Foreign Quarterly Review* (37: 257–80) and *Fraser's Magazine* (34: 105–17).

[3] Probably the Duke of Wellington.

[4] François Ferdinand Philippe Louis Marie, Prince de Joinville, a French expert in military and naval affairs.

To Francis Wedgwood 27 November [1846]

Down Farnborough Kent
Nov. 27th

Dear Frank

Would you be so good (in accordance with enclosed instructions from Jos.) as to send to "M^{ssrs}. Hine & Robinson Charter House Sq^{r}. London", the bundle of parchments which you hold for our Trust.[1] Please to see that they are securely tied up & may I beg you, to send them by secure means to Whitmore Station & have them *booked*; please rather wait than send them insecurely.—

Will you enclose Jos' letter to M^{rss} Hine & Robinson.

Will you *post* mine to Mrs Hine & Robinson on *same* day that the parcel is despatched by Railway.

And lastly be so kind as to send 〈 〉 on same day 〈 〉.—

I am really sorry to give you so much trouble, but the deeds are important & this is the only means to ascertain that they arrive safely at M^{rs}. Hine & Co.—

Believe me, dear Frank, with kind remembrances to Fanny[2] 〈 〉[3]

AL signature excised
Keele University Library (Wedgwood 26783-35)

[1] The marriage settlement funds for CD and Emma were in a trust, of which Josiah Wedgwood III and Erasmus Alvey Darwin were the trustees. Francis (Frank) Wedgwood has endorsed the letter: 'Sent by Rails as directed Nov 30 1846'.

[2] Frances Mosley Wedgwood, Frank Wedgwood's wife.

[3] CD's signature has been excised, which also accounts for the missing text in the letter.

To *Annals and Magazine of Natural History* [December 1846]

Down Farnborough Kent
Saturday

My dear Sir

I fear the enclosed is longer than you wished, but a page in Annals swallows up much M.S. I c^{d}. not make it shorter to do justice to work, or to make my little notice[1] at all interesting.

I have marked one page for extraction,[2] without which my remarks w^{d} be unintelligible.—. I hope you will not think my notice too laudatory; it expresses my *most honest* conviction after *careful* perusal; You can alter it as you like, but I rather hope you won't shorten much. I have inserted some critical objections; as too much praise defeats its end.— Heaven knows whether this notice will suit you,—but I hope so.

Yours very faithfully | C. Darwin

P.S. | If you do not *much* object, I sh^{d} be very glad to correct press, as my style is often very faulty.

American Philosophical Society

[1] CD's unsigned review of George Robert Waterhouse's *A natural history of the Mammalia* vol. 1, *Marsupiata* (Waterhouse 1846–8). It appeared in the January 1847 issue of the *Annals and Magazine of Natural History* 19: 53–6 (see *Collected papers* 1: 214–17).

[2] CD refers, not to his manuscript, but to p. 537 of Waterhouse's volume, which is quoted in the review. CD's copy of Waterhouse 1846–8 is in the Darwin Library–CUL, in the original parts, the earliest numbers of which are dated 1845.

To J. D. Hooker [December 1846]

Down Farnborough Kent
Saturday

My dear Hooker

I sh^{d} have answered your note sooner, only I have been very unwell, with a small abscess at the fangs of one of my grinders, complicated as is invariably the case with me with my wretched stomach: I think, however, I shall be well again tomorrow.—

I am truly delighted to hear that you will have some work & come here in middle of January. In all human probability this time will suit us capitally, as indeed w^{d} almost all times: for the only engagement I know of this winter is a visit for a week to my Father.—

Sorrow take you for wishing me sorrow, merely because I egged you on to make a capital speech, & which I am very glad, if it does not turn you from more valuable work, you are forced to publish.—[1] I do not think you were the *least* "hard" on Bunbury;[2] perhaps he hardly got the usual proportion of "butter" on the occasion.— I suggest & **urge** on you, to find out where the Coal-plants from Melville Is^{d} are deposited & have a good look at them: the case is all important & never has been gone into even with approximate care, *do think of this*.—[3]

When I was drawing with Leonard,[4] I was so delighted with the appearance of the objects, especially with their perspective, as seen through the weak powers of a good compound Microscope that I am going to order one: indeed I often have structures, in which the $\frac{1}{30}$th is not power enough.—

I enclose a prospectus of a lady, our neighbour & friend of Berkeley: I do not know how the book will be done, but her own drawings are in truth *quite wonderfully* beautiful:—

Thanks for the corallines; Heaven knows when I shall begin them; I have been nearly 3 months on Cirripedia & have done only 3 genera!!!

We are both very sorry not to hear a better account of your Sister.

Ever my dear Hooker | Yours | C. Darwin

DAR 114.2: 76–76b

[1] Possibly a reference to Hooker's reading of his paper (J. D. Hooker 1846) to the Linnean Society on 1 and 15 December 1846, or to his having spoken at the Geological Society (see n. 2, below).

[2] Charles James Fox Bunbury read a paper on the coal plants of Cape Breton to the Geological Society on 3 December 1846 (C. J. F. Bunbury 1847). Hooker was present and according to Bunbury had a 'favourable opinion' of the paper (see F. J. Bunbury ed. 1891–3, *Middle life* 1: 196).

[3] Melville island, much further into the Arctic than Cape Breton, was known to possess coal strata. Hooker was at this time preparing his account of the vegetation of the Carboniferous period for the Geological Survey (J. D. Hooker 1848).

[4] Probably Samuel William Leonard, the artist used by W. B. Carpenter (see letter to W. B. Carpenter, [October–December 1846], n. 1). In his Account Book (Down House MS) CD recorded £3 paid to 'Mr Leonard drawing Conia' on 5 December 1846.

To J. D. Hooker [December 1846 – January 1847]

Down Farnborough Kent
Wednesday

My dear Hooker

Will Saturday the 16th. suit you to come here?[1] If so, & I hope so, the Phaeton shall be at *Sydenham* to meet the Train which leaves at 3° 15′. (remember this). Bring some work & do stay as long as you can. Please send me one line by return of Post, for I will ask the old set to come for the Sunday.

I will, also, ask the Lyells, who said they wd. come this winter & bring some work & stay some days with me & I daresay you wd. like to meet them & I am sure they wd to meet you.

Ever yours | C. Darwin

As I must fix, please observe that you are to come in the Phaeton, as earliest asked.—

DAR 114.2: 77

[1] CD refers to 16 January 1847, when Charles Lyell and presumably Hooker visited Down. See *Correspondence* vol. 4, letter to Charles Lyell, [23 January 1847].

To John Higgins 12 December [1846]

Down Farnborough Kent
Dec 12th

My dear Sir

I am much obliged for your note & the draft of £185 " s13 " d1, which I beg to acknowledge.[1]

I am very glad to hear so excellent an account of the Farm House; & I should much like to see it next summer.

You proposed last half year to pay the Rent direct into my Bankers Mrs Robarts & Curtis & I should **greatly** prefer this plan & shd be much obliged if for the future you would endeavour so to arrange it—

Believe me dear Sir | Yours very faithfully | C. Darwin

My Father has been ill, but he is now, thank God, greatly better; he has had much suffering.—

Lincolnshire Record Office

[1] The half-yearly rent for Beesby farm, CD's Lincolnshire property, with deductions for Higgins' commission and a £5 contribution to Beesby school. See CD's Account Book (Down House MS) entries for 9 and 12 December 1846.

To A. C. Ramsay 21 December [1846]

Down Farnborough Kent
Dec. 21st

My dear Sir

I am much obliged by your interesting & friendly letter, which I shd. have answered sooner had I not been unwell during the whole of last week.— I am delighted that you have thought parts of my Book worth reading:[1] it is in my opinion much more difficult to get a geologist to read a book than to write one.—

I have been much interested at what you say about traces of Terraces;[2] I could never see any signs of such in Snowdonia the only part of N. Wales I have of late years visited. This subject has always been a hobby-horse of mine; simple as the subject may be, it was long before I understood it.— I see to the present day, (for instance in Murchison's paper on Scand. drift)[3] that the fundamental error (in my opinion) is made in speaking of the successive terraces as the direct effects of so many elevations, instead of looking at them as the indirect effect of an elevation, & the direct effect of sea's destroying power at those levels. This leads me to a subject, mentioned in your letter; namely whether the old Tertiary beds of S. America were under water during the whole period intervening between their deposition & that of the superficial recent layers: on this point I have no knowledge; but it is certain that they were very slowly uplifted from under the recent sea, & therefore the less high parts, at least, must have been long under water. I declare I think this absence of any considerable recent fossiliferous deposits on both E. & W. coasts, the most remarkable thing I observed there.[4] I hope you will turn this in your mind & take the trouble to reflect over my remarks at p. 135. I have found it a most fertile subject for thought; it helps to explain the breaks in Geological chronology & has disabused my mind of a prejudice that *durable* fossiliferus formations are in most places now accumulating. As it appears to me I see this error prevailing or implied in the writings of most geologists.—

Your trappean deposits, I can well believe are very perplexing; but I do not think I have ever seen anything quite analogous. As far as my memory serves, I think ejected volcanic crystals of glassy feldspar are always broken.— Near the Stiper Stones, I went to look at some of Murchison's thin Silurian lava-streams; but I could not make them out; all that I traced far enough seemed to have been

injected: in my book on Volcanic Isl^{ds} (p. 109. & 103:) I give the measurements of the thinnest aërial streams, I ever heard of. You must have an interesting field, where you are now at work; & Forbes told me of some very curious points.—[5]

Where lavas are vesicular & have become decomposed, I have seen the most marvellous transitions into sedimentary beds; partly caused by the compression & movement of the once solid & *vesicular* lava; so that it was quite impossible to say where lava ended & tuff began, though neither had been in the least metamorphosed. I suspect, however, slight degrees of metamorphic action are usually overlooked.—

I will only allude to one other point in your letter, namely to the long continued action of gravel on the underlying rocks beneath the sea: now this first supposes a very thin layer of gravel, for we well know how great is the inertia from mutual friction of a very thin layer of sand.— Secondly the very frequent presence, as stated in my book, of delicate corallines on gravel in **very** turbulent seas & in not very deep water;[6] this appears to me opposed to much grinding action of the gravel; & the more I think of the subject, the more inclined I am (putting out of the question beds of mud or fine loose sand) to doubt the sea having much destroying power, more than two or three fathoms below low-water mark.

You apologise for the length of your letter, but I fear I have much greater cause to do so.—

I direct this to Charing Cross;[7] thinking it probable that you may have returned from Bala & I beg to remain | My dear Sir | Yours sincerely | C. Darwin

American Philosophical Society

1 *South America.*

2 Ramsay later read a paper to the British Association on the terraces of Cardiganshire (Ramsay 1847).

3 Murchison 1846.

4 CD discussed the absence of geologically recent conchiferous deposits in *South America*, pp. 135–9, and offered his explanation of why, because of denudation during periods of elevation, 'there is but a small chance of *durable* fossiliferous deposits accumulating' (*South America*, p. 139).

5 Both Ramsay and Edward Forbes were employed by the Geological Survey in Wales.

6 *South America*, p. 23.

7 The postal area of London in which the offices of the Geological Survey were located.

To Leonard Horner [23 December 1846 – January 1847][1]

Down. Farnborough. Kent.
Wednesday.

My dear Mr Horner.

I am truly pleased at your approval of my Book[2] & it was very kind of you taking the trouble to tell me so.— I long hesitated whether I would publish it or not, & now that I have done so at a good cost of trouble, it is indeed highly satisfactory to think that my labour has not been quite thrown away.

I entirely acquiesce in your criticism on my calling the Pampean form "recent"; pleistocene w^{d} have been far better. I object however, altogether on principle (whether I have always followed my principle is another question) to designate any epoch after man. It breaks through all principles of classification to take one mammifer as an epoch. And this is presupposing we know something of the introduction of man: how few years ago all beds earlier than the pleistocene were characterized as being before the monkey epoch. It appears to me, that it may often be convenient to speak of an Historical or Human depoisit in the same way as we speak of an Elephant bed, but that to apply it to an Epoch is unsound.

I have expressed myself very ill, & I am not very sure that my notions are very clear on this subject; except that I know that I have often been made wrath (even by Lyell) at the confidence with which people speak of the introduction of man, as if they had seen him walk on the stage, & as if, in a geological chronological sense, it was more important than the entry of any other mammifer.—

You ask me to do a most puzzling thing to point out what is newest in my volume, & I found myself incapable of doing almost the same for Lyell.— My mind goes from point to point without deciding: what has interested oneself or given most trouble is, perhaps quite falsely thought newest.— The elevation of the land is perhaps more carefully treated than any other subject; but it cannot of course be called new. I have made out a sort of index, which will not take you a couple of minutes to skim over, & then you will perhaps judge, what seems newest. The summary at end of book w^{d} also serve same purpose.—[3]

I do not not know where E. de B.[4] has lately put forth on the recent elevation of the Cordillera, He "rapported" favourably on d'Orbigny, who in late times fires off a most Royal salute;[5] every volcano bursting forth in the Andes at the same time with their elevation, the debacle thus caused depositing all the Pampean mud & all the Patagonian shingle! is not this making Geology nice & simple for beginners?

We have been very sorry to hear of Bunbury's severe illness; I believe the measles are often dangerous to grown-up-people. I am very glad that your last account was so much better.[6]

With many thanks | Most truly yours | C. Darwin.

I am astonished that you should have had the courage to go right through my Book. It is quite obvious that most geologists find it far easier to write than to read a book.—

[Enclosure]

Ch. I & II. | *Elevation of the land.*— equability on E. coast as shown by terraces. p 19— length on W. coast p 53— Height at Valparaiso p 32— number of periods of rest at Coquimbo— p 49. elevation within Human period near Lima greater than elsewhere observed— The discussion p 41 on non horizontality of terraces perhaps one of newest features. on formation of terraces rather newish.—

Ch III p 62. Argument of horizontal elevation of Cordillera I believe new.— I think the connection (p 54) between earthquake shocks & insensible rising important.—

Ch. IV The strangeness of the (as strange as Eocene) mammifers, coexisting with recent shells.—

Ch V. Curious pumice— infusorial mudstone p 118 of Patagonia— climate of old Tertiary period p 134— *The subject which has been most fertile in my mind*, is the discussion from p. 135 to end of Chapter on the non-ready-accumulation of fossiliferous deposits.[7]

Ch. VI Perhaps some facts on metamorphism, but chiefly on the layers in mica-slate &c being analagous to cleavage.

Ch VII. The grand up & down movements (& vertical silicified trees) in the Cordillera. see summary p. 204 and p. 240 Origin of the Claystone porphyry formation p. 170

Ch VIII p 224. Mixture of Cretaceous & Oolitic forms— p 226 great subsidence— I think (p 232) there is some novelty in discussion on axes of eruption & injection. p. 247 Continuous volcanic action in the Cordillera. I think the concluding Summary (p 237) w^{d} show what are the most salient features in the Book.

Copy
DAR 145

[1] Dated from the reference to Charles James Fox Bunbury's measles, see n. 6, below. The first possible Wednesday in this period was 23 December.
[2] *South America*.
[3] The concluding pages of chapter eight summarise the entire volume, see *South America*, pp. 237–48.
[4] Jean Baptiste Louis Léonce Élie de Beaumont.
[5] Élie de Beaumont 1843, one in a series of reports on contemporary geology.
[6] Bunbury recorded the onset of measles on 16 December 1846 and the beginning of his recovery on 18 January 1847 (F. J. Bunbury ed. 1891–3, *Middle Life* 1: 205).
[7] CD refers to his explanation for the absence of fossiliferous deposits on the coasts of South America, which leads him to discuss the conditions most favourable for the deposition and preservation of geological formations (*South America*, pp. 135–9).

To John Murray 30 December [1846][1]

Down Farnborough Kent
Dec 30th

My dear Sir

Would you be so kind as to gratify my curiosity, by directing some one in your Office to inform me how many copies of my Journal up to end of the current year have been disposed of.— I heard some months ago from Mr Lyell, that you were pretty well satisfied with the Sale, which I assure you gave me hearty satisfaction.

Is it in your power, without **very** great trouble, to oblige me in another way, viz to send for me for a copy of the N. American Edition of my Journal;[2] I applied to my Bookseller & he told me you alone could effect this. I, of course, would pay for it.—

I should like to have a copy, & am curious to see whether they have altered any part on Slavery or other subjects.—[3]

Pray believe me | My dear Sir | Yours faithfully | C. Darwin

John Murray Archive

[1] Dated from CD's entry in his 'Journal' (Appendix II): '4000 copies of new Edit of Journal sold at Jan. 1. 1847.' The letter has a note, presumably by John Murray: 'nearly 4000'.

[2] *Journal of researches* US ed., published in two volumes by Harper & Brothers, New York, in 1846. CD did not own the copyright of his *Journal* after 1845 (Freeman 1977, p. 38).

[3] There was no change from the English edition, see *Journal of researches* US ed. 2: 302–4. CD is referring to his remarks on slavery in *Journal of researches* 2d ed., pp. 499–500, and possibly to the passage where he criticised Americans, 'with their boastful cry of liberty' (p. 500).

APPENDIX I

Translations of letters

From C. G. Ehrenberg 15 June 1844[1]

Berlin
15 June 1844

Most honoured Sir

The shipment of earth samples of which you so kindly notified me on April 20 arrived soon thereafter in May. I am most thankfully obliged to you for this shipment and especially for the considerate manner in which you fulfilled my requests as communicated by Dr. Dieffenbach.[2] I soon found these materials so very interesting that I immediately devoted myself to them day and night. In a few days I hope to be able to send you a summary of the results of my investigations. I have now received for examination a considerable number of microscopic forms from the Galapagos Islands. But especially important for me was the atmospheric dust or the volcanic ash that came from the area of the Cape Verde Islands. Approximately $\frac{1}{6}$ of the mass consists of siliceous shelled organisms, some of which I have only previously observed (not in Africa) but in Cayenne, although I know numerous forms from Senegal in particular, some of which are peculiar to it, but none of which is characteristic of that dust. I assume that what you say in your Voyage about hazy air on the Cape Verde Islands refers to this dust.[3] How many days did this rain of dust fall? It is especially interesting to keep a record of the circumstances that you know as exactly and in as much detail as possible.— I have not yet been able to get anything from New Zealand. What you sent is inorganic. I already possess a great deal of material from New Holland, but I still have very little from the many islands of the Australian Ocean.[4] I am also writing a few lines of thanks to Mr W. Hooker,[5] since his shipment was also most interesting.

I shall write to you again in about 14 days, in order to provide you with the results of my investigations. Until then I recommend myself to your continued good will and repeat my best thanks.

Ever in greatest respect | Your | thankfully most devoted | Dr C G Ehrenberg | Professor and Secretary of the Academy of Science

DAR 163

[1] For the transcription of this letter in its original German, see pp. 40–1. Ehrenberg's letter was delivered by hand, but not until late July or August. See letter to J. D. Hooker, [25 July – 29 August 1844], and letter to C. G. Ehrenberg, 5 September [1844].
[2] See letter to Ernst Dieffenbach, 25 January 1844.
[3] *Journal of researches*, p. 4.
[4] Ehrenberg probably used the term in a general sense to mean Oceania. See his letter of 11 July 1844 in which he refers to the 'Süd-australischen Archipel von Neu Guinea bis zu den Marquesas I.'
[5] Ehrenberg misread CD's letter of 20 April [1844]. The specimens were provided by Joseph Dalton Hooker.

From C. G. Ehrenberg 11 July 1844[1]

Berlin
11 July 1844

Most honoured Sir

It has been three weeks since I wrote to you through the opportunity presented by the young Mr Gladstone of London, who offered to carry a letter. I hope that this letter is now in your hands.[2] Previously Dr. Dieffenbach had offered to inform you of the fortunate arrival of the very interesting shipment. I have already sent my deepest thanks for your very great favour and I repeat it gladly. You have provided me with such rich material for my investigations that I will feast on them for a long time yet. Yesterday through the good offices of Mr Gibsone of Perth I sent you the printed offprint of my report (in the May issue of the *Monatsberichten* of the Berlin Academy of Science[3]) on a part of your shipment. I had addressed a copy for you and one for Dr Hooker to Mr Francis at the address of Mr Richard Taylor Fleet Street Red Lion Court London.[4] I hope that you receive it.

In answer to your letter of July 4 which came today, let me say that I would very much like to have everything you consider worthy of a microscopic analysis. All kinds of sea-sand and mud or clay-like deposits of the sea and large rivers, even sometimes in small quantities, are rich in forms. What you told me about the rain of dust is extremely interesting with respect to my analysis. I therefore request details that are as complete as possible. Also especially any sediment samples that can be obtained from the great depths of the sea would be very welcome to me, particularly if perhaps the samples from Ross's extensive deep sea-bed soundings are still available from the Admiralty.[5]

Mr Francis, Fleet Street Red Lion Court London at Mr R. Taylor's, has the kindness to forward what you give him for me. That is in case you do not want to send it directly which is perhaps best. In Hamburg the merchant Herr G. Morgenstern am alten Wandrahm takes care to send me safely and inexpensively everything that finds its way to him under my address.

Ever with greatest respect and thankfulness. Your | most devoted | Dr C G Ehrenberg.

The salt deposits in Patagonia are not rich in life, although I have highly concentrated brine from other places.

Isolated islands from every corner of the earth interest me especially with regard to their smallest forms of life. Through your kindness I have good yields from the Galapagos and Ascension.

The sand from Keeling Atoll contains not only dead parts, but also in part small dried animals collected alive, among them many siliceous shells of Polygastric Infusoria. It is not merely a digested excrement but a small bustle of life with many coral fragments from soft coral in it, such as I also found in the Red Sea.

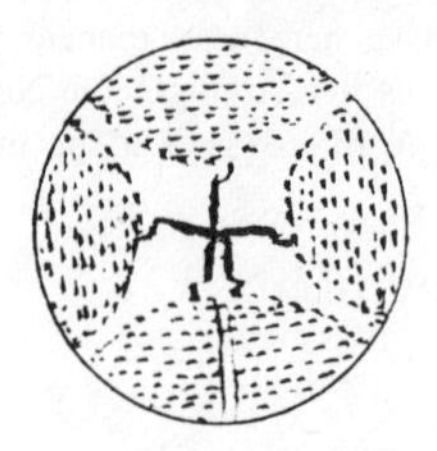

Asteromphalus
Darwinii.

I now possess much material from Terra del Fuego but I lack any from California, the Cape Verde Islands, Sierra Leone, Congo, Angola, or from the South Australian archipelago from New Guinea to the Marquesas I. I possess many from New Holland; nothing yet from New Zealand, since the rock samples sent give no clear result. I possess many from Guinea and Senegal. Still nothing from the Laccadives and Maldives.

Prof. C. G. Ehrenberg. Secret. d. Akad. d. Wissensch. | Berlin. | Unter den Linden 21.

pr. adr. d. Herrn Gustav Morgenstern | Meissner Porzellan Niederlage | am alten Wandrahm | Hamburg.

DAR 163

1 For the transcription of this letter in the original German, see pp. 45–7.
2 Ehrenberg's first letter to CD had not arrived. See letter from C. G. Ehrenberg, 15 June 1844, n. 1.
3 Ehrenberg 1844a. CD's copy of the paper is in the Darwin Library–CUL.
4 William Francis served as an assistant in Richard Taylor's printing firm, located in Red Lion Court, Fleet Street. The paper was being sent to Taylor for publication in English (Ehrenberg 1844b).
5 CD had offered specimens collected by Joseph Dalton Hooker during James Clark Ross's Antarctic expedition, see letters to J. D. Hooker, [27 January 1844] and 23 February [1844], and letter to C. G. Ehrenberg, 20 April [1844].

From Alcide Charles Victor Dessalines d'Orbigny 14 February 1845[1]

Dear Sir,

I had at last come to the end of the work on your fossils when your last letter reached me. I have hurried to draw up the general observations, and I send you the box by way of M. Ballière as you requested.

You will observe that, as I expected, your magnificent collection of fossils shows no contradiction with those that I saw in America. The apparent differences were entirely owing to the incorrect determinations of Mr Sowerby. Fossils are elements of immense value in geology, but they should be taken with every appropriate precaution. Failing that they will give results entirely opposed to the facts. If I kept you waiting a long time for your results, it is because I did not want to give you provisional work. This required very long searches which I would not wish to neglect to avoid giving you faulty information. I hope you will pardon the delay in view of the motive that guided me.

Your interesting collection is without question magnificent, and I would recommend a special publication produced with all due care for it can serve as a basis for the study of South America. I urge you to have the plates made of all the new species so that your work and mine can complete the geology of South America.[2] Above all it would be good if the fossils could be represented from several angles and in many series. Imperfect drawings can only burden science without doing any good whatever.

I owe to your collection the opportunity to correct an error that I committed concerning the age of the sandstone of Concepcion;[3] I had placed it in the Tertiary, but, according to your new *Baculite*, it belongs to the *Cretaceous*.

Furthermore, I often found myself troubled by the names of species. For example when Mr Sowerby had given a familiar name to a species that I considered to be new, I believed I ought to give it a new *designation*, but I attach no importance to whether you keep it or not, change it or do as you please; I wanted only to call your attention to them and to be able to know what you referred to if it became necessary to give you additional information on the details and on the whole collection.

Finally, here is the summary showing the age of the formations which I was able to recognize according to your fossils. It begins with the lowest.

Jurassic Period[4]

A part of the Cordillera of Copiapo, the Alto of Guasco (Chile) and the Cordillera of Coquimbo

Neocomian Period or *lower Cretaceous.*

The central Cordillera of Chile, the central chain of Tierra del Fuego, the Cordillera of Peru near Pasco.

Turonian Period, or chalk chlorite.

Concepcion du Chile. Cordillera of Copiapo, Cordillera of Coquimbo.

Patagonian Period or lower Tertiary.

On the Atlantic Ocean. La Bajada de Santa Fe. St. Joseph's Bay. Port Desire. Port St Julien River of Santa-Cruz.

By the great ocean

Chiloe (east coast) Huafo. Navidad. Coquimbo (lower and middle layers) Payta, La Mocha.

All these fossils are no longer found in the seas today; they belong to a past era.[5]

Diluvial Period or from the modern epoch.

Buenos Ayres. Maldonado. Coquimbo, (upper levels of the plain).

With this rapid summary, the details of which you have in my book, you see that in this manner everything is classified without indecision or confusion.

You have among your fossils the most beautiful fact that I know of to prove the abrupt changes of the period in which they became extinct. It is a group of Crépidules. These shells are still joined together, just as one finds them now alive at Callao. The least change, even the death of individuals, if it occurred before they were covered over would have separated the different shells that compose the group. It is a very curious fact.[6]

I should be happy, Sir, if the work can be useful to you, I insist again on the importance of your collections so that they may be published.

Please, I beg of you, to make use of me under any circumstances and believe me.

Your very humble and very | Obedient Servant | Alcide d'Orbigny

Paris, 14 February 1845

Your box is at M. Baillière's.

DAR 43.1: 62–5

[1] For the transcription of this letter in the original French and CD's annotations, see pp. 143–5.

[2] CD eventually included four plates of shells in *South America*.

[3] Orbigny (1835–47, vol. 3, pt 3: *Géologie*, p. 90) had earlier described the Concepción sandstone as Tertiary. Though CD was inclined to accept the new Cretaceous dating, he noted the possibility that the baculite may simply have been a survivor from an earlier epoch (*South America*, pp. 126–31).

[4] According to CD, the shells distinguished by Orbigny as Jurassic and Turonian (Upper Cretaceous) had all come fom the same deposit. CD, who preferred the term Oolitic to Jurassic, compromised on the expression 'Cretaceo-Oolitic' (*South America*, p. 216).

[5] Orbigny (1835–47, vol. 3, pt 3: *Géologie*, p. 191) had criticised CD's opinion (*Journal of researches*, p. 423) that a few living species were to be found in the lower beds at Coquimbo. Following George Brettingham Sowerby's species determinations, CD qualified his position in *South America*, pp. 128–9, and *Journal of researches* 2d ed., p. 344.

[6] Orbigny believed that the shells were joined together in a way that would have been easily disturbed, and that the molluscs were therefore still living when the sea-bed was elevated. According to CD, the molluscs appeared to have died long before the shells were lifted up (*South America*, pp. 46–8).

From C. G. Ehrenberg 13 March 1845[1]

Berlin
13 March 1845

Most honoured Sir

Unfortunately Dr. Hooker did not come to Berlin and thus there will be further delays in carrying out your wishes.[2] Since I have an opportunity today of returning to you the manuscript on the atmospheric dust, I am taking it although it prevents me from writing to you in more detail. First I want to say a few words about Dr. Dieffenbach, who unfortunately had already left Berlin when your first letter arrived. As you may know, the poor man had to leave his fatherland because he had political connections that were not tolerated. In order to make this less damaging to his future career, he set out on a journey to Giessen in order to appear before a judicial examination where he might clear himself. He hoped to get away soon, but he has already been sitting in Giessen now for half a year without accomplishing the desired result.[3] I myself was travelling for several months in the autumn and was then very busy. I always had your second letter with me as a reminder. I wrote immediately to Dr Dieffenbach and received the enclosed answer from him. I hope you have already received the things you wanted from the book dealer.

As for your excellent reports on the Atlantic atmospheric dust and particularly your material, I can now also inform you of the result of the investigation.[4]

Apart from the 37 microscopic organisms partly constituting the atmospheric dust which have already been listed earlier,[5] another 29 can now be recognised

Cocconema Lunula
Eunotia Argus
—— granulata
—— longicornis
—— Pileus
—— quaternaria
—— tridentula
—— Triodon
Gallionella decussata
Grammatophora oceanica

Gomphonema gracile
Navicula Bacillum
Pinnularia viridula

——

Amphidiscus armatus
—— obtusus
Lithodontium rostratum
—— platyodon
Lithostylidium biconcavum
—— crenulatum
—— Clepsammidium
—— Emblema
—— obliquum

Lithostylidium Rhombus
—— rostratum
—— Rajarum
—— unidentatum
Spongolithis cenocephala
—— Fustis

——

Textilaria globulosa.

Now these are especially remarkable, since there is not one among them of the characteristic coastal forms from Senegambia the Sahara or from Africa generally that I have frequently come across. Lith. Rajarum alone is also known to me from the Isle de France.[6]

Among the 66 forms none is new, since Eunotia longicornis appears in similar fossil form in Hungary.

However just as 2 of the first 37 species were characteristic of America, 4 of the last 29 are common to Guiana and Senegambia, namely Eunotia quaternaria Eunotia tridentula [Eunotia] Pileus [and] Amphidiscus obtusus.

Furthermore all 6 of the dust types you sent me react similarly. They all exhibit a very rich mixture of siliceous shelled infusoria and also others of similar species.

It can now be said with confidence, that all the dust from the 6 samples must come from one and the same source. This source can be determined still more closely.— Only freshwater animal species appeared earlier; these could be from the middle of a continent. Now two exclusively saltwater animals have been found among them, Grammatophora oceanica and Textilaria globulosa. Thus the source of the dust must be a coastal region. To be sure the African coast lies closest, but there are no African forms among them, although many of them are citizens of the world. However there are now 6 South American ones among them.

I have further determined that according to the seamen the wind that carries the dust is never the harmattan, but that it is specifically the trade wind. You have not made a sharp distinction in setting forth your own experience, but this would be highly desirable. You probably still have ship's journals at your disposal that would allow you to check this point. I have not found living things among them, that is any that dried quickly and still exhibit organs.— It would be good to institute direct, lengthy comparative observations on San Jago and along the African Coast concerning this remarkable dust. You seem to have all the necessary means at hand to accomplish this and I am confident you will certainly attempt it.

Your face paint from Tierra del Fuego was also very interesting to me. It also consists of infusoria, the first fossil deposit that has come from there. Thus far I have determined 18 species of which none is new but all are exclusively freshwater forms.[7]

I would consider the white earth in Patagonia a volcanic product.[8] It looks like pumice, that is, like weathered pumice and I have recognised in it a few almost indistinguishable infusorial shells, just a few, yet still clear to me. It may tentatively be identified as a species of the genus Fragilaria. Unfortunately I now have too much else to do to pursue the matter intensively, yet I consider the material very interesting if it is not just a local phenomenon. Is the earth only dusty, no solid rock? How massive do you consider the deposit? How far does it extend? Leagues? Miles? Should it really extend for hundreds of miles, if I understand correctly, it must be accompanied over that distance by clearly volcanic mountain formations.

I cannot give you any further news for today. Merely the warmest thanks for the rich material you and Dr. Hooker have sent. It is unfortunate that he declined to spend some time with us, for we would have received him in a most friendly way.

I would very much like to see even the smallest samples from great depths of the sea, of about 10000 to 20000 feet; after all, I hear depths of 24,000 F. have been measured. These may be either with or without tallow from the plummet.

Remaining with cordial respect | Your | most devoted | Dr C G Ehrenberg[9]

DAR 39.1: 59a–b

[1] For the transcription of this letter in the original German and CD's annotations, see pp. 153–6.
[2] Since Joseph Dalton Hooker intended to visit Berlin, CD had asked Ehrenberg to return by him the manuscript of 'An account of the fine dust which often falls on vessels in the Atlantic Ocean' (*Collected papers* 1: 199–203) and also to urge Ernst Dieffenbach to return the materials CD had sent to him for the German edition of *Journal of researches*. See letter to C. G. Ehrenberg, 23 January [1845], and letter from J. D. Hooker, [late February 1845].
[3] Ernst Dieffenbach had been implicated in the April 1833 storming of the Frankfurt Guardhouse by a militant liberal group and had had to flee Germany. In 1843 he was informed that proceedings against him had been dropped in Prussia; however, he was not able to return permanently to his home in Giessen until he had permission from authorities in Hesse-Darmstadt (Bell 1976, pp. 19–21, 87–9).
[4] Ehrenberg (1845a, pp. 64–6) had reported on these findings before the Berlin Academy on 27 February 1845.
[5] Described in Ehrenberg 1844a, pp. 194–207.
[6] Mauritius.
[7] Described in Ehrenberg 1845a, pp. 63–4.
[8] Described in Ehrenberg 1845b, pp. 143–8.
[9] CD's abstract of this letter is in DAR 39.1: 59c.

From C. G. Ehrenberg 8 April 1845[1]

Berlin
8 April 1845

Most honoured Sir,

I have received your letter of March 23 and I am pleased that your wishes were in some degree satisfied. Enclosed I am sending you the promised excerpt from my last paper at the Berlin Academy of Sciences, in which I took into account the materials you sent me, in so far as I had been able to examine them.[2] May the result agree with your other researches. As for the dust it would be quite interesting if you could ascertain the wind directions exactly. That the dust almost reaches America sometimes does not seem to me to speak decisively for an African origin. The trade winds blow horizontally from Africa to America, but the harmattan hardly ever reaches so far over the ocean. What the trade winds carry must go with them from Africa over the ocean to America again or at least very close to it, even if it had originally been carried from America over to Africa. *But this notion is only a preliminary attempt at explaining the notable absence of genuine African forms.* The dust you collected yourself is presumably No. IA on my table,[3] all the others are from Mr. James.[4] You did not give the package a particular label. It was part of the first shipment. You can make whatever use you wish of my information.

As for the white Patagonian rock from St. Julian I have ascertained the following from it. It is quite clearly a volcanic pumice-like Tuff, with no recognisable admixture of gypsum in the samples sent. If the extension of the mass is that great then it belongs to the most important surface phenomena of the earth known to me. The mass is so rich in microscopic, siliceous shelled organisms, that even the apparently inorganic pumice-like material appears to have arisen only by the fusion of the organic. I estimate that the now still recognisable organic element constitutes $\frac{1}{10}$ of the mass. But most of it is in very small and somewhat altered fragments. This organic admixture is entirely alike at Port St. Julian and Port Desire (Device??), but less so at New Bay, though still recognisable. Up until now I have discovered 28 species in your samples. 21 of them are known, 7 are new ones which hence characterise the country. Your reports have interested me so much that I have resumed the analysis anew with all strength and effort.

The most remarkable thing about this phenomenon is that while the volcanic masses everywhere until now have exhibited only fused *freshwater forms, saltwater forms* are dominant in the Patagonian Tuff, a relationship that I had previously sought in vain, though it was to be expected.[5] I can name the following forms for you:

Those marked with * are new forms.

Siliceous-shelled Polygastrica:	Port St. Julian	Port Desire	New Bay
Actinocyclus Venus?	—	+	—
Coscinodiscus marginatus	+	+	—
radiatus	—	+	—
* spinulosus	—	+	—
Diploneïs didyma	+	—	—
*Discoplea Mam̄illa	+	+	—
Fragilaria rhabdosoma	—	+	—
vulgaris?	—	—	+
*Gallionella ?coronata	+	+	—
* ?plana	+	+	—
sulcata	+	+	—
Goniothecium hispidum	+	—	—
*Hyalodiscus patagonicus	+	+	—
*Mastogonia ?Discoplea	+	+	—
Pinnularia borealis?	+	—	—
Siliceous Phytolitharia:			
Lithasteriscus tuberculatus	+	—	—
Lithosphaera stellata	—	+	—
Lithostylidium amphiodon	+	—	—
articulatum	—	+	—
rostratum	—	—	+
Spongolithis acicularis	+	+	—
appendiculata	+	+	—
aspera	+	—	—
Caput serpentis	+	—	—
Clavus	—	+	—
Fustis	+	+	—
porosa	—	+	—
*Thylaecium hispidum	+	—	—

There were no calcareous forms and therefore it seems that these have formed the glass constituting the pumice of the Tuff by fusion with the adjacent siliceous shells and have lost their organic form. I should like to ask for somewhat more exact information about the occurrence and quantity of gypsum (Sulphate of lime).

It would also be very important to compare exactly the sample from Rio Negro, since the external appearance can easily be misleading. In general considering the significance of the subject I would like to have had a *somewhat larger really solid piece of rock* at least 1 inch in size to examine. Perhaps you don't possess many yourself? What you sent me were small powdery quantities, nothing solid. But, assuming that you selected these small samples yourself and found them characteristic of the whole, I examined them with great care, particularly when I saw results coming.

I am also attaching the note you wanted on the Pampas. Although you have not sent me samples marked Pampas, yet I believe that you mean the samples from Patagonia and the La Plata plains that include mastodon remains. Since I have already examined some of this material, I can now inform you of my chief preliminary result.

To be sure both deposits also contain microscopic organisms. In the Patagonian sample with the label: Earth attached to fossil Bones M. Hermoso in Patagonia, I was able to recognise 3 forms of polygastric animals and 6 forms of Phytolitharia, all but one of which are freshwater forms.[6]

Another sample on which I did a preliminary examination has the label: Mastodon tooth with earth from the Plata estuary Mud.— Banks of the Parana. In this one I have already recognised 7 Polygastrica and 13 Phytolitharia. They are almost equally divided between distinctly sea creatures and freshwater forms with more of the latter than the former. The deposit was therefore semi-salty, brackish.

The Patagonian deposit is perhaps a freshwater formation, but we should be cautious and label it brackish, since the single sea form observed is clearly of that origin.[7]

I have some chalk samples from Peru from Dr. Tschudy,[8] but I would like to compare your samples from the Cordillera of Chile.

Did you not bring incompletely vitrified Obsidians from various places? These could be especially interesting. Non-vitrified obsidians are less interesting.

Ever in greatest respect | Your | most devoted friend | Dr C G Ehrenberg

The dust which has been observed on the high seas between Senegal and Cayenne has unfortunately not been collected and it cannot be currently decided whether or not it was another kind of dust entirely. In Dr. Meyen's Travels (with the Princess Luise) it is reported that the phenomenon occurred closer to the African coast.[9] He considered the reddening of the sails to be the generatio spontanea of a cryptogamous plant, which he named Aerophytum atlanticum. He probably thought the dew pearls were a plant. He was often a very superficial observer who casually helped himself out with his imagination.[10]

DAR 39.1: 60–1b

[1] For the transcription of this letter in the original German and CD's annotations, see pp. 171–4.
[2] Ehrenberg 1845a. CD's copy is in the Darwin Library–CUL.
[3] Ehrenberg enclosed with this letter a chart listing species of infusoria found in Atlantic dust. The chart, in somewhat different form, was printed in Ehrenberg 1845a, pp. 85–7. On the MS version CD noted totals of species, which he cited in 'An account of the fine dust which often falls on vessels in the Atlantic ocean' (*Collected papers* 1: 201).
[4] Robert Bastard James, who had collected dust that had fallen on his vessel in the Atlantic Ocean (see *Correspondence* vol. 2, letter from R. B. James to Charles Lyell, [*c.* 10 March 1838]).
[5] Ehrenberg's results were read to the Berlin Academy on 24 April 1845 and published in Ehrenberg 1845b, pp. 143–8. CD's annotated copy is in the Darwin Library–CUL. For CD's use of Ehrenberg's information see *South America*, pp. 111, 118–19, 248.
[6] Ehrenberg 1845b, p. 147. CD used these results in *South America*, p. 81.
[7] If the Pampas mud had been deposited in fresh or brackish water, it would explain the absence of shells and the presence of large numbers of fossil mammal remains (see *South America*, pp. 98–9, and letter to C. G. Ehrenberg, 23 March [1845]).
[8] Johann Jacob von Tschudi, who had travelled in Peru in 1838.
[9] Meyen 1834–5, 1: 54–5.
[10] The final paragraph was written on the verso of Ehrenberg's chart, see n. 3, above.

From A. C. V. D. d'Orbigny 31 January 1846[1]

Sir,

I have delayed much longer than I would have liked in sending you the list of identifications of your shells from Bahia Blanca,[2] but circumstances beyond my control prevented me from doing so, in spite of my great wish to send you this little work promptly. I beg you to forgive me and not to charge me with negligence.

All the species marked with a + in the margin are described by me in the *Palaeontology of my Voyage*.[3] Those that do not have these marks are likewise described in my *Molluscs of my Voyage*,[4] so you will be able to mention them & dispense with describing them.

All these species are found living on the neighbouring shores and belong to the existing fauna. They are as you thought, of the same age as the fossil shells from Bahia de San-Blas, and have all identical forms living in the same place.

Regarding the species from the Uruguay, they are marine, and also have living analogues on the coast. They are both described in the *Molluscs of my Voyage*, the last part of which is being printed at the moment.[5]

I am very grateful, Sir, for the details that you are willing to give me about people with whom I will be able to correspond in order to obtain English fossils.[6] I am going to see about preparing my duplicates, and I will follow your good advice and direct to them some collections of fossils. If in the meantime you see any of them, I beg you to tell me whether they are disposed to accept my proposal.

If you have any duplicates of fossils from the Falkland Islands, they would be very useful for my future publications.

Believe, Sir, in the most devoted sentiments with which I have the honour to be | Your very humble servant | Alcide d'Orbigny

Paris 31 January 1846.

DAR 43.1: 66–7

[1] For the transcription of this letter in the original French, see pp. 280–1.

[2] The list has not been preserved with the numerous other lists of shells, identified by Orbigny, in DAR 43.1. In *South America*, p. 5, CD cited Orbigny on the shells of Bahia Blanca and San Blas.

[3] Orbigny 1835–47, vol. 3, pt 4.

[4] Orbigny 1835–47, vol. 5, pt 3.

[5] See *South America*, p. 2.

[6] Orbigny was writing Orbigny 1845–7 and planning work on Orbigny 1850–2. See letter from A. C. V. D. d'Orbigny, [June–July 1846].

From C. G. Ehrenberg 11 March 1846[1]

Berlin
11 March 1846

Most honoured Sir

I have received your kind new shipment of mountain samples through the Prussian embassy in London and I would have hastened long ago to give you news of it if the severe illness of my wife had not disturbed my entire correspondence for almost half a year.

Nevertheless the following information may indicate to you that I have not left your wishes unattended.

1. The samples from the River Gallegos in Patagonia also show microscopic organisms, but completely different ones from the large white tuff mass from St. Cruz. They are freshwater—or brackish forms and not fused. Accordingly these deposits are related to those of Mount Hermoso. Incidentally I found organic material only in the 2 friable pumice-like samples especially in the yellowish one.[2]

2 In addition, you want to know whether the word *Fluthgebeit* is identical with the English expression *Estuary deposit*.[3] They both mean much the same thing, but an *Ablagerung (deposit)* in a *Fluthgebeit of the sea* could be a *freshwater deposit*, whereas *estuary deposit* is definitely always a *deposit of the sea in a tidal area*. The *obere Fluthgebeit des Meeres im Festlande* is *the upper district of the tide in a River or Continent*. Here the blocking up of the river water and its mixing [with seawater] always results in brackish deposits, e.g. near Hamburg there is an upper tidal district 18 German miles from the sea, and a lower tidewater district at Cuxhaven. The tide reaches the upper one only on occasion.

3. I immediately analysed the sample of atmospheric dust that was sent to me from Malta, and I reported the remarkable result to our Academy of Science.[4] I observed 43 organisms in it, of which 31 are exactly like those in the Cape Verde dust. You would certainly want to publish your learned and comprehensive

collection of dust observations. The matter seems important to me and even if our views on the origin of the dust and winds differ somewhat, the continued observation that would be stimulated thereby would soon establish the boundaries of probability. It is to be expected that the relatively small number of geographical observations of microscopic organisms made so far cannot give definitive results. Nevertheless the path of observation is the only one to follow and its result alone can give scientific confirmation. You are in the midst of a most magnificent situation for developing the resulting material further.

You will receive my last report on these matters either on this occasion or another one soon. My heartiest greetings close this letter.

I remain with sincere respect | Your most humble | Dr C G Ehrenberg

If you have the opportunity to send me some fragments of grasses now growing on Ascension I should like them for comparison. There must be dependable names on them, but they can be leaf fragments or stem fragments without botanic value, an inch in size.

DAR 39.1: 62–3

[1] For the transcription of this letter in the original German and CD's annotations, see pp. 298–9.

[2] The information in this paragraph was used in *South America*, p. 117. It led CD to conclude that 'the 200 to 300 feet plain at Port Gallegos is of unknown age, but probably of subsequent origin to the great Patagonian tertiary formation.'

[3] CD quoted this letter as his authority in correcting mistranslations of Ehrenberg's term 'Fluthgebiethe'. Previously, it had been translated by Alcide d'Orbigny as a 'flood', lending support to Orbigny's view that the Pampas deposits had been laid down by a debacle. See *South America*, p. 248 n., and letter to C. G. Ehrenberg, 29 October 1845, n. 2.

[4] Ehrenberg 1845c, pp. 377–81.

APPENDIX II

Chronology 1844–6

This appendix contains a transcription of Charles Darwin's 'Journal' for the period 1844–6. Darwin commenced his 'Journal' in August 1838 and continued to maintain it until December 1881. In this small notebook, measuring 3 inches by $4\frac{1}{2}$ inches, Darwin recorded the periods he was away from home, the progress and publication of his work, and important events in his family life.

The version published by Sir Gavin de Beer as 'Darwin's Journal' (1959), was edited before the original 'Journal' had been found and used a transcription made by an unknown copyist. The original, now in the Darwin Archive, Cambridge University Library (DAR 158), reveals that the copyist did not clearly distinguish between the various types of entries it contains.

Initially Darwin kept alternate pages of the 'Journal' blank for comments about particular entries on the opposite page and for additional entries. All the entries on these pages are printed in this transcription within double angle brackets (« »). In many cases it is impossible to tell whether these entries were made later than those on the opposite page and therefore the brackets simply indicate that the enclosed text occupies a different physical position in the manuscript without presupposing when the entry was made.

From 1845 onwards, however, Darwin adopted a different policy. Instead of leaving alternate pages blank, he recorded all that pertained to his work (including his illnesses, since these accounted for time lost from work) on the left hand pages of the 'Journal', whilst the periods he was away from home and family events were noted on the right hand pages. In order to show clearly Darwin's deliberate separation of the types of entries he made in his 'Journal', the transcription will be printed in double columns representing the left and right hand pages of his notebook.

All alterations, interlineations, additions, and the use of a different ink or pencil have been noted. In addition, the editors have inserted relevant information to supplement the 'Journal' entries. These interpolations are enclosed in square brackets to distinguish them from Darwin's own entries.

1844. Jan 5th sent M.S. of Volcanic Islands to Printers.[1]

[31 January. Attended council meeting of the Geological Society.][2]

Feb. 13th: finished correcting. In intervals & previously slowly enlarged[3] & improved pencil sketch in 35 pages (written in midsummer of 1842)[4] of Species Theory.[5]

[21 February. Attended council meeting of the Geological Society.]

[20 March. Attended council meeting of the Geological Society.]
[6 April. 'On the origin of mould' published.][6]
[17 April. Attended council meeting of the Geological Society.]
April 23d to Maer & Shrewsbury; returned May 30th
[8 June. 'Manures, and steeping seeds' published.][7]
[12 June. Attended council meeting of the Geological Society.]
July 5th. Sent a written sketch of species theory (seven years after commencement[8] in about 230[9] pages to Mr. Fletcher to be copied. Corrected it, last week in Sept:[10] Paper on Atlantic Dust:[11] on Planariæ[12] Sorted my collections.
[5 July. Instructions to Emma concerning editing and publication of species theory in case of death.]
[17 July. Went to London with Emma.
18 July. Visited Kew Gardens with Emma.][13]
July 27th.—/44/—/ Began S. America[14]
[14 September. 'Variegated leaves' and 'What is the action of common salt on carbonate of lime?' published.][15]
«October 18th to 29th at Shrewsbury»
[20 November. Attended council meeting of the Geological Society.]
[18 December. Attended council meeting of the Geological Society.]

1845.
[3 February. 'Extracts from letters to the general secretary, on the analogy of the structure of some volcanic rocks with that of glaciers' read to the Royal Society of Edinburgh.][16]
April 24th. Finished first time over S. America.— (9 months)
April 25 Began 2d Edit of Journal.[17]
[4 June. 'An account of the fine dust which often falls on vessels in the Atlantic Ocean' read to the Geological Society.][19]
August 26th. Finished.—d[itt]o [i.e., *Journal of researches* 2d ed.]— (4 months)[20] Rested idle for a fortnight
Oct. 29th. Recommenced Geolog. of S. America

4000 copies of New. Edit of Journal *sold* at Jan 1. 1847[24]

1845.
[8 January. Attended council meeting of the Geological Society.]
[5 February. Attended council meeting of the Geological Society.]
[12 March. Attended council meeting of the Geological Society.]
[2 April. Attended council meeting of the Geological Society.]
April 29th to Shrewsbury, returned on May 10th.—[18]
[28 May. Attended council meeting of the Geological Society.]
[11 June. Attended council meeting of the Geological Society.]
[2 July. Attended council meeting of the Geological Society.]
George Howard born July 9.[21]
Sept 15th. to Shrewsbury; Lincolnshire; York: the Dean of Manchester: Waterton: Chatsworth: Camp. Hill: returned home[22] Oct. 26th.[23]

[1845 *cont.*]

[19 November. Attended council meeting of the Geological Society.]

[24 November. Returned to Down from London.][25]

[10 December. Attended council meeting of the Geological Society.]

[1846]

[25 March. 'On the geology of the Falkland Islands' read to the Geological Society.][26]

[August. 'Origin of saliferous deposits: salt-lakes of Patagonia and La Plata' published.][27]

1846. Oct 1st. Finished last proof of my Geolog. Obser. on S. America; This volume, including Paper in Geolog. Journal on the Falkland Islands took me 18 & ½ months: the M.S., however, was not so perfect as in case of Volcanic Islands. So that my Geology has taken me 4 & ½ years:[28] now it is 10 years since my return to England. How much time lost by illness!

Oct 1st. Paper on new Balanus. Arthro balanus[29] 10 days in London, during 2 visits & visitors here[30] & some days unwell[31]

November, December Conia & Megatrema[32]

1846

[7 January 1846. Attended council meeting of the Geological Society.]

[4 February. Attended council meeting of the Geological Society.]

[20 February. Attended council meeting of the Geological Society.]

Feb 21st. to Shrewsbury.

March 3d. Home

[25 March. Attended council meeting of the Geological Society.]

[31 March. Elizabeth (Bessy) Wedgwood died.]

[22 April. Attended council meeting of the Geological Society.]

[20 May. Attended council meeting of the Geological Society.]

[3 June. Attended council meeting of the Geological Society.]

[17 June. Attended council meeting of the Geological Society.]

July 31st to d[itt]o [i.e., Shrewsbury]

Augt 9th. Home

September 9th with Emma[33] to Brit Assoc: at Southampton, on the 12th to Portsmouth & coast of Isle of Wight on 13th to Winchester & S. Cross,[34] on 14th Netley Abbey & Southampton Common. 17th Home

Sept. 22d. With Em. & Susan to Knole Park

[18 November. Attended council meeting of the Geological Society.]

[2 December. Attended council meeting of the Geological Society.]

1 *Volcanic islands*, the second part of the geology of the *Beagle* voyage.
2 For CD's attendance at the Geological Society council meetings see the Council Minute Books, Geological Society Archives.
3 'slowly enlarged' *above del* 'copied'.
4 'in 35 . . . 1842)' *interl.*
5 See *Foundations* for both the sketch of 1842 and the essay of 1844.
6 *Collected papers* 1: 195–6.
7 *Collected papers* 1: 196–7.
8 '(seven . . . commencement' *added.*
9 '230' *underl in pencil.*
10 'Corrected . . . Sept:' *added.*
11 'An account of the fine dust which often falls on vessels in the Atlantic Ocean', *Collected papers* 1: 199–203.
12 'Brief descriptions of several terrestrial *Planariae*, and of some remarkable marine species, with an account of their habits', *Collected papers* 1: 182–93.
13 See letter to J. D. Hooker, [14 July 1844], and Emma Darwin's diary.
14 *South America*, the third part of the geology of the *Beagle* voyage.
15 *Collected papers* 1: 198, 198–9.
16 *Collected papers* 1: 193–5.
17 *Journal of researches* 2d ed., published by John Murray.
18 CD originally wrote 'May 1st to Shrewsbury for a week', then, in pencil, altered 'a week' to '11 days'. The entire entry was then deleted and the entry, 'April 29th . . . May 10th.—' was made. A second 'on May 10th' was written in pencil below 'returned', presumably earlier.
19 *Collected papers* 1: 199–203.
20 '(4 months)' *added.*
21 'George . . . July 9.' *added in Emma Darwin's hand.*
22 'home' *interl.* CD travelled to Beesby in Lincolnshire to view the farm he had purchased. Following this he visited William Herbert, Dean of Manchester, and Charles Waterton in York, the Chatsworth estate in Derbyshire, and Camp Hill, the home of Sarah Elizabeth (Sarah) Wedgwood.
23 *Followed by a line drawn across the page and* '*1846*— Feb. 21 to Shrewsbury; March 3d. home.—July 31 to Shrewsbury. Home. Augt 9th.' *deleted in pencil.*
24 '4000 . . . 1847' *added to entries for 1845.*
25 Emma Darwin's diary.
26 *Collected papers* 1: 203–12.
27 *Collected papers* 1: 212–13.
28 CD's geology of the *Beagle* voyage comprised *Coral reefs*, *Volcanic islands*, and *South America.*
29 'Arthro balanus' *added.* No paper on 'Arthrobalanus', later named *Cryptophialus minutus*, has been found. See also letter to J. D. Hooker, [26 October 1846].
30 Bartholomew James Sulivan and his wife and two children stayed at Down House in mid-October (see letter to Robert FitzRoy, 28 October [1846]). Joseph Dalton Hooker and John Hensleigh Allen also visited Down at this time (see letter to J. D. Hooker, [8 October 1846]).
31 '10 days . . . 2 visits' *added*; '2' *written in pencil*; '& visitors . . . unwell' *written in ink over pencil.*
32 The cirripedes *Conia* and *Megatrema* of the Balanidae were later described in *Living Cirripedia* (1854).
33 'with Emma' *interl.*
34 St Cross is a suburb of Winchester.

APPENDIX III

Darwin's notes arising from conversations with Joseph Dalton Hooker

Dec. 8.— 1844. J. d. Hooker. Notes[1]

Maccormick[2] says there are blocks of granite &c on Kerguelan land— Icebergs certainly stranded—not known to do so on Auckland Islands. or S. of New Zealand.

An astelia, *Beech*, (N.B the Astelia of Sandwich Is[d] very different) & Oriobolus (*peculiar* genus of grass) common to Mountains of Van Diemen's land, New Zealand & Tierra del Fuego; H thinks other analogous facts—so that the S. Hemisphere is like north in some community. =He mentioned another genus=

In Falklands there is a peculiar grass common to Melville Isl[d] & same identical grass in Upper Peru.

The 14 Plants (& 4 peculiar) common to Kerguelen Land & T: del Fuego, chiefly grasses & insignificant plants[3]—one Auckland Island form there.

Hooker[4] *thinks*, that There is some affinity between St. Helena & J. Fernandez (!!)[5] in two peculiar species of Campanulaciæ; & relations to external conditions in strong caulescent Plumbagos

---

Hooker says there is case of foreign Leguminous plant, having been **reared** from seed picked up on coast of Ireland.—

Cocos Is[d] (N. of Galapagos) has Mexican form of ferns— Galapagos allied to W. Indian Islands—, certainly American character flora, more than to continent.—

A Rhododendron on summit of Ceylon & Java— An Andromeda summit of Sandwich Is[d].— Believe Europæan form on Organ Mountains in Brazil— Schonburgk[6] found Rubi on 6000 ft mountains in Guyana.— What is Flora of Bourbon?[7] Really all mountains, except of *Australia* (a)[8] are allied. (My views require much correcting).[9] Iceland 400 plants all Europæan; believes Greenland 600 plants are also so.— *Heaths*, but none have crossed Baffins Bay.—[10] Have all these been received from Europe by migration. Good case of evidence for migration.

(a) On Van Diemens Land Mountains, An Anemone & Andromeda, & none of these genera on low-land. (The Andromeda very wonderful, because belong to Ericeæ, & these are represented by Epacridæ.)[11] One Rubus, likens it to Arcticus—[12] one lowland species. There is allied Anemone in T. del Fuego.— On New Zealand Mountains a herbaceous Veronica, (& Epilobiums?) in lowlands only shrubby ones, almost other genus.— Believe that on most mountains as Ceylon & E. India, the extra-tropical division of grasses are found. Certainly appears as if all mountains in some degree uniform.— Certainly the Gentianæ are Alpine plants, though some are found in lowlands in most countries—yet great tendency to be on mountains. (All my reasoning will apply to sub-alpine Floras.)[13] One species found on mountains of *Chile*, N. America & Europe. important like the grass, showing a high-Road. The further south gone, on the islands S of New Zealand, the more American Flora becomes. Kerguelen Land not **so** distant in straght line from T. del Fuego

---

Thinks Australia & C. of Good Hope rather more allied than S. America & Australia:—same main divisions as with Mammifers:[14] except w^{d} include all South-part of N. America, & does not know about Madagascar.— {St. Helena versus the world would perhaps be the first division.}

Agrees with me about diffusion of species.

Accounts for difference between East & West Australia from plants being "confined" genera, hence the species have "confined ranges"—

Agrees with Lesson that going to Westward in Pacific, more nutmegs Pandini &c showing more Indian Flora, that it dies away to Eastward & does not get any American character.—[15]

Senecio is a large mundane genus, yet no species has a very wide range. Berkely[16] says most mundane Fungus has some species with most wide range.

I see our cow, which has two abortive mammæ, then these two are unequally developed.[17]

Hooker's case of Polypodium fine case of species varying in one country & not in another.[18]

Australian Orchideous Plant. Pterostylos, which from Spring in Labellum, when touched must exclude insect agency, though possibly, I think, an insect might get crushed between

H. has capital plan of testing number of species by seeing how many species he collected in same time, in two countries. In New Zealand, though looking so much more fertile he got only half as many as near Sydney. At Cape he thinks perhaps even more.

Juan Fernandez. quite S. American Flora—some peculiar genera; many peculiar species of Chilian genera, & some Chilian species— Analogy with St. Helena; such analogy opposed to my views

---

No facts, in Embryology, in Botany: cannot say, which is highest group—most complicated. There is an orchis (Listera 'avis-nidi??) colourless leafless, &c presenting strong analogy with Orobanche

Hooker says, an Impatiens fulva or biflora, he has no doubt is American & has recently overrun Europe & England & taken place of other plants— Railway Cutting.—introduced in A.D not given in Loudon (lately)[19] One species *is* doubtful English; but not this large I. fulva[20]

I compared Maer Heath, unaltered, during 20,000 years. w^{d} perhaps not receive new plants—but plant firs & so alter conditions & see how many new things are introduced: so with a whole country.—[21]

Are not New Zealand leaves related to present Flora of Van Diemen's Land?— V. my volcanic volume. Hooker thinks that they are certainly Eucalypti=[22] The silicified wood there is *Araucaria*..— The fossil trees of Kerguelen Land, probably coeval with the odd Cabbage.—

Does not agree, (but had not thought) of Waterhouses remark on alliances being general & not especial[23]

Believe part, which is normally in a species *abortive* appears often as a *rudiment*— Has lately seen & described this in case of pistil of diœcious *Umbelliferous plant*:[24] does not know anything on Bentham's law of variability of abortive parts.=[25]

DAR 100: 35–40

Manuscript alterations

3.1 (N.B . . . different)] *added; square brackets in MS*
3.4 =He . . . genus= 3.5] *added*
6.1 Hooker *thinks*, that] *added*
6.2 two] *above del* 'one'
6.3 caulescent] *interl*
7.1 foreign] *below del* 'W. Indian'
8.2 certainly American character] *interl*
9.2 in Brazil] *interl*
9.4 (My . . . correcting) 9.5] *square brackets in MS*
9.5 Iceland 400] 'I' *over illeg*; '400' *interl*
9.6 Have all . . . of migration. 9.8] *added pencil*

10.1 Mountains] *interl*
10.2 (The . . . Epacridæ.) 10.3] *added; square brackets in MS*
10.5 (& Epilobiums?)] *added*
10.10 (All . . . Floras.)] *added; square brackets in MS*
10.11 on] 'o' *over* 'i'
11.3 {St. Helena . . . division.} 11.4] *added*
14.2 Flora] *after del* 'Fauna'
15.1 very] *after del* 'w'
16.1 I see . . . developed 16.2] *added after brace in MS*
19.3 half as] *interl*
19.3 as near Sydney] *before del* '—many.'; 'as' *over* '—'
21.1 say, which . . . complicated. 21.2] *interl*
22.1 or biflora] *interl*
22.2 overrun] *before omitted point*
22.2 & taken place of other plants] *interl*
22.3 not given in Loudon] *interl*
24.1 New] 'Newz' *in MS*
24.2 V. my . . . Eucalypti=] *added*

[1] CD made these notes on the occasion of Joseph Dalton Hooker's first visit to Down, 7–8 December 1844. They refer to questions asked during conversation and topics raised in their previous correspondence. Hooker provided CD with further comments in letter from J. D. Hooker, 12 December 1844.

[2] McCormick 1841 and 1842. Hooker soon wrote to tell CD that this information was incorrect, see letter from J. D. Hooker, 12 December 1844.

[3] This information was incorporated into the essay of 1844 (*Foundations*, p. 167). See also letter to J. D. Hooker, 23 February [1844].

[4] The paragraphs transcribed as five and six were written on the verso of the first page in a way that makes the exact sequence of the text hard to determine.

[5] St Helena lies in the middle of the South Atlantic ocean, and Juan Fernandez off the coast of Chile. For Hooker's opinion on their close botanical relationship, see J. D. Hooker 1846, pp. 244, 246. CD queried Hooker's interpretation in letter to J. D. Hooker, [23 November 1846].

[6] Robert Hermann Schomburgk.

[7] Réunion island, Indian ocean.

[8] The '(a)' and a small drawing of a pointing hand indicates the insertion of paragraph nine. To retain the integrity of the text, it has been transcribed following paragraph eight.

[9] In his sketch of 1842, CD had stated that the mountain faunas of eastern South America, the Altai, and southern India were dissimilar (*Foundations*, p. 30). The issue is discussed again in the essay of 1844, see *Foundations*, pp. 162–8. See also letter to J. D. Hooker, 23 February [1844].

[10] CD has scored the passage 'Greenland . . . Baffins Bay.—' with brown crayon.

[11] The Ericaceae are rare in Australia. Instead, they are represented by the closely allied Epacridaceae. Hence the presence of *Andromeda* (Ericaceae) was a striking example of the uniqueness of montane floras.

[12] *Rubus arcticus*, a northern hemisphere plant.

[13] The sentence '(All my reasoning . . . Floras.)' was added in the margin. Its intended position is not made clear.

[14] CD used the geographical distribution of mammals to illustrate regional sub-divisions of the earth in his essay of 1844 (*Foundations*, pp. 151–3). He linked Australia with South America.

[15] Lesson and Garnot 1826–30, 1: 12–14, and CD's essay of 1844 (*Foundations*, p. 162). See letter to J. D. Hooker, [6 March 1844], and letter from J. D. Hooker, 9 March 1844.

[16] Miles Joseph Berkeley. See letter to J. D. Hooker, 23 February [1844], for CD's first formulation of this question about geographical range.

[17] This sentence was added by CD on the verso of the preceding page in such a way that its intended position is not made clear. The topic seems to be an addendum to his remarks on George Bentham's law of abortive parts, see n. 25, below. CD's interest in the unequal development or variability of 'abortive' organs is explained in his essay of 1844 (*Foundations*, pp. 231–8). He describes the normal cow as possessing four teats plus two abortive or rudimentary mammae, and instances a case in which the two rudimentary organs were well-developed and even gave milk (*Foundations*, p. 232).

[18] See letter to J. D. Hooker, [10–11 November 1844], for CD's first formulation of this question.

[19] Loudon 1839.

[20] Hooker describes his views on introduced species in letter from J. D. Hooker, 28 October 1844. See CD's annotations to this letter for the basis of his question to Hooker.

[21] See *Natural selection*, pp. 196, 198, and *Origin*, pp. 71–2. This paragraph was added on the verso of the preceding page. It seems that CD intended it to supplement information given in paragraph twenty-one.

[22] *Volcanic islands*, p. 140, discusses fossil leaves from Tasmania, which Robert Brown said were not the most usual form of the *Eucalyptus*. See also letter from J. D. Hooker, 30 December 1844.

[23] See *Correspondence* vol. 2, letter from G. R. Waterhouse, [*c.* 2 August 1843], and *Origin*, pp. 429–30.

[24] A reference to *Anisotome*. Hooker later wrote to CD to tell him that this information was incorrect, see letter from J. D. Hooker, 12 December 1844.

[25] George Bentham's 'law of abortive parts' is explained in CD's *Questions & experiments notebook*: 21, 'N.B. Bentham remarks, where parts of flower are reduced from normal number, they are apt to vary in number in individuals of same species'.

APPENDIX IV

Darwin's questions on the breeding of animals in captivity[1]

(1) A list of those quadrupeds or birds whether British or foreign which have coupled & bred in the Gardens.[2]

(2) A list of those quads or birds which have coupled, but of which the female has not conceived; stating whether the coupling has been once, seldom, or frequent[3]

(3) A list of those, which have had opportunity (ie. male & female being together) but have never coupled.[4]

(4) A list of those, of which the male has shown desire, but of which either there has been no female, or the female has refused to allow the male to mount—stating which has been the case,

(5) A list of those, of which the female has been 'proud' or appeared ready for the male. but for whom, there has been no male, or the male has shown no desire.

(6) A list of those, of which both male & female have shown desire but notwithstanding have not coupled.[5]

A draft
DAR 206[6]

Manuscript alterations

1.1 whether . . . foreign] *interl*
2.1 quads or birds] *interl*
2.1 coupled,] *before del interl* '(once or often?)'
2.2 stating . . . frequent 2.3] *added*
3.1 (ie. . . . together) 3.2] *interl*
4.2 to allow . . . mount] *interl*
4.3 which] *below del* 'which'
4.3 has] *before del interl* 'each'
4.3 been the] *above del* 'the'
4.3 case,] *before del* 'which has been the alternative'
5.1 been 'proud' or] *interl*
5.1 for the male. 5.2] *interl*
5.2 but] *before del* 'of the males has refused to m'
6.1 but notwithstanding 6.2] *above del* '& yet'

[1] These questions may well have prompted the manuscript report on breeding in captivity that was prepared for CD by the Zoological Society, as reported in *Natural selection*, p. 75, and *Variation* 2: 149. The manuscript has not been found, but the report evidently described successful and unsuccessful breeding attempts by animals in the Society's gardens between 1838 and 1846; it is probable,

therefore, that CD requested these details in 1845 or 1846. The manuscript report is probably the 'Breeding list' that Louis Fraser notes is in progress in letter from Louis Fraser, 23 July 1845. CD used the breeding information in *Natural selection*, pp. 74–81, and in *Variation* 2: 148–58. In both these works CD also used information on breeding in captivity from the Earl of Derby's menagerie at Knowsley Hall, Lancashire, of which Fraser was conservator. Another item in the loose notes associated with the *Questions & experiments notebook* (see n. 6, below) is a draft of a series of questions relating to the breeding of specific birds, headed 'Ld Derbys or Paris', which may date from around the same time as the draft to the Zoological Society. The original aims of the Zoological Society (founded 1826) included the introduction and domestication of new breeds of animals and experimental breeding. These aims were enthusiastically pursued by the Society under the presidency of Lord Derby, 1831–51, particularly in the experimental farm at Kingston Hill (1829–33) and later in the gardens themselves. See Scherren 1905 and A. Desmond 1985.

[2] CD has underlined the word 'coupled' and added in pencil: 'It must be those which have been kept some time alive.'

[3] Beside this question CD has added in pencil: 'Males & females not [known] separate in many cases.'

[4] CD subsequently renumbered the first three questions in pencil, so that question 1 became question 3, question 2 became question 1, and question 3 became question 2. He also deleted in ink the whole of the original question 3, presumably because the content was repeated in question 6.

[5] In the margin alongside questions 5 and 6, CD has noted in pencil, '(Experiments on Fowls)'. This note may relate to breeding experiments he proposed in an entry made in his *Notebook D*: 180, probably at the beginning of October 1838: 'Experimentize on crossing of the several species of wild fowl of India with our common ones in Zoolog. Gardens'.

[6] This draft is one of several scraps that were interleaved in CD's *Questions & experiments notebook* (DAR 206), which itself contains further queries on the breeding of animals in the Zoological Society's gardens (*Questions & experiments notebook*: 20).

MANUSCRIPT ALTERATIONS AND COMMENTS

The alteration notes and comments are keyed to the letter texts by paragraph and line numbers. The precise section of the letter text to which the note applies precedes the square bracket. The changes recorded are those made to the manuscript by CD or his amanuensis; readers should consult the Note on editorial policy in the front matter for details of editorial practice and intent. The following terms are used in the notes as here defined:

del deleted
illeg illegible
interl interlined, i.e., inserted between existing text lines
omitted omitted in transcription
over written over, i.e., superimposing

To Geological Society of London [3 January 1844]
1.2 Evening] 'E' *over* 'e'
1.3 "*Down*] *interl*

To Joseph Dalton Hooker [11 January 1844]
0.2 Thursday] *after del* 'Wednesday'
1.1 to tell] 'to' *over illeg*
1.5 generalize] 'generatize' *in MS*
2.3 & northward] *interl*
2.6 with the] 'the' *interl*
2.11 rap] 'r' *over* 'wr'
2.12 Berkeley] *after del* 'Henslow'
2.16 (though poor)] *interl*
4.2 whether] *before del* 'the'
4.4 number of species] 'number of' *interl*
4.6 fewness] *above del* 'number'
4.6 number of] *interl*
4.6 in *Coral-islets*., 4.7] *interl*
4.8 as I have supposed.—] *added*
7.5 &c &c that] '&c &c' *added*
7.12 from his] *interl*
7.13 the simple] 'the' *over* 'a'
7.14 exquisitely] *interl*
7.15 been] *over* 'was'

To Henry Denny 20 January [1844]
2.6 then] 'n' *over illeg*
3.2 if the] 'the' *altered from* 'they'

To J. D. Hooker [27 January 1844]
1.4 Ehrenberg] *before omitted point*
1.7 plants like] *interl*
2.1 trouble] 'b' *crossed in MS*

To J. D. Hooker [3–17 February 1844]
2.1 to] *after del* 'flo'
6.2 imperfect] 'i' *over illeg*

To Charles Wicksted 13 February [1844?]
1.7 inherit] *altered from* 'inherited'

To J. D. Hooker 23 February [1844]
1.7 iceberg] *after del* 'floating'
1.10 from London] *interl*
3.5 good] *before del* 'to'
3.6 that] *altered from* 'they'
3.8 Archipelago] *after del* 'small'
3.9 separate] *interl*
3.11 separate] *interl*
3.12 representative] *interl*
4.6 Fauna] *above del* 'plants'
4.8 remarkably] *above del* 'very'
4.12 if not all] *interl*
5.6 plants; have not] *interl*
5.7 viz] *interl*
5.10 genera] *above del* 'plants'

To J. D. Hooker [6 March 1844]
1.6 one] 'o' *over* 'of'

1.6 but is not now so called] *interl*
1.7 Byron's] 'y' *over* 'r'
3.1 peculiar] *before omitted point*
3.2 Archipelago] *interl*
6.2 but] *over* '—'
6.6 range from] *before omitted point*
6.8 Or] 'O' *over* 'o'
6.9 whole] *interl*
6.9 will] *added*
6.9 through Europe] *interl*
6.10 in] 'i' *over* ')'

To J. D. Hooker 11 March [1844]
1.6 (which read first) 1.7] *interl*
1.7 my & Andrew Smith's] *above del* 'our'
1.11 a sight of] *interl*
3.1 consult] *before omitted point*
5.1 supposed] *interl*
5.1 had thought] *above del* 'sh[d] think'
5.2 almost] *interl*
6.1 on] *blotted over illeg*
6.2 Lithog.] *interl*
7.4 at the Galapagos] *interl*
7.7 close] *interl*
7.11 question] *above del* 'remark'
9.1 on] *over* 'of'

To J. D. Hooker 16 March [1844]
1.3 in a few days] *interl*
1.5 London] *before del* 'th'
1.8 &] *over* 'as'

To J. D. Hooker 31 March [1844]
1.6 it] *over* 'wh'
1.9 (] *over comma*
2.3 necessary] *interl*
2.4 or more] *interl*
2.5 all or most of] *interl*
2.6 would have] *above del* 'had'
2.6 (ie were found in most parts of)] *interl*
2.7 found] *after del* 'widely'
2.9 hand] *before del* 'mo'
2.9 so large a part of the 2.10] *above del* 'whole'
2.11 that in] 'in' *interl*
2.12 genus of] 'genus' *interl*
3.2 are] *over* 'is'
3.2 certainly] *before del* 'some'
3.2 their] *altered from* 'the'
3.5 It is evident] *after del* 'There is'
3.6 different] *interl*
4.3 this] *interl*

To J. D. Hooker [17 April 1844]
3.5 6 weeks] *after del* 'couple'
3.5 days] *before omitted point*
5.5 forms] *interl*
5.5 those,] *before interl del* 'forms,'
5.5 in] *over interl* 'of'; *above del* 'of'
5.5 variety] *before interl del* 'of forms'

To Christian Gottfried Ehrenberg
20 April [1844]
1.4 immediately] 'im' *over illeg*
2.1 naturalist in . . . expedition 2.2] *interl*

To Josiah Wedgwood III [May 1844]
1.1 will] *before interl del* 'soon'
1.2 from me] *before del* '(which with 900£ part of the former money, ['making' *del interl*] he has got security for)'
1.4 whenever] *after del illeg*
1.5 the difference] 'the' *over* 'it'; 'difference' *interl*
1.9 us] *after del* 'me'

To Julian Jackson 23 May [1844]
1.4 scattered] *over* 'in'

To J. D. Hooker 1 June [1844]
2.1 now on my return] *above del* 'just'

To Emma Darwin [3 June 1844]
4.1 reading] *interl*
4.5 just] *interl*

To Henry Denny 3 June [1844]
2.2 inhabitants of] *interl*
3.1 specimens] *before del* 'f'
3.2 inform] *after omitted point*
5.4 of] *before del* 'all'
5.5 by] *over* 'in'

To G. R. Waterhouse
10 [June 1844 – March 1845]
1.2 what] *altered from* 'whether'
1.2 was] *interl*
1.3 it] *above del* 'me'
1.4 about] *interl*

To J. D. Hooker 29 [June 1844]
1.4 N W.] 'W' *over* 'E.'
3.3 when you] 'you' *interl*
3.4 , if there are none thence,] *interl*
6.1 at London] *before omitted point*
6.2 Croydon] *interl*

To C. G. Ehrenberg 4 July [1844]
1.5 sea-] *interl*
1.6 under the blowpipe] *interl*
1.17 them,] *after del* 'your'
2.3 take me] 'me' *interl*
2.8 or lumps] 'or' *interl*
2.9 from sea-] 'f' *over illeg*
2.9 or corallines] 'or' *interl*

To Emma Darwin 5 July 1844
0.3 My.] *added*
1.2 judge,] *before del* 'I'
2.9 I wish you to . . . Editor. 2.10] *interl*
2.10 hand] *after del* 'have'
2.11 Portfolios:] *before omitted point*
2.11 copied] *interl*
2.11 from various works 2.12] *interl*
2.12 (or some amanuensis) 2.13] *interl*
2.15 these] *after del illeg*
2.15 under] 'u' *over* 'as'
3.4 Professor] *after del* 'M^r^'
3.4 London.] *before del* 'or Mr Lonsdale (if his health w^d^ permit)'
3.5 *Henslow??*] *underl and question marks added*
3.5 Dr Hooker . . . Editor= 3.6] *added in margin*
3.6 probably] *added*
3.6 probably=] *before del* 'possibly'
3.6 Dr Hooker would be **very** good] *interl*
3.7 Strickland.—] *before del* 'Professor Owen w^d^ be very good, but I presume he w^d^ not undertake such a work.'
4.2 500£.] *followed by line drawn across page in MS*
7.1 thoroughily] *before omitted point*
7.4 works] *added* 's'
8.1 especially] *interl*
9.1 Without . . . sum.— 9.2] *added*

To J. D. Hooker [14 July 1844]
1.3 home] 'om' *over illeg*
1.4 certainly] *interl*
2.2 Tuesday] 'd' *over* 'm'
5.1 causing] *after del* 'losing'

To J. D. Hooker 22 July [1844]
1.4 from your voyage] *interl*
1.5 right?] '?' *over* '—'
3.2 letter has] *altered from* 'letters have'
5.1 Sir] *before omitted point*

To John Stevens Henslow [25 July 1844]
1.1 stumbled] 'bled' *over illeg*
1.3 like to] *before del* 'see'
1.3 have] 'h' *over illeg*
2.1 more] 'm' *over illeg*
2.2 **quite**] *interl*
2.2 my] *interl*
2.3 marked] 'm' *over* ')'
5.2 for this] *interl*
5.4 you] *interl*

To J. D. Hooker [25 July – 29 August 1844]
1.2 such] *interl*
3.3 390 & 391] 'o' *over* '1'; '1' *over* '2'
3.5 392] '2' *over* '3'
4.1 fm] *over* 'or'
5.1 ready] *interl*

To The Royal Geographical Society [30 July 1844 – 1 October 1846]
1.2 is] *interl*
2.5 there] *over illeg*

To Adolf von Morlot 9 August [1844]
1.1 replied] 'i' *over* 'y'
2.5 Bookseller] *above del* 'of'
2.8 on] *over illeg*
3.3 my] *over illeg*
3.4 one] *altered from* 'one's'
3.7 & during] *interl*
3.8 in place] *interl*
3.9 shares] *above del* 'shares'
3.9 work] 'w' *over illeg*
3.12 a] *interl*
4.1 bottoms of] *before del illeg*

To Henry Denny 12 August [1844]
1.7 ie.] *after del* 'or'
1.12 Aperea] 'A' *over illeg*

To J. D. Hooker 29 [August 1844]
1.1 for] *altered from* 'from'

To Leonard Horner 29 August [1844]
1.2 verdict of] 'of' *over illeg*
1.3 thoroughily] *before omitted point*
1.4 which small] 's' *over* 'h'
2.1 off] *interl*
2.2 erased] *over illeg*
2.7 & relieved] *interl*
3.1 that] *interl*
3.1 of the volume] *interl*
3.4 natural] *interl*

3.6 of the stretching & consequent elevation] *interl*
3.7 most of 3.8] *above del* 'all'
3.8 a] *interl*
3.8 angle] *altered from* 'angles'
5.6 the great] *after del* 'Lyell's'
5.6 was] *interl*
5.7 by Lyell] *above del* 'him'

To Charles Lyell [1 September 1844]
0.2 Sunday] *below del* 'Saturday'
1.7 The] *over* 'My'
4.1 chiefly] *interl*
5.1 attached] *interl*
6.1 the present] *interl*
6.3 depth.] *interl*
6.3 is it] *altered from* 'it is'
6.5 deposit, is] 'is' *interl*
6.5 that] *over illeg*
6.5 very] *interl*
6.8 Pampæan] *interl*
6.9 Frenchmen] *altered from* 'French men'
6.10 to] 'o' *over* 'h'
6.10 Whiston] *over* 'Wiston'
6.12 much] *interl*
7.2 thinks] *altered from* 'think'
7.3 to know this 7.4] *above del* 'this'
8.3 for you] *over illeg*
8.3 wish] *over illeg*
9.1 of] *interl*

To C. G. Ehrenberg 5 September [1844]
2.1 Taylor's] *above del* 'some English'
2.8 Abstract] 'A' *over* 'a'
4.1 Al:] *interl*
4.8 tooth of the] *interl*

To J. D. Hooker [8 September 1844]
1.8 are] *above del* 'were'
2.2 of] *over illeg*
3.1 in certain] 'in' *over* 'on'
3.2 East] 'ast' *added over full stop*
3.6 Andes)] ')' *over comma*
3.8 short] *before del* 'short'
4.1 those areas, in which] *interl*
4.2 have] *after del* 'in those areas, which'
4.2 other] *altered from* 'others'
4.2 areas,] *interl*
5.3 or] *above del illeg*
6.1 often] *before del* 'introduced'
6.4 consider] *after del* 'doubt,'
6.5 of forms,] *interl*
6.8 or had not been created, 6.9] *interl*
6.13 I] *before omitted* '—'
10.3 no] *altered from* 'not'
10.3 great] *double underl del*
10.3 or **sudden**] *interl*

To E. A. Darwin [before 1 October 1844]
1.3 Monmouths] *before omitted point*
1.4 & date] *interl*
5.1 C.] *before omitted point*

To Adolf von Morlot 10 October [1844]
1.8 their] *over illeg*
1.11 speak] *before omitted point*
1.11 Secretary] 'ary' *added*
1.12 Society] 'iety' *added*
1.15 out,] *interl*
1.16 local] *after del illeg*
1.17 how] *above del* 'why'
2.3 the *reference*] 'the' *added*
2.3 American] *interl*
2.4 beds] *interl*
2.5 *in their present state*] *interl*
3.5 the undersoil] 'the under' *interl*
3.5 , however,] *interl*
3.6 & vegetation] *interl*
3.7 a] *interl*
3.7 Scandinavio-German 3.8] *above del* 'Europæan'
3.8 & then] *above del illeg*
3.10 you seem to] *interl*
4.1 from me,] *interl*
4.4 the illustrious] 'the' *above del* 'your'
4.5 every] *above del* 'each'
4.9 appear] *over illeg*
4.11 one] *after del* 'to'
5.3 but] *before del* 'in'
5.6 at present] *interl*

To James David Forbes 11 October [1844]
1.10 or stream] *interl*
1.11 ice-streams] '-streams' *interl*
2.7 you,] *added over comma*
2.11 Obsidian] 'O' *over* 'o'

To Leonard Jenyns 12 October [1844]
2.3 on the] 'the' *interl*
2.7 & which] *after del* 'nam'
2.9 by] 'b' *over* 'a'
2.11 these] 'se' *over* 'n'
2.11 ie. if free from accidents] *interl*
2.13 how great] *above del* 'what'

2.13 or occasionally] *interl*
2.14 of such destruction] *interl*
3.1 & collecting] '&' *over* 'on'
3.4 a directly] 'a' *interl*

To Henry Denny 7 November [1844]
2.1 any of] 'of' *over illeg*
2.4 comparison] 'i' *over* 'a'
2.4 of the] 'the' *over* 'these'
4.1 countries] *after del* 'st'

To J. D. Hooker [10–11 November 1844]
1.6 homogeneous] [2]'e' *over* 'i'
1.8 **running**] *interl*
1.15 of water,] *interl*
1.18 they were . . . fixed & 1.19] *interl*
2.5 certain] *interl*
2.14 in distant . . . world 2.15] *interl*
2.16 through fossils.] *interl*
3.2 for one] *interl*
3.5 of climate] *interl*
3.5 the creation of a] *interl*
3.9 (p. 342)] *parentheses added*; ')' *over comma*
3.9 South] *altered from* 'S.'
3.11 almost] *above del* 'quite'
3.14 (N.B. . . . different!) 3.16] *square brackets in MS*
3.19 & divided;] *interl*
3.20 does] *interl*
3.21 follow] *altered from* 'follows'
3.22 are] *over* 'is'
3.26 but] *added after del* '&'
3.26 there are] *interl*
3.26 &c,] *added*
3.30 such a book, as] *interl*
3.31 insects] *after del* 'animals'
3.31 should *will*] *after del* 'or plants their'
3.32 so as] *interl*
3.33 The other . . . notion] *above del* 'It'
3.34 viz] *interl*
3.35 woodpecker] 'r' *over* 'rs'
3.37 domestication.] *above del* 'it.'
3.37 was] *over* 'am'
3.39 considerable] *interl*
4.3 same] *interl*
4.5 club?] *before del* ')'

To J. D. Forbes 13 [November 1844]
1.4 by steam-boat . . . University 1.5] *interl*
1.5 at,] *interl*
1.8 perfect] *interl*
1.8 or] *over illeg*
1.8 rather] *interl*
1.9 pile] *above del* 'series'
2.5 : one of . . . interesting. 2.6] *interl*
2.7 to my Brothers House] *interl*
3.1 on] *interl*

To J. D. Hooker [18 November 1844]
2.2 Ever] *before del illeg*

To Leonard Jenyns 25 [November 1844]
1.3 interesting;] *interl*
1.4 lives] 'li' *over illeg*
1.7 thus] 'th' *over* 'so'
1.8 killed your] 'your' *over* 'the'
1.11 young] *before del* 'of these'
1.11 thus far] *interl*
1.11 expelled] *before del* 'thus far'
2.1 & increase 2.2] *interl*
2.2 in the succeeding years] *interl*
2.2 about the] *interl after del* 'now increased' *above del* 'about the'
2.3 with the same 2.4] 'the same' *above del* 'increased'
2.4 but] *above del* '&'
2.5 (as indeed seems probable)] *parentheses over commas*
2.6 & *very*] '&' *over* 'or'
2.6 appeared] *after del* 'been'
3.2 you] 'y' *over* 't'
3.2 I meant to say that] *interl*
3.4 *alone*;] *interl*
3.6 ie] *above del illeg*
3.6 species are] *above del* 'they were'
3.6 (as with] *above del* 'like'
3.11 between those] 'b' *over* 'a'
3.14 common] *altered from* 'commont'
3.14 amongst] *over* 'in'
3.17 am a] 'a' *interl*
3.20 these] *altered from* 'they'
3.20 grounds cannot] 'grounds' *interl*
4.1 not] *over illeg*
4.5 a] *over illeg*
5.1 & ill written] *interl*

To Susan Darwin [27 November 1844?]
1.2 am] *interl*
2.4 it] *interl*
3.1 the China] 'the' *interl*
3.1 releifs] 'ei' *over* 'ie'
3.2 is] *above del* 'was'
5.5 have] *before omitted point*
5.6 yesterday,] *interl*
5.10 estates,] *interl*

5.12 as] *before del* 'Susan'
5.12 you visit] 'you' *above del* 'she'; 'visit' *altered from* 'visits'
5.12 your] *above del* 'her'
5.12 you can] 'you' *interl*
6.1 two] *interl*

To Adolf von Morlot 28 November [1844]
1.6 language,] *above del illeg*
2.4 *volcanic*] *interl*
2.5 plutonic] *interl*
2.6 later] *above del illeg*
3.1 great] *above del* 'ice'
3.3 lived] *after del* 'were'
4.3 on *certain*] 'on' *over illeg*
6.4 therefore] *after omitted comma*

To J. D. Hooker [4 December 1844]
1.4 at the B. Arms 1.5] *interl*

To William Benjamin Carpenter [11 or 18 December 1844]
1.2 to London 1.3] *interl*
1.6 Sq^re] *interl*

To J. D. Hooker 16 [December 1844]
1.1 done for] 'for' *interl*
2.3 first] *interl*
2.5 J:] *interl*

To William Darwin Fox 20 December [1844]
1.9 thus] *interl*
2.3 human] 'hu' *over* 'ev'
2.6 for] *over* 'viz'
2.6 long ago] *interl*
2.11 *this*] *interl*
2.13 3 or 4 years old] *interl*
2.17 stupor] *after del* 'sleep'
2.18 tryed] *after del* 'ty'
3.3 attempt] *over* 'do'
3.6 (Miss . . . complaints.) 3.8] *square brackets in MS*
3.6 Ovaria.] *before del* ']'
3.7 the tumor is reduced greatly] *interl*
3.8 complaints.)] *before del full stop*
4.2 happen to] *interl*
4.3 (R. 389)] *interl*; *square brackets in MS*
4.5 enough] *interl*

To J. D. Hooker 25 December [1844]
1.2 very] *interl*
1.7 botanists] 'o' *over illeg*
1.8 be] *interl*
1.9 the Botanical] 'the' *interl*
2.4 the southern ones] *interl*
2.4 northern] *after del* 'we'
2.4 temperate] *interl*
2.6 relations] 's' *added*
2.6 the Floras] 'the' *altered from* 'these'
2.6 like to] *before del* 'know how many genera, ['of' *del*] *peculiar to & confined to [*interl*] Australia & New Zealand, have species, ['either' *del*] (either the same *species or different ones) [*interl*] on separate or not) in Auckland Is^d. & Campbell Is^d.'
2.7 northern] 'n' *over* 's'
3.1 Australia] *after omitted caret*
3.1 Van Diemens land] *interl*
3.3 do] *above omitted* '----'
3.5 (ie. . . . story.)] *square brackets in MS*
3.7 Tropics;] *before del interl* 'or'
3.8 *or only*] *after del* 'except'
4.1 (as far . . . serve)] *interl*
5.1 analogous] *above del* 'same'
5.1 (& this . . . genera.) 5.2] *interl*
5.2 of] *before del* 'the same'
5.3 prepared] *before del* 'w'
5.4 with the] *interl*
5.5 between . . . Auckland Is^d &c 5.6] *interl*
5.8 A] *after del* 'In'
5.9 not] *after del* 'even'
5.11 ought] *before del* 'f'
7.1 on the one hand . . . Campbell &c 7.2] *interl*
7.2 (*but . . . Zealand*) 7.4] *square brackets in MS*
7.3 *S. tropical*] '*S.*' *interl*
7.4 at the] 'the' *interl*
7.5 arctic] *after del* 'northern'
8.1 Is] *altered from* 'There is'
8.1 might] *above del* 'would'
8.3 Arctic] 'A' *over* 'N'
8.6 about] *interl*
8.8 the East] 'the' *interl*
8.9 both] *interl*
8.10 less distance] *interl*
8.10 4000] *above del* 'one'
8.10 miles, instead] *above del* 'miles out'
8.10 explain] *after del* 'make'
8.11 in the degree of similarity] *interl*
8.12 the two] 'the' *interl*
9.1 have] *interl*
9.1 at end of letter: 9.2] *interl*
9.4 peculiar] *after del* 'quite'
9.7 (bottom . . . letter)] *interl*

9.7 that] *above del* 'common'
9.8 are] *above del* 'were'
9.11 having] *before del* 'on'
9.12 an] *above del* 'th'
9.12 (so called)] *interl*
9.12 the whole of 9.13] *interl*
9.13 (or ought)] *interl*
9.14 in the external conditions] *interl*
9.18 means] 's' *added*
9.19 by attributing . . . conditions. 9.20] *added*; 'by' *over* '—'
11.1 remarks on] *above del* 'answers to'
11.1 let me 11.2] *interl*
11.2 of boulders] *interl*
11.4 subject.] *full stop over semicolon*
11.4 see] *before del* 'Collens'
11.6 the coal] 'the' *interl*
13.6 be] *added over comma*
13.7 which] *interl*
13.8 Paper; just] *semicolon over full stop*; 'just' *over* '—'
13.9 (ie is not related to) 13.10] *interl*; *square brackets in MS*
13.9 is not] *interl*
13.11 with] 'w' *over* 'y'
14.1 that] *over* 'you'

To G. B. Sowerby [1845?]
1.1 (if . . . you)] *parentheses over commas*

To David Thomas Ansted [*c.* January 1845]
2.1 they] *interl*
2.3 or "S.] 'or' *over illeg*
2.4 are] *over* 'is'
2.4 one] *over illeg*
3.3 is] *over illeg*
3.3 *in the upper museum*] *interl*
3.4 must] 'm' *over* 'i'
3.4 to the] 'the' *interl*

To J. D. Hooker [7 January 1845]
1.1 which] *before omitted comma*
1.7 with Bailliere] *interl*
2.1 the Pacific] 'the' *interl*
2.6 them back again] *above del* 'me'
2.6 you may] 'you' *over* 'we'
2.7 other] *interl*
3.1 you make] 'you' *interl*
3.1 out] *interl*
3.2 note] *above del* 'passage for me'
5.3 as,] *over* 'th'
6.2 but] *after omitted point*
6.5 of a mother] *interl*
6.8 *before her confinement*] *interl*
6.9 case] *interl*
6.10 the child] 'the' *altered from* 'their'
6.14 latter's mind] *above del* 'former'
7.7 the Voyage] 'the' *over* 'his'
7.8 if you do not know it,] *interl*
8.3 *subsidence*] *after del* 'action'
11.2 be] 'b' *over* 'v'

To Charles Hamilton Smith 14 January [1845]
1.12 which] *before del* 'can'
1.14 the] *over illeg*
1.14 statements] *interl*
2.1 Hoping] 'H' over 'h'

To J. D. Hooker 22 [January 1845]
2.3 Southern] *interl*
2.4 old] *interl*
2.8 (I intend . . . Parts.) 2.9] *interl; square brackets in MS*
3.3 & last] *interl*

To C. G. Ehrenberg 23 January [1845]
1.4 if] *above del* 'should'
2.7 curious] *after del* 'very'
3.1 might] *after del* 'I'
3.4 which] *after del illeg*
3.5 in preparing . . . Edition.—] *added*

To C. H. Smith 26 January [1845]
1.5 & p. 168] *interl*
1.6 Seniavane] [1]'a' *interl*
1.9 barrier] *after del illeg*
1.10 referred] *interl*
1.10 Every] 'E' *over* 'I'
1.12 account,] *before omitted point*
1.16 moral] *interl*
1.23 accumulation] *after del* 'de'
1.23 trees,] *before del* 'superficial,'
1.24 which] *over illeg*
3.2 believe me] *before omitted comma*

To Emma Darwin [3–4 February 1845]
1.2 overcome] *after del* 'oppressed'
1.4 & had . . . Baby] *interl*
1.5 little] *after del illeg*
1.10 first] *interl*
1.12 have] *altered from* 'had'
2.4 so] *interl*
4.2 I asked] *after del* 'Wil'

To Emma Darwin [7–8 February 1845]
1.4 said] 's' *over* 'I'
1.8 a long] 'a' *over* 'l'
1.9 Baby] *before omitted point*
1.13 eagerly] *interl*

To Trenham Reeks [before 8 February 1845]
1.1 Besides . . . glass.] *followed by line across page in MS*
2.1 Please . . . me] *added; square brackets in MS*
2.1 paper] *over illeg*
2.3 often] *interl*
2.4 *& peculiarly*] *interl*
3.1 or . . . work] *interl*
3.2 mixed] *after del* 'placed'
3.6 the shells] 'the' *interl*
3.7 *does*] *altered from illeg*
3.7 *contain*] *after del* 'compose'
3.8 *four*] *interl*
3.8 *four?*] *followed by line across page in MS*
3.9 it . . . crystals:] *added; followed by line across page in MS*
3.9 pure] *interl*
3.13 which] *after del* 'on'
3.13 know.] *followed by line across page in MS*
3.15 climate] *above del* 'soil'
3.15 muriate of soda, but] *above del* 'common salt &'
3.16 damp.] *followed by line across page in MS*
3.17 : does . . . lime?] *interl*
3.18 country.—] *followed by line across page in MS*
3.19 salt?] *followed by line across page in MS*
3.20 well] *interl*
3.21 *sea-salt?*] *followed by line across page in MS*
3.22 how] *interl*
3.23 Saline matter, abundant.] *interl*
3.24 *it?*] *followed by line across page in MS*
3.26 & even . . . flame:] *interl*
3.27 *contain?*] *followed by line across page in MS*

To Charles Lyell [8 February 1845]
1.3 all] *added*
1.7 *Natica*] *altered from* 'Naticae'
3.7 from the] 'the' *over illeg*
3.9 to you] *interl*
3.11 : he is . . . work. 3.12] *interl*
3.12 make] 'a' *over illeg*
3.14 extinct] *added*

To J. D. Hooker [10 February 1845]
1.10 take] *before del* 'not'
1.13 indifferent] *interl*
1.16 a knot] 'a' *interl*
2.11 Geographical Distribution] *after del* 'the'; 'G' *over* 'g'; 'D' *over* 'd'
3.5 Youth] *above del* 'Boy'
3.7 Ask] *below del* 'Talk to'
4.2 proof-Plates] 'proof-' *interl*

To W. D. Fox [13 February 1845]
1.9 stomach,] *before del* 'is,'
3.1 coming] *before del* 'on'
4.2 dying] 'y' *over illeg*
7.1 sometime] *interl*
7.1 true] *added*

To Leonard Jenyns 14 February [1845]
2.2 that it] *over* 'they w'
2.3 I publish] 'I' *over illeg*
2.4 perhaps] *interl*

To John Murray 17 March [1845]
1.4 bound to] 'to' *over illeg*
1.11 without the appendix 1.12] *interl*
4.4 by you,] *interl*

To J. D. Hooker 19 March [1845]
1.2 & your Pacific MS.] *interl*
1.3 it should] 'should' *interl*
1.3 should arrive] 'arrive' *altered from* 'arrived'
2.4 simple-leaved] *after del* 'single'
3.5 extraordinary] 'ex' *over* 'in'
3.5 is almost] 'is' *interl*
4.6 in] *above del* 'under'
4.6 under] *interl*
4.10 the actuality of 4.11] *above del* 'such probable'

To John Murray 20 March [1845]
2.7 it] *interl*

To C. G. Ehrenberg 23 March [1845]
4.1 stone] *above del* 'rock'
4.2 reasons] *above del* 'causes'
4.3 specimens] *after del* 'white'
4.6 perhaps a *little* softer,] *interl*
4.7 extends] *before del* 'in'
4.8 geographical] *interl*
4.8 probably is] 'is' *interl*
4.9 believe] *after del* 'found layers'
5.3 abound] *before del* 'with'
7.1 this Spring] *interl*
7.2 most] *interl*
8.1 *fresh-water*] *interl*

To J. D. Hooker [26 March 1845]
3.1 will] *above del* 'shall'
3.1 on Thursday or Friday] *interl*

To J. D. Hooker 31 March [1845]
1.3 of a] *interl*
1.6 at your] 'at' *above del* 'of'
1.8 heard] *after del illeg*
1.16 to] *interl*
3.2 better] *before del* 'than'
3.3 whole] *interl*
3.10 I have] 'I' *over illeg*
3.10 a] *interl*

To John Murray [5 April 1845]
1.4 an] *interl*
2.2 must] *after del* 'will'
3.3 must] *after del* 'wo'
3.5 the three] 'the' *interl*
3.5 of the General Work] *interl*
4.1 penny] 'ny' *over illeg*
4.1 some] 'me' *added*
6.4 with] *interl*
7.3 entered] *after del* 'resp'

To Ernst Dieffenbach 8 April [1845]
2.2 them!] '!' *possibly in another hand*

To John Murray [10 April 1845]
2.1 have] *interl*
4.3 one or] *interl*

To John Murray 12 April [1845]
1.2 questions] ²'s' *added*
1.3 they] *over* 'it'
1.3 only to] 'to' *over illeg*
1.7 unwillingly] *before omitted point*
1.19 him] *interl*
1.21 a claim] 'a' *above del* 'some'
2.3 Of] 'f' *over illeg*

To J. D. Hooker [16 April 1845]
1.3 to enquire] 'to' *over* '&'
2.3 else,] *interl*
3.3 & I sh^{d} . . . Journal. 3.4] *interl*
5.3 few] *interl*
5.4 hands if he so liked.] 'hands' *altered from* 'hand'; 'if he' *over* '.—'
6.2 the means] 'the' *over* 'it'
6.2 plusser] 'ss' *over* 's'
7.2 coolly] ²'l' *added*
7.3 from a] *after del* 'a'
10.1 lichens,] *before del illeg*
10.1 San] *after omitted point*
11.2 criticisms or corrections 11.3] *interl*
12.2 *sometime*] *interl*
12.4 physionomy] 'i' *over* 'y'; 'o' *over illeg*
12.5 notion] *after del* 'fact'
13.1 & imperfect] *interl*
13.2 in my Journal] *interl*

To John Murray 17 [April 1845]
1.3 will] *altered from* 'with'
1.8 are] *added*
2.3 whole] *interl*
2.3 the press] *interl*
2.4 whenever] *above del* 'as'
2.4 shall be] *above del* 'am'
2.4 rather] *interl*
2.4 getting] *after del* 'the'
2.5 already] *interl*

To John Murray [23 April 1845]
1.3 line] *before del* 'till'

To W. D. Fox [24 April 1845]
1.1 have] *altered from* 'had'
1.3 have asked] 'have' *interl*
1.6 Miss Gifford that was] *interl*
1.7 of Elston 1.8] *interl*
1.13 we] *altered from* 'I'
4.2 thanks to 4.3] 's to' *added*
5.1 into a] 'a' *interl*

To Thomas Bell [26 April – August 1845]
3.2 Murray's] 'M' *over* 'T'

To J. D. Hooker [28 April 1845]
1.4 double] *after del* 'two'
1.5 confervoid] *interl*

To J. D. Hooker [May 1845]
1.3 390, 391] *interl*
4.2 organic] *interl*
4.5 during these] 'during' *interl*
6.3 isl^{ds}] *interl*
6.4 parts] *before omitted point*
7.2 not very low] *above del* 'true'
8.4 a whole] 'a' *interl*
11.2 *quite*] *interl*
11.3 you] *interl*
11.3 very] *interl*

To G. B. Sowerby [May 1845]
2.2 fossil] *before omitted point*

To Edward Forbes 13 May [1845]
3.1 either . . . homewards] *interl*
4.1 Revillagego] 2'e' *over* 'i'
5.1 NW. &] '&' *over illeg*
5.2 coast] *interl*
5.3 by floating ice] *interl*
5.3 across] *interl*
5.4 in comparison] *interl*
5.4 with] *before del* 'respect to'
9.2 I shall . . . inform me.] *added*
11.1 round] *after del* 'by'

To J. S. Henslow 16 May [1845]
1.3 now] *interl*
2.7 of the] 'of' *over* 'to'
2.7 will] *over* 'have'
2.7 be] *before del* 'been'
3.3 one] *over* 'two'
3.3 on] *added*
3.3 two] *altered from* 'the'
3.3 islands] *after del* 'same'

To C. G. Ehrenberg 21 May [1845]
1.8 as well] 'as' *over* 'is'
1.9 its formation] 'its' *interl*
1.10 further] *interl*
1.10 subject, I] *before omitted point*
1.11 The Bahia . . . Formation. 1.12] *added*
1.12 is] *after del illeg*
3.1 tufaceous] 'a' *over* 'f'
3.2 soon to] *interl*
3.4 by] *over* 'in'
3.4 of those already sent.] *interl*
3.6 S.] *interl*
3.6 & there] '&' *added*; 'there' *interl*
4.2 find the] 'the' *over* 'it'
4.2 wind] *interl*

To Paul Edmund de Strzelecki [25 May 1845]
2.3 are] *over* 'is'

To John Murray [31 May 1845]
1.1 the first] 'the' *interl*
1.3 As] 'A' *over* 'a'
4.2 in] *before del* 'the'
4.4 or not] *interl*
5.1 Map and] *interl*
5.1 say] *after del* 'to'
7.2 published] *interl*

To J. D. Hooker [4 June 1845]
1.4 Species] 'S' *over* 's'

To John Murray [4 June 1845]
4.1 by me] 'me' *interl*
5.1 viz . . . next] *interl*
5.2 about] *interl*

To John Murray [6 June 1845]
1.2 consequently] *interl*
1.4 having . . . 1st number] *interl*

To John Murray 20 [June 1845]
2.2 4th] *interl*
2.3 the three] *above del* 'my'
2.4 on] *added*
2.9 you] *added*
4.2 Beagles] *before omitted point*
4.2 (& . . . else)] *interl*
4.3 at one] 'at' *over* 'in'
4.4 out of] *above del* 'in'

To John Murray [23 June 1845]
1.3 so that . . . less 1.4] *interl*
1.4 prominent] *after del* '&'
2.2 Please . . . point: 2.3] *interl*
2.4 & putting] '&' *added*
3.1 (ie Sunday)] *interl*
3.2 them] *over* 'it'
3.3 Excuse . . . well.— 3.4] *added*
5.1 all the sheets] *interl*
5.1 the type 5.2] 'the' *over* 'it'; 'type' *interl*

To John Murray [26 June 1845]
2.5 & distributing] *interl*
5.2 been . . . Clowes] *above del* 'go nicely'

To J. D. Hooker [27 June 1845]
1.1 having] *interl*
1.1 answered] *altered from* 'answering'
1.3 I finally] 'I' *before omitted point*
2.3 evaporation] *after del illeg*
2.5 Polybirus] 'u' *over* 'i'
2.6 Do] *before del* 'the'
3.7 apparently very] *interl*
3.9 viz] *interl*
3.11 then arctic] *interl*
3.11 must] *above del* 'would'
4.5 extinct] *after del* 'fo'
4.5 (ie without . . . doctored)] *interl*
5.1 soon] *interl*

5.4 else] *interl*
5.5 ever] *interl*
6.6 isolated] *before del* 'points'
6.7 by] *over full stop*

To John Murray [28 June 1845]
2.3 I ask] 'I' *above* 'you'
2.3 ask you] 'you' *over illeg*
2.6 Thursday] *after del* 'Wednes'
2.7 "C. Darwin Esqre.] *interl after del* 'me at'

To John Murray [3 July 1845]
1.1 twelve] *over illeg*

To Charles Lyell [5 July 1845]
1.10 thinking] [1]'i' *over illeg*
2.4 anything] 'thing' *added*, *before del* 'part'
2.9 & undervalued] *interl*

To J. D. Hooker [11–12 July 1845]
1.1 had] *altered from* 'have'
1.8 on its back] *interl*
2.2 shall] *interl*
2.2 these] 'se' *added*
3.4 my error in the range] *interl above del* 'it'
4.10 on Variation] *interl*
4.12 more] *interl*
4.12 than at] *before omitted point*
4.19 I sh^{d} particularly be obliged] *interl*
4.21 from ignorance] *interl*
4.22 (I must . . . on you) 4.24] *square brackets in MS*
5.4 Arctic] 'A' *over* 'a'
5.4 N] *over* 'n'
5.7 (not . . . plant)] *interl*
5.7 or monoœcious] *interl*
6.1 would] *before del* 'would'
6.2 any] *above del* 'the'
6.2 since they grew] *interl*
6.2 once] *interl*
6.3 within Tertiary epochs] *interl*
7.1 in your] 'in' *over illeg*; 'your' *interl*
7.1 that] [2]'t' *over* 'n'
7.6 & . . . distinct] *interl*
7.6 most] *after del* '&'
8.5 information] *before del* 'on'
8.6 the moths . . . cocoons] *interl after del* 'they'
8.8 ie if] *above del* 'supposing that'
8.8 of the same species 8.9] *interl*
8.9 This w^{d}. be . . . me. 8.10] *added*
9.2 that there] 'that' *altered from* 'there'; 'there' *added*
9.2 Forbes] *above del* 'his'
9.3 but] *added*
9.7 on] *over illeg*
13.3 I think I collected] *above del underl*
16.1 are they] 'are' *added*
17.1 *Icaleria*] *above del* '*Icalleria*'
19.2 peculiar Galapageian] *interl*

To J. D. Hooker [22 July – 19 August 1845]
1.1 solitary] 'l' *over* 'i'
1.3 to ask about.] *interl*
1.13 regions] 'r' *over illeg*
1.14 on a wild bank 1.15] *interl*
1.15 up,] *interl*
2.8 **natural**] *above del* '*mule*'
2.8 (nor mule?)] *interl*
3.3 my] *after del* 'th'
3.6 when] *interl*
3.7 Kinnordy,] *before del* '& where afterwar'
4.4 (where . . . nights] *interl*
4.5 Cuming . . . it 4.6] *interl*
5.1 origin] *after del* 'cause'
6.2 Darwin] 'r' *over illeg*

To John Murray [27 July 1845]
2.1 next] *before omitted point*
3.2 (Amblyrhynchus)] *interl*
3.2 (No 229 . . . Geology) 3.3] *square brackets in MS*
4.2 w^{d} you] 'you' *interl*
4.2 to the] 'to' *over illeg*
6.1 my] *after del illeg*

To Charles Lyell [30 July – 2 August 1845]
0.2 (Saturday)] *before del* 'Wednesday'
1.2 has been] *above del* 'was'
1.3 than] *over illeg*
1.6 any] *altered from* 'my'
1.10 ie non-scientific] *interl*
1.16 that] *interl*
2.5 the finger of] *interl*
3.3 to the] 'to' *over illeg*
3.3 chapters,] 's,' *added over comma*
3.3 principal] *interl*
3.4 geographical] *interl*
3.6 Sometimes] 'es' *over* 'e'
3.8 disturbing] *interl*
3.18 has] *over* 'is', *before del* 'known'
3.18 N.] *added*
7.3 You] *after del* 'In'
7.4 In] 'I' *over* 'i'
7.5 mammals] *after del* 'ani'

7.5 have thought] *above del* 'say'
7.6 show] *after del* 'are'
7.6 relations] 's' *added*
7.9 insects] *after del* 'f'
7.9 (N.B. . . . L.) 7.10] *square brackets in MS*
7.10 and . . . usual.] *added*
9.1 paragraph] *interl*
9.1 in fact . . . it. 9.2] *interl*
9.2 () *over comma*
9.5 yields] 'i' *over* 'e'
9.6 *decay*] *before del* 'of'
10.2 knowing] 'n' *over illeg*
10.3 or not,] *added*
12.3 sewed] [1]'e' *over* 'o'
13.2 in effect] *added before del* 'strictly'
13.3 Radack] *before del interl* '[& here northern & torrid zone productions]'
13.6 *northern*] *interl*
13.6 must have travelled] *interl*
13.6 (together . . . Palms) 13.7] *interl*
15.2 carboniferous] *after del* 'N'
17.3 in] *after del illeg*

To J. D. Hooker [15 or 22 August 1845]
2.1 Ch.] *interl*
2.3 tameness of birds] *interl*
3.3 latter place] *interl*
4.1 & any others] *interl*
4.2 separate] *above del* 'duplicate'
4.2 in a row] *interl*
4.3 gladly] *interl*
6.4 if such . . . to you] *interl*
7.2 then] *interl*
7.2 fortnight] *after del* 'wee'
8.2 unrubbed out] *interl*
12.1 deal] 'l' *over* 'd'

To William Jackson Hooker [23 August 1845]
1.2 from] *over illeg*
3.1 We] *altered from* 'I'; *before del* 'a'

To John Murray [23 August 1845]
1.6 p. 486] *interl*
1.8 & I] '&' *above del* 'but'
1.8 think if] *above del* 'fear'
1.8 is in very small] *above del* 'without the'
1.9 this] 'is' *over illeg*
3.1 freely] *interl*
3.2 by] *after del* 'on'
4.2 sent] 't' *over* 'd'
7.1 two] *above del* 'a'

To W. J. Hooker 25 August 1845
1.5 fulfill] 'fill' *over illeg*

To Charles Lyell 25 August [1845]
1.4 viz] *after del* 'w'
1.6 (in Ure)] *interl*
1.7 (& highest)] *interl*
1.7 gold, silver, copper 1.8] *above del* 'metals'
1.8 plumbago 75 and] *interl*
2.4 or] *over comma*
3.7 comparative] *interl*
4.3 Humbolt] *above del* 'he'
4.6 volcano] *after del* 'm'
5.3 another] *after del* 'some'
6.2 a] *interl*

To John Murray 27 August [1845]
0.3 My] 'M' *over* 'D'
2.2 St.; as] *semicolon added*; 'as' *over* '—'
3.1 my Bookseller & *Binder* 3.2] *interl*
3.2 Mr Darwin] *above del* 'me'
3.2 Tuesday] *after del* 'Wednes'
4.3 packed] *interl*
6.4 this] *after del* 'my'

To J. D. Hooker [29 August 1845]
1.6 well] *after del* 'very'

To J. D. Hooker [3 September 1845]
2.3 so great] 'so' *over* 'th'
3.3 now] *added*
4.1 Astrolabe] 'e' *over illeg*
4.2 (NB. . . . stand) 4.3] *square brackets in MS*
5.3 so spelt] 'spelt' *altered from* 'spelled'
5.3 also spelt] 'spelt' *altered from* 'spelled'
6.2 in upheaving] *altered from* 'to upheave'
7.4 both] *after del* 'the'

To J. D. Hooker [10 September 1845]
1.1 Monday] *above del* 'Wednesday'
1.5 Cosmos] *above del* 'it'
1.6 if you will inform me.] *added*
2.4 for] *after del* 'without e'
2.5 I have] *added*
2.5 done so] *interl*
2.11 of] *over illeg*
3.4 that he] *interl*
3.10 at least] *interl*
3.13 attempting . . . speculations] *interl*
3.14 has] *added*

To J. D. Hooker [18 September 1845]
1.1 (NB. . . . Pt I) 1.2] *square brackets in MS*
3.3 before the Beagle's survey 3.4] *interl*
4.1 you] *interl*
4.3 variation] *above del* 'species'

To J. D. Hooker [8 October 1845]
1.13 & my wife] *interl*

To Charles Lyell 8 October [1845]
1.5 negros] *altered from* 'negroes'
1.5 N. America] 'N.' *added*
1.8 these] *altered from* 'this'
3.6 & caught 3.7] *interl*
4.2 few] *interl*
4.4 only] *interl*
4.9 part] *altered from* 'parts'

To J. D. Hooker 28 October [1845]
1.3 read] *interl*
2.1 you] *over* 'I'
4.7 Professors;] *semicolon altered from full stop*, *before del* '—'

To J. D. Hooker [5 or 12 November 1845]
3.6 junction of the] *interl*
4.6 hurry] 'ry' *over* 'y'

To G. B. Sowerby 12 [November 1845]
1.3 take] *before del* 'me'
2.1 Yours] *over illeg*
3.1 20th] *after del* '19th'

To Smith, Elder & Co. 13 November [1845]
1.3 much] *over illeg*
1.5 omitted.] *followed by line drawn across page in MS*
2.1 Owen . . . Bell 2.2] *interl*
3.2 Heading] 'g' *over* 'gs'
3.2 to . . . advertisement 3.3] *interl*
3.3 &c."] *followed by line drawn across page in MS*
4.2 introduced,] *before del* 'either,'
4.3 marked it.] *before del* 'or after the III Part, showing that all these Parts are by me.—'

To J. D. Hooker [17 November 1845]
1.1 more] *interl*
1.2 specimen of] *interl*
1.3 described by Henslow] *interl*
1.4 miserablest] *interl*
1.4 Galapageian] *interl*
3.1 Lycopodium] 'Lyco' *above del* 'Poly'
3.2 am] *above del* 'have'
3.9 more] *after del* 'most'
3.13 large] 'l' *over illeg*
3.20 me.—] *before del* 'To a non-botanist'
4.1 When] 'W' *over* 'w'
4.1 sometimes] *above del* 'often'
4.2 station] *after del* 'lo'
4.2 () *over comma*
4.6 polleniferous] 1'e' *over illeg*

To J. D. Hooker [21 November 1845]
1.5 (& am) 1.6] *added*
2.5 faint] *interl*

To J. D. Hooker [29 November 1845]
1.3 in a Fly 1.4] *interl*
1.5 in] *over* 'at'

To G. B. Sowerby [1 December 1845]
2.3 which will cost me some trouble— 2.4] *added*

To G. B. Sowerby [3 December 1845]
2.2 (also . . . Josef)] *added*; *square brackets in MS*
3.1 the] *over illeg*
4.2 you.] *before del* '&'
4.4 the] *interl*
4.4 other] *altered from* 'others'
4.5 Turritellæ] *interl*
4.7 from two places 4.8] *interl*

To G. B. Sowerby [9? December 1845]
0.3 My] 'M' *over* 'D'

To J. D. Hooker [10 December 1845]
3.3 if . . . them 3.4] *interl*

To John William Lubbock [16 January 1846]
1.3 how] *interl*

To Royal Geographical Society [28 January or 4 February 1846]
1.7 in my writings] *interl*
2.1 my own 2.2] *interl*
2.2 gives] *before del* 'much'
2.3 might have] *above del* 'will'
2.3 found] *over* 'find'
5.1 vol 5. p. 103] *interl*
5.2 something] *before del* '& very'

To J. D. Hooker [31 January 1846]
2.1 packed it] 'it' *above del* 'him'

2.7 in proportional numbers] *interl*
3.1 Berthelot] *above del* 'he'

To J. D. Hooker [5 February 1846]
0.3 My] 'M' *over illeg*

To J. D. Hooker [8? February 1846]
2.1 M.] *after omitted point*
6.1 they] *over illeg*
7.1 seen] *before illeg*

To J. D. Hooker [10 February 1846]
2.3 two] *over illeg*
2.9 & views] *interl*
3.6 ought to] *interl*
3.9 should] *after del illeg*
3.11 one,] *interl*
4.2 across] *after del* 'through'
5.1 20 or] *before del* '25'
6.3 or polymorphous] *interl*
7.1 the same] *above del* 'a'
9.2 follow it,] 'it,' *interl*
9.3 even with the assumption of the] *above del* 'assuming the sea'
9.3 assumption] *before del* 'that'
9.3 having] *after del* 'to', *altered from* 'have'
9.4 simply risen] *above del* 'stood at'
9.4 feet.] *before del* 'lower'
9.5 hopelessly] *after del illeg*

To William Thompson 18 February [1846?]
1.2 more] 'm' *over* 'f'

To J. D. Hooker [25 February 1846]
1.4 service] *after del* 'disposal'
1.8 &c?] '?' *over semicolon*
1.9 could] *above del* 'good'
1.9 spin] *before del interl* 'out'
3.1 the] *interl*
3.1 this] *over* 'it'
3.4 seems a] 'seems' *altered from* 'seem'; 'a' *over* 't'
4.2 subject] *above del* 'ground'
4.3 likes] 'li' *over illeg*
5.1 very] *interl*

To J. D. Hooker [25 February – 2 March 1846]
1.1 send it] 'it' *interl*

To J. D. Hooker [13 March 1846]
0.2 Friday] *below del* 'Wednesday'
1.1 I] *over* '&'
1.3 are] *interl*
1.7 most] *after del* 'the'
1.11 (more recent than the miocene age),] *interl*
2.2 *non-littoral*] *interl*
2.3 sickening] 'e' *over* 'n'
3.3 such] *above del* 'the'
3.3 cases] *altered from* 'case'
3.3 as] *above del* 'of'
3.4 on] *over comma*
3.5 peculiar insular] *interl*
3.6 generally] *interl*
3.7 (where there is a continent near)] *interl*
3.10 such] *above del* 'this [*over illeg*]', *before del interl* '& Sandwich'
3.10 groups] *altered from* 'group'
5.4 different families of the] *interl*
5.6 to tell] 'to' *interl*
5.7 the highness] 'the' *above del* 'its'
5.10 , as legs or wings or teeth,] *interl*
7.1 this] *after del* 'to'

To J. D. Hooker [24 March 1846]
1.3 *named*] *interl*
1.4 He] *altered from* 'His'
1.4 an inch] 'an' *altered from* 'a'; 'inch' *above del* 'inch'
1.7 endeavour to] *interl*
3.1 I] *over illeg*

To C. G. Ehrenberg 25 March [1846]
3.2 , however, 3.3] *interl*
4.2 them] *interl*

To J. D. Hooker [29 March or 5 April 1846]
1.1 by] *after del* 'befo'
1.2 them] *interl*

To Smith, Elder & Co. 30 March [1846]
1.2 it)] ')' *over comma*
1.2 can not] *above del* 'could'
2.4 not more] *interl*
3.2 on] *over illeg*

To Robert Hutton [April 1846]
1.1 the Horticultural] 'the' *over* 'Dr'

To J. D. Hooker 10 April [1846]
0.2 10] *altered from* '9'
1.5 a more] 'a' *over illeg*
1.9 have not] 'have' *above del* 'did'

1.9 found as yet] 'found' *altered from* 'find'
1.11 them] 'th' *over* 'at'
2.3 must be] 'must' *over* 'is'; 'be' *interl*
2.3 to take] *after del* 'for'
2.8 has] *over* 'is'
3.4 & much . . . mammifers do 3.5] *interl*
4.5 produces . . . & thus 4.6] *interl*
5.2 (ie modified ribs)] *interl*
5.3 vertebra] 'a' *over* 'ae'
5.4 & lessening] *interl*
5.5 palpi &c] *before del interl* 'branchiæ &c'; '&c' *over caret*
5.6 but larger] *interl*
5.10 locomotive,] 've' *over* 'on'
5.10 &c;] *added*
5.10 even] *altered from* 'every'
6.2 heard] *after del* 'found'
6.4 I quite forget name 6.5] *interl*

To J. D. Hooker [16 April 1846]
1.8 rough] *interl*
1.9 some] *interl*
1.11 altogether] *interl*
1.15 state] *after del* 'suppose'
1.17 same] *interl*
5.1 when] *after del* 'to append to'
5.1 remark] *before del* 'of'
5.1 the] *interl*
5.2 of univers⟨al⟩] *after del illeg*
5.3 *viz*] *interl*
5.3 the species of] *interl*
5.4 are large in number &] *interl*
5.4 ranges have] *before del* 'the species of which'
5.4 themselves] *before del* 'very'

To Richard Owen [21 April 1846]
1.2 on Thursday] 'on' *over illeg*
1.4 save me] 'me' *interl*

To J. D. Hooker [May 1846]
1.8 as it is in some cases] *interl*
1.9 What] *after del* 'I presume no'
1.10 have acquired] 'have' *interl*
2.3 a reference] 'a' *interl*
2.5 what] *interl*
3.3 (Mem: . . . Galapagos.) 3.4] *square brackets in MS*
3.4 or Mexican] *interl*
4.3 ie not elevated 4.4] *interl*
6.3 him] 'm' *over* 's'
6.3 by letter] *interl*
8.4 read them] *interl*

To the Admiralty 9 [May 1846]
1.2 also] 'al' *over* 'the'

To J. D. Hooker [19 May 1846]
1.3 Elizabeth Is^d. 1.4] *interl*
1.4 (] *over* '—'
1.6 these?] '?' *over colon*
1.7 (if you . . . them) 1.8] *parentheses over commas*
2.4 number of] *interl*
2.5 on W. coast of S. America] *interl*
2.5 bears] *after del illeg*
4.4 know] *altered from* 'known'
4.8 Cycad-] *interl*
4.9 From *simple*] *interl*
4.10 Geological] *altered from* 'Geologically'

To Richard Owen 21 [June 1846]
3.2 them] *altered from* 'the', *before del* 'fossils'

To William Crawford Williamson 23 June [1846]
2.3 almost invariably] *above del* 'generally'
2.3 strata] *after del* 'layers'
2.4 fossil] *altered from* 'fossils'
2.7 such] *above del* 'the'
3.5 microscopical] *interl*
3.7 In] *before del* 'almost'
3.7 many] *above del* 'most'
3.8 sandstone] *interl*

To Emma Darwin [24 June 1846]
1.9 about] 'a' *over* 't'
1.10 come] *over* 'be'
4.4 asked] *after del* 'told'
5.2 can] *above del* 'have'
5.3 doubtfully] 'fully' *over* 'ed'

To Emma Darwin [25 June 1846]
1.4 of] *over* 'in'
1.9 have] *interl*
1.10 been] *altered from* 'be'

To J. D. Hooker [8 or 15 July 1846]
4.2 most] *above del* 'several'
4.2 according to Bunbury] *interl*
4.4 of climate] 'of' *over* '&'
4.5 for] *over* 'to'
4.7 with] *over* '—'
6.2 are] *interl*
6.2 so] *over* 'is'
6.3 ;—] *colon over full stop*; '—' *added*

6.4 Myrtaceæ] 'Myr' *over* 'Mry'
6.9 on] *over* 'in'
6.10 they are] *above del* 'it is'
7.1 or] *over* '&'
7.4 "Habitat,] *above del* 'say,'
7.5 & reading your *remarks*] *interl*

To Reeve Brothers [August 1846]
1.1 are] *interl*
1.1 trifling] *interl*
1.4 No[r]] *interl*
2.1 (as close as I have done on left side)] *interl*

To Charles Lyell [8 August 1846]
1.2 Principles] 's' *added*
1.4 S. American] *interl*
1.5 in France] *above del* 'there'
1.7 much] *after del* 'am'
2.3 subjects] *altered from* 'subject'
2.12 *German* Book] *interl*
3.2 heard] *after del* 'never'
3.2 very few] *above del* 'any one'
3.7 & short discussion] *interl*
5.1 proof-sheets] 'sheets' *over illeg*
5.2 whenever] *before omitted point*

To Leonard Jenyns [14 or 21 August 1846]
1.7 foreign] *interl*
1.8 naturalists] 'na' *over illeg*

To Leonard Horner
[17 August – 7 September 1846]
1.4 persuade] 'er' *over illeg*
2.3 head of the] *interl*
3.1 on] *after del* 'there'
3.2 glacier] *interl*

To Robert Mallet 26 August [1846]
2.1 writing] *interl*
2.2 that of waves] *above del* 'it'
2.3 what I had said,] 'had' *interl*
3.4 elevate] *altered from* 'elevates'

To J. D. Hooker [3 September 1846]
1.3 has] *added*
2.1 but] *interl*
2.2 papers] *interl*
2.2 have] *altered from* 'has'
2.12 (do you . . . there?) 2.13] *square brackets in MS*
4.2 (or two yards square?)] *interl; square brackets in MS*
4.4 if] *over* '—'
5.4 whenever] 'ever' *added*
5.5 districts, saying] *interl*
5.5 America] *before del comma*
5.8 results:] *above del* '['researches.' *del*] table'
5.8 M[r].] *interl*
5.9 in, as] ', as' *over* '.—'
5.9 indeed] *interl*
6.3 noble] *interl*
6.5 it necessary to 6.6] *interl before del* 'of'
6.6 refer] *altered from* 'referring'
6.6 Linnæus,] *before del closing quotation marks*
7.1 the] *interl*
8.6 on] *over* 'in'
9.1 you here] 'you' *over illeg*
12.1 (I do not . . . it.—) 12.3] *square brackets in MS*

To John Gould [*c.* October 1846]
1.7 very] *interl*
2.1 know] *interl*

To W. B. Carpenter
[October–December 1846]
1.4 The] *over illeg*
1.9 artist] *after del* 'at'

To J. D. Hooker [2 October 1846]
1.3 but] *interl*
3.4 stand] *interl*
5.7 look at] *before omitted point*
5.8 he had] *interl*
5.8 in the] *after del* 'was'
5.10 to Forbes] 'to' *interl*
5.10 for him] *above del* 'to him'

To W. D. Fox [before 3 October 1846]
1.3 collected] *after del* 'sent off early'
1.4 (ie autumn of the S. Hemisphere)] *interl*
4.2 & quill] *over* '.—'
5.1 us all] *before omitted point*

To Charles Lyell [3 October 1846]
1.2 is,] *interl*
1.6 p. 327] *interl*; '327' *above interl del* '319, 321,'
1.6 believe that] 'that' *interl*
1.7 main part of his] *interl*
1.8 gratuitously] 'ly' *added*
1.8 slow *Tertiary*] 'slow' *above del* 'su'
1.8 subsequent] *interl*
1.9 slow elevation] 'slow' *interl*
1.9 our] *interl*

1.9 is] *interl*
1.10 for *great* denudation,] *interl*
1.10 the general] 'the' *interl*
1.11 the off-shores] *interl*
1.12 two pages 1.13] *interl*
2.1 viz of one great sudden upheaval,] *interl*
2.2 the odder] 'the' *interl*
2.6 similar] *interl*
2.11 having been] *above del* 'being'
2.13 have been] 'have' *interl*; 'en' *added to* 'be'
3.2 there is] *interl*
3.3 the] *interl*
3.4 ones] *over* '.—'
4.2 blinded] *after del* 'overlooking,'
4.2 supposed] *after del* 'ne'
4.3 the common error, on 4.4] *interl*
4.4 on] *above del* 'ab'
4.4 which] *over illeg*
4.4 being] *after del caret*
4.4 , as I do not doubt, the 4.5] *interl*
4.6 the preservation . . . be] *interl after del* 'not being'
4.8 what] *before del* 'I'
5.1 at the same time 5.2] *interl*
5.4 sudden] *interl*
6.1 Ever] 'E' *over* 'e'

To J. S. Henslow [5 October 1846]
1.9 one,] *interl*
1.12 we] *over* 'I'
3.1 at] *over* 'to'

To J. D. Hooker [8 October 1846]
1.2 to bring] *interl*
1.2 good] *interl*
5.1 M^{r}] *over* 'a'
5.1 a very tall young man] *interl*

To John Lindley [*c.* 10 October 1846]
1.6 well] *after del* 'so'

To A. C. Ramsay 10 October [1846]
1.3 related] ^{1}'e' *over illeg*
1.9 likewise] *after del* '&'
1.9 horizontal] *interl*
2.3 that] *interl*
2.11 is] *after del* 'are'
2.12 after] *after del* 'in'
2.14 & yet producing a symmetrical effect.] *interl*
2.14 a] *interl*
3.2 me] *altered from* 'my'
3.3 forgive] *above del* 'excuse'

To Leonard Jenyns 17 October [1846]
1.6 this] *over* 'I'
3.4 splitting] *above del* 'cracking'
3.5 spherical] *interl*
3.5 maggot] *altered from* 'maggots'
3.8 two] *after del* 'some'
3.13 his acid] *interl*
3.13 Carabi] *over illeg*
4.6 possibly] *interl*

To Joseph Beete Jukes [18 October 1846]
1.3 specimens] *interl*

To J. D. Hooker [18 October 1846]
1.4 oclock] *interl*
1.4 if you will give me some 1.5] *interl*

To Smith, Elder & Co. [19 October 1846]
1.2 Plate] *interl*
1.3 bound] *interl*
1.3 in my] 'in' *over* 'on'
2.1 get] *after del* 'if'

To J. D. Hooker [26 October 1846]
1.5 microscope is a] *interl*
1.7 having] *after del* 'my'
2.3 stony, dentated] *interl*
2.4 valves?] '?' *over semicolon*
2.5 manner] *before del* '?'
2.11 I can . . . remember.—] *added*
2.12 the oblique] *interl*
2.13 that] *over* 'it'
2.19 in same] *after del* 'of'
2.19 my] *above del* 'the'
2.22 legs] *after del* 'shell'
2.23 last] *interl*
2.23 (your fig. 21)] *interl*
2.24 figures,] *comma over colon*
2.25 state on] *interl*
6.2 animal] *before del comma*
6.4 3^{d} & inner] *interl*

To Daniel Sharpe [1 November 1846]
1.4 quite] *after del* 'employed'
1.8 on mountain-flanks] *interl*
1.9 observed it,] 'it,' *interl*
1.10 usual] *after del* 'common '
1.14 with compass in hand & note book] *interl*
1.20 (as might . . . statement) 1.21] *interl*
2.3 precise] 'r' *over illeg*
2.4 Post on] 'on' *over* 'of'
2.5 Monday or] *interl*

3.7 the axis-plane of] *interl*
3.8 ridges] *above del* 'planes'

To J. D. Hooker [6 November 1846]
3.1 Wednesday] *interl*
3.1 for Geolog. Soc] *interl*
3.3 which would suit you] *interl*
4.2 should you] *after del* 'you'

To J. D. Hooker [12 November 1846]
1.2 next] *interl*
1.2 & if] '&' *over illeg*
2.2 splitting] *after del illeg*
2.2 the posterior] 'the' *interl*
2.4 this] *interl*
2.4 (viz two penes)] *interl*

To J. D. Hooker [14 November 1846]
2.4 on] *interl*
2.4 fixed] *interl*
3.1 in note to Park S^t,] *interl*
3.1 on Friday] *interl*
5.1 is] *interl*

To J. D. Hooker [15 November 1846]
1.3 to London.] *added*; 'to' *over full stop*

To Catherine Darwin [22 November 1846]
4.2 & met] *after del* 'with'
4.6 added] *before del* '&'

To J. D. Hooker [23 November 1846]
1.3 geograph.] *after del* 'the'
2.2 about] *interl*
2.6 (which . . . part) 2.7] *square brackets in MS*
2.7 yet often thick] *interl*
2.9 C. Verds] *interl*
2.10 (how . . . Compositæ) 2.11] *square brackets in MS*
2.26 other] *after del* 'following'
2.32 *exclude*] *before del* 'the'
2.33 Panama.—] 'P' *over* 'I'
2.35 parts of both] *interl*
2.35 Americas] *altered from* 'American'
2.35 first] *interl*
2.41 exclude] 'ex' *over* 'in'
2.43 of] *before omitted point*
2.44 be] *interl*
2.45 is] *interl*
2.46 *some*] *interl*

To Richard Owen 25 November [1846]
1.1 Having] 'H' *over illeg*
1.9 **Sessile**] *interl*

To J. L. Stokes [*c*. 26 November 1846]
3.1 you] *interl*
3.2 prouder] 'u' *over illeg*
3.3 extraordinary] *after del* 'not'

To Francis Wedgwood 27 November [1846]
4.1 on same day] *interl*

To *Annals and Magazine of Natural History* [December 1846]
2.1 marked] *interl*
2.6 you] *before omitted point*
4.1 not] *interl*

To J. D. Hooker [December 1846]
3.5 Coal-plants] *after del illeg*

To J. D. Hooker [December 1846 – January 1847]
1.4 for the] 'for' *over* 'on'
2.1 this winter] *interl*

To A. C. Ramsay 21 December [1846]
2.3 been] *interl*
2.5 drift)] ')' *over comma*
2.6 direct] *interl*
2.7 indirect] *before del* 'ele'
2.8 the direct] 'the' *interl*
2.10 during] *interl*
2.11 deposition] *above del* 'age'
2.11 of the] *interl*
2.13 less high] *above del* 'lower'
3.1 but] *after del* 'though'
3.2 as] *interl*
3.3 volcanic] *interl*
3.4 thin] *before del* 'volcanic'
3.6 book on] *interl*
3.6 & 103:] *interl*
4.2 compression &] *interl*
4.3 & *vesicular*] *interl*
5.3 for] *over* '&'
5.3 great] *after del* 'thin'
5.3 from] *over* 'of'
5.3 mutual] *interl*
5.7 subject] *after del* 'more'
5.8 loose] *interl*
5.9 two] 't' *over* 'a'
6.2 so.] *before omitted point*

To John Murray 30 December [1846]
1.2 of my journal] *interl*
2.1 it] *interl*

BIBLIOGRAPHY

The following bibliography contains all books and papers referred to in this volume by author–date reference or by short title. Short titles are used for some standard reference works (e.g., *DNB*, *OED*), for CD's books and papers, and for editions of his letters and manuscripts (e.g., *Descent*, *LL*, *Red notebook*). Other references are given in author–date form.

ADB: *Allgemeine deutsche Biographie.* Under the auspices of the Historical Commission of the Royal Academy of Sciences. 56 vols. Leipzig: Duncker and Humblot. 1875–1912.

Agassiz, Louis. 1842. On the succession and development of organized beings at the surface of the terrestrial globe; being a discourse delivered at the inauguration of the Academy of Neuchâtel. *Edinburgh New Philosophical Journal* 33: 388–99.

Alexander, Richard Chandler. 1846a. Botanical excursion in Lower Styria in 1842. *Annals and Magazine of Natural History* 17: 457–66.

——. 1846b. Excursions in Upper Styria, 1842. *Annals and Magazine of Natural History* 18: 94–102.

Allan, Mea. 1967. *The Hookers of Kew 1785–1911.* London: Michael Joseph.

Alum. Cantab.: *Alumni Cantabrigienses. A biographical list of all known students, graduates and holders of office at the University of Cambridge, from the earliest times to 1900.* Compiled by J. A. Venn. Part II. From 1752 to 1900. 6 vols. Cambridge: Cambridge University Press. 1940–54.

Alum. Oxon.: *Alumni Oxonienses: the members of the University of Oxford, 1715–1886.* By Joseph Foster. 4 vols. Oxford. 1888.

Anderson, Peter John, ed. 1889–98. *Fasti Academiae Mariscallanae Aberdonensis. Selections from the records of the Marischal College and University 1593–1860.* 3 vols. Aberdeen.

Anson, George. 1748. *A voyage round the world, in the years 1740–4.* Compiled by Richard Walter. London.

Arago, Dominique François Jean. 1833. On the ground-ice or the pieces of floating ice observed in rivers during winter. *Edinburgh New Philosophical Journal* 15: 123–37.

Arbuckle, Elisabeth Sanders, ed. 1983. *Harriet Martineau's letters to Fanny Wedgwood.* Stanford, California: Stanford University Press.

Arcet, Jean d', Giroud, A., Lelièvre, C. H., and Pelletier, B. 1794. *Rapport sur les divers moyens d'extraire avec avantage le sel de soude du sel marin.* Paris.

Ashworth, J. H. 1935. Charles Darwin as a student in Edinburgh, 1825–1827. *Proceedings of the Royal Society of Edinburgh* 55: 97–113.

Athenæum. 1844. A few words by way of comment on Miss Martineau's statement. No. 896 (28 December): 1198–9.

Atkins, Hedley. 1974. *Down: the home of the Darwins*. London: Curwen Press for the Royal College of Surgeons of England.

Aust. Dict. Biog.: *Australian dictionary of biography: 1788–1850; 1851–1890*. Edited by Douglas Pike and Bede Nairn. 6 vols. Melbourne: Melbourne University Press. 1966–76.

Autobiography: *The autobiography of Charles Darwin 1809–1882. With original omissions restored.* Edited by Nora Barlow. London: Collins. 1958.

Babington, Charles Cardale. 1841–3. Dytiscidæ Darwinianæ; or, descriptions of the species of *Dytiscidæ* collected by Charles Darwin, Esq., . . . in South America and Australia, during his voyage in H.M.S. Beagle. *Transactions of the Entomological Society of London* 3: 1–17.

Backhouse, James. 1844. *A narrative of a visit to the Mauritius and South Africa.* London.

Barnes, Donald Grove. 1930. *A history of the English corn laws from 1600–1846.* London: George Routledge.

Barnes, James John. 1964. *Free trade in books: a study of the London book trade since 1800.* Oxford: Clarendon Press.

Barrett, Paul H. 1960. A transcription of Darwin's first notebook on 'Transmutation of Species'. *Bulletin of the Museum of Comparative Zoology at Harvard College* 122: 247–96.

BDR: *British diplomatic representatives 1789–1852*. Edited by S. T. Bindoff, E. F. Malcolm Smith, and C. K. Webster. London: Offices of the Royal Historical Society. 1934.

'Beagle' diary: *Charles Darwin's diary of the voyage of H.M.S. "Beagle"*. Edited by Nora Barlow. Cambridge: Cambridge University Press. 1933.

Beechey, Frederick William. 1831. *Narrative of a voyage to the Pacific and Beering's Strait, to co-operate with the Polar expeditions: performed in H.M.S. Blossom . . . in the years 1825, 26, 27, 28.* 2 vols. London.

Bell, Gerda Elizabeth. 1976. *Ernst Dieffenbach: rebel and humanist.* Palmerston North: Dunmore Press.

Bellot, Hugh Hale. 1929. *University College London, 1826–1926*. London: University of London Press.

Bénézit, Emmanuel. 1976. *Dictionnaire critique et documentaire des peintres, sculpteurs, dessinateurs et graveurs de tous les temps et de tous les pays, par un groupe d'écrivains spécialistes français et étrangers.* Revised edition. 10 vols. Paris: Librairie Gründ.

Bentham, George and Hooker, Joseph Dalton. 1844. *The botany of the voyage of H.M.S. Sulphur under the command of Captain Sir Edward Belcher*. London.

Berkeley, Miles Joseph. 1845. On an edible fungus from Tierra del Fuego, and an allied Chilian species. *Transactions of the Linnean Society of London* 19: 37–43.

Bertero, Carlo Guiseppe. 1830. Notice sur l'histoire naturelle de l'île Juan Fernandez. *Annales des Sciences Naturelles* 21: 344–51.

——. 1831–3. List of the plants of Chile; translated from the 'Mercurio Chileno,' by W. S. W. Ruschenberger. *American Journal of Science and Arts* 19 (1831): 63–70, 299–311; 20 (1831): 248–60; 23 (1833): 78–96, 250–71.

BHGW: *Biographisch-literarisches Handwörterbuch zur Geschichte der exacten Wissenschaften*. By Johann Christian Poggendorff. Vols. 1–4, Leipzig: Johann Ambrosius Barth. 1863–1904. Vol. 5, Leipzig and Berlin: Verlag Chemie, G.m.b.H. 1926.

Binney, Edward William. 1846. Description of the Dukinfield Sigillaria. *Quarterly Journal of the Geological Society of London* 2: 390–3.

Birds: Part III of *The zoology of the voyage of H.M.S. Beagle*. By John Gould. Edited and superintended by Charles Darwin. London. 1838–41.

Blume, Carl Ludwig and Fischer, Joann Baptist. 1828[–51]. *Flora Javae nec non insularum adjacentium*. Brussels.

BNB: *Biographie nationale publiée par l'Académie royale des sciences, des lettres et des beaux-arts de Belgique*. 28 vols. Brussels. 1866–1944.

Boott, Francis. 1851. Caricis species novæ, vel minus cognitæ. [Read 3 and 17 June 1845, 17 February 1846.] *Transactions of the Linnean Society of London* 20: 115–47.

Bory de Saint-Vincent, Jean Baptiste Georges Marie. 1804. *Voyage dans les quatres principales îles des mers d'Afrique*. 3 vols. Paris.

Bosanquet, Samuel Richard. 1845. *'Vestiges of the natural history of creation': its argument examined and exposed*. London.

Bravais, Auguste. 1845. On the lines of ancient level of the sea in Finmark. *Quarterly Journal of the Geological Society of London* 1: 534–49.

Brehm, Christian Ludwig. 1831. *Handbuch der Naturgeschichte aller Vögel Deutschlands*. Ilmenau.

Breton, William Henry. 1843. Excursion to the western range, Tasmania. *Tasmanian Journal of Natural Science* 2: 121–41.

British Museum. 1904–6. *The history of the collections contained in the natural history departments of the British Museum*. 2 vols. London.

Brodie, Peter Bellinger. 1845. *A history of the fossil insects in the secondary rocks of England*. London.

Brongniart, Adolphe Théodore. 1828. Considérations générales sur la nature de la végétation qui couvrait la surface de la terre aux diverses époques de formation de son écorce. *Annales des Sciences Naturelles* 15: 225–58.

Bronn, Heinrich Georg. 1841–9. *Handbuch einer Geschichte der Natur*. 3 vols. in 5 and atlas. Stuttgart.

Brown, Robert. 1818. Observations, systematical and geographical, on Professor Christian Smith's collection of plants from the vicinity of the river Congo. Appendix 5 of Tuckey, James Kingston, *Narrative of an expedition to explore the river Zaire*. London.

Browne, Janet. 1983. *The secular ark: studies in the history of biogeography*. New Haven and London: Yale University Press.
Brush, Stephen G. 1979. Nineteenth-century debates about the inside of the earth: solid, liquid or gas? *Annals of Science* 36: 225–54.
Buch, Christian Leopold von. 1813. *Travels through Norway and Lapland . . . Translated . . . by John Black. With notes . . . by Robert Jameson*. London.
——. 1836. *Description physique des îles Canaries, suivie d'une indication des principaux volcans du globe*. Translated by C. Boulanger. Paris.
Bunbury, Charles James Fox. 1846. On some remarkable fossil ferns from Frostburg, Maryland, collected by Mr. Lyell. *Quarterly Journal of the Geological Society of London* 2: 82–91.
——. 1847. On fossil plants from the coal formation of Cape Breton. *Quarterly Journal of the Geological Society of London* 3: 423–38.
Bunbury, Frances Joanna. 1891–3. *Memorials of Sir C. J. F. Bunbury, Bart. Early life* 1 vol.; *Middle life* vols. 1–3; *Later life* vols. 1–5. Mildenhall.
Burke's Landed Gentry: *Burke's genealogical and heraldic history of the landed gentry*. By John Burke. 1–17 editions. London: Burke's Peerage Ltd. 1833–1952.
Burke's Peerage: *Burke's peerage and baronetage*. By John Burke. 1–105 editions. London: Burke's Peerage Ltd. 1826–1980.
Burmeister, Karl Hermann Konrad. 1834. *Beiträge zur Naturgeschichte der Rankenfüsser (Cirripedia)*. Berlin.
Byron, George Anson. 1826. *Voyage of H.M.S. Blonde to the Sandwich Islands, in the years 1824–1825*. London.
Calendar: *A calendar of the correspondence of Charles Darwin, 1821–1882*. Edited by Frederick Burkhardt and Sydney Smith. New York and London: Garland. 1985.
Candolle, Augustin Pyramus de. 1818–21. *Regni vegetabilis systema naturale, sive ordines, genera et species plantarum secundum methodi naturalis normas digestarum et descriptarum*. 2 vols. Paris, Strasbourg, and London.
——. 1839–40. *Vegetable organography; or an analytical description of the organs of plants*. Translated by Boughton Kingdon. 2 vols. London.
Candolle, Augustin Pyramus de and Candolle, Alphonse de. 1824–73. *Prodromus systematis naturalis regni vegetabilis, sive enumeratio contracta ordinum generum specierumque plantarum huc usque cognitarum, juxta methodi naturalis normas digesta*. 17 vols. Paris.
Carlyle, Thomas. 1843. *Past and present*. London.
Carmichael, Dugald. 1818. Some account of the island of Tristan da Cunha and of its natural productions. *Transactions of the Linnean Society of London* 12: 483–513.
Carpenter, William Benjamin. 1844. On the microscopic structure of shells. *Report of the 14th meeting of the British Association for the Advancement of Science held at York*, pp. 1–24.

[Chambers, Robert]. 1844. *Vestiges of the natural history of creation*. London.
——. 1845. *Explanations: a sequel to the 'Vestiges of the natural history of creation'*. London.

Chancellor, Gordon, *et al.* 1987. Charles Darwin's *Beagle* collections in the Oxford University Museum. *Archives of Natural History*.

Clarke, William Branwhite. 1839. A notice of showers of ashes which fell on board the Roxburgh, at sea, off the Cape de Verd islands, February, 1839. [Read 6 November 1839.] *Proceedings of the Geological Society of London* 3 (1838–42): 145–6.

——. 1842. On the occurrence of atmospheric deposits of dust and ashes; with remarks on the drift pumice of the coasts of New Holland. *Tasmanian Journal of Natural Science* 1: 321–41.

Clergy List: *The clergy list . . . containing an alphabetical list of the clergy*. London. 1841–.

Colenso, William. 1843. An account of some enormous fossil bones, of an unknown species of the class Aves, lately discovered in New Zealand. *Tasmanian Journal of Natural Science* 2: 81–107.

Collected papers: *The collected papers of Charles Darwin*. Edited by Paul H. Barrett. 2 vols. Chicago and London: University of Chicago Press. 1977.

Colp, Ralph, Jr. 1977. *To be an invalid: the illness of Charles Darwin*. Chicago and London: University of Chicago Press.

Colwell, Hector A. 1922. *An essay on the history of electrotherapy and diagnosis*. London: William Heinemann.

Complete Peerage: *The complete peerage of England Scotland Ireland Great Britain and the United Kingdom extant extinct or dormant*. By G. E. Cokayne. New edition, revised and much enlarged. Edited by the Hon. Vicary Gibbs and others. 12 vols. London: The St Catherine Press. 1910–59.

Coral reefs: *The structure and distribution of coral reefs. Being the first part of the geology of the voyage of the Beagle, under the command of Capt. FitzRoy, R.N. during the years 1832 to 1836*. By Charles Darwin. London. 1842.

Correspondence: *The correspondence of Charles Darwin*. Edited by Frederick Burkhardt and Sydney Smith. Cambridge: Cambridge University Press. 1985–.

Course, Edwin. 1962. *London railways*. London: B. T. Batsford.

Couthouy, Joseph Pitty. 1844. Remarks upon coral formations in the Pacific; with suggestions as to the causes of their absence in the same parallels of latitude on the coast of South America. *Boston Journal of Natural History* 4: 66–105, 137–62.

Cowper, William, trans. 1791. *The Iliad and Odyssey of Homer translated into English blank verse*. 2 vols. London.

Crisp, D. J. 1983. Extending Darwin's investigations on the barnacle life-history. *Biological Journal of the Linnean Society* 20: 73–83.

Cruz, Luis de la. 1835. *Viage a su costa, del alcalde provincial del muy ilustre cabildo de la Concepción de Chile . . . desde el fuerte de Ballenar, frontera de dicha Concepción, por tierras desconocidas . . . hasta la cuidad de Buenos-Aires*. Buenos Aires.

Cunningham, Allan. 1827. A few general remarks on the vegetation of certain coasts of *Terra Australis*. Vol. 2, pp. 497–565, of King, Philip Parker, *Narrative of a survey of the intertropical and western coasts of Australia . . . in the years 1818 and 1822*. 2 vols. London.

DAB: *Dictionary of American biography*. Under the auspices of the American Council of Learned Societies. 20 vols., index, and 7 supplements. New York: Charles Scribner's Sons; London: Oxford University Press. 1928–81.

Darwin, Erasmus. 1791. *The economy of vegetation*. Part 1 of *The botanic garden*. London.

——. 1800. *Phytologia; or the philosophy of agriculture and gardening. With the theory of draining morasses, and with an improved construction of the drill plough*. London.

Darwin Pedigree: *Pedigree of the family of Darwin*. Compiled by H. Farnham Burke. Privately printed. 1888. Reprinted in Freeman, Richard Broke, *Darwin pedigrees*. London: Printed for the author. 1984.

Davies, Gordon L. 1969. *The earth in decay: a history of British geomorphology 1578–1878*. New York: American Elsevier.

Dawson, Llewellyn Styles. 1885. *Memoirs of hydrography including brief biographies of the principal officers who have served in H.M. naval surveying service between the years 1750 and 1885*. 2 pts. Eastbourne. Facsimile reprint. London: Cornmarket Press. 1969.

DBF: *Dictionnaire de biographie française*. Under the direction of J. Balteau, M. Barroux, M. Prevost, and others. 17 vols., (A–Halicka). Paris: Librairie Letouzey et Ané. 1933–86.

DBI: *Dizionario biografico degli Italiani*. Edited by Alberto M. Ghisalberti. 29 vols., (A–Corvo). Rome: Instituto della Enciclopedia Italiana. 1960–83.

DBL: *Dansk biografisk leksikon*. Founded by Carl Frederick Bricka, edited by Povl Engelstoft and Svend Dahl. 26 vols. and supplement. Copenhagen: J. H. Schultz. 1933–44.

Dean, Dennis R. 1969. Hitchcock's dinosaur tracks. *American Quarterly* 21: 639–44.

Dease, Peter Warren and Simpson, Thomas. 1838. An account of the recent Arctic discoveries. *Journal of the Royal Geographical Society of London* 8: 213–25.

de Beer, Gavin, ed. 1958. *Charles Darwin and Alfred Russel Wallace. Evolution by natural selection*. Cambridge: Cambridge University Press. Reprint. New York and London: Johnson Reprint Corporation. 1971.

——, ed. 1959. Darwin's Journal. *Bulletin of the British Museum (Natural History) Historical series* 2: 3–21.

——, ed. 1960. Darwin's notebooks on transmutation of species. *Bulletin of the British Museum (Natural History) Historical Series* 2: 25–183.

de Beer, Gavin and Rowlands, M. J., eds. 1961. Darwin's notebooks on transmutation of species. Addenda and corrigenda. *Bulletin of the British Museum (Natural History) Historical Series* 2: 185–200.

de Beer, Gavin, Rowlands, M. J., and Skramovsky, B. M., eds. 1967. Darwin's notebooks on transmutation of species. Pages excised by Darwin. *Bulletin of the British Museum (Natural History) Historical Series* 3: 129–76.

Decaisne, Joseph. 1845. Recherches sur les anthéridies et les spores de quelque *Fucus*. [Read 11 November 1844.] *Annales des Sciences Naturelles. Botanique* 3d ser. 3: 5–15.

——. 1853. *Plantes vasculaires* (*Botanique*, vol. 2). In Dumont d'Urville, J. S. C. *Voyage au pôle sud et dans L'Océanie sur les corvettes L'Astrolabe et La Zélée*. 23 vols. Paris. [1841–54].

Dejean, Pierre François Marie Auguste. 1825–38. *Species général des Coléoptères de la collection de M. le Comte Dejean*. 6 vols. Paris.

Descent: *The descent of man, and selection in relation to sex*. By Charles Darwin. 2 vols. London. 1871.

Desmond, Adrian. 1982. *Archetypes and ancestors: palaeontology in Victorian London, 1850–1875*. London: Blond & Briggs.

——. 1985. The making of institutional zoology in London 1822–1836. *History of science* 23: 153–85, 223–50.

Desmond, Ray. 1977. *Dictionary of British and Irish botanists and horticulturists, including plant collectors and botanical artists*. 3d ed. London: Taylor and Francis.

Dick, Thomas Lauder. 1823. On the parallel roads of Lochaber. [Read 2 March 1818.] *Transactions of the Royal Society of Edinburgh* 9: 1–64.

Dieffenbach, Ernst. 1843. *Travels in New Zealand; with contributions to the geography, geology, botany, and natural history of that country*. 2 vols. London.

——, trans. 1844. *Naturwissenschaftliche Reisen* . By Charles Darwin. Braunschweig.

DNB: *Dictionary of national biography*. Edited by Leslie Stephen and Sydney Lee. 63 vols. and 2 supplements (6 vols.). London: Smith, Elder and Co. 1885–1912. *Dictionary of national biography 1912–70*. Edited by H. W. C. Davis, J. R. H. Weaver, and others. 6 vols. London: Oxford University Press. 1927–81.

DNZB: *A dictionary of New Zealand biography*. Edited by G. H. Scholefield. 2 vols. Wellington: Department of Internal Affairs New Zealand. 1940.

Don, David. 1841. An account of the Indian species of *Juncus* and *Luzula*. *Transactions of the Linnean Society of London* 18: 317–26.

Doubleday, Edward. 1845. Remarks on the genus *Argynnis* of the *Encyclopédie Méthodique*, especially in regard to its subdivision by means of characters drawn from the neuration of the wings. *Transactions of the Linnean Society of London* 19: 477–85.

Drury, Robert. 1729. *Madagascar: or, Robert Drury's journal, during fifteen years captivity on that island*. London.

DSB: *Dictionary of scientific biography*. Edited by Charles Coulston Gillispie. 14 vols., supplement, and index. New York: Charles Scribner's Sons. 1970–80.

Duchesne, Antoine Nicolas. 1766. *Histoire naturelle des fraisiers*. Paris.

Duméril, André Marie Constant and Bibron, Gabriel. 1834–54. *Erpétologie générale ou histoire naturelle complète des reptiles.* 9 vols. in 10 and atlas. Paris.

Dumont d'Urville, Jules Sébastien César. 1826. Flore des Malouines. *Mémoires de la Société Linnéenne de Paris* 4: 573–621.

——. 1832–3. *Voyage de découvertes autour du monde et à la recherche de La Pérouse.* 5 vols. Paris.

——. [1841–54]. *Voyage au pôle sud et dans L'Océanie sur les corvettes L'Astrolabe et La Zélée, 1837–40.* 23 vols. Paris.

Du Petit-Thouars, Abel Aubert. 1840–3. *Voyage autour du monde sur la frégate La Vénus, pendant les années 1836–1839. Relation.* 4 vols. and atlas. Paris.

EB: *Encyclopaedia Britannica.* 11th ed. 29 vols. Cambridge: Cambridge University Press. 1910–11.

Egerton, Frank N. 1970–1. Refutation and conjecture: Darwin's response to Sedgwick's attack on Chambers. *Studies in History and Philosophy of Science* 1: 176–83.

Ehrenberg, Christian Gottfried. 1844a. Vorläufige resultate seiner Untersuchungen der ihm von der Südpolreise des Capitain Ross, so wie von den Herren Schayer und Darwin zugekommenen materialien über das verhalten des kleinsten Lebens in den Oceanen und den grössten bisher zugänglichen Tiefen des Weltmeers vor. *Bericht über die zur Bekanntmachung geeigneten Verhandlungen der Königlichen Akademie der Wissenschaften zu Berlin*, pp. 182–207.

——. 1844b. On microscopic life in the ocean at the South Pole, and at considerable depths. *Annals and Magazine of Natural History* 14: 169–81.

——. 1844c. Einfluss des unsichtbar kleinen organischen Lebens . . . auf die Massenbildung von Bimstein, Tuff, Trass, vulkanischem Conglomerat und auch auf das Muttergestein des Nordasiatischen Marekanits. *Bericht über die zur Bekanntmachung geeigneten Verhandlungen der Königlichen Akademie der Wissenschaften zu Berlin*, pp. 324–44.

——. 1845a. Neue Untersuchungen über das kleinste Leben als geologisches Moment. *Bericht über die zur Bekanntmachung geeigneten Verhandlungen der Königlichen Akademie der Wissenschaften zu Berlin*, pp. 53–87.

——.1845b. Vorläufige zweite Mittheilung über die . . . Beziehungen des kleinsten organischen Lebens zu den vulkanischen Massen der Erde. *Bericht über die zur Bekanntmachung geeigneten Verhandlungen der Königlichen Akademie der Wissenschaften zu Berlin*, pp. 133–57.

——. 1845c. Über einen am 15 Mai 1830 in Malta gefallenen atmosphärischen Staub. *Bericht über die zur Bekanntmachung geeigneten Verhandlungen der Königlichen Akademie der Wissenschaften zu Berlin*, pp. 377–81.

——. 1846. Über die vulkanischen Phytolitharien der Insel Ascension. *Bericht über die zur Bekanntmachung geeigneten Verhandlungen der Königlichen Akademie der Wissenschaften zu Berlin*, pp. 191–202.

——. 1853. Über das mikroskopische Leben der Galapagos-Inseln. *Bericht über die zur Bekanntmachung geeigneten Verhandlungen der Königlichen Akademie der Wissenschaften zu Berlin*, pp. 178–94.

Élie de Beaumont, Jean Baptiste Armand Louis Léonce. 1838. Recherches sur la structure et sur l'origine du Mont Etna. In Dufrénoy, Pierre Armand and Élie de Beaumont, J. B. A. L. L., *Mémoires pour servir à une description géologique de la France* 4: 1–226. Paris.

——. 1843. Rapport sur un mémoire de M. Alcide d'Orbigny, intitulé: Considérations générales sur la géologie de l'Amérique méridionale. *Comptes rendus hebdomadaires des séances de l'Académie des Sciences Paris* 17: 379–417.

Elliott, Clark A. 1979. *Biographical dictionary of American science: the seventeenth through the nineteenth centuries.* Westport, Conn. and London: Greenwood Press.

Emma Darwin: *Emma Darwin: a century of family letters 1792–1896.* Edited by Henrietta Litchfield. Revised edition. 2 vols. London: John Murray. 1915.

Enciclopedia Italiana: *Enciclopedia Italiana di scienze, lettere ed arti.* 35 vols. Rome: 1929–39.

Endlicher, Stephan Ladislaus. 1833. *Prodromus florae Norfolkicae sive catalogus stirpium quae in insula Norfolk annis 1804 et 1805 a Ferdinando Bauer collectae et depictae nunc in Museo Caesareo Palatino rerum naturalium Vindobonae servantur.* Vienna.

Erman, Georg Adolph. 1838. Extrait d'une lettre . . . à M. Arago, sur la température de la terre en Sibérie. *Comptes rendus hebdomadaires des séances de l'Académie des Sciences Paris* 6: 501–3.

Eschscholtz, Johann Friedrich. 1829–33. *Zoologischer Atlas, enthaltend Abbildungen und Beschreibungen neuer Thierarten während des Flottcapitains von Kotzebue zweiter Reise um die Welt, auf der Russisch-Kaiserlichen Kriegsschlupp Predpriaetië in den Jahren 1823–1826.* 5 vols. in 1. Berlin.

Falconer, Hugh. 1846. Abstract of a discourse by Dr. Falconer, on the fossil fauna of the Sewalik Hills. *Journal of the Royal Asiatic Society* 8: 107–11.

Fish: Part IV of *The zoology of the voyage of H.M.S. Beagle.* By Leonard Jenyns. Edited and superintended by Charles Darwin. London. 1840–2.

FitzRoy, Robert. 1846. *Remarks on New Zealand, in February 1846.* London.

Flinders, Matthew. 1814. *A voyage to Terra Australis.* 2 vols. and atlas. London.

Flower, William Henry. 1879–91. *Catalogue of the specimens illustrating the osteology and dentition of vertebrated animals, recent and extinct, contained in the museum of the Royal College of Surgeons of England.* 3 vols. London.

Forbes, Edward. 1843. Report on the Mollusca and Radiata of the Ægean Sea, and on their distribution, considered as bearing on geology. *Report of the 13th meeting of the British Association for the Advancement of Science held at Cork*, pp. 130–93.

——. 1844. On the light thrown on geology by submarine researches; being the substance of a communication made to the Royal Institution of Great Britain, Friday evening, the 23d February 1844. *Edinburgh New Philosophical Journal* 36: 318–27.

——. 1845. On the distribution of endemic plants, more especially those of the British Islands, considered with regard to geological changes. *Report of the 15th meeting of the British Association for the Advancement of Science held at Cambridge* Transactions of the sections, pp. 67–8.

——. 1846. On the connexion between the distribution of the existing fauna and flora of the British Isles, and the geological changes which have affected their area, especially during the epoch of the northern drift. *Memoirs of the Geological Survey of Great Britain, and of the Museum of Economic Geology in London* 1: 336–432.

Forbes, James David. 1843. *Travels through the alps of Savoy and other parts of the Pennine chain with observations on the phenomena of glaciers.* Edinburgh.

——. 1844. Sixth letter on glaciers. Addressed to the Right Hon. Earl Cathcart. *Edinburgh New Philosophical Journal* 37: 231–44.

Fossil Cirripedia (1851): *A monograph on the fossil Lepadidæ, or, pedunculated cirripedes of Great Britain.* By Charles Darwin. London.

Fossil Cirripedia (1854): *A monograph on the fossil Balanidæ and Verrucidæ of Great Britain.* By Charles Darwin. London.

Fossil Mammalia: Part I of *The zoology of the voyage of H.M.S. Beagle.* By Richard Owen. Edited and superintended by Charles Darwin. London. 1838–40.

Foundations: *The foundations of the Origin of Species. Two essays written in 1842 and 1844 by Charles Darwin.* Edited by Francis Darwin. Cambridge: Cambridge University Press. 1909. Reprint. New York: Kraus Reprint Co. 1969. Also reprinted in de Beer, ed. 1958.

Freeman, Richard Broke. 1977. *The works of Charles Darwin: an annotated bibliographical handlist.* 2d ed., revised and enlarged. Folkestone: W. Dawson and Sons; Hamden, Conn.: Archon Books, Shoe String Press.

——. 1978. *Charles Darwin: a companion.* Folkestone: W. Dawson and Sons; Hamden, Conn.: Archon Books, Shoe String Press.

Freeman, Richard Broke and Gautrey, Peter J. 1969. Darwin's *Questions about the breeding of animals*, with a note on *Queries about expression. Journal of the Society for the Bibliography of Natural History* 5: 220–5.

Gardner, George. 1846. The vegetation of the Organ mountains of Brazil. *Journal of the Horticultural Society of London* 1: 273–93.

Gaudichaud, Charles Beaupré-. 1826. *Botanique.* Vol. 4 of Freycinet, L. C. D. de, *Voyage autour du monde, entrepris par ordre du Roi . . . exécuté sur les corvettes de S. M. l'Uranie et la Physicienne, pendant les années 1817, 1818, 1819 et 1820.* 9 vols. Paris. 1824–44.

Gay, Claude. 1833. Aperçu sur les recherches d'histoire naturelle faites dans l'Amérique du Sud, et principalement dans le Chili, pendant les années 1830 et 1831. *Annales des Sciences Naturelles* 28: 369–93.

Geoffroy Saint-Hilaire, Étienne. 1830. *Principes de philosophie zoologique.* Paris.

Geoffroy Saint-Hilaire, Isidore. 1832–7. *Histoire générale et particulière des anomalies de l'organisation chez l'homme et les animaux . . . ou traité de tératologie.* 3 vols. and atlas. Paris.

Geoffroy Saint-Hilaire, Isidore. 1841. *Essais de zoologie générale.* Paris.

Gérard, Frédéric. 1844. *De l'espèce dans les corps organisés.* Extract from d'Orbigny, Alcide Charles Victor Dessalines, ed., *Dictionnaire universel d'histoire naturelle.* 16 vols. Paris. 1841–9.

Ghiselin, Michael T. 1969. *The triumph of the Darwinian method.* Berkeley and Los Angeles: University of California Press.

Gilbert, Pamela. 1977. *A compendium of the biographical literature on deceased entomologists.* London: British Museum (Natural History).

Gloger, Constantin Wilhelm Lambert. 1833. *Das Abändern der Vögel durch Einfluss des Klimas. Nach zoologischen, zunächst von den Europäischen Landvögeln entnommenen Beobachtungen dargestellt, mit den entsprechenden Erfahrungen bei den Europäischen Säugthieren verglichen.* Breslau.

Gmelin, Johann Georg. 1747–69. *Flora Sibirica sive historia plantarum Sibiriae.* 4 vols. St Petersburg.

Goodsir, Harry D. S. 1844. Notice of observations on the development of the seminal fluid and organs of generation in the Crustacea. *Edinburgh New Philosophical Journal* 36: 183–6.

Gould, John. 1837. Remarks on a group of ground finches from Mr. Darwin's collection, with characters of the new species. *Proceedings of the Zoological Society of London* 5: 4–7.

——. 1843. On nine new birds collected during the voyage of H.M.S. *Sulphur. Proceedings of the Zoological Society of London* 11: 103–8.

Gray, John Edward. 1846. *Gleanings from the menagerie and aviary at Knowsley Hall.* Knowsley.

——. 1850. *Gleanings from the menagerie and aviary at Knowsley Hall. Hoofed quadrupeds.* Knowsley.

——, ed. 1852. *Molluscorum Britanniæ Synopsis. A synopsis of the Mollusca of Great Britain, arranged according to their natural affinities and anatomical structure.* By William Elford Leach. London.

Grey, George. 1841. *Journals of two expeditions of discovery in north-west and western Australia, during the years 1837, 38, and 39, under the authority of Her Majesty's Government. Describing many newly discovered, important, and fertile districts, with observations on the moral and physical condition of the aboriginal inhabitants.* 2 vols. London.

Grove, George. 1980. *The new Grove dictionary of music and musicians.* Edited by Stanley Sadie. 20 vols. London: Macmillan.

Gruber, Howard Ernest and Barrett, Paul H. 1974. *Darwin on man. A psychological study of scientific creativity . . . Together with Darwin's early and unpublished notebooks.* New York: E. P. Dutton and Co.; London: Wildwood House.

GSE: *Great Soviet encyclopedia* (translation of 3d ed. of *Bol'shaia Sovetskaia entsiklopediia.* Moscow. 1970–9). 31 vols. New York: Macmillan; London: Collier Macmillan. 1973–82.

Haidinger, Wilhelm Karl. 1848. Ueber Herrn v. Morlot's Sendschreiben an Herrn Élie de Beaumont die Bildung des Dolomits betreffend. *Sitzungsberichte der Mathematisch-naturwissenschaftlichen classe der Kaiserlichen Akademie der Wissenschaften in Wien* 1: 171–3.

Hamilton, William Richard. 1843. Address to the Royal Geographical Society of London; delivered at the anniversary meeting on the 22nd May, 1843. *Journal of the Royal Geographical Society of London* 13: xli–cv.

Hawkesworth, John. 1773. *An account of the voyages undertaken by . . . Commodore Byron, Captain Wallis, Captain Carteret and Captain Cook.* 3 vols. London.

Hayes, John Lord. 1844. Probable influence of icebergs upon drift. *Boston Journal of Natural History* 4: 426–52.

HDA: *Historial dictionary of Argentina.* By Ione S. Wright and Lisa M. Nekhom. Latin American Historical Dictionaries, no. 17. Metuchen, N. J. and London: The Scarecrow Press. 1978.

Hearne, Samuel. 1795. *A journey from Prince of Wales's Fort in Hudson's Bay, to the Northern Ocean, undertaken by order of the Hudson's Bay Company, for the discovery of copper mines, a north west passage, &c. in the years 1769, 1770, 1771, & 1772.* London.

Henslow, John Stevens. 1837. Description of two new species of *Opuntia*; with remarks on the structure of the fruit of Rhipsalis. *Magazine of Zoology and Botany* 1: 466–9.

——. 1838. Florula Keelingensis. An account of the native plants of the Keeling Islands. *Annals of Natural History* 1: 337–47.

——. 1841a. Report on the diseases of wheat. *Journal of the Royal Agricultural Society of England* 2: 1–25.

——. 1841b. On the specific identity of the fungi producing rust and mildew. *Journal of the Royal Agricultural Society of England* 2: 220–24.

——. 1844. *Suggestions towards an enquiry into the present condition of the labouring population of Suffolk.* Hadleigh.

——. 1845a. On concretions in the Red Crag at Felixstow, Suffolk. *Quarterly Journal of the Geological Society of London* 1: 35–7.

——. 1845b. On nodules, apparently coprolitic, from the Red Crag, London Clay, and Greensand. *Report of the 15th meeting of the British Association for the Advancement of Science held at Cambridge* Transactions of the sections, pp. 51–2.

——. 1847. On detritus derived from the London Clay and deposited in the Red Crag. *Report of the 17th meeting of the British Association for the Advancement of Science held at Oxford* Transactions of the sections, p. 64.

Herbert, Sandra. 1980. The red notebook of Charles Darwin. *Bulletin of the British Museum (Natural History) Historical Series* 7: 1–164. Published as a separate volume, Ithaca: Cornell University Press. 1980.

Herbert, William. 1846. Local habitation and wants of plants. *Journal of the Horticultural Society of London* 1: 44–9.

Herschel, John Frederick William, ed. 1849. *A manual of scientific enquiry; prepared for the use of Her Majesty's Navy: and adapted for travellers in general.* London.

Hinds, Richard Brinsley. 1842. Remarks on the vegetation of the Feejee Islands, Tanna, New Ireland, and New Guinea. *London Journal of Botany* 1: 669–76.

——. 1843. The regions of vegetation; being an analysis of the distribution of vegetable forms over the surface of the globe in connexion with climate and physical agents. Appendix to vol. 2 of Belcher, Edward, *Narrative of a voyage round the world, performed in H.M.S. Sulphur, 1836–42*. 2 vols. London.

——. 1845. Memoirs on geographic botany. *Annals and Magazine of Natural History* 15: 11–30, 89–104.

Hitchcock, Edward. 1841. *Final report on the geology of Massachusetts*. 2 vols. Northampton, Mass.

Hooker, Joseph Dalton. 1844–7. *Flora Antarctica*. Part 1 of *The botany of the Antarctic voyage of H.M. Discovery Ships Erebus and Terror in the years 1839–1843, under the command of Captain Sir James Clark Ross*. 2 vols. London.

——. 1845a. On the Huon pine, and on *Microcachrys*, a new genus of Coniferæ from Tasmania; together with remarks upon the geographical distribution of that order in the southern hemisphere. *London Journal of Botany* 4: 137–57.

——. 1845b. Note on some marine animals, brought up by deep-sea dredging, during the Antarctic voyage of Captain Sir James C. Ross, R.N. *Annals and Magazine of Natural History* 16: 238–9.

——. 1845c. *Testimonials in favour of Joseph Dalton Hooker R.N., M.D., F.L.S. as a candidate for the vacant chair of botany in the University of Edinburgh in four series*. 4 vols. in 1. Edinburgh.

——. 1845d. An enumeration of the plants of the Galapagos Archipelago; with descriptions of those which are new. [Read 4 March, 6 May, and 16 December 1845.] *Transactions of the Linnean Society of London* 10 (1851): 163–233.

——. 1846. On the vegetation of the Galapagos Archipelago, as compared with that of some other tropical islands and of the continent of America. [Read 1 and 15 December 1846.] *Transactions of the Linnean Society of London* 10 (1851): 235–62.

——. 1848. On the vegetation of the Carboniferous period, as compared with that of the present day. *Memoirs of the Geological Survey of Great Britain, and of the Museum of Practical Geology in London* 2, pt II: 387–430.

——. 1853–5. *Flora Novæ-Zelandiæ*. Part 2 of *The botany of the Antarctic voyage of H.M. Discovery Ships Erebus and Terror, in the years 1839–1843, under the command of Captain Sir James Clark Ross*. 2 vols. London.

——. 1860. *Flora Tasmaniæ*. Part 3 of *The botany of the Antarctic voyage of H.M. Discovery Ships Erebus and Terror, in the years 1839–1843, under the command of Captain Sir James Clark Ross*. 2 vols. London.

Hooker, William Jackson. 1843. Notes on the botany of H.M. Discovery Ships, Erebus and Terror in the Antarctic voyage; with some account of the Tussac grass of the Falkland Islands. *London Journal of Botany* 2: 247–329.

Hopkins, William. 1836. An abstract of a memoir on physical geology; with a further exposition of certain points connected with the subject. *London and Edinburgh Philosophical Magazine* 8: 227–36, 272–81, 357–66.

——. 1838. Researches in physical geology. *Transactions of the Cambridge Philosophical Society* 6: 1–84.

——. 1845. On the motion of glaciers. *London, Edinburgh and Dublin Philosophical Magazine* 26: 1–16, 146–69.

——. 1845–56. On the geological structure of the Wealden district and of the Bas Boulonnais. [Read 3 February 1841.] *Transactions of the Geological Society of London* 2d ser. 7: 1–51.

——. 1849. On the motion of glaciers. [Read 1 May 1843.] *Transactions of the Cambridge Philosophical Society* 8: 50–74.

Hopkirk, Thomas. 1817. *Flora anomoia. A general view of the anomalies in the vegetable kingdom.* Glasgow.

Horner, Leonard. 1846. Anniversary address of the president. *Quarterly Journal of the Geological Society of London* 2: 145–221.

Humboldt, Alexander von. 1814–29. *Personal narrative of travels to the equinoctial regions of the New Continent, during the years 1799–1804, by Alexander de Humboldt, and Aimé Bonpland . . . written in French by Alexander de Humboldt, and translated into English by Helen Maria Williams.* 7 vols. London.

——. 1843. *Asie centrale. Recherches sur les chaînes de montagnes et la climatologie comparée.* 3 vols. Paris.

——. 1845–8. *Kosmos; a general survey of the physical phenomena of the universe.* Translated by Augustin Prichard. 2 vols. London.

——. 1845–62. *Kosmos. Entwurf einer physischen Weltbeschreibung.* 5 vols. Stuttgart and Tübingen.

——. 1846–8. *Cosmos: sketch of a physical description of the universe.* Translated [by Elizabeth Julia Sabine] under the superintendence of Edward Sabine. 2 vols. London.

Hume, Abraham. 1845. *Examination of the theory contained in the 'Vestiges of creation'.* London.

Huxley, Leonard, ed. 1918. *Life and letters of Sir Joseph Dalton Hooker.* 2 vols. London: John Murray.

Jenyns, Leonard. 1837. Some remarks on the study of zoology, and on the present state of the science. *Magazine of Zoology and Botany* 1: 1–31.

——, ed. 1843. *The natural history of Selborne by the late Rev. Gilbert White, M.A. A new edition, with notes.* London.

——. 1846. *Observations in natural history: with an introduction on habits of observing, as connected with the study of that science. Also a calendar of periodic phenomena in natural history; with remarks on the importance of such registers.* London.

——. 1858. *Observations in meteorology.* London.

——. 1862. *Memoir of the Rev. John Stevens Henslow.* London.

Johnston, William. 1917. *Roll of commissioned officers in the medical service of the British Army, 1727–1898.* Aberdeen: Aberdeen University Press.

Journal of researches: *Journal of researches into the geology and natural history of the various countries visited by H.M.S. Beagle.* By Charles Darwin. London. 1839.

Journal of researches 2d ed.: *Journal of researches into the natural history and geology of the countries visited during the voyage of H.M.S. Beagle round the world. Corrected with additions.* London. 1845.

—— US ed.: *Journal of researches into the natural history and geology of the countries visited during the voyage of H.M.S. Beagle round the world.* 2 vols. New York. 1846.

Jussieu, Adrien de. 1842. *Botanique.* In Jussieu, Adrien de, Milne-Edwards, Henri, and Beudant, F. S., eds., *Cours élémentaire d'histoire naturelle à l'usage des collèges et des maisons d'éducation, rédigé conformément au programme de l'université du 14 Septembre 1840.* 3 vols. Paris. 1841–8.

Kane, Robert John. 1841. *Elements of chemistry.* Dublin.

Keilhau, Baltazar Mathias. 1838–40. Theory of granite, and the other massive rocks; together with that of crystalline slate; proposed in lectures in geology, in the University of Christiania in Norway, in the year 1836. *Edinburgh New Philosophical Journal* 24 (1838): 387–403; 25 (1838): 80–101, 263–72; 28 (1840): 366–71.

——. 1844. On the mode of formation of crystalline limestone . . . with preliminary observations on the present state of geology, and on the methods of investigation pursued in that science. *Edinburgh New Philosophical Journal* 36 (1844): 341–62; 37 (1844): 143–76.

Keith, Arthur. 1955. *Darwin revalued.* London: Watts.

Kirby, William. 1825. A description of some insects which appear to exemplify Mr. William S. MacLeay's doctrine of affinity and analogy. *Transactions of the Linnean Society of London* 14: 93–110.

Knight, Thomas Andrew. 1799. An account of some experiments on the fecundation of vegetables. *Philosophical Transactions of the Royal Society of London*, pp. 195–204.

Kölreuter, Joseph Gottlieb. 1761–6. *Vorläufige Nachricht von einigen das Geschlecht der Pflanzen betreffenden Versuchen und Beobachtungen.* 4 vols. Leipzig.

Kotzebue, Otto von. 1821. *Entdeckungs-Reise in die Süd-See und nach der Berings-Strasse zur Erforschung einer nordöstlichen Durchfahrt. Unternommen in den Jahren 1815 . . . 1818.* 3 vols. Weimar.

Krusenstern, Adam Johann von. 1824–7a. *Atlas de l'Océan Pacifique.* 2 vols. St Petersburg.

——. 1824–7b. *Recueil de mémoires hydrographiques, pour servir d'analyse et d'explication à l'Atlas de l'Océan Pacifique.* 2 vols. St Petersburg.

——. 1835. *Supplémens au recueil de mémoires hydrographiques, pour servir d'analyse et d'explication à l'Atlas de l'Océan Pacifique.* St Petersburg.

Laing, Samuel. 1836. *Journal of a residence in Norway during the years 1834, 1835 and 1836.* London.

Lamarck, Jean Baptiste Pierre Antoine de Monet de. 1815–22. *Histoire naturelle des animaux sans vertèbres.* 7 vols. Paris.

——. 1830. *Philosophie zoologique, ou exposition des considérations relatives à l'histoire naturelle des animaux.* 2d ed. 2 vols. Paris.

——. 1835–45. *Histoire naturelle des animaux sans vertèbres.* 2d ed. 11 vols. Paris.
Larousse XX: *Larousse du XXe siècle.* Under the direction of Paul Augé. 6 vols. Paris: Librairie Larousse. 1928–33.
Leach, William Elford. 1811–16. On the genera and species of Eproboscideous insects. *Memoirs of the Wernerian Natural History Society* 2: 547–66.
Leighton, William Allport. 1841. *A flora of Shropshire.* London and Shrewsbury.
Lesson, René Primevère and Garnot, Prosper. 1826–30. *Zoologie.* In Duperrey, Louis Isidore, *Voyage autour du monde, sur la corvette de sa majesté La Coquille, pendant les années 1822–5.* 2 vols. in 3 and atlas. Paris. 1826–30.
Liebig, Justus von. 1840. *Organic chemistry in its applications to agriculture and physiology.* Edited by Lyon Playfair. London.
——. 1842. *Chemistry in its application to agriculture and physiology.* Edited by Lyon Playfair. 2d ed., with very numerous additions. London.
Liebmann, Frederik Michael. 1843. Nachrichten über die Reise des Dänischen Botanikers Liebmann in Mejico. *Flora oder allgemeine botanische Zeitung* 26: 108–18.
Lindley, John. 1844. Editorial. *Gardeners' Chronicle and Agricultural Gazette*, no. 33 (17 August): 555.
——. 1845a. Editorial. *Gardeners' Chronicle and Agricultural Gazette*, no. 35 (30 August): 591.
——. 1845b. Do the races of plants wear out? *Gardeners' Chronicle and Agricultural Gazette*, no. 50 (13 December): 833.
——. 1846. *The vegetable kingdom; or, the structure, classification, and uses of plants, illustrated upon the natural system.* London.
Linnaeus, Carolus (Carl von Linné). 1741. Rön om växters plantering, grundat på naturen. *Kongliga Swenska Wetenskaps Academiens handlingar* 1: 5–24.
——. 1768. *De coloniis plantarum.* Defended by Johannes Flyggare. Uppsala. Also in *Amoenitates academicae.* 8 vols. Leyden, Amsterdam, and Erlangen. 1749–85.
Living Cirripedia (1851): *A monograph on the sub-class Cirripedia, with figures of all the species. The Lepadidæ; or, pedunculated cirripedes.* By Charles Darwin. London.
Living Cirripedia (1854): *A monograph on the sub-class Cirripedia, with figures of all the species. The Balanidæ (or sessile cirripedes); the Verrucidæ*, etc. By Charles Darwin. London.
LL: *The life and letters of Charles Darwin, including an autobiographical chapter.* Edited by Francis Darwin. 3 vols. London. 1887.
Lockhart, John Gibson. 1837–8. *Memoirs of the life of Sir Walter Scott.* 7 vols. Edinburgh and London.
[Long, Edward]. 1774. *The history of Jamaica . . . with reflections on its situation, settlements, inhabitants, climate, products, commerce, laws, and government.* 3 vols. London.
Loudon, John Claudius. 1839. *Hortus Britannicus. A catalogue of all the plants, indigenous, cultivated in, or introduced to Britain.* 2d ed. London.
Lund, Peter Wilhelm. 1839a. Coup d'œil sur les espèces éteintes de mammifères du Brésil. *Annales des Sciences Naturelles (Zoologie)* 2d ser. 11: 214–34.

Lund, Peter Wilhelm. 1839b. Nouvelles observations sur la faune fossile du Brésil. *Annales des Sciences Naturelles (Zoologie)* 2d ser. 12: 205–8.
——. 1841–2. Blik paa Braziliens Dyreverden för sidste Jordomvæltning. *Det Kongelige Danske Videnskabernes selskabs naturvidenskabelige og mathematiske Afhandlinger* 8 (1841): 61–144, 217–72, 273–96; 9 (1842): 137–208, 361–4.
Lütke, Fedor Petrovich. 1835–6. *Voyage autour du monde, exécuté par ordre de Sa Majesté l'empereur Nicolas 1er, sur la corvette le Séniavine, dans les années 1826, 1827, 1828 et 1829*. 4 vols. and 2 atlases. Paris.
Lyell, Charles. 1830–3. *Principles of geology, being an attempt to explain the former changes of the earth's surface, by reference to causes now in operation.* 3 vols. London.
——. 1834. Observations on the loamy deposit called "Loess" of the basin of the Rhine. *Edinburgh New Philosophical Journal* 17: 110–22.
——. 1837. *Principles of geology: being an inquiry how far the former changes of the earth's surface are referable to causes now in operation.* 5th ed. 4 vols. London.
——. 1838. *Elements of geology.* London.
——. 1840a. *Principles of geology: or, the modern changes of the earth and its inhabitants, considered as illustrative of geology.* 6th ed. 3 vols. London.
——. 1840b. On the boulder formation, or drift and associated freshwater deposits composing the mud-cliffs of eastern Norfolk. *London and Edinburgh Philosophical Magazine* 16: 345–80.
——. 1842. On the geological position of the *Mastodon giganteum* and associated fossil remains at Bigbone Lick, Kentucky, and other localities in the United States and Canada. *Proceedings of the Geological Society of London* 4: 36–9.
——. 1845a. *Travels in North America; with geological observations on the United States, Canada, and Nova Scotia.* 2 vols. London.
——. 1845b. On the Miocene Tertiary strata of Maryland, Virginia, and of North and South Carolina. *Quarterly Journal of the Geological Society of London* 1: 413–29.
——. 1846a. On foot-marks discovered in the coal-measures of Pennsylvania. *Quarterly Journal of the Geological Society of London* 2: 417–20.
——. 1846b. Observations on the fossil plants, of the Coal field of Tuscaloosa, Alabama; with a description of some species by C. T. F. Bunbury. *American Journal of Science and Arts* 2d ser. 2: 228–33.
——. 1846c. On the delta and alluvial deposits of the Mississippi, and other points in the geology of North America, observed in the years 1845–46. [Read 14 September 1846.] *Report of the 16th Meeting of the British Association for the Advancement of Science held at Southampton* Transactions of the sections, pp. 117–19.
——. 1847. *Principles of geology; or, the modern changes of the earth and its inhabitants considered as illustrative of geology.* 7th ed. London.
——. 1849. *A second visit to the United States of North America.* 2 vols. London.
Lyell, Katharine Murray, ed. 1881. *Life, letters and journals of Sir Charles Lyell, Bart.* 2 vols. London.
McCormick, Robert. 1841. Geological remarks on Kerguelen's Land. *Abstracts of the papers printed in the Philosophical Transactions of the Royal Society of London* 4: 299.

——. 1842. Geological remarks on Kerguelen's Land. *Tasmanian Journal of Natural Science* 1: 27–34.

MacCulloch, John. 1837. *Proofs and illustrations of the attributes of God, from the facts and laws of the physical universe.* 3 vols. London.

Macleay, William Sharp. 1819–21. *Horæ entomologicæ: or essays on the annulose animals.* Vol. 1, pts 1 and 2 (no more published). London.

——. 1830. On the dying struggle of the dichotomous system . . . in a letter to N. A. Vigors. *Philosophical Magazine* 7: 431–45; 8: 53–7, 134–40, 200–7.

Macquart, Pierre Justin Marie. 1834–5. *Histoire naturelle des Insectes. Diptères.* 2 vols. Paris.

Mallet, Robert. 1848a. On the dynamics of earthquakes; being an attempt to reduce their observed phenomena to the known laws of wave motion in solids and fluids. [Read 9 February 1846.] *Transactions of the Royal Irish Academy* 21: 51–105.

——. 1848b. On the objects, construction, and use of certain new instruments for self-registration of the passage of earthquake shocks. [Read 22 June 1846.] *Transactions of the Royal Irish Academy* 21: 107–13.

Mammalia: Part II of *The zoology of the voyage of H.M.S. Beagle.* By George Robert Waterhouse. Edited and superintended by Charles Darwin. London. 1838–9.

Martineau, Harriet. 1844. On mesmerism. *Athenæum* (23 and 30 November, 7, 14, and 21 December): 1070–2, 1093–4, 1117–18, 1144–6, 1173–4.

——. 1845. *Letters on mesmerism.* London.

MDCB: *The Macmillan dictionary of Canadian biography.* Edited by W. Stewart Wallace. 4th ed., revised, enlarged, and updated by W. A. McKay. Toronto: Macmillan of Canada. 1978.

Mellersh, Harold Edward Leslie. 1968. *FitzRoy of the Beagle.* London: Rupert Hart-Davis.

Meteyard, Eliza. 1875. *The Wedgwood handbook.* London.

Meyen, Franz Julius Ferdinand. 1834–5. *Reise um die Erde ausgeführt auf dem Königlich Preussischen Seehandlungs-Schiffe Prinzess Louise, commandirt von Capitain W. Wendt, in den Jahren 1830, 1831 und 1832.* 3 vols. Berlin.

Miers, John. 1826. *Travels in Chile and La Plata.* 2 vols. London.

Mill, John Stuart. 1845. The claims of labour: an essay on the duties of the employers to the employed. *Edinburgh Review* 81: 498–525.

Mitford, William. 1784–1818. *History of Greece.* 5 vols. London.

ML: *More letters of Charles Darwin: a record of his work in a series of hitherto unpublished letters.* Edited by Francis Darwin and Albert Charles Seward. 2 vols. London: John Murray. 1903.

Modern English Biography: *Modern English biography containing many thousand concise memoirs of persons who have died since the year 1850.* By Frederic Boase. 3 vols. and supplement (3 vols.). Truro: Netherton and Worth. 1892–1921.

Montagne, Jean François Camille. 1844. Mémoire sur le phénomène de la coloration des eaux de la Mer Rouge. *Comptes rendus hebdomadaires des séances de l'Académie des Sciences Paris* 19: 171–4.

Moquin-Tandon, Horace Bénédict Alfred. 1841. *Éléments de tératologie végétale, ou histoire abrégée des anomalies de l'organisation dans les végétaux.* Paris.

Morris, John and Sharpe, Daniel. 1846. Description of eight species of brachiopodous shells from the palæozoic rocks of the Falkland Islands. *Quarterly Journal of the Geological Society of London* 2: 274–8.

Murchison, Roderick Impey. 1846. On the superficial detritus of Sweden, and on the probable causes which have affected the surface of the rocks in the central and southern portions of that kingdom. *Quarterly Journal of the Geological Society of London* 2: 349–81.

Murchison, Roderick Impey, Verneuil, Edouard de, and Keyserling, Alexander von. 1845. *The geology of Russia in Europe and the Ural mountains.* 2 vols. London and Paris.

Murray, John. 1845. *Strictures on morphology: its unwarrantable assumptions, and atheistical tendency.* London.

Narrative: *Narrative of the surveying voyages of His Majesty's Ships Adventure and Beagle, between the years 1826 and 1836.* Edited by Robert FitzRoy. 3 vols. and appendix. London. 1839.

Natural selection: *Charles Darwin's Natural Selection; being the second part of his big species book written from 1856 to 1858.* Cambridge: Cambridge University Press. 1975.

Navy List: *The navy list.* London. 1815–70.

NBL: *Norsk biografisk leksikon.* Edited by Edvard Bull, Anders Krogvig, Gerhard Gran, and others. 19 vols. Oslo: H. Aschehoug & Co. 1923–83.

NBU: *Nouvelle biographie universelle.* Edited by Jean Chrétien Ferdinand Hoefer. 46 vols. Paris. 1852–66.

NDB: *Neue deutsche Biographie.* Under the auspices of the Historical Commission of the Bavarian Academy of Sciences. 14 vols., (A–Locher-Freuler). Berlin: Duncker and Humblot. 1953–85.

NDBA: *Nuevo Diccionario Biografico Argentino (1750–1930).* Edited by Vicente Osvaldo Cutolo. 6 vols., (A–SA). Buenos Aires: Editorial Elche. 1968–83.

Nelson, Richard John. 1840. On the geology of the Bermudas. *Transactions of the Geological Society of London* 2d ser. 5: 103–23.

New Zealand encyclopaedia: *An encyclopaedia of New Zealand.* Edited by A. H. McLintock. 3 vols. Wellington: R. E. Owen. 1966.

Nicol, William. 1834. Observations on the structure of recent and fossil Coniferæ. *Edinburgh New Philosophical Journal* 16: 137–58, 310–14.

NNBW: *Nieuw Nederlandsch Biografisch Woordenboek.* Edited by P. C. Molhuysen, P. J. Blok, and K. H. Kossmann. 10 vols. Leiden: A. W. Sijthoff. 1911–37.

Notebook B. See Barrett 1960; de Beer 1960; de Beer and Rowlands 1961; de Beer, Rowlands, and Skramovsky 1967; *Notebooks.*

Notebook C. See de Beer 1960; de Beer and Rowlands 1961; de Beer, Rowlands, and Skramovsky 1967; *Notebooks.*

Notebook D. See de Beer 1960; de Beer and Rowlands 1961; de Beer, Rowlands, and Skramovsky 1967; *Notebooks.*

Notebooks: *Charles Darwin's notebooks*. Edited by Paul H. Barrett, Sandra Herbert, David Kohn, Sydney Smith, and Peter Gautrey. London: British Museum (Natural History); Ithaca: Cornell University Press. 1987.

NUC: *The National Union Catalog*. Pre-1956 Imprints. 685 vols. and supplement (vols. 686–754). London and Chicago: Mansell. 1968–81.

O'Byrne, William R. 1849. *A naval biographical dictionary: comprising the life and services of every living officer in Her Majesty's Navy, from the rank of admiral of the fleet to that of lieutenant, inclusive*. London.

OED: *The Oxford English dictionary. Being a corrected re-issue with an introduction, supplement, and bibliography of A new English dictionary*. Edited by James A. H. Murray, Henry Bradley, W. A. Craigie, C. T. Onions. 12 vols. and supplement. Oxford: Clarendon Press. 1970. Supplement. Edited by R. W. Burchfield. 4 vols. Oxford: Clarendon Press. 1972–86.

Orbigny, Alcide Charles Victor Dessalines d'. 1835–47. *Voyage dans l'Amérique Méridionale (le Brésil, la République orientale de l'Uruguay, la République Argentine, la Patagonie, la République du Chili, la République de Bolivia, la République du Pérou), exécuté pendant les années 1826 . . . 1833*. 9 vols. Paris and Strasbourg.

——. 1845–7. *Mollusques vivants et fossiles, ou description de toutes les espèces de coquilles et de mollusques classées suivant leur distribution géologique et géographique*. Vol. 1 (no more published) and atlas. Paris.

——. 1850–2. *Prodrome de paléontologie stratigraphique universelle des animaux mollusques et rayonnés faisant suite au cours élémentaire de paléontologie et de géologie stratigraphiques*. 3 vols. Paris.

Origin: *On the origin of species by means of natural selection, or the preservation of favoured races in the struggle for life*. By Charles Darwin. London. 1859.

Ornithological notes: Darwin's ornithological notes. Edited by Nora Barlow. *Bulletin of the British Museum (Natural History) Historical Series* 2 (1963): 201–78.

Owen, Richard. 1840–5. *Odontography; or, a treatise on the comparative anatomy of the teeth; their physiological relations, mode of development, and microscopic structure*. 2 vols. London.

——. 1841. Report on British fossil reptiles. *Report of the 11th meeting of the British Association for the Advancement of Science held at Plymouth*, pp. 60–204.

——. 1843a. *Lectures on the comparative anatomy and physiology of the invertebrate animals . . . from notes taken by William White Cooper*. London.

——. 1843b. On Dinornis Novæ Zealandiæ. *Proceedings of the Zoological Society of London* 11: 8–10, 144–6.

——. 1845. Appendix to Professor Henslow's paper, consisting of a description of the fossil tympanic bones referable to four distinct species of Balæna. *Quarterly Journal of the Geological Society of London* 1: 37–40.

——. 1846a. Memoir on the Dinornis. *Proceedings of the Zoological Society of London* 14: 46–9.

——. 1846b. Report on the archetype and homologies of the vertebrate skeleton. *Report of the 16th Meeting of the British Association for the Advancement of Science held at Southampton*, pp. 169–340.

Owen, Richard. 1846c. *A history of British fossil mammals, and birds.* London.

——. 1846d. *Lectures on the comparative anatomy and physiology of the vertebrate animals. Part I—Fishes.* London.

Owen, Richard Startin. 1894. *The life of Richard Owen. With the scientific portions revised by C. Davies Sherborn, also an essay on Owen's position in anatomical science by the Right Hon. T. H. Huxley.* 2 vols. London.

Palissy, Bernard. 1636. *Le moyen de devenir riche, et la maniere veritable, par laquelle tous les hommes de la France pourront apprendre à multiplier & augmenter leurs thresors & possessions.* Paris.

Pernety, Antoine Joseph. 1769. *Journal historique d'un voyage fait aux Iles Malouines en 1763 & 1764.* 2 vols. Berlin.

Perrottet, Samuel. 1838. Extrait d'une lettre sur la végétation des montagnes dites Nelligherries, dans les Indes-Orientales. *Annales des Sciences Naturelles (Botanique)* 2d ser. 9: 288–90.

Physicians: *The roll of the Royal College of Physicians of London.* By William Munk. 2d ed., revised and enlarged. Vol. 3 (1801–25). London. 1878. Vol. 4, *Lives of the fellows of the Royal College of Physicians* (1826–1925), compiled by G. H. Brown. London: Published by the Royal College of Physicians. 1955.

Plarr, Victor Gustave. 1930. *Plarr's lives of the fellows of the Royal College of Surgeons of England.* Revised by Sir D'Arcy Power. 2 vols. London: Simpkin Marshall.

Pöppig, Eduard Friedrich. 1835. *Reise in Chile, Peru und auf dem Amazonenstrome während der Jahre 1827–1832.* 2 vols. in 1. Leipzig.

Porter, David. 1823. *A voyage in the south seas, in the years 1812, 1813, and 1814. With particular details of the Gallipagos and Washington Islands.* London.

Porter, Duncan M. 1980a. Charles Darwin's plant collections from the voyage of the *Beagle*. *Journal of the Society for the Bibliography of Natural History* 9: 515–25.

——. 1980b. The vascular plants of Joseph Dalton Hooker's *An enumeration of the plants of the Galapagos Archipelago; with descriptions of those which are new. Botanical Journal of the Linnean Society* 81: 79–134.

——. 1981. Darwin's missing notebooks come to light. *Nature* 291: 13.

Post Office directory of the six home counties: *Post Office directory of the six home counties, viz., Essex, Herts, Kent, Middlesex, Surrey and Sussex.* London. 1845–.

Post Office London directory: *Post Office London directory.* London. 1802–.

Prescott, William Hickling. 1843. *History of the conquest of Mexico, with a preliminary view of the ancient Mexican civilization and the life of Fernando Cortes.* 3 vols. London.

Prichard, James Cowles. 1836–7. *Researches into the physical history of mankind.* 3d ed. 2 vols. London.

Prideaux, J. 1845. Potatoes: storing and preserving. *Gardeners' Chronicle and Agricultural Gazette*, no. 39 (27 September): 655–6.

Provincial Medical Directory: *The provincial medical directory.* London. 1847–.

Questions & experiments notebook. See *Notebooks.*

Ramsay, Andrew Crombie. 1846. On the denudation of South Wales and the adjacent counties of England. *Memoirs of the Geological Survey of Great Britain, and of the Museum of Economic Geology in London* 1: 297–335.

——. 1847. On the origin of the existing physical outline of a portion of Cardiganshire. *Report of the 17th meeting of the British Association for the Advancement of Science held at Oxford* Transactions of the sections, pp. 66–7.

[Rathbone, Hannah Mary]. 1844–8. *So much of the diary of Lady Willoughby as relates to her domestic history, and to the eventful period of the reign of Charles the First.* 2 vols. London.

Red notebook. See S. Herbert 1980; *Notebooks.*

Rehbock, Philip F. 1983. *The philosophical naturalists: themes in early nineteenth-century biology.* Madison: University of Wisconsin Press.

Reptiles: Part V of *The zoology of the voyage of H.M.S. Beagle.* By Thomas Bell. Edited and superintended by Charles Darwin. London. 1842–3.

Richardson, E. W. 1916. *A veteran naturalist: being the life and work of W. B. Tegetmeier.* London: Witherby and Co.

Richardson, John. 1829–37. *Fauna Boreali-Americana; or the zoology of the northern parts of British America. Assisted by William Swainson and the Reverend William Kirby.* 4 vols. London.

Richardson, John and Gray, John Edward. 1844–75. *The zoology of the voyage of H.M.S. Erebus and Terror.* 2 vols. London.

Ripa, Matteo. 1844. *Memoirs of Father Ripa, during thirteen years' residence at the Court of Peking . . . with an account of the foundation of the college for the education of young Chinese at Naples.* London.

Robberds, John Warden. 1843. *A memoir of the life and writings of the late William Taylor of Norwich.* 2 vols. London.

Robertson, John. 1841. Catalogue of geological specimens procured from Kerguelen's Land during the months of May, June, and July, 1840. *Abstracts of the papers printed in the Philosophical Transactions of the Royal Society of London* 4: 305. [Title only printed.]

Robinson, Murrell R. 1844. On the town of Carmen and the Rio Negro. *Journal of the Royal Geographical Society of London* 14: 130–41.

Rogers, Henry Darwin. 1844. An address on the recent progress of geological research in the United States.

——. 1846. On cleavage of slate-strata. *Edinburgh New Philosophical Journal* 41: 422–3.

Rogers, William Barton and Rogers, Henry Darwin. 1843. On the physical structure of the Appalachian chain, as exemplifying the laws which have regulated the elevation of great mountain chains, generally. *Transactions of the Association of American Geologists and Naturalists*, pp. 474–531.

Ross, Helena C. G. 1979. Recently discovered correspondence between Charles Darwin and William Thompson in the Ulster Museum. *Irish Naturalists' Journal* 19: 364–5.

Ross, James Clark. 1847. *A voyage of discovery and research in the southern and Antarctic regions, during the years 1839–43*. 2 vols. London.

Rowbottom, Margaret and Susskind, Charles. 1984. *Electricity and medicine: history of their interaction*. London: Macmillan.

Rowlinson, J. S. 1971. The theory of glaciers. *Notes and Records of the Royal Society of London* 26: 189–204.

Royal College of Surgeons. 1854. *Descriptive catalogue of the fossil organic remains of Reptilia and Pisces contained in the museum of the Royal College of Surgeons of England*. London.

Rudwick, Martin J. S. 1974. Darwin and Glen Roy: a 'great failure' in scientific method? *Studies in the History and Philosophy of Science* 5: 97–185.

Russell-Gebbett, Jean. 1977. *Henslow of Hitcham: botanist, educationalist and clergyman*. Lavenham, Suffolk: Terence Dalton.

Rutherford, H. W. 1908. *Catalogue of the library of Charles Darwin now in the Botany School, Cambridge . . . with an introduction by Francis Darwin*. Cambridge: Cambridge University Press.

Saint-Hilaire, Auguste de. 1841. *Leçons de botanique comprenant principalement la morphologie végétale, la terminologie, la botanique comparée, l'examen de la valeur des charactères dans les diverses familles naturelles*. Paris.

Salaman, Redcliffe N. 1985. *The history and social influence of the potato*. Revised edition by J. G. Hawkes. Cambridge: Cambridge University Press.

Sarjeant, William A. S. 1980. *Geologists and the history of geology. An international bibliography from the origins to 1978*. 5 vols. London: Macmillan Press.

Sartorius, Christian. 1855–8. *Mexico: landscapes and popular sketches . . . Edited by Dr. Gaspey. With steel engravings by distinguished artists, from original sketches by Moritz Rugendas*. London.

Sauer, Gordon C. 1982. *John Gould: the bird man. A chronology and bibliography*. London: H. Sotheran.

SBL: *Svenskt Biografiskt Lexikon*. Edited by Bertil Boëthius, Erik Grill, and Birgitta Lager-Kromnow. 24 vols. Stockholm: Albert Bonnier and P. A. Norstedt. 1918–84.

Scherren, Henry. 1905. *The Zoological Society of London. A sketch of its foundation and development and the story of its farm, museum, gardens, menagerie and library*. London, Paris, New York and Melbourne: Cassell and Company.

Schouw, Joakim Frederik. 1816. *Dissertatio de sedibus plantarum originariis*. Copenhagen.

——. 1823. *Grundzüge einer Allgemeinen pflanzengeographie*. Berlin.

——. 1845. Les conifères d'Italie, sous les rapports géographiques et historiques. *Annales des Sciences Naturelles (Botanique)* 3d ser. 3: 230–72.

Schwartz, Joel S. 1980. Three unpublished letters to Charles Darwin: the solution to a "geometrico-geological" problem. *Annals of Science* 37: 631–7.

Secord, James A. 1986. The Geological Survey of Great Britain as a research school, 1839–1855. *History of Science* 24: 223–75.

Sedgwick, Adam. 1835. Remarks on the structure of large mineral masses, and especially on the chemical changes produced in the aggregation of stratified rocks during different periods after their deposition. *Transactions of the Geological Society of London* 2d ser. 3: 461–86.

——. 1845. Vestiges of the natural history of creation. *Edinburgh Review* 82: 1–85.

Selwyn, George Augustus. 1844. *New Zealand. Part I, Letters from the Bishop . . . with extracts from his visitation journal, from July 1842, to January 1843.* Edited by C. B. Dalton. London.

Seymour, W. A., ed. 1980. *A history of the ordnance survey.* Folkestone: Dawson.

Sharpe, Daniel. 1847. On slaty cleavage. *Quarterly Journal of the Geological Society of London* 3: 74–105.

Sheets-Pyenson, Susan. 1981. Darwin's data: his reading of natural history journals, 1837–1842. *Journal of the History of Biology* 14: 231–48.

Sloan, Phillip. 1985. Darwin's invertebrate program, 1826–1836. In Kohn, David, ed., *The Darwinian heritage.* Princeton, New Jersey: Princeton University Press.

Sloane, Hans. 1696. An account of four sorts of strange beans, frequently cast on shoar on the Orkney Isles, with some conjectures about the way of their being brought thither from Jamaica, where three sorts of them grow. *Philosophical Transactions of the Royal Society of London* 19: 298 (misprinted as 398)–300.

Smith, Andrew. 1838–49. *Illustrations of the zoology of South Africa, consisting chiefly of figures and descriptions of the objects of natural history collected during an expedition into the interior of South Africa in the years 1834–1836; fitted out by the Cape of Good Hope Association for exploring Central Africa.* 5 pts. London.

Smith, Charles Hamilton. 1845. On the original population of America, and the modes of access from the old to the new continent, with preliminary observations on the recently published *Travels in North America*, of Prince Maximilian of Wied. *Edinburgh New Philosophical Journal* 38: 1–20.

Smith, James. 1846. On the geology of Gibraltar. [Read 20 November 1844]. *Quarterly Journal of the Geological Society of London* 2: 41–51.

Smith, Sydney. 1960. The origin of 'The Origin' as discerned from Charles Darwin's notebooks and his annotations in the books he read between 1837 and 1842. *The Advancement of Science* 16: 391–401.

South America: *Geological observations on South America. Being the third part of the geology of the voyage of the Beagle, under the command of Capt. FitzRoy, R.N. during the years 1832 to 1836.* By Charles Darwin. London. 1846.

Southward, A. J. 1983. A new look at variation in Darwin's species of acorn barnacles. *Biological Journal of the Linnean Society* 20: 59–72.

Sprengel, Christian Konrad. 1793. *Das entdeckte Geheimniss der Natur im Bau und in der Befruchtung der Blumen.* Berlin.

Stafleu, Frans A. and Cowan, Richard S. 1976–9. *Taxonomic literature.* 2d ed. 2 vols. (A–L). Utrecht: Bonn, Scheltema & Holkema.

Stanley, Arthur Penrhyn. 1844. *The life and correspondence of Thomas Arnold, D.D., late head master of Rugby School.* 2 vols. London.

Steenstrup, Johannes Japetus Smith. 1842. *Om Forplantning og Udvikling gjennem vexlende Generationsraekker, en saeregen form for Opfostringen i de lavere Dyreklasser.* Copenhagen.

Stenton, Michael. 1976. *Who's who of British members of Parliament, 1832–1885.* Brighton: Harvester Press.

Stokes, John Lort. 1846. *Discoveries in Australia; with an account of the coasts and rivers explored and surveyed during the voyage of H.M.S. Beagle, in the years 1837 . . . 43.* 2 vols. London.

Strzelecki, Paul Edmund de. 1845. *Physical description of New South Wales and Van Diemen's Land.* London.

Studer, Bernhard. 1842. General view of the geological structure of the Alps. *Edinburgh New Philosophical Journal* 33: 144–65.

Sulivan, Henry Norton, ed. 1896. *Life and letters of the late Admiral Sir Bartholomew James Sulivan, K.C.B. 1810–1890.* London.

Sulloway, Frank J. 1982. Darwin's conversion: the *Beagle* voyage and its aftermath. *Journal of the History of Biology* 15: 325–96.

——. 1984. Darwin and the Galapagos. *Biological Journal of the Linnean Society* 21: 29–59.

Swainson, William. 1832–3. *Zoological illustrations, or original figures and descriptions of new, rare, or interesting animals.* 2d ser. 3 vols. London.

Taylor, Richard. 1845. The Arctic expedition under the command of Sir John Franklin. *Annals and Magazine of Natural History* 16: 163–6.

Taylor, William Cooke. 1840. *The natural history of society in the barbarous and civilized state: an essay towards discovering the origin and course of human improvement.* 2 vols. London.

Tegetmeier, William Bernhard. 1866–7. *The poultry book.* 15 pts in 1 vol. London.

Thackray, John C. 1978. R. I. Murchison's *Geology of Russia* (1845). *Journal of the Society for the Bibliography of Natural History* 8: 421–33.

Theoretical notebooks. See *Notebooks.*

Thom, Alexander. 1845. *An inquiry into the nature and course of storms in the Indian Ocean south of the equator . . . with suggestions on the means of avoiding them.* London.

Torrey, John and Gray, Asa. 1838–43. *A flora of North America: containing abridged descriptions of all the known indigenous and naturalized plants growing north of Mexico; arranged according to the natural system.* 2 vols. New York.

Turton, William. 1840. *A manual of the land and fresh-water shells of the British Islands.* Edited by J. E. Gray. London.

Ure, Andrew. 1823. *A dictionary of chemistry.* 2d ed. London.

Vallemont, Pierre Le Lorrain, Abbé de. 1707. *Curiosities of nature and art in husbandry and gardening.* Translated by William Fleetwood. London.

Variation: *The variation of animals and plants under domestication.* By Charles Darwin. 2 vols. London. 1868.

Volcanic islands: *Geological observations on the volcanic islands, visited during the voyage of H.M.S. Beagle, together with some brief notices on the geology of Australia and the Cape of Good Hope. Being the second part of the geology of the voyage of the Beagle, under the command of Capt. FitzRoy, R.N. during the years 1832 to 1836*. By Charles Darwin. London. 1844.

Vorzimmer, Peter J. 1977. The Darwin reading notebooks (1838–1860). *Journal of the History of Biology* 10: 107–53.

Waterhouse, George Robert. 1839. Observations on the Rodentia, with a view to point out the groups, as indicated by the structure of the crania, in this order of mammals. *Magazine of Natural History* 3: 90–6, 184–8, 274–9, 593–600.

——. 1843. Observations on the classification of the Mammalia. *Annals and Magazine of Natural History* 12: 399–412.

——. 1845a. Descriptions of coleopterous insects collected by Charles Darwin, Esq., in the Galapagos Islands. *Annals and Magazine of Natural History* 16: 19–41.

——. 1845b. Descriptions of some new genera and species of Heteromerous coleoptera. *Annals and Magazine of Natural History* 16: 317–24.

——. 1846–8. *A natural history of the Mammalia*. 2 vols. London.

Watson, Hewett Cottrell. 1835. *Remarks on the geographical distribution of British plants*. London.

——. 1835–7. *The new botanist's guide to the localities of the rarer plants in Britain*. 2 vols. London.

——. 1843. *The geographical distribution of British plants*. London.

——. 1843–7. Notes of a botanical tour in the western Azores. *London Journal of Botany* 2 (1843): 1–9, 125–31, 394–408; 3 (1844): 582–617; 6 (1847): 380–97.

——. 1845. On the theory of 'progressive development,' applied in explanation of the origin and transmutation of species. *Phytologist* 2: 108–13, 140–7, 161–8, 225–8.

——. 1847–59. *Cybele Britannica; or British plants, and their geographical relations*. 4 vols. London.

Waugh, Francis Gledstanes. 1888. *Members of the Athenæum Club, 1824 to 1887*. London.

Webb, Philip Barker and Berthelot, Sabin. 1836–50. *Histoire naturelle des Iles Canaries*. 3 vols. (9 pts) and atlas. Paris.

Webb, Robert Kiefer. 1960. *Harriet Martineau; a radical Victorian*. New York: Columbia University Press.

Wedgwood, Josiah C. 1908. *A history of the Wedgwood family*. London.

Wells, William Charles. 1815. *An essay on dew, and several appearances connected with it*. 2d ed. London.

Westwood, John Obadiah. 1836. On the modern nomenclature of natural history. *Magazine of Natural History* 9: 561–6.

——. 1841. Illustrations of the relationships existing amongst natural objects, usually termed affinity and analogy, selected from the class of insects. *Transactions of the Linnean Society of London* 18: 409–21.

Whately, Richard. 1831. *Introductory lectures on political economy*. London.

Whately, Thomas. 1785. *Remarks on some of the characters of Shakespeare*. London.

White, Gilbert. 1789. *The natural history and antiquities of Selborne, in the county of Southampton*. London.

Wickham, John Clements. 1838. Outline of the survey of part of the N. W. coast of Australia, in H.M.S. *Beagle* in 1838. *Journal of the Royal Geographical Society of London* 8: 460–6.

Wiedemann, Christian Rudolph Wilhelm 1828–30. *Aussereuropäische zweiflügelige Insekten*. 2 vols. Hamm.

Wilkes, Charles. 1845. *Narrative of the United States Exploring Expedition during the years 1838, 1839, 1840, 1841, 1842*. 5 vols. Philadelphia.

Williamson, William Crawford. 1848. On some of the microscopical objects found in the mud of the Levant, and other deposits; with remarks on the mode of formation of calcareous and infusorial siliceous rocks. [Read 4 November 1845.] *Memoirs of the Literary and Philosophical Society of Manchester* 2d ser. 8: 1–128.

Wilson, George and Geikie, Archibald. 1861. *Memoir of Edward Forbes, F.R.S.* Edinburgh, Cambridge, and London.

Wilson, Leonard G. 1972. *Charles Lyell. The years to 1841: the revolution in geology*. New Haven and London: Yale University Press.

Wilson, Patrick. 1788. Experiments and observations upon a remarkable cold which accompanies the separation of hoar-frost from a clear air. *Transactions of the Royal Society of Edinburgh* 1: 146–77.

Wiltshear, F. G. 1913. The botany of the Antarctic voyage. *Journal of Botany: British and Foreign* 51: 355–8.

Winsor, Mary Pickard. 1976. *Starfish, jellyfish, and the order of life: issues in nineteenth-century science*. New Haven and London: Yale University Press.

Woodward, Horace B. 1907. *The history of the Geological Society of London*. London: Geological Society.

Wrangel, Ferdinand Petrovich. 1840. *Narrative of an expedition to the Polar Sea, in the years 1820–3*. Translated by Elizabeth Julia Sabine. Edited by Edward Sabine. London.

Zoology: *The zoology of the voyage of H.M.S. Beagle, under the command of Captain FitzRoy, during the years 1832 to 1836. Published with the approval of the Lords Commissioners of Her Majesty's Treasury*. Edited and superintended by Charles Darwin. 5 pts. London. 1838–43.

BIOGRAPHICAL REGISTER AND INDEX TO CORRESPONDENTS

This list includes all persons mentioned in the letters and notes that the editors have been able to identify, and all correspondents. Dates of letters to and from correspondents are given in chronological order; letters to the correspondent are listed in roman type and letters from in italic. Following the register a list of all biographical sources referred to in the entries is given. These works are also listed in the main bibliography.

Adie, Alexander James (1775–1858). Optician and instrument maker in Edinburgh. Invented the sympiesometer in 1818. (*Modern English Biography.*)

The Admiralty.
9 [May 1846]

Agassiz, Elizabeth Cabot Cary (1822–1907). Educator. A founder of the educational establishment for women which later became Radcliffe College, Cambridge, Massachusetts. President of Radcliffe College, 1894–1902. Married Louis Agassiz in 1850. (*DAB.*)

Agassiz, Jean Louis Rodolphe (1807–73). Swiss geologist and zoologist. Professor of natural history, Neuchâtel, 1832–46. Emigrated to the United States in 1846. Professor of natural history, Harvard University, 1847–73. Established the Museum of Comparative Zoology, Harvard, 1859. Foreign member, Royal Society, 1838. (*DAB*, *DSB.*)

Agassiz, Louis. *See* Agassiz, Jean Louis Rodolphe.

Ainslie, Robert. Farmer of Pond House, Down. (*Post Office directory of the six home counties 1851.*)

Alexander, Richard Chandler. *See* Prior, Richard Chandler Alexander.

Allen, Catherine (1765–1830). Daughter of John Bartlett Allen. Married James Mackintosh in 1798. (*Emma Darwin.*)

Allen, Emma (1780–1866). Daughter of John Bartlett Allen. Emma Darwin's aunt. (*Emma Darwin.*)

Allen, Frances (Fanny) (1781–1875). Daughter of John Bartlett Allen. Emma Darwin's aunt. (*Emma Darwin.*)

Allen, John Bartlett (1733–1803). Of Cresselly, Pembrokeshire. Emma Darwin's grandfather. (*Burke's Landed Gentry 1952.*)

Allen, John Hensleigh (1818–68). Emma Darwin's cousin. (*Alum. Oxon.*, *Emma Darwin.*)

Angelis, Pedro de (1784–1859). Portuguese historian, resident in Buenos Aires. Edited collections of manuscripts relating to South American history. (*NDBA.*)

***Annals and Magazine of Natural History*.**
[December 1846]

Anson, George (1697–1762). Circumnavigated the globe, 1740–4. First lord of the Admiralty, 1751–6, 1757–62. Admiral of the fleet, 1761. (*DNB*.)

Ansted, David Thomas (1814–80). Professor of geology, King's College, London, 1840–53. Assistant-secretary of the Geological Society, 1844–7. Consultant geologist and mining engineer. FRS 1844. (*DNB*, Sarjeant 1980.)
[*c*. January 1845]

Arago, Dominique François Jean (1786–1853). French physicist, astronomer, and politician. An editor of the *Annales de chimie et physique*, 1816–40. Perpetual secretary of the Académie des Sciences, 1830. Foreign member, Royal Society, 1818. (*DBF*, *DSB*.)

Arcet, Jean d' (1725–1801). French chemist. Professor of chemistry at the Collège de France, 1774. Director of Sèvres porcelain works. (*DBF*, *DSB*.)

Arnold, Thomas (1795–1842). Clergyman. Headmaster of Rugby, 1828–42. Added mathematics, modern history, and modern languages to the ordinary school course. Regius professor of history, Oxford University, 1841. (*DNB*.)

Austin, Sarah (1793–1867). Editor and translator of French and German works. (*DNB*.)

Babington, Charles Cardale (1808–95). Botanist and archaeologist. Professor of botany, Cambridge University, 1861–95. An expert on plant taxonomy. FRS 1851. (*DNB*, *DSB*.)

Backhouse, James (1794–1869). Naturalist and Quaker missionary. Travelled in Tasmania, 1832–4; Australia, 1835–7; South Africa, 1838–40. Collected Australian plants for Kew Gardens. Returned to England in 1841 and maintained a nursery in York, devoting much time to missionary and temperance work. (*Aust. Dict. Biog.*, R. Desmond 1977.)

Baillière, Hippolyte (d. 1867). Bookseller and publisher in London who specialised in French medical and scientific texts. (*Modern English Biography*.)

Bain, James (d. 1866). Bookseller at 1 Haymarket, London, 1831–66. (*Modern English Biography*.)

Balfour, John Hutton (1808–84). Physician and botanist. Professor of botany, Glasgow University, 1841–5. Professor of botany and regius keeper of the Royal Botanic Garden, Edinburgh, 1845–79. FRS 1856. (*DNB*, *DSB*.)

Banks, Joseph (1743–1820). Naturalist and patron of science. Accompanied James Cook on his circumnavigation of the globe, 1768–71. President of the Royal Society, 1778–1820. FRS 1766. (*DNB*, *DSB*.)

Basket, Fuegia (1821?–83?). A Fuegian girl brought to England in 1830 by Robert FitzRoy. In 1833 she returned to Tierra del Fuego in the *Beagle*. (Freeman 1978.)

Beaufort, Francis (1774–1857). Naval officer; retired as rear-admiral in 1846. Hydrographer to the Admiralty, 1832–55. One of the founders of the Royal Astronomical Society and of the Royal Geographical Society. FRS 1814. (*DNB*.)

Beck, Henrick Henricksen (1799–1863). Danish conchologist. (*BHGW*, *DBL*.)

Beechey, Frederick William (1796–1856). Naval officer and geographer. Participated in exploration and surveying voyages to the Arctic, Africa, South America, and Ireland. President, Royal Geographical Society, 1855. FRS 1824. (*DNB*.)

Belcher, Edward (1799–1877). Naval officer engaged in hydrographic surveys, 1830–47. Commander of the expedition to the Arctic in search of John Franklin, 1852. Admiral, 1872. (*DNB*.)

Bell, Thomas (1792–1880). Dental surgeon at Guy's Hospital, London, 1817–61. Professor of zoology, King's College, London, 1836. President, Linnean Society, 1853–61. Described the reptiles from the *Beagle* voyage. FRS 1828. (R. Desmond 1977, *DNB*.)
[26 April – August 1845]

Bentham, George (1800–84). Botanist. Honorary secretary of the Horticultural Society, 1829–40. President of the Linnean Society, 1861–74. Published *Genera plantarum* (1862–83) with J. D. Hooker. FRS 1862. (*DNB*, *DSB*.)

Berkeley, Miles Joseph (1803–89). Clergyman and botanist. B.A., Christ's College, Cambridge, 1825. An expert on British fungi. FRS 1879. (*DNB*, *DSB*.)

Bertero, Carlo Guiseppe Luigi (1789–1831). Italian physician, botanist, and traveller. (*DBI*.)

Berthelot, Sabin (1794–1880). French naturalist and traveller who studied in the Canary Islands, 1820–30, and directed the botanic garden at La Orotava, Tenerife. Agent in Santa Cruz, Tenerife, 1847; consul, 1867. (*DBF*.)

Bibron, Gabriel (1806–48). French zoologist who described the reptiles collected by Alcide d'Orbigny in South America. (*DBF*.)

Binney, Edward William (1812–81). Solicitor in Manchester. Palaeobotanist. A founder of the Manchester Geological Society; president, 1857–9, 1865–7. FRS 1856. (R. Desmond 1977, *DNB*.)

Black, Adam (1784–1874). Scottish publisher. Lord provost of Edinburgh, 1843–7. (*DNB*.)

Blume, Carl Ludwig (1796–1862). Dutch botanist. (*NNBW*.)

Bonpland, Aimé Jacques Alexandre (1773–1858). French traveller and botanist. Accompanied Alexander von Humboldt on his South American travels, 1799–1804. (*DBF*, *EB*.)

Boott, Francis (1792–1863). Physician and botanist. Lecturer on botany at the Webb Street School of Medicine, London, 1825. Secretary, Linnean Society, 1832–9; treasurer, 1856–61. (R. Desmond 1977, *DNB*.)

Bory de Saint-Vincent, Jean Baptiste Georges Marie (1778–1846). French army officer and naturalist. Leader of several botanical collecting expeditions. Editor of the *Dictionnaire classique d'histoire naturelle*, 1822–31. (*DBF*, *DSB*.)

Bosanquet, Samuel Richard (1800–82). Barrister and writer. Published *Vestiges of the natural history of creation, its arguments examined and exposed* (1845). (*DNB.*)

Botfield, Thomas (1762–1843). Magistrate and deputy-lieutenant of Shropshire. (*Burke's Landed Gentry 1858.*)

Bowerbank, James Scott (1797–1877). London distiller and geologist with a special interest in London Clay fossils; devoted his later career to the study of sponges. FRS 1842. (*DNB*, Sarjeant 1980.)

Bravais, Auguste (1811–63). French naval officer, botanist, and physicist. Travelled with the Commission Scientifique du Nord in Norwegian Lapland, 1838–9, and published an account of the expedition. Professor of physics at the École Polytechnique, 1845–57. (*DBF*, *DSB.*)

Brayley, Edward William (1802–70). Writer on science. One of the editors of *Annals of Philosophy*, *Zoological Journal*, and *Philosophical Magazine*, 1822–45. FRS 1854. (*DNB*, Sarjeant 1980.)
7 February 1845

Brehm, Christian Ludwig (1787–1864). German clergyman and ornithologist. His bird collection formed the basis of the Rothschild museum at Tring. (*NDB.*)

Breton, William Henry. Naval officer. Police magistrate in Launceston, Van Diemen's Land (Tasmania). (O'Byrne 1849.)

Brewster, David (1781–1868). Scottish physicist who specialised in optics. Invented the kaleidoscope, 1816. Assisted in organising the British Association for the Advancement of Science, 1831. FRS 1815. (*DNB*, *DSB.*)

Broderip, William John (1789–1859). Magistrate and naturalist. A founder of the Zoological Society, 1826. FRS 1828. (*DNB*, Sarjeant 1980.)

Brodie (d. 1873). The Darwin childrens' nurse at 12 Upper Gower Street and Down House, 1842–51. (Freeman 1978.)

Brodie, Benjamin Collins (1783–1862). Surgeon. Professor of comparative anatomy and physiology, Royal College of Surgeons, 1816. Sergeant-surgeon to Queen Victoria. President of the Royal Society, 1858–61. Created baronet, 1834. FRS 1810. (*DNB*, *DSB.*)

Brodie, Peter Bellinger (1815–97). Clergyman and palaeontologist. Founded the Warwickshire Naturalists and Archaeologists' Field Club, 1854. A specialist on fossil insects. (*Modern English Biography*, Sarjeant 1980.)

Brongniart, Adolphe Théodore (1801–76). French palaeobotanist and taxonomist. A founder of the *Annales des Sciences Naturelles*, 1824. Professor of botany, Muséum d'Histoire Naturelle, 1833. Foreign member, Royal Society, 1852. (*DBF*, *DSB.*)

Bronn, Heinrich Georg (1800–62). German palaeontologist. Professor of natural science, Heidelberg, 1833. Translated and superintended the first German edition of the *Origin* (1860). (*DSB*, *NDB.*)

Brown, Robert (1773–1858). Botanist. Librarian to Joseph Banks, 1810–20. Keeper of the botanical collections, British Museum, 1827–58. FRS 1811. (*DNB*, *DSB*.)

Buch, Christian Leopold von (1774–1853). Widely travelled German geologist and geographer. Foreign member, Royal Society, 1828. (*DSB*, *NDB*.)

Buckland, William (1784–1856). Reader in geology, Oxford University, 1818–49. President of the Geological Society, 1824–5 and 1840–1. Dean of Westminster from 1845. Author of the Bridgewater treatise on geology (1836). FRS 1818. (*DNB*, *DSB*.)

Buffon, Georges Louis Leclerc, Comte de (1707–88). French naturalist and scientific administrator. (*DBF*, *DSB*.)

Bunbury, Charles James Fox, 8th Baronet (1809–86). Botanist and palaeobotanist. Collected plants in South America, 1833–4; South Africa, 1838–9. Accompanied Charles Lyell to Madeira in 1853. Married Frances Horner, Charles Lyell's sister-in-law, in 1844. FRS 1851. (R. Desmond 1977, Sarjeant 1980.)

Bunbury, Frances Joanna. *See* Horner, Frances Joanna.

Bunsen, Christian Karl Josias Freiherr von (1791–1860). Prussian diplomat and scholar. Ambassador in London, 1841. (*EB*, *NDB*.)

Burmeister, Karl Hermann Konrad (1807–92). German zoologist, ethnographer, and geologist. Professor of zoology, University of Halle, 1842. Travelled in Brazil, 1850–2. Director of the museum in Buenos Aires, 1861–80. (*NBU*, Sarjeant 1980.)

Burnet, Thomas (1635–1715). Cambridge Platonist whose *The sacred theory of the earth* (1684) attempted to reconcile scripture and geology. (*DNB*, *DSB*.)

Byron, George Anson, 7th Baron (1789–1868). Naval officer. Admiral, 1862. Succeeded to the peerage, 1824. (*Complete Peerage*, *Modern English Biography*.)

Candolle, Alphonse de (1806–93). Swiss botanist whose home was a centre of botanical activity. Professor of botany and director of the botanic gardens, Geneva, 1835–50. Son of Augustin Pyramus de Candolle. Foreign member, Royal Society, 1869. (*DSB*.)

Candolle, Augustin Pyramus de (1778–1841). Swiss botanist. Professor of natural history, Academy of Geneva, 1816–35. Foreign member, Royal Society, 1822. (*DSB*.)

Carlyle, Thomas (1795–1881). Essayist and historian. (*DNB*.)

Carmichael, Dugald (1772–1827). Army officer. Collected plants on Mauritius and Bourbon, 1810–14; Tristan da Cunha, 1817; India, 1815–17. (R. Desmond 1977.)

Carpenter, William Benjamin (1813–85). Physician and naturalist. Fullerian professor at the Royal Institution, professor of forensic medicine at University College London, and lecturer at the London Hospital, 1845. FRS 1844. (*DNB*, *DSB*.)

[11 or 18 December 1844]; *21 December 1844*; *2 January [1845]*; *5 May 1845*; [October–December 1846]

Carus, Julius Victor (1823–1903). German zoologist. Professor of comparative anatomy, Leipzig, 1853. Translated CD's works into German. (*DSB*, *NDB*.)

Cavendish, William George Spencer, 6th Duke of Devonshire, (1790–1858.) Lord lieutenant of Derbyshire, 1811–58. President of the Horticultural Society of London, 1838–58. Employed Joseph Paxton as manager of his estate at Chatsworth, Derbyshire. Played a major role in the establishment of Kew as a national botanic garden. (R. Desmond 1977, *DNB*.)

Chambers, Robert (1802–71). Publisher, writer, and geologist. Anonymous author of *Vestiges of the natural history of creation* (1844). (*DNB*, *DSB*.)

Chamisso, Adelbert von (1781–1838). German poet and naturalist who accompanied the expedition of Otto von Kotzebue to the Pacific, 1815–18. Curator of the Berlin Royal Botanic Gardens, 1833. (*DSB*, *NDB*.)

Charlesworth, Edward (1813–93). Naturalist and palaeontologist. Honorary curator, Ipswich Museum, 1835–7; Museum of the Zoological Society of London, 1837–40. Curator of the Yorkshire Philosophical Society Museum, 1844–58. (*Modern English Biography*, Sarjeant 1980.)

Charlton, Isaac. House steward of the Geological Society, 1841–91.

Christopher, William (1814–48). Officer in the Indian Navy who carried out surveys in the Red Sea and Indian Ocean. (Dawson 1885.)

Clarke, William Branwhite (1798–1878). Geologist and clergyman. Emigrated to Australia in 1839. Discovered gold, tin, and diamonds in New South Wales. FRS 1876. (*DNB*, *DSB*.)

Clausen, Peter (1804–?55). Danish exile in Brazil. Served in the Brazilian army. Sold his extensive natural history collections in Europe in the 1840s.

Clowes, William (1779–1847). Prominent London printer. (*DNB*.)

Cockell, Edgar. Surgeon and apothecary of Down, Kent, *c*. 1840–55. Member of the Royal College of Physicians.

Colburn, Henry (d. 1855). London publisher. Published *Narrative* and *Journal of researches*. (*DNB*.)

Colby, Thomas Frederick (1784–1852). Army officer and director of the Ordnance Survey, 1820–47. FRS 1820. (*DNB*.)

Coldstream, John (1806–63). Physician. M.D. Edinburgh, 1827. Practitioner in Leith, 1829–47. Friend of CD at Edinburgh University. (*DNB*.)

Cole, William Willoughby, 3d Earl of Enniskillen, (1807–86). Naturalist who collected fossil fish. M.P. for Fermanagh, 1831–40. Succeeded to the peerage, 1840. FRS 1829. (*Modern English Biography*.)

Colenso, William (1811–99). Missionary and explorer in New Zealand with an extensive knowledge of Maori subjects and the natural history of New Zealand. FRS 1886. (R. Desmond 1977, *DNZB*.)

Collie, Alexander (1793–1835). Surgeon in the Royal Navy. Surgeon on board H.M.S. *Blossom*, 1825–8. Settled in Albany, Western Australia, 1829. (*Aust. Dict. Biog.*, R. Desmond 1977.)

Cook, James (1728–79). Commander of several voyages of discovery. Circumnavigated the world, 1768–71 and 1772–5. FRS 1776. (*DNB*, *DSB*.)

Cooley, William Desborough (d. 1883). Geographer. Published papers on African geography, 1841–74, and a manual, *Physical geography* (1876). (*DNB.*)

Copleston, Edward (1776–1849). Professor of poetry, Oxford University, 1802–12. Provost of Oriel College, Oxford, 1814–28. Bishop of Llandaff and dean of St Paul's, 1828–49. (*DNB.*)

Couthouy, Joseph Pitty (1808–64). American sailor, conchologist, and invertebrate palaeontologist. (Sarjeant 1980.)

Covington, Syms (1816?–61). Became CD's servant in the *Beagle* in 1833 and remained with him as assistant, secretary, and servant until 1839. Emigrated to Australia, 1839. (Freeman 1978.)

Cowper, William (1731–1800). Poet. Translated the works of Homer. (*DNB.*)

Cresy, Edward (1792–1858). Architect and civil engineer. A neighbour of CD in Down, Kent. (*DNB.*)

Crosse, Thomas. Captain. Resident of Down, Kent. (*Post office directory of the six home counties 1845.*)

Crozier, Francis Rawdon Moira (1796?–1848). Naval officer. Commander of the *Terror* on the Antarctic voyage, 1839–43. Commander of the *Terror* on the Arctic voyage of the *Terror* and *Erebus* under the orders of John Franklin, 1845–8, during which all hands perished. FRS 1843. (*DNB.*)

Cruz, Luis de la (1768–1828). Author of works on South America. (*NUC.*)

Cuming, Hugh (1791–1865). Naturalist and traveller. Collected shells and living orchids in the Pacific, on the coast of Chile, and in the Philippine Islands. Returned to England in 1839. (R. Desmond 1977, *DNB.*)
28 July 1845

Cunningham, Allan (1791–1839). Botanist and explorer. Collected plants for Kew Gardens in Brazil, 1814–16; New South Wales, 1816–26, 1827–30; New Zealand, 1826. Superintendent of the botanic garden, Sydney, 1836–8. (*Aust. Dict. Biog.*, R. Desmond 1977.)

Darwin, Anne Elizabeth (Annie) (1841–51). CD's oldest daughter. (*Darwin Pedigree.*)

Darwin, Caroline Sarah. *See* Wedgwood, Caroline Sarah.

Darwin, Catherine. *See* Darwin, Emily Catherine.

Darwin, Catty. *See* Darwin, Emily Catherine.

Darwin, Doddy. *See* Darwin, William Erasmus.

Darwin, Elizabeth de St Croix (1790–1868). Wife of William Brown Darwin and mother of Robert Alvey Darwin. (*Darwin Pedigree.*)

Darwin, Emily Catherine (1810–66). CD's sister. Became Charles Langton's second wife in 1863. (*Darwin Pedigree.*)
[22 November 1846]

Darwin, Emma (1808–96). Youngest daughter of Bessy and Josiah Wedgwood II. Married CD, her cousin, in 1839. (*Emma Darwin.*)
[3 June 1844]; 5 July 1844; [20 or 27 October 1844];
[3–4 February 1845]; [7–8 February 1845]; [24 June 1846];
[25 June 1846]

Darwin, Erasmus (1731–1802). Physician, botanist, and poet. Advanced an evolutionary theory similar to that subsequently expounded by Lamarck. CD's grandfather. FRS 1761. (*DNB, DSB.*)

Darwin, Erasmus Alvey (1804–81). CD's brother. Attended Shrewsbury School, 1815–22. Matriculated Christ's College, Cambridge, 1822; M.B., 1828. At Edinburgh University, 1825–6. Qualified but never practised as a physician. Lived in London from 1829 to his death. (*Alum. Cantab.*)

[May 1844 – 1 October 1846]; 25 July 1844; [before 1 October 1844]

Darwin, Etty. *See* Darwin, Henrietta Emma.

Darwin, Francis (1848–1925). CD's son. B.A., Trinity College, Cambridge, 1870. Collaborated with CD on several botanical projects, 1875–82. Lecturer in botany, Cambridge University, 1884; reader, 1888–1904. Edited CD's letters. FRS 1882. (*DNB, DSB.*)

Darwin, George Howard (1845–1912). CD's son. B.A., Trinity College, Cambridge, 1868. Plumian professor of astronomy and experimental philosophy, Cambridge University, 1883–1912. FRS 1879. (*DNB, DSB.*)

Darwin, Henrietta Emma (Etty) (1843–1927). CD's daughter. Married Richard Buckley Litchfield (*Alum. Cantab.*) in 1871. (*Burke's Landed Gentry 1952.*)

Darwin, Katty. *See* Darwin, Emily Catherine.

Darwin, Mary Eleanor (September–October 1842). CD's third child. (*Darwin Pedigree.*)

Darwin, Robert Alvey (1826–47). Of Elston Hall. Oldest son of William Brown Darwin. Matriculated, Exeter College, Oxford, 1845. (*Alum. Oxon., Darwin Pedigree.*)

Darwin, Robert Waring (1766–1848). Physician. M.D. Leiden, 1785. Had a large practice in Shrewsbury, and resided at The Mount which he built *c.* 1796–8. Third son of Erasmus Darwin by his first wife, Mary Howard. Married Susannah, daughter of Josiah Wedgwood I, in 1796. CD's father. FRS 1788. (Freeman 1978.)

Darwin, Susan Elizabeth (1803–66). CD's sister. Lived at The Mount, Shrewsbury, until her death. (*Darwin Pedigree.*)

[27 November 1844?]; 3[–4] September 1845

Darwin, William Brown (1774–1841). Of Elston Hall. Barrister. (*Alum. Cantab., Darwin Pedigree.*)

Darwin, William Erasmus (1839–1914). CD's oldest child. B.A., Christ's College, Cambridge, 1862. Banker in Southampton. (*Alum. Cantab.*)

Davy, Humphry (1778–1829). Professor of chemistry at the Royal Institution, 1802–13. President of the Royal Society, 1820–7. FRS 1803. (*DNB, DSB.*)

Dease, Peter Warren (1788–1863). Fur-trader and explorer. Commander, with Thomas Simpson, of the Hudson's Bay Company's expedition that explored the north-western coast of North America, 1836–9. Retired from the Hudson's Bay Company in 1842 and settled near Montreal, Canada. (*MDCB.*)

Decaisne, Joseph (1807–82). French botanist at the Jardin des plantes, Paris, 1824. Named professor of statistical agriculture, Collège de France, 1845. Foreign member, Royal Society, 1877. (*DBF*.)

Dejean, Pierre François Marie Auguste (1780–1845). French general and entomologist. A specialist on Coleoptera. (*DBF*, Gilbert 1977.)

De la Beche, Henry Thomas (1796–1855). Geologist. First director of the Geological Survey of Great Britain, 1835–55. Established the Museum of Economic Geology and the School of Mines. FRS 1819. (*DNB*, *DSB*.)

Delessert, Jules Paul Benjamin (1773–1847). French botanist, industrialist, banker, and philanthropist. (*DBF*.)

Denny, Henry (1803–71). Entomologist who studied parasitic insects. Curator of the Museum of the Literary and Philosophical Society of Leeds. (*DNB*, Gilbert 1977.)

20 January [1844]; 3 June [1844]; 12 August [1844]; *30 October 1844*; 7 November [1844]

Derby, Lord. *See* Stanley, Edward Smith.

Devonshire, Duke of. *See* Cavendish, William George Spencer.

Dick, Thomas Lauder. *See* Lauder, Thomas Dick.

Dieffenbach, Ernst (1811–55). German physician, naturalist, and geologist. Surgeon and naturalist to the New Zealand Company, 1839–41. Supernumerary professor of geology at Giessen, 1850–5. Translated *Journal of researches* into German (1844). (*DNZB*, Sarjeant 1980.)

25 January 1844; 14 March 1844; 11 June [1844]; 6 April 1845; 8 April [1845]; [before 9 July 1845]; 6 April [1846]

Don, David (1800–41). Professor of botany, King's College, London, 1836–41. Linnean Society librarian, 1822–41. (R. Desmond 1977, *DNB*.)

Doubleday, Edward (1810–49). Entomologist. Collected insects in the United States, 1835–7. Assistant in the British Museum with special charge of the collections of butterflies and moths, 1839–49. (R. Desmond 1977, *DNB*.)

Douglas, David (1799–1834). Sent to America in 1823 by the Horticultural Society of London to collect plants. (R. Desmond 1977.)

Drury, Robert (*fl.* 1729). Traveller and writer on Madagascar. (*DNB*.)

Duchesne, Antoine Nicolas (1747–1827). French botanist. (*DBF*.)

Duméril, André Marie Constant (1774–1860). French physician and naturalist. Professor of medicine, Paris, 1801–57. Professor of zoology specialising in reptiles and fish, Muséum d'Histoire Naturelle, 1825. (*DBF*, *DSB*.)

Dumont d'Urville, Jules Sébastien César (1790–1842). French navigator. Commander of the *Astrolabe* expeditions, 1826–9 and 1837–41. (*DBF*, *DNZB*.)

Duncan, John (1805–49). Scottish explorer of West Africa. (*DNB*.)

Du Petit-Thouars, Abel Aubert (1793–1864). French naval officer and administrator. Carried out hydrographic surveys and voyages of exploration. Vice-admiral, 1846. (*DBF*.)

Edmondston, Thomas (1825–46). Botanist. Elected professor of botany and natural history, Andersonian Institution, Glasgow, 1845. Naturalist aboard H.M.S. *Herald*, 1845–6. Accidentally shot in Peru, 1846. (R. Desmond 1977, *DNB*.)

Egerton, Philip de Malpas Grey- (1806–81). Of Oulton Park, Cheshire. Tory M.P. for South Cheshire, 1835–68. Palaeontologist who specialised in fossil fish. FRS 1831. (*DNB*, Sarjeant 1980.)
5 May [1844]

Ehrenberg, Christian Gottfried (1795–1876). German naturalist, microscopist, and traveller. Studied the development of coral reefs and worked extensively on Infusoria. Foreign member, Royal Society, 1837. (*DSB*, *NDB*.)
20 April [1844]; *15 June 1844*; 4 July [1844]; *11 July 1844*; 5 September [1844]; 23 January [1845]; *13 March 1845*; 23 March [1845]; *8 April 1845*; 21 May [1845]; 29 October [1845]; *11 March 1846*; 25 March [1846]

Élie de Beaumont, Jean Baptiste Armand Louis Léonce (1798–1874). French mining engineer and geologist. Propounded a catastrophist theory of mountain elevation. (*DBF*, *DSB*.)

Endlicher, Stephan Ladislaus (1804–49). German botanist. (*NDB*.)

Enniskillen, Earl of. *See* Cole, William Willoughby.

Erman, Georg Adolph (1806–77). German physicist and geologist. Travelled round the world, 1828–30. Assistant professor of physics, Berlin, 1834. Foreign member, Royal Society, 1873. (*DSB*, *NDB*.)

Eschscholtz, Johann Friedrich (1793–1831). Russian physician and zoologist who accompanied the expedition of Otto von Kotzebue to the Pacific, 1815–18. Professor of anatomy, Dorpat University, 1828. (*DSB*, *NDB*.)

Falconer, Hugh (1808–65). Palaeontologist and botanist. Superintendent of the botanic garden, Saharanpur, India, 1832–42. Superintended arrangement of Indian fossils for the British Museum, 1844. Superintendent of the botanic garden, Calcutta and professor of botany, Calcutta Medical College, 1848–55. FRS 1845. (*DNB*, *DSB*.)

Fischer, Johann Baptist (d. 1832). Botanist who collaborated with C. L. Blume. (*NUC*.)

Fitch, Walter Hood (1817–92). Botanical artist at Kew Gardens. (R. Desmond 1977.)

Fitton, William Henry (1780–1861). Physician and geologist. President of the Geological Society, 1827–9; vice-president, 1831–46. FRS 1815. (*DNB*, *DSB*.)

FitzRoy, Mary Henrietta (d. 1852). Married Robert FitzRoy in 1836. (*Burke's Peerage 1980*.)

FitzRoy, Robert (1805–65). Naval officer, hydrographer, and meteorologist. Commander of the *Beagle*, 1828–36. Author of a narrative of the surveying voyages of the *Adventure* and *Beagle*, 1839. Tory M.P. for Durham, 1841–3. Governor of New Zealand, 1843–5. Chief of the meteorological department of the board of trade, 1854. Vice-admiral, 1863. FRS 1851. (*DNB*, *DSB*.)

1 October 1846; 28 October [1846]; 23 November [1846]

Flaxman, John (1755–1826). Noted sculptor and draughtsman. In his early years he executed cameos and bas-reliefs for the Wedgwood factory. Professor of sculpture at the Royal Academy from 1810. (*DNB.*)

Flinders, Matthew (1774–1814). Naval officer, hydrographer, and explorer. Lieutenant, 1798, commander, 1801. Surveyed a large part of the Australian coast. (*Aust. Dict. Biog.*, *DNB.*)

Forbes, Edward (1815–54). Zoologist, botanist, and invertebrate palaeontologist. Naturalist on board H.M.S. *Beacon*, 1841–2. Professor of botany, King's College, London, 1842. Palaeontologist with the Geological Survey, 1844–54. Professor of natural history, Edinburgh University, 1854. FRS 1845. (*DNB*, *DSB.*)

28 May 1844; *[after 14 February 1845]*; *[March? 1845]*; *[9 May 1845]*; 13 May [1845]; *[25 February 1846]*; *[7 August 1846]*; *[September 1846]*

Forbes, James David (1809–68). Professor of natural philosophy, Edinburgh University, 1833–60. Secretary of the Royal Society of Edinburgh, 1840–51. FRS 1832. (*DNB*, *DSB.*)

11 October [1844]; [November? 1844]; 13 [November 1844]

Fox, Ellen Sophia (1820–87). Second wife of William Darwin Fox, 1846. (*Darwin Pedigree.*)

Fox, Harriet (1799–1842). Married William Darwin Fox in 1834. (*Darwin Pedigree.*)

Fox, William Darwin (1805–80). Clergyman. Matriculated Christ's College, Cambridge, 1824; B.A., 1829. CD's second cousin. A close friend at Cambridge who shared CD's enthusiasm for entomology. Maintained an active interest in natural history throughout his life and provided CD with much information. Rector of Delamere, Cheshire, 1838–73. Spent the last years of his life at Sandown, Isle of Wight. (*Alum. Cantab.*)

20 December [1844]; [13 February 1845]; [24 April 1845]; [before 3 October 1846]

Francis, William. Printer. Assistant and later partner of Richard Taylor of the printing firm of Taylor and Francis.

Franklin, John (1786–1847). Naval officer and Arctic explorer. Lieutenant-governor of Van Diemen's Land (Tasmania), 1837–43. Leader of the 1845 expedition, in search of a north-west passage, during which all hands perished. FRS 1823. (*DNB.*)

Fraser, Louis. Naturalist to the Niger expedition, 1841–2. Curator of the Zoological Society Museum, 1843–5. Conservator of Lord Derby's menagerie at Knowsley until 1850. (*DNB.*)

23 July 1845; *[24? July 1845]*

Freycinet, Louis Claude Desaulses de (1779–1842). French navigator. (*DBF*, *EB.*)

Gardeners' Chronicle and Agricultural Gazette.
[27 March 1844]; [before 8 June 1844]; [before 14 September 1844]; [before 14 September 1844]

Gardner, George (1812–49). Botanist. Collected plants in Brazil, 1836–41. Superintendent of the botanic garden, Peradeniya, Ceylon, 1844. (R. Desmond 1977, *DNB*.)

Garnot, Prosper (1794–1838). French naval surgeon. Naturalist on Duperry's voyage round the world in the *Coquille*, 1822–4. (*DBF*, *DSB*.)

Gaudichaud, Charles Beaupré- (1789–1854). French botanist and pharmacist. Botanist on board the *Uranie* on the voyage round the world, 1817–19. Travelled in South America, 1830–3, and undertook a second voyage round the world in 1836. Published the botanical results of his voyages. (*DBF*.)

Gay, Claude (1800–73). French naturalist and traveller who surveyed the flora and fauna of Chile. Professor of physics and chemistry, Santiago College, 1828–42. (*DBF*.)

Geoffroy Saint-Hilaire, Étienne (1772–1844). French zoologist. Professor of zoology, Muséum d'Histoire Naturelle, 1793. Devoted much attention to teratology. (*DBF*, *DSB*.)

Geoffroy Saint-Hilaire, Isidore (1805–61). French zoologist. Replaced his father, Étienne Geoffroy Saint-Hilaire, as professor at the Muséum d'Histoire Naturelle, 1841. Continued his father's work in teratology. (*DBF*, *DSB*.)

Geological Society of London.
[3 January 1844]

Gérard, Frédéric. French writer on botanical and horticultural topics, a follower of Étienne Geoffroy Saint-Hilaire.

Gibson, Alexander (1800–67). Botanist. Surgeon to the East India Company. Superintendent of the Dapuri botanic gardens, 1838–47; conservator of forests in Bombay, 1847–60. (R. Desmond 1977, *DNB*.)

Gloger, Constantin Wilhelm Lambert (1803–63). German zoologist and ornithologist. (*NDB*.)

Gmelin, Johann Georg (1709–55). Naturalist and explorer. Professor of chemistry and natural history at the Academy of Sciences, St Petersburg, 1731–47. Professor of medicine, botany, and chemistry at Tübingen University, 1749. (*DSB*, *NDB*.)

Goethe, Johann Wolfgang von (1749–1832). German poet and naturalist. (*DSB*, *NDB*.)

Goodsir, Harry (d. 1845). Physician. Brother of John Goodsir (*DNB*). Assistant surgeon in H.M.S. *Erebus*; perished in John Franklin's 1845 expedition in search of a north-west passage.

Gould, John (1804–81). Self-taught ornithologist and artist. Taxidermist to the Zoological Society of London, 1826–81. Travelled in Australasia, 1838–40. Described the birds collected on the *Beagle* and *Sulphur* expeditions. FRS 1843. (*DNB*, *DSB*.)
[*c*. October 1846]

Graham, Robert (1786–1845). Physician and botanist. Regius professor of botany, Edinburgh University, 1820–45. Physician to the Edinburgh Infirmary. (*DNB.*)

Graham, Thomas (1805–69). Chemist. Professor of chemistry at the Andersonian Institution, Glasgow, 1830–7; University College London, 1837–55. Master of the mint, 1855–69. FRS 1836. (*DNB, DSB.*)

Gray, Asa (1810–88). American botanist. Fischer professor of natural history, Harvard University, 1842–73. Devoted much time to the Harvard botanic garden and herbarium. Foreign member, Royal Society, 1873. (*DAB, DSB.*)

Gray, George Robert (1808–72). Zoologist; an expert on insects and birds. Assistant in the zoological department of the British Museum, 1831–72. Brother of John Edward Gray. FRS 1865. (*DNB*, Gilbert 1977.)

Gray, John Edward (1800–75). Naturalist. Assistant zoological keeper at the British Museum, 1824; keeper, 1840–74. FRS 1832. (R. Desmond 1977, *DNB.*)

Greenough, George Bellas (1778–1855). Geologist, geographer, and politician. A founder of the Geological Society of London. First president of the Geological Society, 1811. FRS 1807. (*DNB, DSB.*)

Grey, George (1812–98). Army officer and explorer. Captain, 1839. Governor of South Australia, 1841–5; New Zealand, 1845–53, 1861–8; Cape Colony, 1854–61. Settled in New Zealand, 1870–94. Prime minister of New Zealand, 1877–9. Knighted 1848. (*Aust. Dict. Biog.*, *DNB.*)
10 May 1846; 10 November 1846

Groom, Henry. Nurseryman and florist, Walworth, London. Acquired business of Curtis, Milliken and Co. Moved to Clapham Rise, London, 1842–3. (R. Desmond 1977.)

Gunn, Ronald Campbell (1808–81). Botanist, public servant, and politician in Van Diemen's Land (Tasmania). FRS 1854. (*Aust. Dict. Biog.*, *DNB.*)

Haidinger, Wilhelm Karl (1795–1871). Austrian mineralogist and geologist who studied pseudomorphs and light absorption in crystals. Inspector of mines, Vienna, 1840. (*DSB, NDB.*)

Hall, Basil (1788–1844). Naval captain and geographer. FRS 1816. (*DNB.*)

Hamilton, William Richard (1777–1859). Antiquary and diplomatist. Under-secretary of state for foreign affairs, 1809–22; minister at Naples, 1822–5. A founder of the Royal Geographical Society, 1833, and served as president. Trustee of the British Museum, 1838–58. FRS 1813. (*DNB.*)

Hamond, Robert Nicholas (1809–83). Naval officer; lieutenant, 1827. (*Burke's Landed Gentry 1952*, O'Byrne 1849.)

Harding, Elizabeth. Nursery maid at Down House. (Freeman 1978.)

Harper, James (1795–1869). American printer and publisher. (*DAB.*)

Harvey, William Henry (1811–66). Irish botanist. Colonial treasurer in Cape Town, 1836–42. Keeper of the herbarium, Trinity College, Dublin, 1844. Professor of botany, Trinity College, Dublin, 1856–66. (*DNB, DSB.*)

Hawkesworth, John (1715?–73). Author. Published an account of voyages in the South Seas, 1773. A director of the East India Company. (*DNB*.)

Hayes, John Lord (1812–87). American lawyer and geologist. (*DAB*.)

Hearne, Samuel (1745–92). Explorer and colonial administrator in Canada. (*DNB, MDCB*.)

Hemmings, Henry. A servant to Sarah Elizabeth (Sarah) Wedgwood. (Freeman 1978.)

Henslow, George (1835–1925). Clergyman and teacher. Lecturer in botany at St Bartholomew's Medical School, 1886–90. Younger son of John Stevens Henslow. (*Alum. Cantab.*, R. Desmond 1977.)

Henslow, Harriet (1797–1857). Daughter of George Leonard Jenyns and sister of Leonard Jenyns. Married John Stevens Henslow in 1823. (*Burke's Landed Gentry 1879*.)

Henslow, John Stevens (1796–1861). Clergyman, botanist, and mineralogist. Professor of mineralogy, Cambridge University, 1822–7; professor of botany, 1825–61. Extended and remodelled the Cambridge botanic garden. Curate of Little St Mary's Church, Cambridge, 1824–32; vicar of Cholsey-cum-Moulsford, Berkshire, 1832–7; rector of Hitcham, Suffolk, 1837–61. CD's teacher and friend. (*DNB, DSB*.)
[25 July 1844]; 16 May [1845]; 25 July 1845; 28 October [1845]; [5 October 1846]

Herbert, Edward, 2d Earl of Powis, (1785–1848). Tory M.P. for Ludlow, 1806–39. Assumed the surname Herbert, in lieu of Clive, in 1807. Succeeded to the earldom in 1839. (*Complete Peerage, DNB*.)

Herbert, John Maurice (1808–82). B.A., St John's College, Cambridge, 1830; Fellow, 1832–40. Barrister, 1835. County court judge, South Wales, 1847–82. (*Alum. Cantab., Modern English Biography*.)
[3 September? 1846]

Herbert, Mary-Anne. Married John Maurice Herbert in 1840.

Herbert, William (1778–1847). Naturalist, classical scholar, linguist, politician, and clergyman. Noted for his work on plant hybridisation. Rector of Spofforth, Yorkshire, 1814–40. Dean of Manchester, 1840–7. (*DNB, DSB*.)

Herschel, John Frederick William (1792–1871). Astronomer, mathematician, chemist, and philosopher. Member of many learned societies. Carried out astronomical observations at the Cape of Good Hope, 1834–8. Master of the mint, 1850–5. Created baronet, 1838. FRS 1813. (*DNB, DSB*.)

Hervey, Arthur Charles (1808–94). Fourth son of the first Marquis of Bristol. Rector of Ickworth with Horringer, 1832–69. Bishop of Bath and Wells, 1869–94. (*Alum. Cantab., DNB*.)

Higgins, John (1796–1872). Land agent. Born in Shrewsbury but settled in Alford, Lincolnshire, in 1819. Agent for R. W. Darwin's estates in that county and later for CD's farm at Beesby.
15 March 1845; *2 October 1845*; 27 May [1846]; 12 December [1846]

Hinds, Richard Brinsley (1812–47?). Served as surgeon and naturalist aboard H.M.S. *Sulphur*, 1836–42. (R. Desmond 1977, Plarr 1930.)

Hitchcock, Edward (1793–1864). American geologist and clergyman. Professor of chemistry and natural history, Amherst College, 1825–45; president, 1845–55; professor of geology and natural theology, 1855–64. Served on the Massachusetts, New York, and Vermont geological surveys. (*DAB*, *DSB*.)
6 November [1845]

Hobson, William (1793–1842). Irish naval officer. As lieutenant-governor of New Zealand he drew up the treaty of Waitangi, 1840. Governor of New Zealand, 1841–2. (*Aust. Dict. Biog.*, *DNZB*.)

Holland, Edward (1806–75). CD's second cousin. B.A., Trinity College, Cambridge, 1829. Liberal M.P. for East Worcestershire, 1835–7; Evesham, 1855–68. President of the Royal Agricultural Society. (*Alum. Cantab.*, Stenton 1976.)

Holland, George Henry (d. 1891). B.A., Trinity College, Cambridge, 1839. Barrister and sportsman. Younger brother of Edward Holland. (*Alum. Cantab.*)

Holland, Henry (1788–1873). Physician. Distant cousin of the Darwins and Wedgwoods. Physician in ordinary to Queen Victoria, 1852. President of the Royal Institution. Created baronet, 1853. FRS 1815. (*DNB*.)

Hombron, Jacques Bernard (b. 1800). French naturalist. (*NUC*.)

Hooker, Elizabeth (1820–98). Sister of Joseph Dalton Hooker. (Huxley ed. 1918.)

Hooker, Joseph (1754–1845). Merchant in Norwich. Grandfather of Joseph Dalton Hooker. (Allan 1967.)

Hooker, Joseph Dalton (1817–1911). Botanist. Accompanied James Clark Ross on the Antarctic expedition, 1838–43, and published the botanical results of the voyage. Botanist to the Geological Survey, 1845. Travelled in the Himalayas, 1848–50. Assistant director, Royal Botanic Gardens, Kew, 1855–65; director, 1865–85. Worked chiefly on taxonomy and plant geography. Son of William Jackson Hooker. Friend and confidant of CD. FRS 1847. (*DNB*, *DSB*.)
[11 January 1844]; [27 January 1844]; *29 January 1844*;
[3–17 February 1844]; 23 February [1844]; *[23 February – 6 March 1844]*;
[6 March 1844]; *9 March 1844*; 11 March [1844]; 16 March [1844];
31 March [1844]; *5 April 1844*; [17 April 1844]; 1 June [1844];
29 [June 1844]; [14 July 1844]; 22 July [1844];
[25 July – 29 August 1844]; 29 [August 1844]; *[c. 3 September 1844]*;
[8 September 1844]; *28 October 1844*; *8 November 1844*;
[10–11 November 1844]; *14 November 1844*; [18 November 1844];
29 November 1844; [2 December 1844]; [4 December 1844];
12 December 1844; 16 [December 1844]; 25 December [1844];
30 December 1844; [7 January 1845]; 22 [January 1845];
[22–30 January 1845]; [10 February 1845]; *[late February 1845]*;

Hooker, J. D., cont.

19 March [1845]; *[23] March 1845*; [26 March 1845]; 31 March [1845]; *[2–6 April 1845]*; [16 April 1845]; [28 April 1845]; *[28 April 1845]*; [May 1845]; [4 June 1845]; [27 June 1845]; *5 July 1845*; [11–12 July 1845]; *[after 12 July 1845]*; *[mid-July 1845]*; [22 July – 19 August 1845]; [15 or 22 August 1845]; [29 August 1845]; *1 September [1845]*; [3 September 1845]; *[4–9 September 1845]*; [10 September 1845]; *14 September 1845*; [18 September 1845]; [8 October 1845]; 28 October [1845]; [5 or 12 November 1845]; [17 November 1845]; *[19 November 1845]*; [21 November 1845]; [25 November 1845]; [29 November 1845]; [10 December 1845]; [31 January 1846]; *1 February 1846*; [5 February 1846]; [8? February 1846]; [10 February 1846]; [15 February 1846]; [25 February 1846]; [25 February – 2 March 1846]; *2 [March] 1846*; [13 March 1846]; [24 March 1846]; *[25 March 1846]*; [29 March or 5 April 1846]; 10 April [1846]; *[11–15 April 1846]*; [16 April 1846]; [May 1846]; [19 May 1846]; [8 or 15 July 1846]; *[before 3 September 1846]*; [3 September 1846]; *28 September 1846*; [2 October 1846]; [6 October 1846]; [8 October 1846]; [18 October 1846]; [26 October 1846]; [6 November 1846]; [12 November 1846]; [14 November 1846]; [15 November 1846]; [17 November 1846]; [23 November 1846]; *[24 November 1846]*; [December 1846]; [December 1846 – January 1847]

Hooker, Maria (1797–1872). Wife of William Jackson Hooker. (Allan 1967.)

Hooker, Maria (1819–89). Sister of Joseph Dalton Hooker. Married Walter McGilvray in 1846. (Huxley ed. 1918.)

Hooker, William Jackson (1785–1865). Botanist. Regius professor of botany, Glasgow University, 1820. Established the Royal Botanic Gardens at Kew, 1841, and served as first director. Father of Joseph Dalton Hooker. FRS 1812. (*DNB*, *DSB*.)

[23 August 1845]; [25 August 1845]; 25 August 1845

Hope, Frederick William (1797–1862). Entomologist and clergyman. Gave his collection of insects to Oxford University and founded a professorship of zoology, 1849. FRS 1834. (*DNB*, Gilbert 1977.)

Hopkins, Evan (1810?–67). Welsh geologist who supervised a number of mining projects in South America. (Sarjeant 1980.)

Hopkins, William (1793–1866). Mathematician and geologist. A highly successful mathematics tutor at Cambridge. Specialised in quantitative studies of geological and geophysical questions. FRS 1837. (*DNB*, *DSB*.)

3 March 1845; *27 April 1846*; *5 May 1846*

Hopkirk, Thomas (1785–1841). Scottish botanist. Founded the Botanical Institution, Glasgow. (R. Desmond 1977, *DNB*.)

Horner, Anne Susan (1789–1862). Married Leonard Horner in 1806. (Freeman 1978.)

Horner, Frances Joanna (1814–94). Daughter of Leonard Horner. Married Charles James Fox Bunbury in 1844. (*Burke's Peerage 1980*, Freeman 1978.)

Horner, Katharine Murray. *See* Lyell, Katharine Murray.

Horner, Leonard (1785–1864). Geologist and educationist. A founder of the *Edinburgh Review*, 1802. Warden of University College London, 1827–31. President of the Geological Society, 1845–7 and 1860–2. Father-in-law of Charles Lyell. FRS 1813. (*DNB*, *DSB*.)
29 August [1844]; [17 August – 7 September 1846];
[23 December 1846 – January 1847]

Horner, Mary Elizabeth. *See* Lyell, Mary Elizabeth.

Horsfield, Thomas (1773–1859). Naturalist. Keeper of the East India Company's museum, London, 1820–59. (R. Desmond 1977, *DNB*.)

Hullmandel, Charles Joseph (1789–1850). Lithographer in London. (*DNB*.)

Humboldt, Friedrich Wilhelm Heinrich Alexander von (1769–1859). Eminent Prussian naturalist and traveller. Official in the Prussian mining service, 1792–6. Explored equatorial South America, 1799–1804. Travelled in Siberia, 1829. Foreign member, Royal Society, 1815. (*DSB*, *NDB*.)

Hume, Abraham (1814–84). Antiquary and clergyman. Curate of St Augustine's, Liverpool, 1843–7; vicar of Vauxhall, Liverpool, 1847. Took an active part in many of the public, scientific, educational, and ecclesiastical movements in Liverpool. (*DNB*.)

Hunter, William (1718–83). Anatomist. First professor of anatomy, Royal Academy, 1768. FRS 1767. (*DNB*, *DSB*.)

Hutton, Robert (1784–1870). Of Putney Park, Surrey. Formerly a merchant in Dublin. M.P. for Dublin, 1837–41. A secretary of the Geological Society, 1837. Served on the council of University College London, and of the British Association for the Advancement of Science. (*Modern English Biography*, Sarjeant 1980.)
[April 1846]

Jackson, Julian (1790–1853). Army officer and geographer. Secretary of the Royal Geographical Society, 1841–7. FRS 1845. (*DNB*.)
23 May [1844]

James, Robert Bastard. Naval officer. Lieutenant-commander of the brig *Spey* from 1833. (*Navy List 1838*.)

Jameson, Robert (1774–1854). Geologist and mineralogist. Regius professor of natural history and keeper of the museum at Edinburgh University, 1804–54. Editor of the *Edinburgh Philosophical Journal*, 1824–54. FRS 1826. (*DNB*, *DSB*.)

Jamieson, Robert (d. 1861). Philanthropist and merchant in London. Explorer of West African rivers. (*DNB*.)

Jardine, William, 7th Baronet (1800–74). Naturalist. A founder of the *Annals and Magazine of Natural History*, 1841. Commissioner on salmon fisheries of England and Wales, 1860. FRS 1860. (*DNB*.)

Jenyns, George Leonard (1763–1848). Vicar of Swaffham Prior, Cambridgeshire, 1787–1848; prebendary of Ely, 1802–48. Inherited Bottisham Hall, Cambridgeshire, from his second cousin in 1787. (*Alum. Cantab.*)

Jenyns, Leonard (1800–93). Naturalist and clergyman. Son of George Leonard Jenyns. Brother-in-law of John Stevens Henslow. Vicar of Swaffham Bulbeck, Cambridgeshire, 1828–49. Member of many scientific societies. Described the *Beagle* fish specimens. Adopted the name Blomefield in 1871. (R. Desmond 1977, *DNB*.)
12 October [1844]; 25 [November 1844]; 14 February [1845]; [14 or 21 August 1846]; 17 October [1846]

Joinville, François Ferdinand Philippe Louis Marie, Prince de (1818–1900). French naval officer. Third son of Louis Philippe, Duc d'Orléans. Exiled after the 1848 revolution. Returned to France and a seat in the National Assembly, 1871. (*EB*, *NBU*.)

Jones, Thomas Heron, Viscount Ranelagh, (1812–85). Army officer. Promoter of the volunteer movement in England. Chairman of the Conservative Land Society. (*Complete Peerage*.)

Jordan, John. Servant in the Darwin household at 12 Upper Gower Street and Down House. (Freeman 1978.)

Jukes, Joseph Beete (1811–69). Geologist. Geological surveyor of Newfoundland, 1839–40. Naturalist aboard H.M.S. *Fly*, 1842–6. Worked in North Wales for the Geological Survey, 1846–50. Director, Irish branch of the Geological Survey, 1850–69. FRS 1853. (*DNB*, *DSB*.)
[18 October 1846]

Jussieu, Adrien Henri Laurent de (1797–1853). French botanist. Succeeded his father, Antoine Laurent de Jussieu, as professor of botany at the Muséum National d'Histoire Naturelle, 1826. (*DSB*.)

Jussieu, Antoine Laurent de (1748–1836). French botanist whose *Genera plantarum* (1789) established a system of plant taxonomy. Professor of botany, Muséum National d'Histoire Naturelle, 1793; director, 1800–26. Foreign member, Royal Society, 1829. (*DSB*.)

Kane, Robert John (1809–90). Irish chemist. Professor of chemistry, Apothecaries' Hall, Dublin, 1831–45, and of natural philosophy to the Royal Dublin Society, 1834–47. Editor of the *Philosophical Magazine*, 1840–90. President of Queen's College, Cork, 1845–73. Knighted 1846. FRS 1849. (*DNB*, *DSB*.)

Kay-Shuttleworth, James Phillips, 1st Baronet (1804–77). Physician and educationist. First secretary of the committee of the council on education, 1839–49. Created baronet, 1849. (*DNB*.)

Keilhau, Baltazar Mathias (1797–1858). Norwegian stratigrapher and petrologist. (*NBL*, Sarjeant 1980.)

Keyserling, Alexandr Andreevich (1815–91). Russian geologist, palaeontologist, and botanist. Accompanied Roderick Impey Murchison on his expedition to Russia, 1841. (*DSB*, Sarjeant 1980.)

King, Phillip Parker (1793–1856). Naval officer and hydrographer. Commander of the *Adventure* and *Beagle* on the first surveying expedition to South America, 1826–30. Settled in Australia. Rear-admiral, 1855. FRS 1824. (*Aust. Dict. Biog.*, *DNB*.)

Kingdon, Boughton (1816–96). Of Ryde, Isle of Wight. Apothecary who practised in Exeter, Croydon, and Sydney. Translated Augustin Pyramus de Candolle's *Organographie végétale*. (R. Desmond 1977.)

Kirby, William (1759–1850). Clergyman and entomologist. Vicar of Barham, Suffolk, 1797–1850. One of the first fellows of the Linnean Society, 1788. FRS 1818. (*Alum. Cantab.*, *DNB*.)

Klotzsch, Johann Friedrich (1805–60). German botanist. (*ADB*.)

Knight, Thomas Andrew (1759–1838). Botanist interested in improving the culture and yield of farm and garden produce. President of the Horticultural Society, 1811–38. FRS 1805. (*DNB*, *DSB*.)

Kölreuter, Joseph Gottlieb (1733–1806). German botanist. Professor of natural history and director of the gardens at Karlsruhe. Carried out hybridisation experiments on plants. (*DSB*, *NDB*.)

Kotzebue, Otto von (1787–1846). Russian explorer and navigator. Published accounts of his voyages. (*EB*.)

Krusenstern, Adam Johann von (1770–1846). Russian navigator, hydrographer, and admiral. Circumnavigated the world, 1803–6. Foreign member, Royal Society, 1837. (*BHGW*, Dawson 1885, *EB*.)

Laing, Samuel (1780–1868). Army officer, traveller, and author. (*DNB*.)

Lamarck, Jean Baptiste Pierre Antoine de Monet de (1744–1829). French naturalist. Held various botanical positions at the Jardin du Roi, 1788–93. Professor of zoology, Muséum d'Histoire Naturelle, 1793. Believed in spontaneous generation and the progressive development of animal types; propounded a theory of transmutation. (*DSB*.)

Lankester, Edwin (1814–74). Physician, lecturer, and author. An active supporter of the British Association for the Advancement of Science. Professor of natural history, New College, London, 1850. FRS 1845. (*DNB*.)

Lansdowne, Lord. *See* Petty-Fitzmourice, Henry.

La Pérouse, Jean François de Galaup, Comte de (1741–88). French navigator. Commander of a voyage of discovery to the Pacific, 1785–8, during which all hands perished. (*EB*, *Larousse XX*.)

Laslett, Isaac. Bricklayer in Down, Kent. (*Post Office directory of the six home counties 1845*.)

Lauder, Thomas Dick, 7th Baronet (1784–1848). Scottish writer. Author of novels, works on Scottish topography, and several scientific papers. Secretary to the Board of Scottish Manufactures, 1839. (*DNB*, Sarjeant 1980.)

Leach, William Elford (1790–1836). Assistant keeper of the natural history department, British Museum, 1813–21. FRS 1816. (*DNB*, Gilbert 1977.)

Lee, James (*fl.* 1800–75). Son of John Lee (*DNB*). Engraver on wood of Kennington, London. (*Post Office London directory 1846*.)

Lefebvre, Alexandre Louis (1798–1867). French lepidopterist. (Gilbert 1977, Sarjeant 1980.)

Leighton, William Allport (1805–89). Botanist, clergyman, and antiquary. Edited the *Transactions of the Shropshire Archaeological Society*. Published *The Flora of Shropshire* (1841). (R. Desmond 1977, *DNB*.)

Leonard, Samuel William. A member of the Microscopical Society from 1840.

Lennox, Charles Gordon-, 5th Duke of Richmond, (1791–1860). Army officer and politician. Provincial grand master of Freemasons, Sussex, 1819–60. Postmaster-general, 1830–4. President of the Royal Agricultural Society, 1845–60. FRS 1840. (*Complete Peerage*, *DNB*.)

Leslie, John (1766–1832). Mathematician and natural philosopher. Professor of natural philosophy, Edinburgh University, 1819. (*DNB*, *DSB*.)

Lesson, René Primevère (1794–1849). French naval surgeon and naturalist. Medical officer on board the *Coquille*, 1822–5. (*DSB*.)

Lewis, John. Carpenter in Down, Kent. (*Post Office directory of the six home counties 1845*.)

Lewis, William. Brewer in Down, Kent. (*Post Office directory of the six home counties 1845*.)

Lhotsky, Johann (b. 1800). Polish physician, traveller, and natural history collector. Claimed to have discovered gold in Australia but was forced to leave Australia and Tasmania owing to financial troubles. Settled in England in 1838. (*Aust. Dict. Biog.*, R. Desmond 1977.)

Liebig, Justus von (1803–73). German organic chemist. Professor of chemistry, Giessen, 1825–51; Munich, 1852–73. Developed artificial agricultural fertilizers. Foreign member, Royal Society, 1840. (*ADB*, *DSB*.)

Liebmann, Frederik Michael (1813–56). Danish botanist. (*DBL*.)

Lindley, John (1799–1865). Botanist and horticulturist. Assistant secretary to the Horticultural Society, 1822–41; vice-secretary, 1841–58. Professor of botany at London University (later University College London), 1828–60. Editor of the *Gardeners' Chronicle* from 1841. FRS 1828. (*DNB*, *DSB*.)
[*c*. 10 October 1846]

Linnaeus (Carl von Linné) (1707–78). Swedish botanist and zoologist. Proposed a system for the classification of the natural world and reformed scientific nomenclature. (*DSB*.)

Litchfield, Henrietta Emma. *See* Darwin, Henrietta Emma.

Litke, Fyodor Petrovich (1797–1882). Russian naval officer, explorer, and geographer. (*DSB*.)

Lockhart, John Gibson (1794–1854). Biographer and novelist. Editor of the *Quarterly Review*, 1825–53. (*DNB*.)

Long, Edward (1734–1813). English judge and historian in Jamaica. (R. Desmond 1977, *DNB*.)

Lonsdale, William (1794–1871). Geologist. Served the Geological Society from 1829 to 1842, first as curator and librarian, and after 1838 as assistant secretary and librarian. (*DNB*, *DSB*.)

Loudon, John Claudius (1783–1843). Landscape gardener and horticultural writer. Travelled in north Europe, 1813–15; in France and Italy, 1819–20. Founded and edited the *Gardener's Magazine*, 1826–43, and *Magazine of Natural History*, 1828–36. (R. Desmond 1977, *DNB*.)

Lubbock, Harriet (1810–73). Married John William Lubbock in 1833. (Freeman 1978.)

Lubbock, John William, 3d Baronet (1803–65). Astronomer, mathematician, and banker. First vice-chancellor of London University, 1837–42. CD's neighbour in Down, Kent. FRS 1829. (*DNB*, *DSB*.)
[16 January 1846]

Lumley, Edward. Bookseller and publisher at 56 Chancery Lane, London. (*Post Office London directory 1846*.)

Lund, Peter Wilhelm (1801–80). Danish naturalist and palaeontologist. Studied fossils of South American caves. (*DBL*, Sarjeant 1980.)

Lütke, Fedor Petrovich. *See* Litke, Fyodor Petrovich.

Lyell, Charles (1797–1875). Uniformitarian geologist. Professor of geology, King's College, London, 1831–3. President of the Geological Society, 1834–6, and 1849–50. Scientific mentor and friend of CD. FRS 1826. (*DNB*, *DSB*.)
[1 September 1844]; [8 February 1845]; [5 July 1845];
[30 July – 2 August 1845]; *[after 2 August 1845]*; 25 August [1845];
8 October [1845]; [8 August 1846]; [3 October 1846]

Lyell, Charles, Sr (1769–1849). Of Kinnordy, Forfarshire. Botanist and Dante scholar. Father of Charles Lyell. (R. Desmond 1977, *DNB*.)

Lyell, Katharine Murray (1817–1915). Daughter of Leonard Horner. Married Henry Lyell, brother of Charles Lyell, in 1848. Collected plants in India. Edited *Life, letters and journals of Sir Charles Lyell* (1881) and memoirs of Charles James Fox Bunbury and Leonard Horner. (R. Desmond 1977.)

Lyell, Marianne (1801–81). Sister of Charles Lyell. (Wilson 1972.)

Lyell, Mary Elizabeth (1808–73). Oldest child of Leonard Horner. Married Charles Lyell in 1832. (Freeman 1978.)

Macaulay, Thomas Babington, 1st Baron (1800–59). Historian and politician. Created Baron Macaulay of Rothley, 1857. FRS 1849. (*DNB*.)

McCormick, Robert (1800–90). Naval surgeon, explorer, and naturalist. Published accounts of his voyages. Surgeon in the *Beagle*, 1831–2. Accompanied James Clark Ross's Antarctic expedition, 1839–43. (*DNB*.)

MacCulloch, John (1773–1835). Physician, chemist, and geologist. President of the Geological Society, 1816–17. FRS 1820. (*DNB*, *DSB*.)

McGilvray, Walter (1807–80). Scottish Free Church minister. Minister of Hope Street Gaelic Church, Glasgow, 1842–6. Gaelic minister in Canada, 1846–8. Minister of St Marks, Glasgow, 1848. Married Maria Hooker in 1846. (*Modern English Biography*.)

Mackintosh, James (1765–1832). Philosopher and historian. Professor of law and general politics at the East India Company College, Haileybury, 1818–24. Married Catherine Allen in 1798. (*DNB*.)

Maclean, John. Merchant in Lima, 1832–54. Sent plants to William Jackson Hooker and William Herbert. (R. Desmond 1977).

Macleay, William Sharp (1792–1865). Zoologist and diplomat. Emigrated to New South Wales in 1839. Originator of the quinary system of taxonomy. (*Aust. Dict. Biog.*, *DNB.*)

Macquart, Pierre Justin Marie (1778–1855). French entomologist. (Gilbert 1977, *NUC.*)

Macrae, James (d. 1830). Collected plants for the Horticultural Society of London on the Sandwich and Galápagos Islands and in Chile and Brazil. Superintendent, Ceylon botanic gardens, 1827–30. (R. Desmond 1977.)

Mallet, Robert (1810–81). Civil engineer and seismologist. Carried out many engineering projects in Ireland. Consulting engineer in London, 1861. FRS 1854. (*DNB*, *DSB.*)
26 August [1846]

Martineau, Harriet (1802–76). Author, reformer, and traveller. (*DNB.*)

Meyen, Franz Julius Ferdinand (1804–40). Prussian botanist and physician. Naturalist on board the *Prinzess Louise*, 1830–2. Published the results of the voyage, 1834–43. Carried out microscopic investigations of plants and animals. (*ADB*, *DSB.*)

Miers, John (1789–1879). Botanist and engineer. Travelled and worked in South America, 1819–38. Author of many papers describing South American plants. FRS 1843. (R. Desmond 1977, *DNB.*)

Mill, John Stuart (1806–73). Philosopher and political economist. (*DNB*, *DSB.*)

Miller, William Hallowes (1801–80). Mineralogist and crystallographer. Professor of mineralogy, Cambridge University, 1832–80. FRS 1838. (*DNB*, *DSB.*)
8 March [1845]

Milman, Henry Hart (1791–1868). Professor of poetry, Oxford University, 1821–31. Rector of St Margaret's, Westminster, 1835. Dean of St Paul's, 1849. (*DNB.*)

Miquel, Friedrich Anton Wilhelm (1811–71). Dutch botanist. Described flora of Dutch East Indies. Director, Rotterdam botanic garden, 1835–46. Professor of botany, Amsterdam, 1846–59; Utrecht, 1859–71. (*DSB.*)

Mitford, William (1744–1827). Historian and politician. (*DNB.*)

Montagne, Jean François Camille (1784–1866). French army surgeon and botanist. (*Larousse XX*, *NBU.*)

Moquin-Tandon, Horace Bénédict Alfred (1804–63). French naturalist. Professor of zoology, Marseilles, 1829; professor of botany, Toulouse, 1833; professor of natural history, Faculty of Medicine, Paris, 1853. (*Larousse XX*, *NBU.*)

Morlot, Adolf von (1820–67). Swiss stratigrapher and archaeologist. Professor of geology, Lausanne University. (*ADB*, Sarjeant 1980.)
9 August [1844]; 10 October [1844]; 28 November [1844]

Morris, John (1810–86). Geologist. Originally a pharmaceutical chemist in Kensington. Professor of geology, University College London, 1854–77. (*DNB*, Sarjeant 1980.)

Morton, Samuel George (1799–1851). American physician and ethnologist, known for research on human skulls. (*DAB*, *DSB*.)

Murchison, Roderick Impey (1792–1871). Geologist noted for his work on the Silurian system. A leading figure in the Geological Society, British Association for the Advancement of Science, and Royal Geographical Society. Director of the Geological Survey of Great Britain, 1855. Knighted 1846; created baronet, 1866. FRS 1826. (*DNB*, *DSB*.)

[1846?]

Murray, John (1786?–1851). Scientific writer and lecturer. (R. Desmond 1977, *DNB*.)

Murray, John (1808–92). Publisher and author of guide-books. CD's publisher from 1845. (*DNB*.)

17 March [1845]; 20 March [1845]; [5 April 1845]; [10 April 1845]; 12 April [1845]; 17 [April 1845]; [23 April 1845]; [31 May 1845]; [4 June 1845]; [6 June 1845]; 20 [June 1845]; [23 June 1845]; [26 June 1845]; [28 June 1845]; [3 July 1845]; 16 [July 1845]; [27 July 1845]; [23 August 1845]; 27 August [1845]; 2 September [1845]; 30 December [1846]

Nelson, Richard John (1803–77). Officer in the Royal Engineers and geologist. Studied the geology of the Bermudas. (*DNB*.)

Nicol, William (1768–1851). Investigated the structure of crystals and invented the Nicol prism. Developed a technique for preparing transparent slivers of rock for viewing directly through a microscope and used the method in the study of fossil woods. (R. Desmond 1977, *DSB*.)

Orbigny, Alcide Charles Victor Dessalines d' (1802–57). French palaeontologist who travelled widely in South America, 1826–34. Professor of palaeontology, Muséum d'Histoire Naturelle, 1853. (*DSB*.)

14 February 1845; *[14 February 1845?]*; *31 January 1846*; *[June–July 1846]*

Owen, Richard (1804–92). Comparative anatomist. Assistant-conservator of the Hunterian Museum, Royal College of Surgeons, 1827; Hunterian professor, 1836–56. Superintendent of the Natural History departments, British Museum, 1856–84. Described the *Beagle* fossil mammal specimens. FRS 1834. (*DNB*, *DSB*.)

[21 April 1846]; 21 [June 1846]; 25 November [1846]

Palgrave, Elizabeth (1799–1852). Joseph Dalton Hooker's aunt. Married Francis Palgrave in 1823. Lived in Hampstead, London. (Huxley ed. 1918.)

Palgrave, Francis (1788–1861). Historian. Author of several works on the history of England and the English Commonwealth. Deputy-keeper of the Queens's records, 1838–61. Married Elizabeth Turner, Joseph Dalton Hooker's aunt, in 1823. Knighted in 1832, FRS 1821. (*DNB*.)

Palissy, Bernard (1510?–90?). French potter and naturalist. (*DSB*.)

Pallas, Pyotr Simon (1741–1811). German naturalist and geographer who travelled extensively in the Russian empire. FRS 1763. (*ADB*, *DSB*.)

Parish, Woodbine (1796–1882). Diplomat. Chargé d'affaires in Buenos Aires, 1825–32. Chief commissioner in Naples, 1840–5. FRS 1824. (*DNB*.) *[1845?]*

Parslow, Joseph (1809/10–98). CD's manservant at 12 Upper Gower Street *c.* 1840 and butler at Down House until 1875. (Freeman 1978.)

Paxton, Joseph (1801–65). Gardener and architect. Head gardener to the Duke of Devonshire at Chatsworth, 1826. Designed the Crystal Palace. Knighted 1852. (R. Desmond 1977, *DNB*.)

Pernety, Antoine Joseph (1716–1801). French writer. In 1763 he accompanied Louis Antoine de Bougainville's expedition to the Falkland Islands. (*Larousse XX*, *NBU*.)

Perrottet, G. Samuel (1793–1860). French traveller and botanist. (*Larousse XX*, *NBU*.)

Petty-Fitzmourice, Henry, 3d Marquis of Lansdowne, (1780–1863). Statesman. (*DNB*.)

Phillips, Richard (1778–1851). Chemist and curator of the Museum of Practical Geology, London, 1839–51. President of the Chemical Society, 1849–50. FRS 1822. (*DNB*, Sarjeant 1980.)

Phillpotts, Henry (1778–1869). Clergyman known for his outspoken support of conservative causes. Bishop of Exeter, 1830–69. (*DNB*.)

Pöppig, Eduard Friedrich (1798–1868). German naturalist who travelled in South America. Professor of zoology, Leipzig University, 1833. (*ADB*, *Larousse XX*.)

Porter, David (1780–1843). American naval officer. Consul-general in Algiers, 1830. Minister in Turkey, 1839. (*DAB*.)

Prescott, William Hickling (1796–1859). American historian of Spain and Spanish America. (*DAB*.)

Price, Edward. Resident of Down, Kent. (*Post Office directory of the six home counties 1845*.)

Prichard, Augustin (1818–98). M.D. Berlin, 1841. Specialised in eye surgery. Lecturer on anatomy at the Bristol Medical School, 1843–54. Translated Alexander von Humboldt's *Cosmos*, 1845–8. Son of James Cowles Prichard. (Plarr 1930.)

Prichard, James Cowles (1786–1848). Physician and ethnologist who maintained that the races of man are varieties of one species. A commissioner of lunacy in London, 1845. FRS 1827. (*DNB*, *DSB*.)

Prior, Richard Chandler Alexander (1809–1902). Physician and botanist. Collected plants from around the world. Took the name Prior in 1859. (R. Desmond 1977.)

Ramsay, Andrew Crombie (1814–91). Geologist with the Geological Survey, 1841; senior director of the Geological Survey for England and Wales, 1862; director-general, 1871–81. Professor of geology, University College London, 1847–52; lecturer at the Government School of Mines, 1852–71. Knighted, 1881. FRS 1849. (*DNB*, *DSB*.)
10 October [1846]; 21 December [1846]

Ranelagh, Lord. *See* Jones, Thomas Heron.

Rathbone, Hannah Mary (1798–1878). Writer and artist. Author of the *Diary of Lady Willoughby* (1844–7). (*DNB*.)

Reeks, Trenham (1823/4–79). Mineralogist. Employed in the laboratory of the Museum of Economic Geology, 1840. Curator and librarian, Museum of Economic Geology, 1851. (*Modern English Biography*.)
[before 8 February 1845]; *8 February 1845*; *25 February 1845*; *14 March 1845*

Reeve, Lovell Augustus (1814–65). Conchologist. Set up a shop in King William Street, Strand, for the sale of natural history specimens and the publication of conchological works. (*DNB*, Sarjeant 1980.)

Reeve Brothers. Publishing company founded by Lovell Augustus Reeve.
[August 1846]

Reinwardt, Caspar Georg Carl (1773–1854). German naturalist. Professor of natural philosophy, University of Leiden, 1823–45. (*BHGW*, *NNBW*.)

Rich, William. Botanist with the United States exploring expedition under the command of Charles Wilkes, 1838–42.

Richardson, John (1787–1865). Arctic explorer, physician, and naturalist. FRS 1825. (*DNB*, Sarjeant 1980.)

Richmond, Duke of. *See* Lennox, Charles Gordon-.

Ripa, Matteo (1682–1746). Italian missionary and painter. (*Enciclopedia Italiana*.)

Rivera, Fructuoso (1788–1854). Uruguayan political and military leader. President of Uruguay, 1830–4 and 1839–45. Exiled to Brazil in 1845. (*HDA*.)

Robberds, John Warden. Author of works on Norfolk and biographer of William Taylor. (*NUC*.)

Robertson, John. Surgeon in H.M.S. *Terror* on the Antarctic voyage under the command of James Clark Ross, 1839–43.

Robinson, George (1824?–93). Botanist. Rector of Tartaraghan, County Armagh until 1892. (R. Desmond 1977.)

Rogers, Henry Darwin (1808–66). American geologist. Professor of geology and mineralogy, University of Pennsylvania, 1835. Director of the geological survey of New Jersey, 1835; Pennsylvania, 1836. Regius professor of natural history, Glasgow University, 1857–66. FRS 1858. (*DAB*, *DSB*.)

Rogers, William Barton (1804–82). American geologist. Professor of natural philosophy, University of Virginia, 1835–53. State geologist for the geological survey of Virginia, 1835. Established the Massachusetts Institute of Technology, 1861; president, 1862–70 and 1878–81. Brother of Henry Darwin Rogers. (*DAB*, *DSB*.)

Rosas, Juan Manuel de (1793–1877). Governor of Buenos Aires, 1829–32 and 1835–52, who ruled as dictator of Argentina. (*HDA*.)

Ross, James Clark (1800–62). Naval officer and polar explorer. Discovered the north magnetic pole in 1831. Commander of an expedition to the Antarctic, 1839–43. FRS 1828. (*DNB*, *DSB*.)

Royal Geographical Society of London.
[30 July 1844 – 1 October 1846]; [28 January or 4 February 1846]

Rugendas, Johann Moritz (1802–58). German landscape painter who travelled widely in South America. (Bénézit 1976.)

Sabine, Edward (1788–1883). Geophysicist and army officer. Promoted and carried out studies of geomagnetism. General secretary of the British Association for the Advancement of Science, 1838–59. President of the Royal Society, 1861–71. FRS 1818. (*DNB*, *DSB*.)

Sabine, Elizabeth Juliana (1807–79). Scientific translator. Wife of Edward Sabine. (*Modern English Biography*.)

Saint-Hilaire, Augustin François César Prouvençal (Auguste de) (1779–1853). French naturalist. Surveyed the flora and fauna of Brazil, 1816–22. (*DSB*.)

Sales, William. Publican and grocer in Down, Kent. (*Post Office directory of the six home counties 1845*.)

Salt, Thomas. Partner in the Shrewsbury law firm of Dukes and Salt.

Sartorius, Christian (1796–1872). German traveller. (*ADB*.)

Schlegel, Herman (1804–84). German naturalist. Director of the National Museum of the Netherlands in Leiden. (*ADB*, Sarjeant 1980.)

Schomburgk, Robert Hermann (1804–65). German-born traveller and naturalist. Explored British Guiana under the direction of the Royal Geographical Society, 1831–5. Government commissioner for surveying, British Guiana, 1841–3. British consul, San Domingo, 1848–57; Bangkok, 1857–64. Knighted 1844. FRS 1859. (R. Desmond 1977, *DNB*.)

Schönbein, Christian Friedrich (1799–1868). German chemist. Professor of physics, University of Basel, 1835–52; professor of chemistry, 1852–68. (*ADB*, *DSB*.)

Schouw, Joachim Frederik (1789–1852). Danish botanist. Extraordinary professor of botany, Copenhagen University, 1820; professor, 1845–53. Curator of the Copenhagen botanic gardens, 1841. (*DBL*, *DSB*.)

Scott, Walter (1771–1832). Scottish novelist and poet. (*DNB*.)

Scouler, John (1804–71). Surgeon and naturalist on the Hudson's Bay Company's voyage to the Columbia River, 1824–5. Professor of geology, Andersonian Institution, Glasgow, 1829. Professor of geology, zoology, and botany, Royal Dublin Society, 1833–54. (R. Desmond 1977, *DNB*.)

Sedgwick, Adam (1785–1873). Geologist and clergyman. Woodwardian professor of geology, Cambridge University, 1818–73. Canon of Norwich, 1834–73. FRS 1821. (*DNB*, *DSB*.)

Selwyn, George Augustus (1809–78). Bishop of New Zealand, 1841–67. Travelled extensively in New Zealand and the Pacific Islands. Bishop of Lichfield, 1868–78. (*DNB*, *DNZB*.)

Selys Longchamps, Michel Edmond, Baron de (1813–1900). Belgian naturalist and statesman. (*BNB*.)

Sharpe, Daniel (1806–56). Geologist. Merchant in Portugal, 1835–8. Studied slaty cleavage and wrote several papers on the cause of cleavage and foliation. President of the Geological Society, 1856. FRS 1850. (*DNB*, Sarjeant 1980.)
[1 November 1846]

Silliman, Benjamin (1779–1864). American chemist, geologist, and mineralogist. Professor of chemistry and natural history, Yale University, 1802–53. Founder and first editor of the *American Journal of Science and Arts*, 1818. (*DAB*, *DSB*.)

Simpson, Thomas (1808–40). Arctic explorer. Second in command, under Peter Warren Dease, on the Hudson's Bay Company's expedition to the north-western coast of North America, 1836–9. Killed by gunshot wound. (*DNB*, *MDCB*.)

Sinclair, Andrew (1796–1861). Naval surgeon and botanist. Surgeon on board H.M.S. *Sulphur*, 1834–42. Collected plants in Australia, New Zealand, Mexico, and Central America. Colonial secretary in New Zealand, 1844–56. (*DNB*, *DNZB*.)

Sismondi, Jean Charles Léonard Simonde de (1773–1842). Swiss historian. Married Jessie Allen in 1819. (*EB*.)

Sismondi, Jessie (1777–1853). Daughter of John Bartlett Allen. Married Jean Charles Léonard Simonde de Sismondi in 1819. Emma Darwin's aunt. (*Emma Darwin*).

Sloane, Hans (1660–1753). Physician and natural historian. His collections formed the basis of the British Museum. (*DNB*, *DSB*.)

Smith, Andrew (1797–1872). Army surgeon stationed in South Africa, 1821–37. An authority on South African zoology. Principal medical officer at Fort Pitt, Chatham, 1837; deputy inspector-general, 1845. Director-general, army medical department, 1853–8. FRS 1857. (*DNB*.)

Smith, Charles Hamilton (1776–1859). Army officer and writer on natural history. FRS 1824. (*DNB*.)
14 January [1845]; *22 January 1845*; 26 January [1845]

Smith, Elder, and Company. Partnership of George Smith (1789–1846) and Alexander Elder (1790–1876). Publishers and East India agents. (*DNB*, *Modern English Biography*.)
13 November [1845]; 30 March [1846]; [19 October 1846]

Smith, James (1782–1867). Of Jordanhill. Geologist, yachtsman, and Biblical scholar. Studied the geology of Madeira, Gibraltar, Lisbon, and Malta, 1839–46. Studied glacial geology. FRS 1830. (*DNB*, Sarjeant 1980.)

Smith, John. Farmer. Resident of Down, Kent. (*Post Office directory of the six home counties 1846*.)

Smith, Sydney (1771–1845). Essayist and wit. A founder of the *Edinburgh Review*, 1802. Canon of St Paul's, London, 1831. (*DNB*.)

Smyth, William. Admiralty mate on board H.M.S. *Blossom*, 1825–8.

Snow, George. London carrier in Down, Kent. (*Post Office directory of the six home counties 1845*.)

Sowerby, George Brettingham (1788–1854). Conchologist and artist. Produced catalogues of shells and molluscs. (*DNB*, Sarjeant 1980.)

28 May 1844; [1845?]; [May 1845]; 12 [November 1845]; [1 December 1845]; [3 December 1845]; [9? December 1845]; *[1846]*; *17 January 1846*; *7 February 1846*

Sowerby, George Brettingham, Jr (1812–84). Eldest son of George Brettingham Sowerby. Conchologist and scientific illustrator. (*DNB*, Sarjeant 1980.)

31 [March 1846]

Sowerby, James de Carle (1787–1871). Naturalist and scientific illustrator. Founder member of the Royal Botanic Society and Gardens, Regents Park, London, 1838; secretary, 1839–69. Brother of George Brettingham Sowerby. (R. Desmond 1977, *DNB*.)

Sprengel, Christian Konrad (1750–1816). German botanist who studied insect-aided fertilisation of flowers. (*ADB*, *DSB*.)

Stanley, Arthur Penrhyn (1815–81). Fellow of University College, Oxford, 1838–50. Professor of ecclesiastical history, Oxford University, 1856. Dean of Westminster, 1864–81. FRS 1863. (*DNB*.)

Stanley, Edward Smith, 13th Earl of Derby, (1775–1851). Whig M.P. for Preston, 1796–1812; Lancashire, 1812–32. President of the Linnean Society, 1828–33. President of the Zoological Society, 1831–51. Formed a private menagerie at Knowsley. (*DNB*.)

Steenstrup, Johannes Japetus Smith (Japetus) (1813–97). Danish zoologist. Professor of zoology, University of Copenhagen, 1846–85. (*DBL*, *DSB*.)

Stokes, Charles (1783–1853). Stockbroker in London, collector of coins and drawings, naturalist, and geologist. Collected fossil woods. FRS 1821. (R. Desmond 1977, *Modern English Biography*, Sarjeant 1980.)

Stokes, John Lort (1812–85). Naval officer. Served in H.M.S. *Beagle* as midshipman, 1826–31; mate and assistant surveyor, 1831–7; lieutenant, 1837–41; commander, 1841–3. Admiral, 1877. (R. Desmond 1977, *DNB*.)

[November–December 1845]; 3 November 1846; *6 November 1846*; [*c.* 26 November 1846]

Strickland, Hugh Edwin (1811–53). Geologist and zoologist. An advocate of reform in zoological nomenclature. FRS 1852. (*DNB*, Sarjeant 1980.)

Strzelecki, Paul Edmund de (1797–1873). Polish-born explorer and geologist. Explored the Australian interior, 1839–40. Became a British subject in 1845. FRS 1853. (*Aust. Dict. Biog.*, *DNB*.)

[25 May 1845]

Studer, Bernhard (1794–1887). Swiss geologist. Professor of geology and mineralogy, University of Berne, 1834–73. (*ADB*, *DSB*.)

Sulivan, James Young Falkland (b. 1842). Naval officer. Oldest son of Bartholomew James Sulivan (*DNB*).

Sulivan, Bartholomew James (1810–90). Naval officer and hydrographer. Lieutenant in H.M.S. *Beagle*, 1831–6. Surveyed the Falkland Islands in H.M.S. *Arrow*, 1838–9. Commander of H.M.S. *Philomel* surveying vessel, on the south-east coast of South America, 1842–6. Resided in the Falkland Islands, 1848–51. Admiral, 1877. (*DNB*.)
13 January – 12 February 1845; 4 July 1845

Surtees, Harriet (1776–1847). Daughter of John Bartlett Allen. Emma Darwin's aunt. (*Darwin Pedigree*.)

Swainson, William (1789–1855). Naturalist and illustrator who collected in Sicily and Brazil. Adopted the quinary system of William Sharp Macleay. Emigrated to New Zealand in 1837. FRS 1820. (*DNB*, *DSB*.)

Swale, Ralph and William Henry. Booksellers at 21 Great Russell Street, London. (*Post Office London directory 1846*.)

Taylor, Richard (1781–1858). Printer, editor, and naturalist. Editor of the *Philosophical Magazine*, 1822. Established *Annals of Natural History* in 1838. (*DNB*.)

Taylor, Thomas (d. 1848). M.D. Dublin, 1814. Professor of botany and natural history, Cork Scientific Institution. An expert on mosses, liverworts, and lichens. Collaborated with William Jackson Hooker and Joseph Dalton Hooker. (R. Desmond 1977, *DNB*.)

Taylor, William (1765–1836). Man of letters. An enthusiast for German literature and the French revolution. (*DNB*.)

Taylor, William Cooke (1800–49). Miscellaneous writer. A contributor to and member of staff of the *Athenæum*, 1829–49. A promoter of the British Association for the Advancement of Science. (*DNB*.)

Tegetmeier, William Bernhard (1816–1912). Editor, journalist, lecturer, and naturalist. Pigeon fancier and an expert on fowl. (E. W. Richardson 1916.)

Thierry, Charles Philip Hippolytus, Baron de (1793–1864). Colonist. Hoped to create a large colony of his own in New Zealand but was reduced, by lack of means, to an ordinary settler. (*DNB*.)

Thompson, William (1805–52). Irish naturalist. President of the Belfast Natural History Society, 1843–52. (R. Desmond 1977, *DNB*.)
18 February [1846?]

Thomson, Thomas (1817–78). East India Company surgeon and botanist in India. Collaborated with Joseph Dalton Hooker on various publications and joined him on his Himalayan expedition in 1849. Curator of the Asiatic Society's museum, Calcutta, 1840. Superintendent of the Calcutta botanic garden and professor of botany at the Calcutta Medical College, 1854–61. Son of Thomas Thomson Sr. FRS 1855. (R. Desmond 1977, *DNB*.)

Thomson, Thomas, Sr (1773–1852). Regius professor of chemistry, Glasgow University, 1818–52. Edited his own journal, *Annals of Philosophy*, 1813–20. FRS 1811. (*DNB*, *DSB*.)

Tollet, Ellen Harriet (d. 1890). Daughter of George Tollet. A close friend of the Wedgwood family. (*Burke's Landed Gentry 1846*.)

Tollet, George (1767–1855). Of Betley Hall, Staffordshire. Justice of the peace and deputy lieutenant of Staffordshire. Agricultural reformer. A close friend of Josiah Wedgwood II. (*Burke's Landed Gentry 1879*.)

Torrey, John (1796–1873). American botanist and chemist. Professor of chemistry at the College of Physicians and Surgeons, New York City, 1827–55. Professor of chemistry and natural history at the College of New Jersey (later Princeton), 1830–54. New York State botanist, 1836. (*DAB*, *DSB*.)

Tschudi, Johann Jacob von (1818–89). Swiss explorer in America. (*ADB*, *BHGW*.)

Turton, William (1762–1835). Conchologist. Practised medicine in Swansea. Author and editor of works on medicine and natural history. (R. Desmond 1977, *DNB*.)

Tyndall, John (1820–93). Irish natural philosopher, lecturer, and populariser of science. Professor of natural philosophy at the Royal Institution, 1853; superintendent of the Royal Institution, 1867–87. FRS 1852. (*DNB*, *DSB*.)

Ure, Andrew (1778–1857). Professor of chemistry and natural philosophy at the Andersonian Institution, Glasgow, 1804–30. Chemical consultant in London, 1830–57. Editor of the *Dictionary of chemistry*. FRS 1821. (*DNB*, *DSB*.)

Vallemont, Pierre Le Lorrain, Abbé de (1649–1721). French writer and naturalist. (*NBU*.)

Verneuil, Philippe Édouard Poulletier de (1805–73). French geologist and palaeontologist. Foreign member, Royal Society, 1860. (*DSB*.)

Vieweg, Eduard (1797–1869). German publisher and politician. (*ADB*.)

Vriese, Willem Hendrik de (1806–62). Dutch botanist. (*NUC*.)

Wakefield, Edward Gibbon (1796–1862). Colonial statesman. Imprisoned, 1826–9, for abducting an heiress. A founder of the New Zealand Association, 1837. Director of the New Zealand Land Company, 1840. Settled in New Zealand in 1853. (*DNB*, *DNZB*.)

Ward, William Robert. Chargé d'affaires in Mexico in 1843. (*BDR*.)

Waterhouse, George Robert (1810–88). Naturalist. A founder of the Entomological Society, 1833. Curator, Zoological Society, 1836–43. On the staff of the mineralogical and geological branch of the natural history department of the British Museum, 1843–80. Described the Mammalia and entomological specimens from the *Beagle* voyage. (*DNB*, Gilbert 1977.)

26 April 1844; *[after 26 April 1844]*; 10 [June 1844 – March 1845]; *21[–22] May 1845*; *[*c. *June 1845]*; *[*c. *June 1845]*; *[*c. *June 1845]*; *[11 July 1845]*; *[30 March 1846]*

Waterton, Charles (1782–1865). Naturalist and traveller. (*DNB*, *DSB*.)

Watson, Hewett Cottrell (1804–81). Botanist, phytogeographer, and phrenologist. Published various guides on the distribution of British plants. Edited the *Phrenological Journal*, 1837–40. An early supporter of the idea of the progressive development of plant species. (*DNB*, *DSB*.)

Webb, Philip Barker (1793–1854). Botanist and traveller. Published several natural history works in various languages. FRS 1824. (R. Desmond 1977, *DNB*.)

Wedgwood, Bessy. *See* Wedgwood, Elizabeth (Bessy).

Wedgwood, Caroline Sarah (1800–88). CD's sister. Married Josiah Wedgwood III, her cousin, in 1837. (*Darwin Pedigree.*)

Wedgwood, Eliza. *See* Wedgwood, Sarah Elizabeth (Eliza).

Wedgwood, Elizabeth. *See* Wedgwood, Sarah Elizabeth (Elizabeth).

Wedgwood, Elizabeth (Bessy) (1764–1846). Oldest daughter of John Bartlett Allen. Married Josiah Wedgwood II in 1792. (*Emma Darwin.*)

Wedgwood, Emma. *See* Darwin, Emma.

Wedgwood, Frances Mackintosh (Fanny) (1800–89). Daughter of James and Catherine Mackintosh. Married Hensleigh Wedgwood in 1832. (*Burke's Peerage 1980*, Freeman 1978.)

Wedgwood, Frances Mosley (d. 1874). Married Francis Wedgwood in 1832. (*Burke's Peerage 1980.*)

Wedgwood, Francis (Frank) (1800–88). Master-potter and partner in the works at Etruria until 1876. Son of Bessy and Josiah Wedgwood II. Married Frances Mosley in 1832. (*Alum. Cantab.*)
27 November [1846]

Wedgwood, Harry. *See* Wedgwood, Henry Allen (Harry).

Wedgwood, Henry Allen (Harry) (1799–1885). B.A., Jesus College, Cambridge, 1821. Barrister. Son of Bessy and Josiah Wedgwood II. Married Jessie Wedgwood in 1830. (*Alum. Cantab.*)

Wedgwood, Hensleigh (1803–91). B.A., Christ's College, Cambridge, 1824; fellow, 1829–30. Philologist and barrister. Metropolitan police magistrate at Lambeth, 1832–7. Registrar of metropolitan carriages, 1838–49. Son of Bessy and Josiah Wedgwood II. Married Frances Mackintosh in 1832. (*DNB.*)

Wedgwood, Jessie (1804–72). CD's cousin. Married Henry Allen Wedgwood in 1830. (*Emma Darwin.*)

Wedgwood, Josiah, II (1769–1843). Of Maer Hall, Staffordshire. Master-potter of Etruria. Whig M.P. for Stoke-on-Trent, 1832–4. Married Elizabeth (Bessy) Allen in 1792. (*Burke's Peerage 1980, Emma Darwin.*)

Wedgwood, Josiah, III (Jos) (1795–1880). Of Leith Hill Place, Surrey. Son of Bessy and Josiah Wedgwood II. Married Caroline Sarah Darwin in 1837. (*Burke's Peerage 1980.*)
[May 1844]; 25 July 1844

Wedgwood, Lucy Caroline (1846–1919). Daughter of Caroline and Josiah Wedgwood III. (*Darwin Pedigree.*)

Wedgwood, Sarah Elizabeth (Eliza) (1795–1857). Oldest daughter of John and Louisa Jane Wedgwood. (*Emma Darwin.*)

Wedgwood, Sarah Elizabeth (Elizabeth) (1793–1880). Oldest daughter of Bessy and Josiah Wedgwood II. (*Burke's Peerage 1980.*)

Wedgwood, Sarah Elizabeth (Sarah) (1778–1856). Youngest daughter of Sarah and Josiah Wedgwood I. CD's aunt. (*Emma Darwin.*)

Wellesley, Arthur, 1st Duke of Wellington, (1769–1852). Army officer and statesman. Field marshal, 1813. Chancellor of Oxford University, 1834–52. Conservative leader in the House of Lords, 1835–41. FRS 1847. (*DNB.*)

Wellington, Duke of. *See* Wellesley, Arthur.

Wells, William Charles (1757–1817). Natural philosopher and physician. FRS 1793. (*DNB*, *DSB.*)

Westwood, John Obadiah (1805–93). Entomologist and palaeographer. Hope professor of zoology, Oxford University, 1861–93. Entomological referee for the *Gardeners' Chronicle.* (*DNB*, Gilbert 1977.)

Whately, Richard (1787–1863). Clergyman, logician, and political economist. Drummond professor of political economy, Oxford University, 1829–31. Archbishop of Dublin, 1831–63. (*DNB*, *DSB.*)

Whately, Thomas (d. 1772). Politician and man of letters. (*DNB.*)

Whewell, William (1794–1866). Mathematician and historian and philosopher of science. Tutor at Trinity College, Cambridge, 1823–38; master, 1841–66. Professor of mineralogy, Cambridge University, 1828–32; professor of moral philosophy, 1838–55. FRS 1820. (*DNB*, *DSB.*)

Whiston, William (1667–1752). Theologian, mathematician, and cosmologist. (*DNB*, *DSB.*)

White, Adam (1817–79). Naturalist. Employed in the zoological department of the British Museum, 1835–63. (*DNB*, Gilbert 1977.)

White, Gilbert (1720–93). Naturalist and clergyman. Author of *The natural history and antiquities of Selborne* (1789). (*DNB*, *DSB.*)

Wickham, John Clements (1798–1864). Naval officer and magistrate. First-lieutenant in H.M.S. *Beagle*, 1831–6. Commander of the *Beagle*, 1837–41, surveying the Australian coast. Emigrated to Australia in 1842. Police magistrate in New South Wales, 1843–57; Government resident, 1857. (*Aust. Dict. Biog.*, R. Desmond 1977.)

Wicksted, Charles (1796–1870). Justice of the peace. High-sheriff of Cheshire, 1822. Son of George Tollet; took the name Wicksted in 1814. (*Alum. Oxon.*, *Burke's Landed Gentry 1879.*)

13 February [1844?]

Wiedemann, Christian Rudolph Wilhelm (1770–1840). German naturalist, gynaecologist, historian, and man of letters. (*ADB*, *BHGW.*)

Wight, Robert (1796–1872). East India Company surgeon and botanist. Superintendent of the Madras botanic garden, 1826–8. In charge of an experimental cotton farm at Coimbatore, India, 1842–50. FRS 1855. (R. Desmond 1977, *DNB.*)

Wilkes, Charles (1798–1877). American naval officer and explorer. Head of the Depot of Charts and Instruments, Washington, 1833–6. Commander of the Pacific exploring expedition, 1838–42. (*DAB.*)

Williams, Henry (1782–1867). Naval officer. Missionary in New Zealand, 1823. Archdeacon of Waimate, 1844. (*DNB, DNZB.*)

Williamson, William Crawford (1816–95). Surgeon and naturalist. Professor of natural history and geology, Owens College, Manchester, 1851. FRS 1854. (*DNB, DSB.*)
23 June [1846]

Wilson, Patrick (1743–1811). Son of Alexander Wilson (*DNB*). Professor of astronomy, Glasgow University, 1784.

Wood, Searles Valentine (1798–1880). Geologist. Studied the fossils of the East Anglian Crag. (*DNB*, Sarjeant 1980.)
5 June 1846

Woodd, Ellen Sophia. *See* Fox, Ellen Sophia.

Wordsworth, William (1770–1850). Poet. (*DNB.*)

Wrangel, Ferdinand Petrovich (1796–1870). Russian navigator who investigated northern waters. Director of the Russian American Company, 1840–9. Naval minister, 1855–7. (*GSE.*)

Yarrell, William (1784–1856). Zoologist. Engaged in business as newspaper agent and bookseller in London. Wrote standard works on British birds and fishes. (*DNB.*)
29 July 1845

Yorke, James Whiting. Of Wybarton, Lincolnshire. Matriculated at Trinity College, Oxford in 1815 aged 18. (*Alum. Oxon.*)

Unidentified.
1 October 1844; 27 October [1846 or 1848?]

Bibliography of biographical sources

ADB: *Allgemeine deutsche Biographie.* Under the auspices of the Historical Commission of the Royal Academy of Sciences. 56 vols. Leipzig: Duncker and Humblot. 1875–1912.

Allan, Mea. 1967. *The Hookers of Kew 1785–1911.* London: Michael Joseph.

Alum. Cantab.: *Alumni Cantabrigienses. A biographical list of all known students, graduates and holders of office at the University of Cambridge, from the earliest times to 1900.* Compiled by J. A. Venn. Part II. From 1752 to 1900. 6 vols. Cambridge: Cambridge University Press. 1940–54.

Alum. Oxon.: *Alumni Oxonienses: the members of the University of Oxford, 1715–1886.* By Joseph Foster. 4 vols. Oxford. 1888.

Aust. Dict. Biog.: *Australian dictionary of biography: 1788–1850; 1851–1890.* Edited by Douglas Pike and Bede Nairn. 6 vols. Melbourne: Melbourne University Press. 1966–76.

BDR: *British diplomatic representatives 1789–1852.* Edited by S. T. Bindoff, E. F. Malcolm Smith, and C. K. Webster. London: Offices of the Royal Historical Society. 1934.

Bénézit, Emmanuel. 1976. *Dictionnaire critique et documentaire des peintres, sculpteurs, dessinateurs et graveurs de tous les temps et de tous les pays, par un groupe d'écrivains spécialistes français et étrangers.* Revised edition. 10 vols. Paris: Librairie Gründ.

BHGW: *Biographisch-literarisches Handwörterbuch zur Geschichte der exacten Wissenschaften.* By Johann Christian Poggendorff. Vols. 1–4, Leipzig: Johann Ambrosius Barth. 1863–1904. Vol. 5, Leipzig and Berlin: Verlag Chemie, G.m.b.H. 1926.

BNB: *Biographie nationale publiée par l'Académie royale des sciences, des lettres et des beaux-arts de Belgique.* 28 vols. Brussels. 1866–1944.

Burke's Landed Gentry: *Burke's genealogical and heraldic history of the landed gentry.* By John Burke. 1–17 editions. London: Burke's Peerage Ltd. 1833–1952.

Burke's Peerage: *Burke's peerage and baronetage.* By John Burke. 1–105 editions. London: Burke's Peerage Ltd. 1826–1980.

CDEL: *A critical dictionary of English literature, and British and American authors, living and deceased.* By S. Austin Allibone. 3 vols. Philadelphia and London. 1859–71. Supplement by John Foster Kirk, 2 vols. Philadelphia and London. 1891.

Clergy List: *The clergy list . . . containing an alphabetical list of the clergy.* London. 1841–.

Complete Peerage: *The complete peerage of England Scotland Ireland Great Britain and the United Kingdom extant extinct or dormant.* By G. E. Cokayne. New edition, revised and much enlarged. Edited by the Hon. Vicary Gibbs and others. 12 vols. London: The St Catherine Press. 1910–59.

DAB: *Dictionary of American biography.* Under the auspices of the American Council of Learned Societies. 20 vols., index, and 7 supplements. New York: Charles Scribner's Sons; London: Oxford University Press. 1928–81.

Darwin Pedigree: *Pedigree of the family of Darwin.* Compiled by H. Farnham Burke. Privately printed. 1888. Reprinted in Freeman, Richard Broke, *Darwin pedigrees.* London: Printed for the author. 1984.

Dawson, Llewellyn Styles. 1885. *Memoirs of hydrography including brief biographies of the principal officers who have served in H.M. naval surveying service between the years 1750 and 1885.* 2 pts. Eastbourne. Facsimile reprint. London: Cornmarket Press. 1969.

DBF: *Dictionnaire de biographie française.* Under the direction of J. Balteau, M. Barroux, M. Prevost, and others. 17 vols., (A–Halicka). Paris: Librairie Letouzey et Ané. 1933–86.

DBI: *Dizionario biografico degli Italiani*. Edited by Alberto M. Ghisalberti. 29 vols., (A–Corvo). Rome: Instituto della Enciclopedia Italiana. 1960–83.

DBL: *Dansk biografisk leksikon*. Founded by Carl Frederick Bricka, edited by Povl Engelstoft and Svend Dahl. 26 vols. and supplement. Copenhagen: J. H. Schultz. 1933–44.

Desmond, Ray. 1977. *Dictionary of British and Irish botanists and horticulturists, including plant collectors and botanical artists*. 3d ed. London: Taylor and Francis.

DNB: *Dictionary of national biography*. Edited by Leslie Stephen and Sydney Lee. 63 vols. and 2 supplements (6 vols.). London: Smith, Elder and Co. 1885–1912. *Dictionary of national biography 1912–70*. Edited by H. W. C. Davis, J. R. H. Weaver, and others. 6 vols. London: Oxford University Press. 1927–81.

DNZB: *A dictionary of New Zealand biography*. Edited by G. H. Scholefield. 2 vols. Wellington: Department of Internal Affairs New Zealand. 1940.

DSB: *Dictionary of scientific biography*. Edited by Charles Coulston Gillispie. 14 vols., supplement, and index. New York: Charles Scribner's Sons. 1970–80.

EB: *Encyclopaedia Britannica*. 11th ed. 29 vols. Cambridge: Cambridge University Press. 1910–11.

Elliott, Clark A. 1979. *Biographical dictionary of American science: the seventeenth through the nineteenth centuries*. Westport, Conn. and London: Greenwood Press.

Emma Darwin: *Emma Darwin: a century of family letters 1792–1896*. Edited by Henrietta Litchfield. Revised edition. 2 vols. London: John Murray. 1915.

Enciclopedia Italiana: *Enciclopedia Italiana di scienze, lettere ed arti*. 35 vols. Rome: 1929–39.

Freeman, Richard Broke. 1978. *Charles Darwin: a companion*. Folkestone: W. Dawson and Sons; Hamden, Conn.: Archon Books, Shoe String Press.

Gilbert, Pamela. 1977. *A compendium of the biographical literature on deceased entomologists*. London: British Museum (Natural History).

Grove, George. 1980. *The new Grove dictionary of music and musicians*. Edited by Stanley Sadie. 20 vols. London: Macmillan.

GSE: *Great Soviet encyclopedia* (translation of 3d ed. of *Bol'shaia Sovetskaia entsiklopediia*. Moscow. 1970–9). 31 vols. New York: Macmillan; London: Collier Macmillan. 1973–82.

HDA: *Historial dictionary of Argentina*. By Ione S. Wright and Lisa M. Nekhom. Latin American Historical Dictionaries, no. 17. Metuchen, N. J. and London: The Scarecrow Press. 1978.

Huxley, Leonard, ed. 1918. *Life and letters of Sir Joseph Dalton Hooker*. 2 vols. London: John Murray.

Johnston, William. 1917. *Roll of commissioned officers in the medical service of the British Army, 1727–1898*. Aberdeen: Aberdeen University Press.

Larousse XX: *Larousse du XXe siècle*. Under the direction of Paul Augé. 6 vols. Paris: Librairie Larousse. 1928–33.

MDCB: *The Macmillan dictionary of Canadian biography*. Edited by W. Stewart Wallace. 4th ed., revised, enlarged, and updated by W. A. McKay. Toronto: Macmillan of Canada. 1978.

Modern English Biography: *Modern English biography containing many thousand concise memoirs of persons who have died since the year 1850*. By Frederic Boase. 3 vols. and supplement (3 vols.). Truro: Netherton and Worth. 1892–1921.

Navy List: *The navy list*. London. 1815–70.

NBL: *Norsk biografisk leksikon*. Edited by Edvard Bull, Anders Krogvig, Gerhard Gran, and others. 19 vols. Oslo: H. Aschehoug & Co. 1923–83.

NBU: *Nouvelle biographie universelle*. Edited by Jean Chrétien Ferdinand Hoefer. 46 vols. Paris. 1852–66.

NDB: *Neue deutsche Biographie*. Under the auspices of the Historical Commission of the Bavarian Academy of Sciences. 14 vols., (A–Locher-Freuler). Berlin: Duncker and Humblot. 1953–85.

NDBA: *Nuevo Diccionario Biografico Argentino (1750–1930)*. Edited by Vicente Osvaldo Cutolo. 6 vols., (A–SA). Buenos Aires: Editorial Elche. 1968–83.

NNBW: *Nieuw Nederlandsch Biografisch Woordenboek*. Edited by P. C. Molhuysen, P. J. Blok, and K. H. Kossmann. 10 vols. Leiden: A. W. Sijthoff. 1911–37.

NUC: *The National Union Catalog*. Pre-1956 Imprints. 685 vols. and supplement (vols. 686–754). London and Chicago: Mansell. 1968–81.

O'Byrne, William R. 1849. *A naval biographical dictionary: comprising the life and services of every living officer in Her Majesty's Navy, from the rank of admiral of the fleet to that of lieutenant, inclusive*. London.

Physicians: *The roll of the Royal College of Physicians of London*. By William Munk. 2d ed., revised and enlarged. Vol. 3 (1801–25). London. 1878. Vol. 4, *Lives of the fellows of the Royal College of Physicians* (1826–1925), compiled by G. H. Brown. London: Published by the Royal College of Physicians. 1955.

Plarr, Victor Gustave. 1930. *Plarr's lives of the fellows of the Royal College of Surgeons of England*. Revised by Sir D'Arcy Power. 2 vols. London: Simpkin Marshall.

Post Office directory of the six home counties: *Post Office directory of the six home counties, viz., Essex, Herts, Kent, Middlesex, Surrey and Sussex*. London. 1845–.

Post Office London directory: *Post Office London directory*. London 1802–.

Provincial Medical Directory: *The provincial medical directory*. London. 1847–.

Richardson, E. W. 1916. *A veteran naturalist: being the life and work of W. B. Tegetmeier*. London: Witherby and Co.

Sarjeant, William A. S. 1980. *Geologists and the history of geology. An international bibliography from the origins to 1978*. 5 vols. London: Macmillan Press.

SBL: *Svenskt Biografiskt Lexikon*. Edited by Bertil Boëthius, Erik Grill, and Birgitta Lager-Kromnow. 24 vols. Stockholm: Albert Bonnier and P. A. Norstedt. 1918–84.

Stenton, Michael. 1976. *Who's who of British members of Parliament, 1832–1885.* Brighton: Harvester Press.

Wilson, Leonard G. 1972. *Charles Lyell. The years to 1841: the revolution in geology.* New Haven and London: Yale University Press.

INDEX

The dates of letters to and from the correspondents are listed in the Biographical register and index to correspondents and are not repeated here. CD's works are indexed under the short titles used throughout this volume and listed in the bibliography.

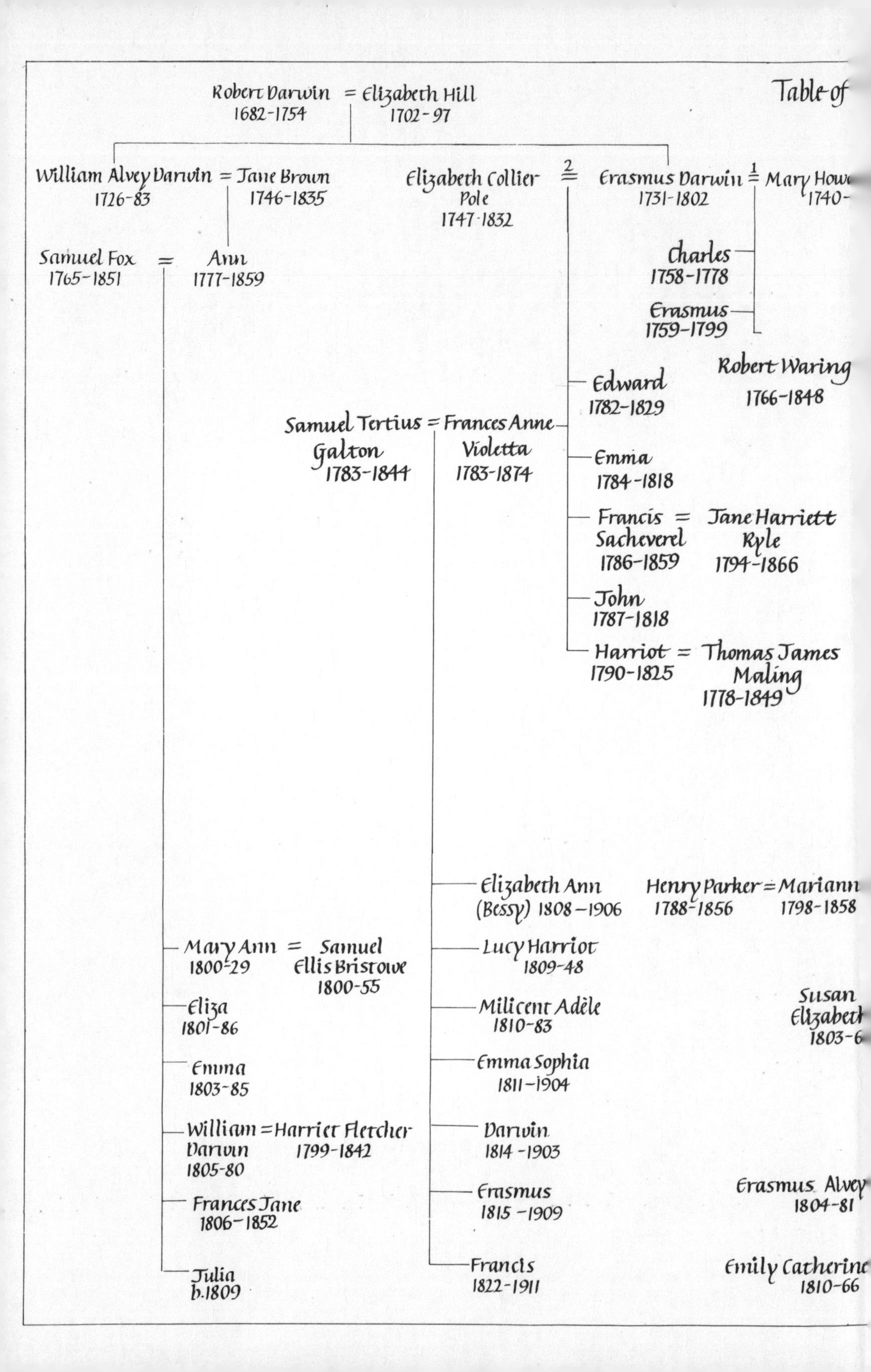

Table of
Robert Darwin = Elizabeth Hill
1682-1754
1702-97
William Alvey Darwin = Jane Brown
1726-83
1746-1835
Elizabeth Collier Pole
1747-1832
2
Erasmus Darwin
1731-1802
1
Mary Howe
1740-
Samuel Fox = Ann
1765-1851
1777-1859
Charles
1758-1778
Erasmus
1759-1799
Robert Waring
1766-1848
Edward
1782-1829
Samuel Tertius Galton = Frances Anne Violetta
1783-1844
1783-1874
Emma
1784-1818
Francis Sacheverel = Jane Harriett Ryle
1786-1859
1794-1866
John
1787-1818
Harriot = Thomas James Maling
1790-1825
1778-1849
Elizabeth Ann (Bessy) 1808-1906
Henry Parker = Mariann
1788-1856
1798-1858
Mary Ann = Samuel Ellis Bristowe
1800-29
1800-55
Lucy Harriot
1809-48
Eliza
1801-86
Milicent Adèle
1810-83
Susan Elizabeth
1803-6
Emma
1803-85
Emma Sophia
1811-1904
William Darwin = Harriet Fletcher
1805-80
1799-1842
Darwin
1814-1903
Frances Jane
1806-1852
Erasmus
1815-1909
Erasmus Alvey
1804-81
Julia
b.1809
Francis
1822-1911
Emily Catherine
1810-66